Soft and Biological Matter

"Soft and Biological Matter" is a series of authoritative books covering established and emergent areas in the realm of soft matter science, including biological systems spanning from the molecular to the mesoscale. It aims to serve a broad interdisciplinary community of students and researchers in physics, chemistry, biophysics and materials science.

Pure research monographs in the series as well as those of more pedagogical nature, will emphasize topics in fundamental physics, synthesis and design, characterization and new prospective applications of soft and biological matter systems. The series will encompass experimental, theoretical and computational approaches. Topics in the scope of this series include but are not limited to: polymers, biopolymers, polyelectrolytes, liquids, glasses, water, solutions, emulsions, foams, gels, ionic liquids, liquid crystals, colloids, granular matter, complex fluids, microfluidics, nanofluidics, membranes and interfaces, active matter, cell mechanics and biophysics.

Both authored and edited volumes will be considered.

More information about this series at http://www.springer.com/series/10783

Isamu Kaneda
Editor

Rheology of Biological Soft Matter

Fundamentals and Applications

Springer

Editor
Isamu Kaneda
Department of Food Science and Wellness
Rakuno Gakuen University
Ebetsu, Hokkaido
Japan

ISSN 2213-1736 ISSN 2213-1744 (electronic)
Soft and Biological Matter
ISBN 978-4-431-56078-4 ISBN 978-4-431-56080-7 (eBook)
DOI 10.1007/978-4-431-56080-7

Library of Congress Control Number: 2016957508

Printed on acid-free paper

This Springer imprint is published by Springer Nature
The registered company is Springer Japan KK
The registered company address is: Chiyoda First Bldg. East, 3-8-1 Nishi-Kanda, Chiyoda-ku, Tokyo 101-0065, Japan

Preface

Rheology is a research field that involves learning about transformation and flow over a wide range. In the twentieth century, the main subject of rheological study was the melt or the solution of the synthetic macromolecule, and excellent results were obtained both experimentally and theoretically. Therefore, many technical books in the field have been published. Although "hard material"—steel, concrete, and kinds of plastics—were the characteristic words of material science through the twentieth century, "soft matter" became the main target of material science in the twenty-first century.

There are many "interfaces" between humans and materials in various areas. Therefore, the importance of rheology for soft material has increased. For example, the quantitative estimation of food texture is the most important issue in the food science field. In particular, special foods for the elderly have been actively developed due to the progression of aging in the population. Because elderly people lose mastication and swallowing capability, developers have to consider the rheological properties of their products seriously. Cosmetics also work in such "interfaces". Because cosmetic products are used every day, the feeling of the products when applied strongly affects their palatability to consumers. Regenerative medicine has also been focused on in recent years. Artificial organs or tissue are in direct contact with living tissue; therefore, their rheological properties are key points for their development.

Not coincidentally, newly developed rheology equipment, which can detect quite low torque by using an air-bearing system and novel methods such as microrheology techniques, may drive such a trend. Specifically, in the fields of food, cosmetic, and personal care products, rheological properties are quite important for both consumers and the related industries. Namely, the dynamic characteristic values such as food texture or the feeling of lubrication with cosmetics are the most important as such product characteristics. In those cases, rheological parameters can become the benchmark of the development of these products. The topics presented in this book deal not only with application aspects but also with selected fundamental research.

Department of Food Science and Wellness
Rakuno Gakuen University, Ebetsu, Japan

Isamu Kaneda

Contents

Part I
Fundamentals

Chapter 1
Bridging the Gap Between Single-Molecule Unbinding Properties and Macromolecular Rheology

Makoto Takemasa, Andrew N. Round, Marit Sletmoen, and Bjørn Torger Stokke

Abstract Conformation and interactions between biological macromolecules are crucial for the mechanical properties of biological soft matter. In this chapter, the method and applications of the mechanical characteristics at the single-molecule level, from a fundamental point of view, are described as basis for understanding aspects of rheology. Atomic force microscope (AFM) and optical tweezers can be applied to investigate mechanical properties and interactions of molecules in the single molecular level. The force between two molecules as a result of specific and/or non-specific interactions can be determined as a function of distance between two molecules. Selected examples for interactions in macromolecules were highlighted based on observations by AFM-based force spectroscopy. This includes polysaccharide pairs such as interactions among hydrophobically modified hydroxyethyl cellulose (HMHEC), between protein polysaccharides and mucin–alginate. The mechanism of physically cross-linked hydrogel formation, HMHEC–amylose gel and alginate gels was also discussed based on single molecular pair interactions. For slower bond formation systems, which may not be capable with normal dynamic force spectroscopy, slide contact force spectroscopy can be applied. For slower dissociation rate, Dudko–Hummer–Szabo model and Friddle–Noy–De Yoreo model can be used for the analysis as an extension of the Bell–Evans model. The relation between characteristic timescale of interaction estimated in the single

M. Takemasa (✉)
School of Creative Science and Engineering, Waseda University, 3-4-1, Okubo Shinjuku, Tokyo, Japan
e-mail: takemasa@physics.soft-matter.org

A.N. Round
School of Pharmacy, University of East Anglia, Norwich Research Park, Norwich, NR4 7TJ UK

M. Sletmoen
Department of Biotechnology, The Norwegian University of Science and Technology, NO-7491 Trondheim, Norway

B.T. Stokke
Department of Physics, The Norwegian University of Science and Technology, NO-7491 Trondheim, Norway

I. Kaneda (ed.), *Rheology of Biological Soft Matter*, Soft and Biological Matter,
DOI 10.1007/978-4-431-56080-7_1

molecular study and relaxation spectra in the mechanical properties obtained at the macroscopic scale is presented as a possible way forward in understanding the gap between the mechanical properties in macroscopic and microscopic scale.

Keywords Rheology • Molecular interactions • Dynamic force spectroscopy • Atomic force microscope (AFM)

1.1 Introduction

What is molecular scale rheology, and why is it important? During deformation of macroscopic objects in response to an applied force, it is reasonable to assume that each molecule in the object deforms as an element. However, the actual origins of many rheological phenomena have not yet been understood at the single molecular level, such as shear thickening, assumption of affine transformation and many others. What really happens at the molecular scale? This is one of the motivations for the mechanical investigation of rheology at the single molecular level. The relationship between the end-to-end distance of each polymer and the deformation of the whole object would be helpful for material design, since the deformation and the stiffness of each molecule are essential for macroscopic materials. Characterization of each molecule is essential for a full understanding and control of the rheological properties of soft materials at the macroscopic scale. In this section, we introduce the characteristic force behaviour in response to the deformation of molecules, i.e. the end-to-end distance of single molecule.

The question 'What is the relationship between the end-to-end distance of a single polymer chain and force?' has been posed as a thought experiment to illustrate the fundamentals of polymer physics, rubber elasticity, or entropy-driven elasticity in many polymer textbooks [1]. Although actual deformation of large-scale objects may not be simply understood based on the superposition of deformation of single molecules, single-molecule characteristics and single molecular pair interactions are definitely important for macroscopic characteristics.

Not only for theoretical understandings but also for actual applications, consideration of single molecular mechanisms is an attractive route to understanding what really happens. For instance, to elucidate each contribution to the gel strength (i.e. network structure, unbending force of cross-link, stiffness of each polymer chain), we need to focus on each factor. From macroscopic experiments using gels, focusing on each factor is essentially impossible, while single molecular experiments offer the opportunity to study these interactions in isolation, in well-defined systems.

Recent developments of experimental techniques and equipment make possible experiments focusing on manipulating a single molecule, i.e. by changing the end-to-end distance of a single polymer by stretching, from random coil to elongated state while measuring the force applied to the molecule. In this section, the principles behind these methods and the practical implementation of manipulation and force measurements of single molecules in specific rheologically relevant applications are described.

1.2 Methods

Nowadays, a series of tools enables both the manipulation and force measurement and the investigation of mechanical properties of single molecules. Here, the two most commonly used techniques, atomic force microscopy (AFM) and optical tweezers (OT), will be introduced. The principles for stretching a single polymer molecule and the simultaneous force measurement using AFM and OT are essentially the same.

For an investigation of the relation between the end-to-end distance of a single polymer chain and the force between both ends, one simple setup is as follows: (i) one end of a polymer is immobilized onto a fixed wall; (ii) another end of the polymer is immobilized onto the manipulator, enabling spatial control in nm precision; (iii) the distance between the surfaces of the manipulator and the wall is measured; and (iv) the force applied to the manipulator, which equals the force required to stretch the molecule, is calculated.

Selective immobilization of a molecule onto the wall or manipulator's surface is possible using an end-selective chemical reaction. For DNA this can be achieved using thiol-modified DNA and a substrate made of Au. Position control of one end of the polymer can be realized precisely using a piezoelectric scanner in sub-nanometre scale, in principle. The magnitude of force applied to a single polymer chain ranges from pN to μN, and an estimation of force in this scale is also possible by precise position (displacement) measurement of a reference spring having a soft spring constant. When the target molecule is immobilized onto the surface of a piezo-manipulator via 'reference spring', the displacement of this reference spring directly reflects the force applied on the target polymer chain. The details of each procedure are different in AFM and OT and will be introduced in more detail below.

AFM is probably the most commonly used example of a scanning probe microscope (SPM), well established and widely used for imaging at the nanometre scale. A schematic diagram of AFM is shown in Fig. 1.1a. A piezoelectric transducer is commonly used for spatial manipulation of the sample stage in SPM and also for optical tweezers. Piezoelectric transducers can move a mechanical stage attached to the target molecule in 3D directions. Using AFM, (1) precise spatial manipulation ability of the target molecule to several nm precision based on piezo-stage and (2) high sensitivity of force, which is enough to sense the force required to deform single molecule, are achieved. AFM imaging is based on a feedback loop that maintains the force between probe tip and sample molecule in different ways, depending on the operation mode (e.g. contact and non-contact modes) used. For this feedback procedure, (a) highly sensitive force sensors for less damage

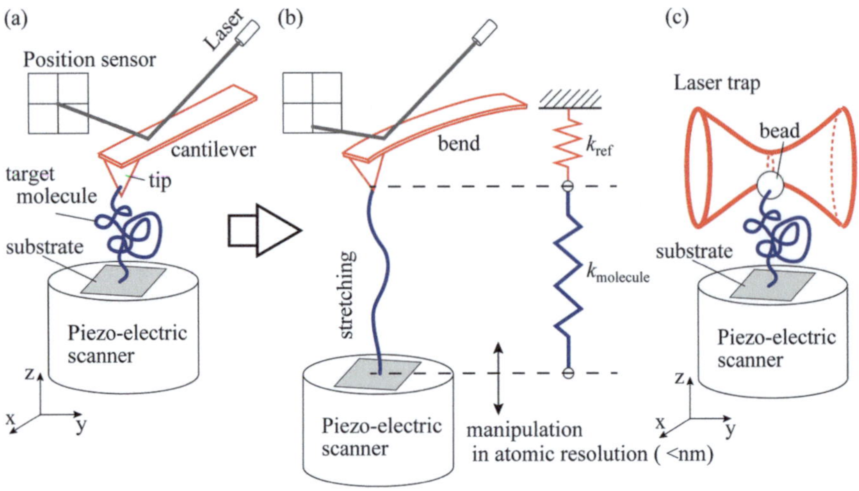

Fig. 1.1 Schematic diagram of typical setup of (**a**) atomic force microscope (AFM) and (**c**) optical tweezers (OT). The end-to-end distance of the target molecule, immobilized on AFM tip and interacting with substrate in this example, will be manipulated by piezoelectric scanner. (**b**) The cantilever can be modelled as reference spring, and the deformation of the target molecule will be controlled by piezo-scanner. (**c**) In OT, target molecule immobilized on bead can be deformed, which is fixed by reference spring, as spring constant, k_{ref}, using laser trapping

on the sample surface and (b) precise spatial control of the target molecule are required.

1.2.1 Principle of Single-Molecule Measurement by AFM or Optical Tweezers

The atomic force microscope (AFM), like other SPMs, scans the surface of the sample by using physical probe to make surface images. The force applied to the physical probe reflects the interaction between the probe and the sample surface, and atomic resolution can be realized when appropriate probe and stage scatter were employed. For high spatial resolution, a sharp tip and soft cantilever are required for the detection of small differences. The most commonly used method for detecting the probe–substrate interactions is the optical detection of cantilever deflection, as shown in Fig. 1.1.

The degree of cantilever bending can be estimated from the position of the laser spot reflected from the backside of the cantilever. Using a soft cantilever, extremely small force differences, such as the height gap due to the existence of single molecule, can be estimated. This high sensitivity is useful not only for molecular scale imaging but also for estimation of the force in response to the mechanical deformation of a single molecule, as shown in Fig. 1.1a and modelled in Fig. 1.1b. The raw measurement data is a position reflecting the displacement of the AFM cantilever. By multiplying the spring constant of the cantilever, k_{ref}, by this displacement, the force applied to the cantilever can be estimated. By using a soft cantilever, i.e. low k_{ref} probe as a reference spring (spring constant of AFM cantilever, k_{ref} in Fig. 1.1b), in comparison to the target molecule, $k_{molecule}$ in Fig. 1.1b, the sensitivity of the force increases due to the larger deflection of the cantilever. For more detailed information about the principle and instruments, refer review papers [2–5].

Micro- or nanofabrication techniques, i.e. silicon etching, are used to produce cantilevers soft enough to sense the deformation of a target single molecule (Fig. 1.1b). In fact, the limit of the actual sensitivity of the molecular deformation comes from thermal fluctuation, rather than the design of the softness of the cantilever ($=k_{ref}$ in Fig. 1.1b).

Optical tweezers can also be used for single molecular study. Highly focused light can hold a dielectric particle, such as a polystyrene bead, in 3D space via constant spring constant [6, 7]. As shown in Fig. 1.1c, a laser trap was employed for trapping a bead by a spring constant k_{ref}, which is determined by the laser power used for laser trapping. The bead position is determined using optical imaging, and the force applied to the target molecule will be estimated from the k_{ref} and the displacement of the bead, which is essentially the same as the case in AFM. By adjusting the laser power, k_{ref} can be varied in the OT system, and a lower k_{ref}, i.e. soft trapping, can be achieved in OT compared to AFM. This potentially enables weaker forces to be applied and measured, although again the force detection limit comes not from the softness of the reference spring but from thermal noise around room temperature (see review papers for more detailed information [7–9]).

1.2.2 Rheology

We focus on viscoelastic objects made of polymers in the broad research area of rheology. In single-molecule experiments on polymers, changes in physical characteristics in response to the mechanical stimulation, such as the force in response to the elongation of single molecule and force caused by intermolecular interactions, are measured. This is not directly the same information as that obtained from conventional rheology experiments in the macroscopic scale, since some physical quantities in the macroscopic scale cannot simply be estimated in the single molecular level. For instance, stress σ cannot be simply evaluated, due to the uncertainty of the cross-sectional area of single molecule. Later in this chapter, Sect. 1.4.4, we will discuss the relationship between single-molecule unbinding properties and the frequency dependence of dynamic shear moduli.

1.3 Information Obtained Using Force Probes and Rheology in Studies of Polymer-Based Systems

Mechanical properties of single molecules can be investigated by using the methods mentioned above. The first applications of single-molecule force probe to investigate interactions aimed at rupture force determination of various types of chemical bonds [10] and force measurements between complementary strands of DNA [11].

Here, a series of typical results will be shown as examples. A force–distance curve shows the measured force applied on the single molecule as a function of distance from the substrate, i.e. the end-to-end distance of the molecule. The data presented as force is originally recorded as the change in position of the laser spot on the optical detector, measured in Volts. This voltage is then converted to the change in distance between AFM probe and substrate by measuring the deflection in nm corresponding to a particular voltage. Finally, this distance is converted to an applied force by multiplying it by the cantilever spring constant k_{ref}.

Typical examples of force–distance curves are shown in Fig. 1.2a. In the approach process of AFM tip to the substrate surface, no force will be detected until direct contact when using an intact tip and substrate having no long-range interaction force between them. After contacting the substrate, a linear force increase is observed due to the deflection of the cantilever (it is the slope of this deflection that is used to convert V to nm in the process described above). In the reverse direction (retrace process), essentially the same trace is reproduced as in the approach curve. For an AFM tip immobilized with target polymers, a characteristic force curve is sometimes observed in the retrace process, reflecting the stretching of single polymer chain, as shown in Fig. 1.2b.

The force–distance profiles obtained in such studies allow the determination of polymer chain stiffness at the single polymer level. The chain stiffness of poly(methacrylic acid) (PMAA) has been estimated by fitting with the wormlike chain model (WLC) or the FJC model [5, 12], and force-induced transitions from boat to chair conformations have been reported for polysaccharides [13]. From the detailed analysis of unbinding forces of single molecular pair interactions, as obtained from force curves of the type displayed in Fig. 1.2b, information about the energy landscape map reflecting molecular interactions in the single molecular scale can be obtained. This information includes the width and height of energy barriers and dissociation rate constant (lifetime of bound state). This type of experiment and analysis is called dynamic force spectroscopy (DFS) [14] and has become established over the last 20 years as an important method for understanding the interactions between binding partners in ligand–receptor pairs [15] as well as many other cases [16]. Among the candidate molecular pairs that may be studied usefully with this technique are those that form cross-links in polymeric solutions and gels and whose properties influence the rheology of those states [17].

Sometimes, multiple stretching events can be observed in the single force–distance measurement (Fig. 1.2c), especially when immobilization density is high. In this case, it is thought that each stretching type force curve reflects stretching of different polymer chain. A different type of force–distance curve is observed

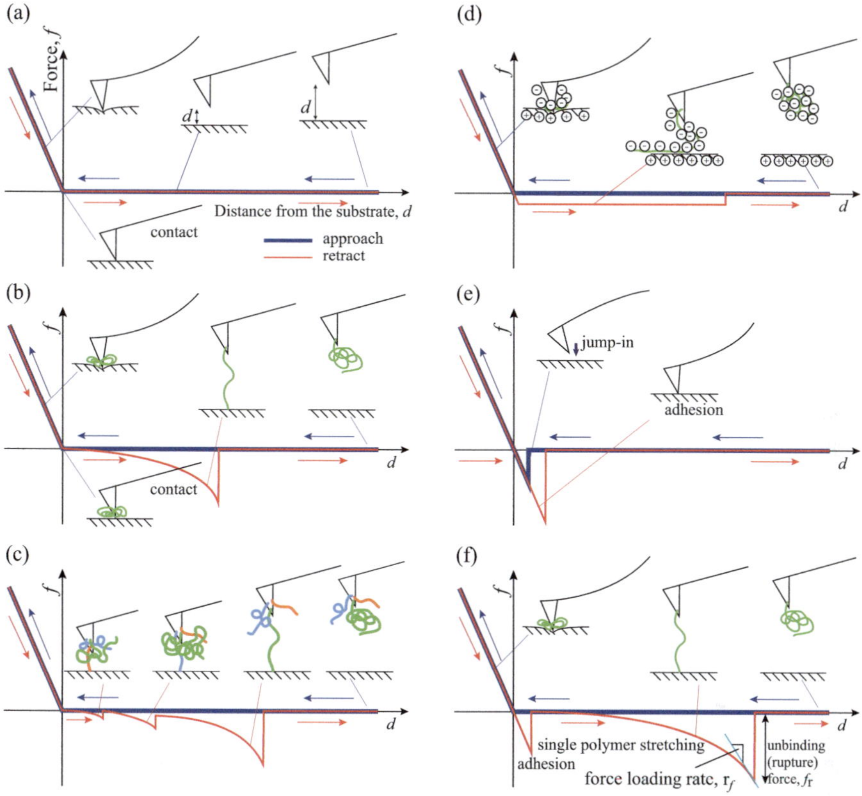

Fig. 1.2 (**a**) No molecule on tip/substrate. No interaction is observed when tip is apart from the substrate and the cantilever bended after contacting (**b**) single-molecule stretching. No interaction during approach to the subset, and the fig shows characteristic behaviour reflecting polymer chain stretching and unbinding during retrace curve. (**c**) Multiple molecules were stretched and unbound at different distances from the substrate. (**d**) Peeling, constant force is sometimes observed to pull the molecule bound to the substrate, i.e. polyelectrolyte. (**e**) Tip suddenly was pulled into the substrate (jump-in) during approach curve, due to attractive force, and adhesion during retrace process due to the attractive force between tip and substrate after the contact. (**f**) Estimation of characteristic properties of unbinding event of single molecular pair, definition of unbinding (rupture) force f_r, force loading rate r_f, and unbinding distance *d*.

for different types of interaction, as is the case with polyelectrolytes peeled from a substrate.

As shown in Fig. 1.2d, a plateau-like constant force is observed in the retrace curve for a polyelectrolyte—oppositely charged substrate interaction. As the polyelectrolyte is pulled away from the substrate, a constant force is observed as the charge–charge interactions between polyelectrolyte and substrate are broken at a constant rate, in a process that can be imagined to be similar to the peeling of a banana. Some observations of constant force plateaus with twice the value of the fundamental level were also reported. Constant force plateaus are also reported to reflect forced desorption of polymers from solid surfaces [18, 19]. This peeling is characterized by equilibrium of the pullout of the polyelectrolyte within an electrostatic attractive field. A special case in which plateau-like force–distance curves are observed is described later in this work.

Interactions between tip and substrate can be unfavourably observed in some cases rather than interactions of target molecules immobilized on tip and/or substrate, as shown in Fig. 1.2c. Jump-in events, sudden jumps of the AFM tip from the non-contact state into contact due to long-range attractive forces between tip and substrate, and adhesion events during retrace process, reflecting the sticking of the tip on the substrate after contacting, are often observed, depending on the materials of tip/substrate and solvent conditions. These jump-in and adhesion events may overlap the stretching event reflecting single-molecule characteristics and are unfavourable in many cases. The characteristic properties of an unbinding event that are used to analyse the interaction are taken from the force–distance curve, as shown in Fig. 1.2f, i.e. unbinding (rupture) force f_{r}, force loading rate r_f, and unbinding distance d. These are described later in more detail for definition and interpretations.

1.4 Examples of Interactions Within Macromolecules at the Single-Molecule Level and Their Relation to Rheological Properties

1.4.1 Dynamic Force Spectroscopy Applied for HMHEC–Amylose Interactions (In Comparison to HMHEC/HMHEC)

One characteristic property of a polymer gel is gel strength, from the viewpoint of rheology, which is a macroscopic value. The origin of the magnitude of gel strength is not clear in general, since it will be affected by a series of factors, i.e. the strength of cross-linking and lifetime of the cross-linking for physically cross-linked gels.

Using the DFS method introduced above, the lifetime of the cross-link and strength of the cross-linking among molecules can be separately estimated. Takemasa et al. reported that in the HMHEC–amylose aqueous system, the gelation mechanism originates from the short lifetime of physical cross-linking rather than enhancement of the strength of the cross-linking compared with self-association among HMHEC (hydrophobically modified hydroxyethyl cellulose) and between HMHEC and amylose investigated by dynamic force spectroscopy [20], as shown in Fig. 1.3. Cross-links between HMHEC and amylose, forming physical gels (Fig. 1.3b, d), were characterized by DFS and compared with HMHEC–HMHEC hydrophobic interactions among hydrophobe C_{16} alkyl side chains in HMHEC (Fig. 1.3a, c).

HMHEC and water-soluble cellulose derivatives are produced by introducing a hydroxyethyl group and grafting longer alkyl chains to the hydroxyethyl group. These alkyl side chains are referred to as hydrophobes. The viscosity of solutions of

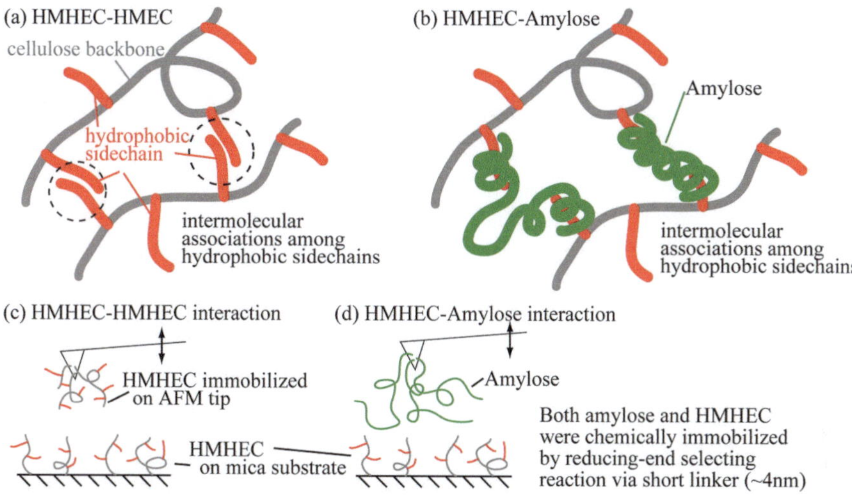

Fig. 1.3 (**a**) Model of HMHEC–HMHEC interactions in aqueous solution, (**b**) model of HMHEC–amylose interaction and schematic representations of immobilization nature of target molecules for (**c**) HMHEC–HMHEC or (**d**) HMHEC–amylose interaction studies. Molecules were chemically immobilized on mica substrate or AFM tip made of SiN, respectively, via short (∼4 nm) linker using end-selective chemical reaction (Takemasa et al. [20])

HMHEC is much higher than expected for a non-associating polymer at the same M_w and chain flexibility, due to hydrophobic interactions among the hydrophobes. It is known that the hydrophobes of HMHEC make complexes with a series of molecules, such as amylose, which can act as a cross-linker among hydrophobes.

Amylose is a water-soluble polysaccharide. It is reported that amylose makes a complex with a variety of molecules, such as alkyl chains [21]. Amylose adopts a helical conformation (V-amylose), and it is suggested that the guest molecule locates inside the helix. By complexation, the steady shear viscosity of a HMHEC aqueous solution increases drastically by addition of amylose, and this is explained by cross-linking due to the complexation of hydrophobes of HMHEC and amylose [22].

Figure 1.4 shows typical force–distance curves obtained for (a) HMHEC immobilized on the AFM tip/HMHEC on mica substrates and (b) HMHEC on mica interacting with amylose molecules on the tip.

A few molecules immobilized on the tip may simultaneously interact with molecules on substrate, which results in complex results due to superposition of stretching type force–distance curves originating from each molecule, as seen in (a-1 to a-5) and (b-1 to b-10). In the retrace curves of (a-1 to a-5) and (b-1 to b-5), short-range adhesion peaks arising from non-specific adhesion events were observed, and these should be eliminated from the detailed analysis of the single molecular pair interaction. Well-separated peaks were occasionally observed (a-8 to a-10, b-3 to b-10) at distances far from the substrate surface, which can be interpreted as

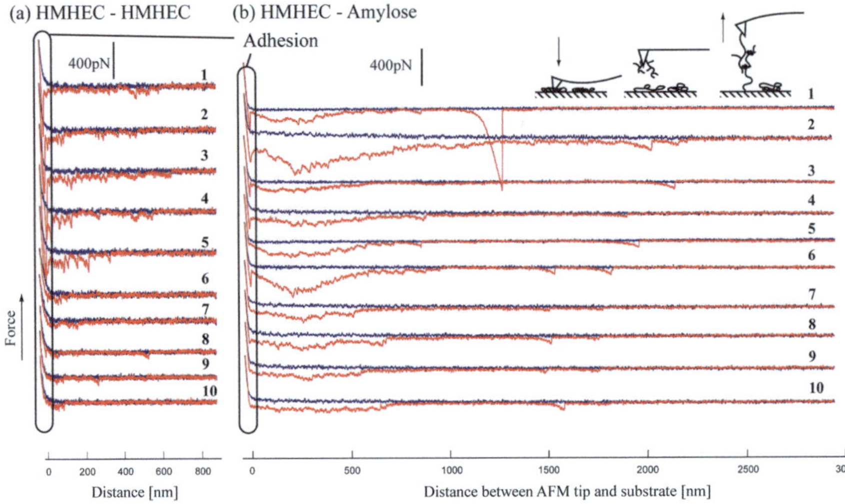

Fig. 1.4 Typical force–distance curves for (**a**) HMHEC–HMHEC and (**b**) HMHEC–amylose interactions. The blue line represents force behaviour when approaching to the substrate, and the red does retrace curve (Figure was reproduced based on data from reference [20])

stretching and unbinding events with rapture force f_{r} ~100pN originating from single molecular pair interactions.

Unbinding events occurred at longer distances from the substrate, more than 2,000 nm for HMHEC–amylose, compared to only up to 600 nm for the HMHEC–HMEC interactions. This difference reflects the difference in length of the two polymers, with HMHEC chains being up to ~300 nm in length, whereas amylose chains may extend up to 2 μm. The observation that the total contour lengths of two molecules agreed with the above-mentioned observed rupture distances for each case constitutes strong evidence that the unbinding events observed originated from single-molecule interactions. The total areas of overlapped peaks are larger for HMHEC–amylose pairs than those for HMHEC–HMHEC (Fig. 1.4).

According to Bell and Evans et al. [23–25], the intermolecular potential is affected by the external pulling force applied to the molecules. This results in the decrease of the activation energy and a subsequent increase of the dissociation rate. Under the constant force loading rate r_f, the probability density $P(f_{\mathrm{r}})$ to observe a bond rupture as a function of pulling force f is given by [26]

$$P\left(f_r\right) = \frac{1}{\tau}\exp\left(\frac{x_\beta f_r}{k_B T}\right)\exp\left(\frac{k_B T}{\tau x_\beta r_f}\left(1 - \exp\left(\frac{x_\beta f_r}{k_B T}\right)\right)\right) \tag{1.1}$$

where τ is the lifetime of the bound state in the absence of external force ($\tau = 1/k^0$, where k^0 is the dissociation rate constant), f_r is the rupture force, and x_β is the thermally averaged distance from the bound complex to the transition state projected

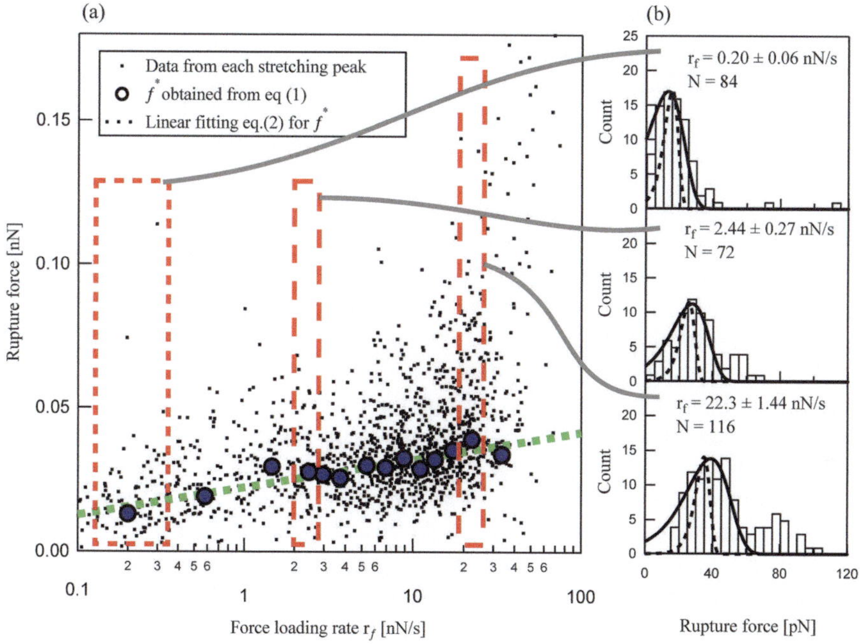

Fig. 1.5 (**a**) Rupture force plotted against force loading rate, r_f, observed for HMHEC immobilized on mica/amylose on the AFM tip. The rupture force and force loading rate were estimated for each stretching peak. (**b**) Histograms of the rupture force distribution for HMHEC–amylose interactions calculated for subintervals based on r_f indicated as *dotted box* in the figure (**a**). The solid and *dotted curves* represent *lines* fitted by Eq. 1.1 for the distribution of the histogram in the limited r_f, and the calculated curve obtained with x_β and τ estimated from the linear fitting for all the data with r_f in all the range (see text). The peak of this line corresponds to the most probable rupture force, f^*, predicted by Eq. 1.1. Figure was made based on data in the literature [20]

along the direction of the applied force. The force loading rate $r_f = dF/dt$, the actual force increase on the molecules as they are being forced apart. r_f can be estimated for each unbinding event as the slope of the unbinding stretching event just before the rupture in force–distance curve, dF/dx, and the retraction speed of the AFM cantilever, dx/dt, where x is the distance between the substrate and the AFM tip.

The unbinding force ensembles (Fig. 1.5) were grouped into subsets based on r_f. Histograms of the f_r were calculated for each subset and fitted by Eq. 1.1 (Fig. 1.5b). The distribution of f_r was well accounted for by Eq. 1.1, and the peak value of the f_r, f^*, increased with increasing r_f.

The values, x_β and τ, were estimated for each subset with the mean force loading rate, $<r_f>$, and plotted in Fig. 1.5. The bond strength f^* increased with increasing r_f, as expected based on the relation,

$$f^* = \frac{k_B T}{x_\beta} \log\left(\frac{\tau x_\beta}{k_B T}\right) + \frac{k_B T}{x_\beta} \log\left(r_f\right) \quad (1.2)$$

where f^* is most probable rupture force obtained from Eq. 1.1 where $P'(f) = 0$. Equation 1.2 means that a single energy barrier empirical model results in the linear relation between f^* and log r_f. From the experimentally obtained f^* as a function of r_f, x_β and τ can be estimated based on Eq. 1.2.

The experimental data can be reproduced well by Eq. 1.2. The linear relation between f^* and log r_f was basically satisfied. Based on assumption of a one barrier model, determination of x_β and τ by applying Eq. 1.2 is possible, and the estimated values of x_β and τ were 0.99 nm and 0.89 s, respectively, for HMHEC–amylose and 0.31 nm and 0.075 s for HMHEC–HMHEC interactions.

The obtained x_β and τ are shorter and smaller for HMHEC–HMHEC interactions than those for the HMHEC–amylose. The longer interaction lifetime of HMHEC–amylose pairs is consistent with larger macroscopic viscosity and longer related relaxation time for the properties at the ensemble level [22].

The observed behaviours in the single molecular study, such as (1) high probability of unbinding event observed in the single molecular pair interaction study and (2) larger peak area (i.e. larger work), are also consistent with the macroscopic behaviour reported, i.e. an increase of the viscosity originating from molecular associations among hydrophobes in HMHEC [27] and gelation of HMHEC in the presence of amylose [22].

1.4.1.1 Alternative Models for DFS Analysis

While in the work described above and many others, the linear relationship between f^* and log r_f means that the Bell–Evans model may be successfully employed to estimate the values of the parameters x_β and τ; in other more complex cases, the estimation relies on the assumption that a departure from a linear relationship between f^* and ln(r) implies more than one energy barrier. Recently, it is argued [28] that the Bell–Evans model is too simplified and alternative models, such as Dudko–Hummer–Szabo model [29] and Friddle–Noy–De Yoreo model (Eq. 1.3a–c) [30], have been introduced. The latter model aims to reconcile the regimes whereby (i) the loading rate r is sufficiently slow that a bond that is broken by the applied force may rebind before the probe has moved away (known as the near-equilibrium regime) and (ii) the regime where the loading rate is sufficiently fast that the probe has moved away from the substrate before the broken bond has time to reform (known as the kinetic regime). The Bell–Evans model is valid in the kinetic regime and attributes non-linearity in the resultant dynamic force spectrum to the existence of multiple energy barriers in the forced rupture pathway. The Friddle–Noy–De Yoreo model recognizes that rebinding at slow loading rates represents an equilibrated situation and so allows for equilibrium free energy of the bond to be extracted alongside the kinetic parameters.

The DFS data is treated in the same way as described above for the Bell–Evans model and fitted to (Eq. 1.3a) [30]. Values for parameters that characterize the bond rupture ($f_\beta = k_B T/x_\beta$, the thermal force scale where x_β is the distance to the transition state, k_{off} the rate of bond dissociation (derived from $k_{\mathrm{u(feq)}}$ using Eq. 1.3b) and f_{eq} the equilibrium unbinding force) are then extracted.

$$\langle f \rangle \cong f_{\mathrm{eq}} + f_{\beta}.\ \ln\left(1 + e^{-0.577}.\frac{r}{f_{\beta}.k_{u\left(f_{\mathrm{eq}}\right)}}\right) \tag{1.3a}$$

$$k_{\mathrm{off}} = k_{u\left(f_{\mathrm{eq}}\right)} \Big/ \exp\left[\frac{\left(f_{\mathrm{eq}} - \frac{1}{2}k_{\mathrm{eff}}.x_{\beta}\right)}{f_{\beta}}\right] \tag{1.3b}$$

$$\Delta G_{\mathrm{bu}} = \frac{\left(f_{\mathrm{eq}}\right)^2}{2k_{\mathrm{eff}}} \tag{1.3c}$$

The particular value of the Friddle–Noy–De Yoreo model is that it allows the estimation of the free energy of unbinding ΔG_{bu} for the single-molecule interaction from the values of f_{eq} and k_{eff}, the effective spring constant (Eq. 1.3c).

1.4.2 Slide Contact Force Spectroscopy

Two salient features of DFS as it has usually been conducted to date are (i) the bond to be broken is formed transiently during the approach of the probe to the surface, since in a typical DFS experiment, binding between two molecules can only occur during the short time (typically 10^{-3}–1 s) that probe and surface are in proximity, and (ii) the bond thus formed is between two 'fixed' molecules. Consequences arising from these conditions include a suboptimal sampling of slowly formed bonds (due to the limited time available for bonds to be formed) and a lack of control over the geometry of the binding partners (since it is not trivial to arrange the binding partners into specific orientations relative to each other), both of which may be highly relevant in complex cases such as alginate gelation. An alternative iteration of dynamic force spectroscopy which is just beginning to be applied to rheological problems may be called 'sliding contact' (dynamic) force spectroscopy (hereafter SCFS).

The SCFS method (Fig. 1.6) uses a cyclodextrin (CD)-based macrocycle tethered to the AFM probe, with the polymer to be interrogated tethered likewise to another surface and induced to form a host-guest complex with the macrocycle to form a polyrotaxane [31] or molecular ring-thread complex. The experiment has parallels with the nanopore sequencing approach that is under continuous development as a DNA sequencing tool [32, 33] and which has recently been shown to discriminate between different PEG polymers on the basis of their molecular weight, with monomer resolution [34]. The most common terminology [35] for describing the processes occurring in a rotaxane depicts a macrocyclic 'bead' shuttling between 'stations' on the polymer axle before finally 'dethreading' from the axle. In the

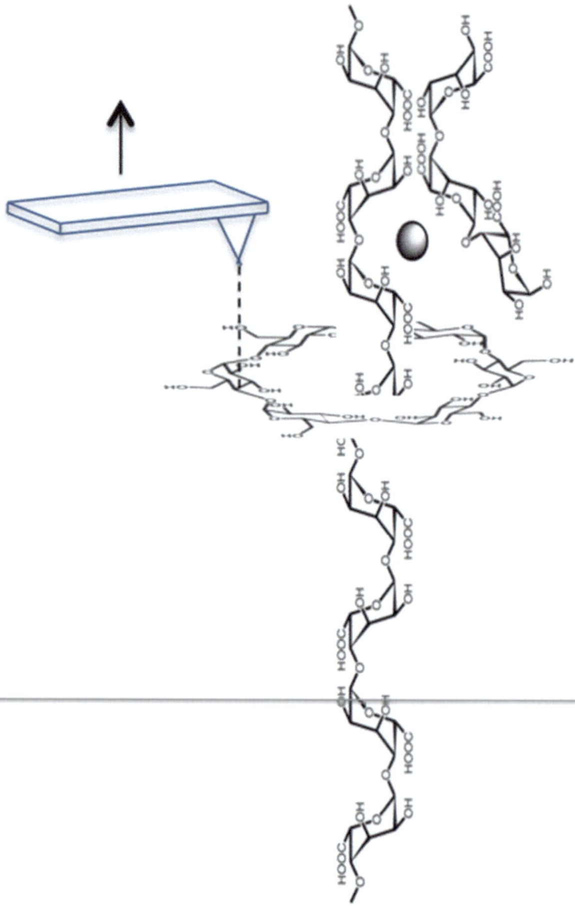

Fig. 1.6 Sliding contact force spectroscopy may be used to 'unzip' a junction zone

SCFS method described here [36, 37], the bead is α- or β-cyclodextrin (α-CD and β-CD hereafter), and the stations are the individual monomers comprising the polymer axle, while the AFM probe supplies the driving force for shuttling the bead between stations under a load generated by the controlled separation of probe and sample. The concept of manipulating a rotaxane using an AFM probe has been explored before: Komiyama and co-workers [38] used STM (Scanning Tunnelling Microscopy) to manipulate α-CD beads forming a polyrotaxane with PEG back and forth along the PEG axle, while Stoddart et al. [39] and Lussis et al. [40] have used AFM to measure the force required to drive a bead between two stations in a rotaxane. It has also been considered as a potential sequencing tool for DNA and other polymers [37, 41, 42].

While SCFS offers the opportunity to measure the difference in interaction between the sliding CD and the monomers in the polymer axle during sliding, the

CD may also be used as a molecular 'zipper' to unzip molecules bound to sites incorporated into the polymer. From the point of view of rheological experiments, this 'unzipping' application of SCFS offers an alternative that potentially addresses the two drawbacks of conventional DFS highlighted above. Firstly, because the AFM probe is functionalized to pick up a tether attached to the functionalized CD and does not need to form the bond to be broken during the approach cycle of the experiment, SCFS offers the freedom to allow the bond of interest to be formed under ideal conditions and timescales. Secondly, once the CD has been picked up by the AFM probe, the retraction of the probe drives the CD along the polymer, so the direction of the unzipping action is controlled by the sequence and orientation of binding sites along the polymer.

Knowledge of the kinetic (x_β, the distance to the transition state, and k_{off}, the rate of bond dissociation) and thermodynamic (f_{eq}, the equilibrium unbinding force and the free energy of unbinding ΔG_{bu}) parameters characteristic of the cross-links and/or junction zones in a polymer solution or gel will aid our understanding of the formation and rupture of these materials under deformation and may be used to guide design of new ones. Specific examples include the complex rheological behaviours of alginate and pectin gels [43–45], which are proposed to be due to a hierarchy of non-covalent cross-links of different strengths and lifetimes. If we can probe these characteristic parameters over a range of timescales from milliseconds to minutes, then we will gain a clearer understanding of the processes occurring during gelation and deformation of a gel. An example of particular relevance here is the oligo(guluronate)-Ca^{2+} junction zone that acts as a cross-link in alginate gels [46]. A recent study [47] used both DFS and SCFS to characterize the Ca^{2+}-mediated interaction between oligogalacturonates ('oligo-Gs'), with the aim of establishing at the single-molecule level the minimum length of galacturonate oligomer required to form an 'egg-box' junction zone in alginate. Oligo-Gs with lengths from 6 to 18 monomers [48] were coupled to short PEG spacers and attached to both an AFM probe and a mica surface. Because the coupling reaction involved the opening of one of the uronide rings and because the stoichiometry of the uronic acid–Ca^{2+}–uronic acid complex is 2:1:2, the series of oligo-Gs used represented examples of junction zones formed between pairs of oligo-Gs by complexation with two, three, four, five and up to eight Ca^{2+} ions. Hereafter we call this complex between a pair of diuronic acids and a Ca^{2+} ion a junction zone unit.

An analysis of the work required to separate pairs of oligo-Gs in a DFS experiment (Work of separation, W_{sep}) revealed a complex picture (Fig. 1.7a). The work distributions showed that even for the shortest oligo-Gs, strong bonds, requiring more than 1 nN·nm to separate, could be formed between individual pairs in the presence of Ca^{2+} ions. In the absence of Ca^{2+}, or in the presence of excess EDTA, very few non-specific interactions were observed. A discontinuity in the magnitudes of W_{sep} between junction zones made by three junction zone units and those made by four units suggested that when at least eight guluronic acids were present in each of the paired sequences, a much stronger bond was formed (mean W_{sep} per Ca^{2+} ion increased from 0.2 to 0.6 nN·nm). However, for longer oligo-Gs, the values of W_{sep} reduced; until for sequences 16–18 residues long (potentially

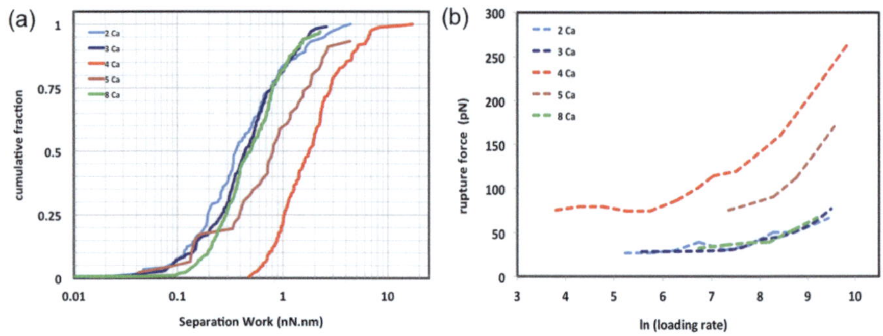

Fig. 1.7 (**a**) Cumulative fraction of work of separation for the oligo-G series with capacities from 2 to 8 Ca^{2+} ions. (**b**) Dynamic force spectra for the same oligo-G series

Table 1.1 Values of parameters fitted to Friddle–Noy–de Yoreo model for oligo-G series with capacities from two to eight Ca^{2+} ions

	2	3	4	5	8
x_β (nm)	0.15 ± 0.10	0.10 ± 0.10	0.05 ± 0.11	0.05 ± 0.03	0.10 ± 0.01
k_{off} (s^{-1})	31.9 ± 23.3	53.4 ± 31.8	6.3 ± 3.4	20.9 ± 6.7	45.8 ± 8.1
f_{eq} (pN)	26.3 ± 3.3	24.8 ± 4.7	73.0 ± 6.9	59.0 ± 6.0	25.7 ± 2.9
ΔG_{bu}(kJ·mol^{-1})	17.4 ± 4.8	11.1 ± 6.5	143.9 ± 9.2	41.7 ± 8.8	13.9 ± 3.7

capable of forming eight-junction zone units), the distribution was identical to that seen for the shortest oligomers. Analysis of the dynamic force spectra using the Friddle–Noy–De Yoreo model [30] provided a similar picture (Fig. 1.7b). Table 1.1 presents the values extracted from the Friddle–Noy–De Yoreo model (Eq. 1.3a–c) for the fitted parameters x_β, the distance to the transition state; k_{off}, the rate of bond dissociation; and f_{eq}, the equilibrium unbinding force. The values for the minimum free energy difference between the bound and unbound states (ΔG_{bu}, calculated from the values of f_{eq} obtained at the low loading rate limit of the dynamic force spectrum) show a maximum for the four junction zone units, increasing from 11 to 140 kJ/mol (equivalent to approximately 0.5–5 times the strength of a hydrogen bond [49–51]).

Such a result is counterintuitive: The strength of the interaction between the pairs of the junction zone, as reflected by W_{sep} and f_{eq}, would be expected to increase with length of junction zone. NMR analysis confirmed the sequence lengths and purities of the oligo-Gs used, but a consideration of the nature of the DFS experiment suggests a possible reason for the decrease in W_{sep} for the longer junction zones: the probe and surface spend only a short time in sufficiently close proximity to each other for bonds to form (in the data discussed here, the time is no more than 50 ms). Thus, it is possible that on this timescale, larger junction zones do not have enough time to form. However, this only explains why the eight-unit oligo-G does not form a stronger bond than the four-unit oligo-G. To account for a decrease in bond strength, we must propose that the weaker interaction is competing with the

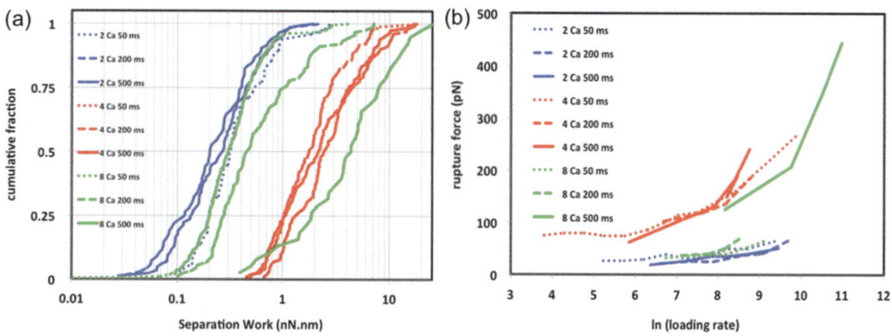

Fig. 1.8 (**a**) Cumulative fraction of work of separation for the oligo-G series with capacities of two, four and eight Ca^{2+} ions with increased surface dwell times of 50, 200 and 500 ms. (**b**) Dynamic force spectra for the same oligo-G series

junction zone interaction for the available Ca^{2+}-binding sites and that the weak interaction is more kinetically favourable and so forms more rapidly. When this is the case, for long (>5 units) oligo-Gs, the increased availability of binding sites for Ca^{2+} favours the weaker interaction over the entropically disfavoured junction zone.

The hypothesis that junction zones longer than four units take longer than 50 ms to form was tested for two-, four- and eight-junction zone unit oligo-Gs by increasing the dwell time of the AFM probe at the surface to 200 and 500 ms. Figure 1.8a, b shows how the work distributions and dynamic force spectra changed as the dwell time was increased. For the two- and four-unit oligo-Gs, both the work distributions and dynamic force spectra for single unbinding events showed little to no change when the interaction time at the sample surface was increased from 50 to 500 ms, suggesting that for these short oligo-Gs, 50 ms is the sufficient time for the maximum possible junction zone to displace the weaker Ca^{2+}-dependent interaction. However, for the longest oligo-G, consisting of up to eight-junction zone units, the work distribution and dynamic force spectrum changed significantly, from being identical to the two-unit oligo-G at 50 ms to extending to higher loading rates, rupture forces and W_{sep} than the four-unit oligo-G at 500 ms. From this we conclude that the largest junction zone that can form between oligo-Gs of 16 residues takes longer than 200 ms to displace the initial, weak bond(s) that is (are) formed and may not yet have reached equilibrium after 500 ms. A more general point here is that long dwell times are necessary in conventional DFS experiments in order to capture all the interactions that may be critical to the performance of a gel formed from them.

In the corresponding SCFS experiment, free oligo-G was allowed to bind to oligo-G tethered to the surface by PEG, as in the DFS experiment, but with the addition of α-CD which was threaded over the PEG polymer to form a pseudorotaxane prior to binding the polymer to the surface. A tether on the α-CD was terminated with an amine group, and an AFM probe was functionalized with another PEG spacer, this time terminated with a succinimide group. At neutral pH,

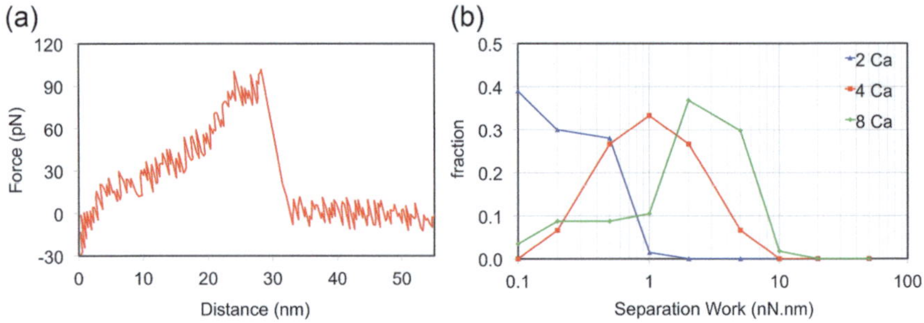

Fig. 1.9 (**a**) Typical force vs. separation curve for a SCFS experiment. After an initial stretch of the PEG spacer, the cyclodextrin slides over the oligo-G, unzipping the junction zone. Here the plateau is 3.2 nm long at an unzipping force of 62 pN. (**b**) Histogram of work distributions for unzipping junction zones by SCFS in the oligo-G series with capacities of two, four and eight Ca^{2+} ions

a strong bond was rapidly formed [36] between the amine and succinimide groups, and this bond allowed the AFM probe to manipulate the α-CD along the polymer chain and over the oligo-G. Where the oligo-G was participating in a junction zone interaction, the free oligo-G was displaced by the α-CD acting as a zipper, separating the two oligo-Gs residue by residue. The resulting force–distance spectrum showed an initial stretch of the PEG tether followed by a short plateau as the two oligo-Gs separated (Fig. 1.9a). The forces at which the plateaus occur, multiplied by their length, gave the work done to unzip the interaction: $W_{sep} = F_{plat} \times d_{plat}$. Figure 1.9b shows the distributions of the unzipping work for the two-, four- and eight-unit oligo-Gs. Similar to the case of the DFS experiment using long dwell times, the unzipping work depended on junction zone length, as expected since the free oligo-G had at least minutes to bind to the tethered oligo-G before the α-CD disrupted it. Thus, this simple modification of the DFS experiment extended the temporal range over which measurements of individual junction zones could be made from milliseconds to minutes.

While both experiments permit the measurement of the work done to separate the two oligo-Gs that make up the junction zone, the values obtained differ significantly. This may be attributed to the different modes of unbinding the oligo-G pair, as illustrated in Fig. 1.10a–c where the unbinding modes are shown in the context of a deforming gel. In the SCFS experiment (c), the junction zone is unzipped one unit at a time as the CD slides along the polymer, whereas in the DFS experiment (b), junction zone failure occurs instantaneously when the bond is subject to sufficient tension. In the latter case, sufficient work must be done to overcome every junction zone unit in the bond simultaneously, whereas in the unzipping case, each unit may be separated individually in turn. The difference in observed forces is magnified by the fact that the instantaneous rupture of the junction zone in the DFS experiment may occur far from equilibrium. The bond's longer lifetime can mean that the loading rate at which force is applied to the bond causes the work done on the bond to increase rapidly before it finally ruptures, so that more rupture events occur away

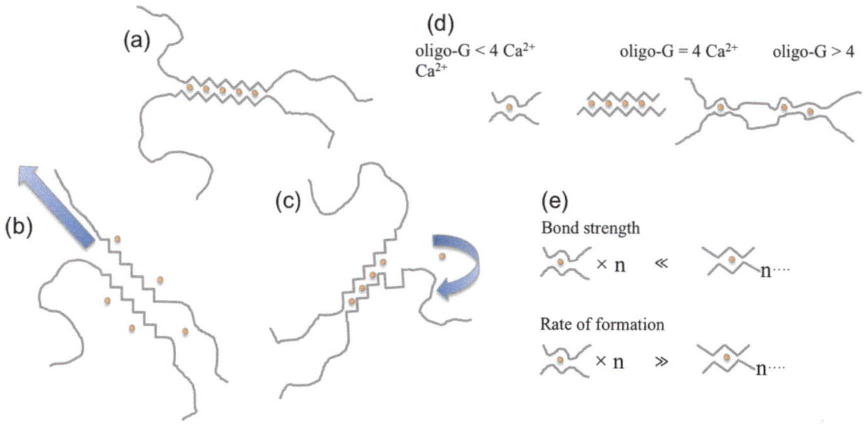

Fig. 1.10 (**a**) A junction zone in alginate, subject to (**b**) shearing, as in dynamic force spectroscopy, and (**c**) unzipping, as in sliding contact force spectroscopy. The *arrows* indicate both the direction an AFM probe acts to separate the junction zone and the direction of deformation in a gel required to rupture the junction zone by each mode. (**d**) The predominating interaction between pairs of oligo-Gs and Ca^{2+} at oligo-G lengths below, equal to and greater than 4. (**e**) The relative bond strength and rate of formation of the two interactions. A *dot* represents a Ca^{2+} ion, freehand lines represent oligo-G segments forming weak interactions, and *straight lines* represent oligo-G segments taking part in egg-box-like interactions

from, and at larger forces than, the equilibrium condition described by the value of f_{eq}. These factors combine to account for the large difference in values for W_{sep} for the two methods, where the mean work measured in the DFS experiment is $\sim 3\times$ the mean work measured in the SCFS experiment.

Therefore, experimental analysis of the interactions of the individual components of the junction zones responsible for the rheological behaviour of gels suggests that the DFS experiment is reflecting a dynamic hierarchy of binding events between oligo-Gs in presence of Ca^{2+}, limited by the time available for binding to occur (0–50 ms dwell times). Within that timeframe, weak bonds are formed between the shortest oligo-G pairs in the presence of Ca^{2+}. For oligo-Gs with four junction zone units, a stronger bond is able to form within this timeframe, reflecting the onset of 'egg-box' junction zone formation, but for longer oligo-Gs, this stronger bond does not form within 50 ms. Instead, the weaker, rapidly forming bond predominates in these interactions as it does for the shortest oligo-Gs. In order to account for the decrease in bond strength observed for oligo-Gs longer than five units, the increased availability of Ca^{2+} binding sites must favour the formation of the weak interaction to an extent that outcompetes the formation of junction zones from this initial state. A candidate for the identity of the interaction forming at short times is the 'tilted egg-box' proposed by Massimiliano-Borgnani et al. [52], as it is Ca^{2+} dependent. At present, the progression from this initial chelated state to the final junction zone is not fully understood [53], and it is this process that the work discussed here sheds light upon. If bonds are allowed time to form (and here we have explored the changes in binding behaviour out to 500 ms in the DFS experiment and out to many minutes in the SCFS experiment), the longer junction zones can form and so more work is required to separate them. At these longer timescales, the strength of interactions in the eight-unit oligo-Gs catches up with and surpasses those of the four-unit oligo-Gs, which themselves do not increase significantly in strength with longer dwell times. Thus, the expected dependence of the interaction strength on junction zone size is restored. Figure 1.10d, e depicts the various bonds formed and their relative strengths and association/dissociation rates. Larobina et al. [43] have recently shown that it is a hierarchy of different interactions, occurring over different timescales, that determine the relaxation behaviour of alginate gels, and experiments like the ones described in the present work provide a way to gain insight into these interactions at the molecular level.

1.4.3 Interactions Between Mucin and Other Biomacromolecules

Mucins are large extracellular proteins that are heavily glycosylated with complex oligosaccharides that potentially engage in numerous interactions. They are a principal constituent of the mucous membranes that lines bodily cavities and canals that lead to the outside, e.g. the respiratory, digestive and urogenital tracts. These complex molecules have been extensively studied, and the molecular properties and functional attributes of mucins were recently reviewed [54]. The documented functions of the mucosal surfaces include gaseous exchange especially in the respiratory tree, nutrient and cofactor adsorption in the gut, transparency at the ocular surface and general roles such as lubrication and chemical sensing. In addition, mucins are believed (through their interaction with other biomolecules) to control many normal and pathological processes in the body [55]. The understanding of how mucins influence these complex processes at the molecular level is in its infancy. However, several papers published over the last decade address the involvement of mucins in intermolecular recognition processes. In the following, we review and compare some of these papers, and we discuss how rheology and single-molecule approaches have been used to obtain increased understanding of the systems and how the data obtained using these two complementary techniques are compared.

Complex oligosaccharides including mucins are involved in the control of many normal and pathological processes through their interaction with other biomacromolecules. The interaction of mucins with lectins has been extensively studied, but this topic is outside of the scope of the present paper. Additionally, reviewed in the following, mucins interact with other biomacromolecules. Improved understanding of these interactions is important for the area of targeted drug delivery to mucosal surfaces. Such targeted delivery often relies on coupling of the drug to functional building blocks that interact with the surface of interest. The resulting polymer-based particles can be designed to facilitate either mucoadhesion or mucopenetration [56]. However, the incomplete understanding of the interactions occurring between mucin and other macromolecules is a challenge for the development of drug delivery vehicles that penetrate rapidly through or establish strong adhesion to the mucosal surface [57]. In order to aid the development of the understanding within this field, we thus argue that it is beneficial to meet the challenges with a broad range of experimental techniques, including both single-molecule techniques and more traditional approaches including rheology. Below we summarize some of the recent insight that has been gained by the combined use of these two approaches.

Chitosan and alginate are among the polymers suggested to mediate mucoadhesive properties in the context of drug delivery applications. Given the fundamental role played by mucins in the mucus gel, it is natural to consider mucoadhesion of such polymers in terms of their intermolecular interactions with mucins. The associative interaction of purified ovine and porcine submaxillary mucins with alginate has been evaluated by comparing the rheological properties of mixtures against those of pure alginate or pure mucin solutions [58]. Based on the obtained rheological data, alginate was found to have a significant interaction with both mucins. The study further concluded that the nature of the binding sites remained unclear, but concluded that non-covalent interactions other than electrostatic play an important role. The interaction of mucins with alginate was later addressed using single-molecule techniques [59], as shown in Fig. 1.11. In addition to alginate, this study also considered the interaction between mucins and chitosan. Two chitosans of high molecular weight with different degrees of acetylation—thus possessing different solubility profiles in aqueous solution as a function of pH—and two alginates with different fractions of α-guluronic acid were employed. The interaction strength was determined using AFM-based forced dissociation of the macromolecular interactions in aqueous solutions. When probed against immobilized

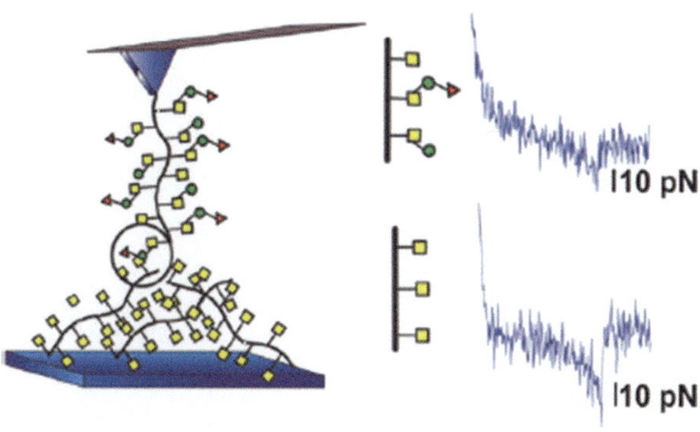

Fig. 1.11 Schematic illustration of measurement setup for mucin interaction studies. Mucin interactions are studied using AFM-based force spectroscopy employing AFM tips and sample surfaces onto which mucins are covalently linked. When studying heterotypic mucin interactions, the mucins immobilized on the AFM tip and the sample surface have differing glycan compositions (Data from reference [59] with permission)

mucin, both chitosans and alginates revealed unbinding profiles characteristic of localized interactions along the polymers. The investigation of the chitosan–mucin interaction revealed signatures arising both from an electrostatic mechanism and hydrophobic interaction.

The topic of molecular interactions between polysaccharides and mucin has also been addressed using microviscosimetry and SAXS [60]. The magnitude of deviation between the viscosity of mixed polysaccharide–mucin solutions and the corresponding individual stock solutions was taken as an indication of underlying molecular interactions. Furthermore, SAXS and fluorescence spectroscopy were used to study the mechanism of interaction in two systems (Alg4-mucin and Alg400-mucin where Alg4 refers to an alginate with M_w 4 kDa and M/G ratio 1.42 and Alg400 alginate characterized by $M_w = 406$ kDa and M/G ratio 0.95) of which one had shown a gain in viscosity and the other a loss relative to the viscosity of the pure mucin solution. The data obtained showed that the charge neutral polysaccharides investigated, i.e. dextran and *Streptococcus thermophilus* exopolysaccharide, did not interact with mucin, whereas most of the charged polysaccharides tested showed a greater or lesser tendency to interact with mucin. The microviscosimetry, scattering and spectroscopy data obtained in this study formed the basis for a proposed model of interaction between mucin and anionic polysaccharides as a function of M_w, charge and degree of contraction, as shown in Fig. 1.12. The mucin molecules are in this model described as two globular protein regions connected by highly glycosylated linkers. The glycosylated regions are considered to be extended due to a high density of negatively charged groups at pH > 2.5. The model further assumes, despite the net negative charge of mucin, the existence of positively charged patches in the nonglycosylated globular regions, forming interaction sites with negatively charged polysaccharides. Low-M_w and stiff polyanions are in the model assumed to interact preferentially with these globular regions without influencing the conformation of the mucin, thus having a negligible impact on bulk properties such as size and viscosity. This interpretation is in accordance with the fluorescence quenching data which indicated exposure of

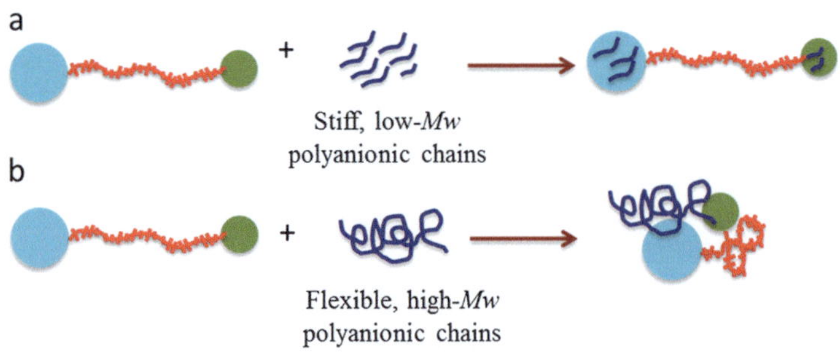

Fig. 1.12 Proposed model for the interaction between alginate and the double globular comb structure of mucin as a function of alginate M_w and chain flexibility (Data from reference [60] with permission)

aromatic residues from the globular protein domains, consistent with penetration of the polysaccharide and successive rearrangement of the globular protein structure. This rearrangement may also favour cross-linking of mucin, consistent with the observed slight increase in viscosity. In contrast, high-M_w polyanions are believed to bridge distant sites and thus influence the conformation of the mucin, with a resulting reduction in the overall hydrodynamic volume and thus a reduced viscosity of the mixture.

The study by Menchicchi and co-workers [60] was performed on dilute mucin and alginate solutions, characterized by a relative viscosity of ~2.0 in water. At higher mucin and alginate concentrations than applied by Menchicchi and co-workers, high-M_w alginates are observed to increase the viscosity and elasticity of mucus [61]. Interestingly, rheological studies have shown that a low-M_w guluronate fraction of alginate (oligoguluronates) can conversely reduce the viscosity and elasticity of mucus matrices including those containing alginate [62]. The rheological data presented by Nordgard and co-workers were by the authors taken to be compatible with the theory that low-molecular-weight guluronate fraction of alginate (oligoguluronates) is able to compete with polymeric alginates at mucin-binding sites, inhibiting the mucin–alginate interactions, reducing intermolecular cross-link density and long-range order and thereby weakening and eventually collapsing the gel [62]. These results lead to the suggestion that pulmonary administration of oligoguluronates may be a viable strategy for improving sputum mobility and thus affording symptom relief in cystic fibrosis (CF) lung disease, which is characterized by a build-up of thick mucus that is unable to be cleared by the mucociliary clearance system, resulting in mucus stasis and severely reducing lung function. As this was the case, it became imperative to demonstrate that these types of interactions occurred through molecular coupling and decoupling rather than through indirect physical changes in the mixtures. The interaction between the oligoguluronates, alginate and mucin was therefore later investigated at the single-molecule level through a determination of the forced unbinding properties between mucin and alginate molecular pairs in aqueous solution and how these properties are affected by the presence of oligoguluronate at different concentrations [63]. More precisely, the effect of added oligoguluronates on the applied work needed to separate two opposing surfaces, one coated with mucin and the other with alginate molecules, was determined. The data obtained by this experimental approach demonstrated that interactions on a molecular level do indeed take place in this system. Furthermore, an analysis of interaction patterns between alginate and mucin immobilized on the AFM tip probed independently against immobilized oligoguluronate suggests that oligoguluronate interaction with mucin is the mechanism underlying the observed suppression of the alginate–mucin interactions. The oligoguluronate concentration-dependent suppression of mucin–alginate interactions increases the molecular understanding of oligoguluronates' action as an apparent mucolytic agent of potential relevance for application to improve lung function in patients suffering from cystic fibrosis.

A similar experimental approach as described in Sletmoen et al. was later used to determine the interaction force between a mucin macromolecule and a sodium alginate gel layer [64]. The study addressed changes in the work of adhesion between mucin and alginate hydrogels depending on the time of contact, the G/M ratio of the alginate and the cross-link ratio of the gel. The results obtained showed that the maximum work of adhesion and the strength of the adhesive force increased with the decrease of the guluronic content of the alginate and increased with increasing contact time up to a saturation level. This was taken to indicate that the mannuronic acid, part of the alginate, plays an important role for mucoadhesion.

In addition to the documented interaction between mucins and other biomacromolecules, mucins may also influence normal and pathological processes through direct carbohydrate-based self-interactions. Complementary binding between carbohydrates on neighbouring cells has for a long time been considered either non-specific or non-canonical, but the recognition of carbohydrates by complementary carbohydrates could, however, be biologically relevant [65]. We have addressed the topic of carbohydrate–carbohydrate interactions by quantifying the strength and lifetimes of self-associations between single molecular pairs of mucin molecules with different carbohydrate decoration patterns [66, 67]. The observations revealed that the probabilities for self-interactions, as well as their strengths and lifetimes, depend on the precise decoration pattern. The results provided in these studies document the ability of complex carbohydrates found on glycoconjugates to modulate interaction strengths through variation in lifetime of low-affinity binding sites created by oligosaccharide groups of varying structure and relative location. The results also document that sensitive force probes make possible direct observations of the rupture of carbohydrate-mediated intermolecular bonds and, therefore, have the potential to lead to important advances in our understanding of the relationship between structure and function in carbohydrate interactions. A more system-specific conclusion that was drawn is related to the suggested role of the Tn carbohydrate cancer antigen in modulating the self-association and signalling properties of membrane-bound mucins in normal and transformed cells [66, 67]. The self-association ability of the mucins was found to be relatively small for fully glycosylated mucins but was more pronounced for mucins displaying the truncated glycans including Tn and STn glycan epitopes, which are the most frequent aberrant O-glycosylation found in epithelial cancer cells.

The findings described above cannot easily be compared to rheological data since to our knowledge, similar well-defined mucin samples with modified carbohydrate composition have not been studied using rheology. Another challenge faced when addressing the topic of self-interactions using rheology is connected to the inability to compare with a mixture where such interactions cannot form. Also, rheological studies of pig gastric mucus gels have revealed that for high mucin densities, both transient and non-transient interactions are responsible in maintaining the gel matrix [68]. In conclusion, it seems that much is still to be discovered related to the precise identity and relative importance of the interactions occurring between mucins. Carbohydrate–carbohydrate interactions as observed in the AFM studies might be one of the several interactions contributing to the viscoelastic properties of mucins, in addition to entanglement of sugar side chains, electrostatic interactions between charged groups and hydrophobic interactions between the protein core.

1.4.4 Combined Single-Molecule and Rheological Approach to Studies of Polyelectrolyte Complexes

Complexes between macromolecular anions and cations are one group of entities where the correspondence between properties acquired by single-molecule techniques and rheology was recently explored for a particular polyelectrolyte pair [69]. In general, the mixing of polyanions and polycations yields polyelectrolyte complexes driven by interactions where the counterion exchange is often a major contribution. The polyelectrolyte complexation may be described as a counterion exchange reaction when blending polyanions (PA) and polycations (PC) with their respective counterions (c^+, c^-) in a solvent [70, 71]:

$$\left(c^+PA^-\right)_p + \left(PC^+c^-\right)_m \rightarrow \left(PC^+PA^-\right)_y + \left(c^+PA^-\right)_{p-y} + \left(PC^+c^-\right)_{m-y} + yc^- + yc^+ \quad (1.4)$$

It is assumed that the charge stoichiometry involved in the interaction is 1:1, but there can be an excess of one of the components. Depending on the mixing ratio of polyanions (p) and polycations (m), either one or both polyions may be present as uncomplexed polyelectrolytes following the mixing.

The widespread abundance of charge-driven macromolecular interactions in biology and man-made materials illustrates the importance of understanding their molecular behaviour at various length scales. The fundamental polyampholyte character of proteins is one reason for the abundance of biological polyelectrolyte complexes in biology. The chromatin fibre composed of DNA and histones is one example, where the presence of charge and consequent electrostatic interactions is pivotal in their organization. The wrapping of the DNA around the histone complex is guided by charged patches of the histone octamer [72], and millions of octamers along its DNA chain with a repeat distance of about 200 bp resembling an overall pearl-necklace structure wrap the DNA to largely reduce the volume occupancy in a cell [73]. Nucleosome–nucleosome interactions appear to be important in the further wrapping of the chromatin, where one histone tail is suggested to play a crucial role [74–76]. Various applications of polyelectrolyte complexes have recently been summarized [77]. Design, synthesis and characterization of polycations as non-viral delivery vehicles for gene therapy, for coding either DNA sequences or siRNA sequences, are an active research area, e.g. [78] among these examples. Another application domain is within drug delivery [79].

The mixing of polyanions and polycations may lead to dispersed, geometrically well-defined structures such toroids and rods as reported in the case of complexes involving semi-flexible polyanions such as DNA [80] or xanthan [81]. Under other well-defined conditions, the mixing of polyanions and polycations in aqueous solution leads to phase separation into a polymer-rich phase and polymer-depleted phase [82]. The polymer-rich phase, known as the coacervate, is sensitive to molecular and aqueous solution parameters due to its balance of electrostatic-mediated interactions, polymer features and solvation. The linear viscoelasticity

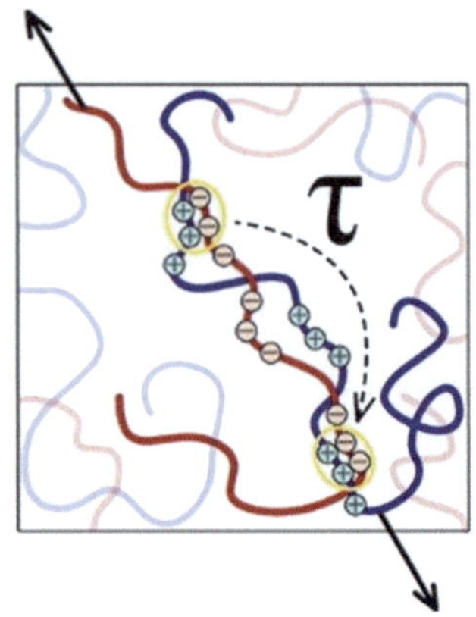

Fig. 1.13 Schematic illustration of aspects of molecular relaxation in polyanion–polycation blends in the complex coacervate state. The lifetime of the localized electrostatically mediated interaction sites along the polyanion (*red chain*) and polycation (*blue chain*) affects the overall relaxation of the dynamic coacervate state (Data from reference [69] with permission)

of the complex coacervate state of PDMAEMA (poly-N,N-dimethylaminoethyl methacrylate) and PAA (polyacrylic acid) was recently reported with emphasis on polymer molecular parameters, mixing ratios of the blends and solvent conditions in the resulting relaxation spectra [69]. The localized, electrostatically based interactions were foreseen to represent 'sticky points' in the relaxation behaviour (Fig. 1.13) tremendously slowing down the relaxation process compared to classical Rouse- or Zimm-like relaxation.

The rheological properties of the complex coacervates were obtained on samples sedimented in the aqueous solution and loaded in the sample cell while due consideration was taken to prevent adverse effects of solvent evaporation during the data acquisition. Following sample relaxation, shear modulus $G(t)$, shear storage G' and loss moduli G'' were determined experimentally. The estimate of the relaxation spectra $H(\tau)$ was calculated based on the assumption that the relaxation of the complex coacervation could be approximated by the sum of the Maxwell elements,

$$G(t) = \sum_{i}^{M} G_i e^{-t/\tau_i} \tag{1.5}$$

$$G'(\omega) = \sum_{i}^{M} \frac{G_i \omega^2 \tau_{i^2}}{1 + \omega^2 \tau_{i^2}} \tag{1.6}$$

$$G''(\omega) = \sum_{i}^{M} \frac{G_i \omega \tau_i}{1 + \omega^2 \tau_i} \tag{1.7}$$

$$H(\tau) = \sum_{i}^{M} G_i \delta \left(1 - {}^{t}\!/_{\tau_i}\right) \tag{1.8}$$

where M are the number of relaxation modes, each with relaxation strength G_i and associated relaxation time τ_i. The relaxation times were assumed to be equally spaced on a logarithmic scale in estimating the relaxation time spectra $H(\tau)$. The estimates of $H(\tau)$ were obtained by iterative fitting of the logarithm of both the time-dependent relaxation data and frequency dependent storage and loss moduli.

The experimentally determined stress relaxation (Fig. 1.14a) and the derived relaxation spectra (Fig. 1.14b) revealed features strongly depending on the chain length of the molecular partners and ionic strength of the solution. Such features should also be considered in view of the various concentrations in the coacervates at different chain lengths. The polymer volume fraction within the complexes is reported to vary between 5 % and 40 % for the decreasing chain lengths. The stress relaxation data (Fig. 1.14a) were reported for nearly equal degree of polymerization of the polyanion and polycation as well as chain length mismatches. There are clear trends in the slowing down of the stress relaxation with increasing chain lengths and decreasing ionic strength.

The time constants derived from the mechanical spectra were compared to those extracted from single-molecule interactions recently reported by the same group [83]. The interaction between a cationic block copolymer (poly(N-methyl-2-vinylpyridum)-block-polyethylene oxide) covalently attached to triangular SiN_3 AFM probes and a polyanionic brush was determined at ionic strengths from 0.5 to 3 M, which is a range similar to that employed in the rheological characterization. Facilitation of electrostatic interactions was achieved by covalently end linking the

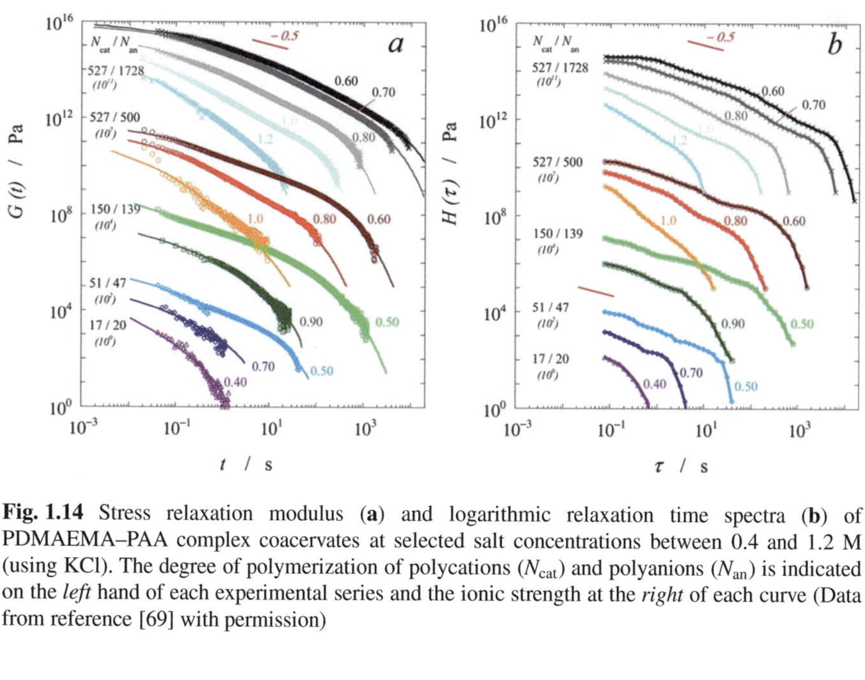

Fig. 1.14 Stress relaxation modulus (**a**) and logarithmic relaxation time spectra (**b**) of PDMAEMA–PAA complex coacervates at selected salt concentrations between 0.4 and 1.2 M (using KCl). The degree of polymerization of polycations (N_{cat}) and polyanions (N_{an}) is indicated on the *left* hand of each experimental series and the ionic strength at the *right* of each curve (Data from reference [69] with permission)

cationic block copolymer to the AFM tip through its PEO block. Numerous force–distance profiles were recorded and used as a basis for analysis of unbinding forces and particular unbinding profiles. The speed of approach and retraction controlled by the z-piezo was kept constant at 500 nm/s. Among the two observed unbinding profiles for this system, the first one (depicted I) was characterized by a constant force plateau in an extended retraction distance followed by an abrupt transition to a lower force (Fig. 1.15a), which is essentially the same as Fig. 1.2d. Thus, it was suggested a peeling-type desorption mechanism accounts for the type I force–distance profiles where there was a parallel chain orientation between the polycation attached to the AFM tip and polyanion of the interacting surface.

In general, the type I unbinding plateau force profiles can be expected when the individual interaction sites, ionic pairs in this case, associate and dissociate at timescales faster than that of the pulling process [19]. The additional type of unbinding profiles, referred to as type II, was characterized by a polymer stretching signature preceding the unbinding event, classified into Fig. 1.2b. The slower association–dissociation rate underpinning the occurrence of the stretching signature was suggested to arise from polycationic chains penetrating the anionic brush and possessing an antiparallel chain orientation. Both type I and II occurred each in about 8 % of the total number of force–distance profiles. Unbinding signatures similar to that identified as type II for the mentioned polycation–polyanion system are reported for a wide variety of macromolecular pairs and occur in general when the imposed force ramp occurs on a timescale that is faster than the characteristic lifetime of the interaction.

Salt concentration was used as a key parameter for affecting the relaxation spectra and single-molecule unbinding features. For type I interactions, Spruijt and co-workers reported that the mean rupture force observed at about 110 pN at 0.5 M NaCl decreased upon increasing salt concentration to 75 pN at 1.2 M NaCl. The decreasing mean unbinding force in this range of ionic strength was interpreted based on salt effect of single ionic bonds. Based on the theory for the lowering of the energy barrier due to an external force f affecting the dissociation rate $k(f)$ of a single rupture process is given by,

$$k(f) = \omega_0 g(f) \exp\left(\frac{E_a - f x_\beta}{k_B T}\right) = k_0 g(f) e^{f/f_\beta} \tag{1.9}$$

In Eq. 1.9, $k_0 = \omega_0 \exp(E_a/k_B T)$ is the dissociation rate in the absence of an external force, x_β is the distance from the energy minimum in the bound complex to the transition state described earlier in Sect. 1.4.1, E_a is the activation energy for the dissociation, $k_B T$ is the thermal energy, $g(f)$ is of order 1, and $f_\beta = k_B T / x_\beta$. The activation energy is calculated as the difference between the effective electrostatic energy of the cation–anion pairs in contact and the situation where both types of

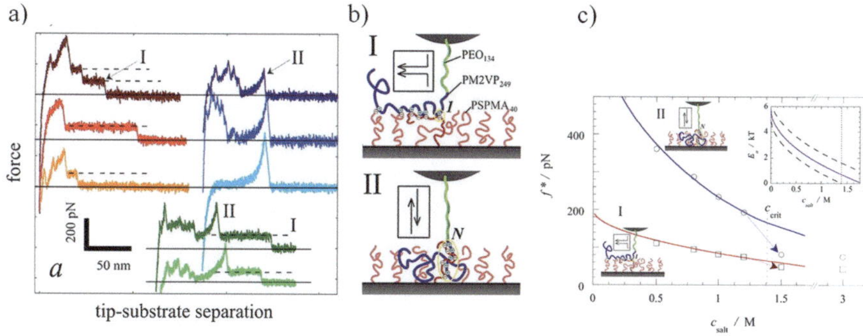

Fig. 1.15 (**a**) Experimentally determined force–distance profiles for forced dissociation between polycation linked to an AFM tip and a polycation brush observed in aqueous 1 M NaCl solution. Ladder-like desorption profiles or stretching of polymer chains just prior to the unbinding events, referred to as type I and II, respectively, was observed among the trials with successful anchoring events. (**b**) Interpretation of type I desorption profiles as peeling of ionically bound train-like segments and type II polymer chain stretching and dissociation originating in polycation–polyanion pair interactions embedded within the brush. (**c**) Most probable unbinding force for type I and II unbinding profiles versus salt concentration. The symbols depict the experimental data, whereas the lines are fit of the theory to the data. Activation energy for dissociation of hydrated ionic bonds versus ionic strength (*inset*) (Data from reference [83] with permission)

ions are surrounded by salt corresponding to the separated polymer chains:

$$E_a = k_B T\left(-\kappa l_B + l_B/d\right) = k_B T\left(-a\sqrt{c_{\text{salt}}} + b\right) \tag{1.10}$$

In Eq. 1.10, κ is the inverse Debye length, l_B is the Bjerrum length (0.71 nm in pure water at room temperature) and d is the contact distance of the ion pair. The inverse Debye length depends on the square root of the salt concentration, and the constants a and b follow from the rearrangement. Application of this theory yields estimates of activation energies in the range 1–6 k_BT for the single ion pair bonds when fitted to the single-molecule data (Fig. 1.15). The lifetime of single anion–cation pairs in a hydrated environment was thus found to range from milliseconds at high salt concentrations to much more extensive lifetimes, e.g. up to years for complexes held together by multiple interactions and at low ionic strengths.

It is interesting to compare the information extracted from the single-molecule data. The relaxation spectra revealed a continuous decrease of $H(\tau)$ up to the terminal relaxation time, and a general appearance reported to be consistent with Rouse or Zimm dynamics. Gucht and co-workers proceeded the analysis by rescaling the relaxation spectra obtained at various ionic strengths (Fig. 1.16).

The shift factor τ_c is proportional to the inverse of k_O (Eq. 1.9) and is estimated to depend on the salt concentration, analogous with the single-molecule experiments (Fig. 1.16, inset). The rescaled relaxation spectra reveal a plateau at short times ($\tau_1 < \tau < \tau_2$) and a regime where the system undergoes Rouse-like relaxation ($\tau_2 < \tau < \tau_3$). The transition from the plateau to the Rouse-like relaxation, τ_2, was

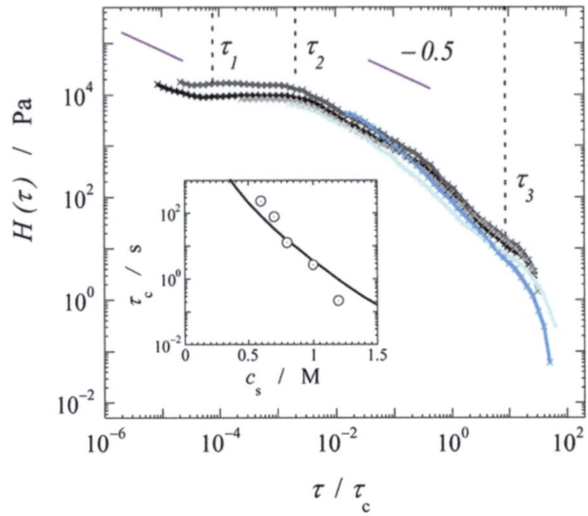

Fig. 1.16 Rescaled relaxation spectra for complex coacervates prepared using N_{cat}/N_{an} = 527/1728 at different salt concentrations. The *inset* depicts the time scaling factors versus the ionic strength, where fit to the theory (*line*) adequately accounts for the data (*symbols*) (Data from reference [69] with permission)

attributed to the lifetime of a single ionic pair making up a 'sticky point'. The lifetime of a single ionic bond was thus estimated from the relaxation spectra to 200 ms at 0.6 M salt concentration, in the same order of magnitude as that obtained from the single-molecule data. The overall relaxation of the coacervate state (Fig. 1.13) involving multiple bonds was additionally considered by including the number of possible interactions supported by the actual chain lengths. In such a framework, the data from such studies can be combined to obtain an improved understanding on the importance of properties at the single-molecule level also on the rheological properties.

1.5 Concluding Remarks

The past decades have seen substantial progress in the understanding of single-molecule interactions, with respect to both the theoretical foundation and the application to numerous molecular groups. During the same time period, macromolecular rheology has further matured, but it is only recently that the interrelation between these approaches is pursued. Intuitively, one would expect that the characteristic features of molecular interactions, such as strength and lifetime, among macromolecules will affect the rheological properties. In this context, one can describe covalently connected polymer networks or gels as molecular entities connected by covalent bonds of (practically) infinite lifetime and strength larger than imposed by external forces. Such systems can be expected to store deformation energies that are regained upon relaxing external forces. In mechanical spectroscopy, these systems will yield frequency-independent storage moduli. At the other extreme, polymers that do not show interactions mediated by specific domains along the chain (e.g. zero lifetime for the single molecular tests) yield long-term rheology with flow behaviour (although depending on parameters such as concentrations, wtemperature and solvent viscosity). Many polymers, both naturally occurring and man-made, include molecular groups hosting interaction capacities, with respect to interaction strengths and lifetime. These systems can, generally, be expected to be intermediate between noninteracting and covalently connected ones. The challenge is to build a quantitative understanding of the rheological properties that include the understanding of the interaction parameters of the sites within the macromolecular constituents. While the single-molecule approach is essential in building a molecular understanding of the nature and energy landscape of these interactions (including also cofactors), the connection to the rheological properties also needs information on the density of interaction sites hosted by individual macromolecules and their overall concentration. In the present overview, we have discussed some examples characterized by both differences in macromolecular origin and energy landscape parameters. The literature is not yet abundant with examples where both the energy landscape parameters of interaction sites of the macromolecules and rheological properties have been determined. Although it is difficult, for this reason, to present a comprehensive overview of how to put this puzzle together, the approaches we review here show potential (e.g. included in the polyelectrolyte complex example) that could represent a possible way forward for other systems as well.

References

1. Gennes, P., *Scaling Concepts in Polymer Physics*. 1979: Cornell University Press
2. Eaton, P. and P. West, *Atomic Force Microscopy*. 2010: Oxford University Press
3. Janshoff, A., et al., *Force spectroscopy of molecular systems – Single molecule spectroscopy of polymers and biomolecules*. Angewandte Chemie-International Edition, 2000. **39**(18): p. 3213–3237
4. Sarid, D. and V. Elings, *Review of Scanning Force Microscopy*. Journal of Vacuum Science & Technology B, 1991. **9**(2): p. 431–437
5. Giannotti, M.I. and G.J. Vancso, *Interrogation of single synthetic polymer chains and polysaccharides by AFM-based force spectroscopy*. Chemphyschem, 2007. **8**(16): p. 2290–2307
6. Ashkin, A., *Acceleration and Trapping of Particles by Radiation Pressure*. Physical Review Letters, 1970. **24**(4): p. 156–159
7. Ashkin, A., et al., *Observation of a Single-Beam Gradient Force Optical Trap for Dielectric Particles*. Optics Letters, 1986. **11**(5): p. 288–290
8. Moffitt, J.R., et al., *Recent advances in optical tweezers*. Annual Review of Biochemistry, 2008. **77**: p. 205–228
9. Williams, M.C., *Optical Tweezers: Measuring Piconewton Forces*, in *Single Molecule Techniques*, P. Schwille, Editor. 2002, Biophysical Society

10. Grandbois, M., et al., *How strong is a covalent bond?* Science, 1999. **283**(5408): p. 1727–1730.
11. Lee, G.U., L.A. Chrisey, and R.J. Colton, *Direct Measurement of the Forces between Complementary Strands of DNA.* Science, 1994. **266**(5186): p. 771–773.
12. Ortiz, C. and G. Hadziioannou, *Entropic elasticity of single polymer chains of poly(methacrylic acid) measured by atomic force microscopy.* Macromolecules, 1999. **32**(3): p. 780–787
13. Marszalek, P.E., et al., *Polysaccharide elasticity governed by chair-boat transitions of the glucopyranose ring.* Nature, 1998. **396**(6712): p. 661–664.
14. Noy, A. and R.W. Friddle, *Practical single molecule force spectroscopy: How to determine fundamental thermodynamic parameters of intermolecular bonds with an atomic force microscope.* Methods, 2013. **60**(2): p. 142–150.
15. Kienberger, F., et al., *Molecular recognition imaging and force spectroscopy of single biomolecules.* Accounts of Chemical Research, 2006. **39**(1): p. 29–36.
16. Rackham, B.D., et al., *Non-covalent duplex to duplex crosslinking of DNA in solution revealed by single molecule force spectroscopy.* Organic & Biomolecular Chemistry, 2013. **11**(48): p. 8340–8347
17. Morris, S., S. Hanna, and M.J. Miles, *The self-assembly of plant cell wall components by single-molecule force spectroscopy and Monte Carlo modelling.* Nanotechnology, 2004. **15**(9): p. 1296–1301
18. Hugel, T., et al., *Elasticity of single polyelectrolyte chains and their desorption from solid supports studied by AFM based single molecule force spectroscopy.* Macromolecules, 2001. **34**: p. 1039–1047
19. Friedsam, C., H.E. Gaub, and R.R. Netz, *Probing surfaces with single-polymer atomic force microscope experiments.* Biointerphases, 2006. **1**(1): p. MR1-MR21
20. Takemasa, M., M. Sletmoen, and B.T. Stokke, *Single Molecular Pair Interactions between Hydrophobically Modified Hydroxyethyl Cellulose and Amylose Determined by Dynamic Force Spectroscopy.* Langmuir, 2009. **25**(17): p. 10174–10182.
21. Egermayer, M., M. Karlberg, and L. Piculell, *Gels of hydrophobically modified ethyl (hydroxyethyl) cellulose cross-linked by amylose: Effects of hydrophobe architecture.* Langmuir, 2004. **20**(6): p. 2208–2214.
22. Chronakis, I.S., M. Egermayer, and L. Piculell, *Thermoreversible gels of hydrophobically modified hydroxyethyl cellulose cross-linked by amylose.* Macromolecules, 2002. **35**(10): p. 4113–4122
23. Bell, G.I., *Models for Specific Adhesion of Cells to Cells.* Science, 1978. **200**(4342): p. 618–627.
24. Evans, E. and K. Ritchie, *Strength of a weak bond connecting flexible polymer chains.* Biophysical Journal, 1999. **76**(5): p. 2439–2447
25. Friedsam, C., M. Seitz, and H.E. Gaub, *Investigation of polyelectrolyte desorption by single molecule force spectroscopy.* Journal of Physics-Condensed Matter, 2004. **16**(26): p. S2369-S2382
26. Evans, E. and K. Ritchie, *Dynamic strength of molecular adhesion bonds.* Biophysical Journal, 1997. **72**(4): p. 1541–1555
27. Karlson, L., K. Thuresson, and B. Lindman, *A rheological investigation of the complex formation between hydrophobically modified ethyl (hydroxy ethyl) cellulose and cyclodextrin.* Carbohydrate Polymers, 2002. **50**(3): p. 219–226.
28. Hane, F.T., S.J. Attwood, and Z. Leonenko, *Comparison of three competing dynamic force spectroscopy models to study binding forces of amyloid-beta (1–42).* Soft Matter, 2014. **10**(12): p. 1924–1930
29. Dudko, O.K., G. Hummer, and A. Szabo, *Intrinsic rates and activation free energies from single-molecule pulling experiments.* Physical Review Letters, 2006. **96**(10): 108101.
30. Friddle, R.W., A. Noy, and J.J. De Yoreo, *Interpreting the widespread nonlinear force spectra of intermolecular bonds.* Proc. Natl. Acad. Sci. U. S. A., 2012. **109**(34): p. 13573–13578.
31. Harada, A. and M. Kamachi, *Complex-Formation between Poly(Ethylene Glycol) and Alpha-Cyclodextrin.* Macromolecules, 1990. **23**(10): p. 2821–2823.

32. Wanunu, M., *Nanopores: A journey towards DNA sequencing.* Physics of Life Reviews, 2012. **9**(2): p. 125–158
33. Laszlo, A.H., et al., *Decoding long nanopore sequencing reads of natural DNA.* Nature Biotechnology, 2014. **32**(8): p. 829–833.
34. Baaken, G., et al., *High-Resolution Size-Discrimination of Single Nonionic Synthetic Polymers with a Highly Charged Biological Nanopore.* Acs Nano, 2015. **9**(6): p. 6443–6449
35. Wenz, G., B.H. Han, and A. Muller, *Cyclodextrin rotaxanes and polyrotaxanes.* Chemical Reviews, 2006. **106**(3): p. 782–817.
36. Dunlop, A., et al., *Mapping the positions of beads on a string: dethreading rotaxanes by molecular force spectroscopy.* Nanotechnology, 2008. **19**(34): 345706
37. Ashcroft, B.A., et al., *An AFM/rotaxane molecular reading head for sequence-dependent DNA structures.* Small, 2008. **4**(9): p. 1468–1475
38. Shigekawa, H., et al., *The molecular abacus: STM manipulation of cyclodextrin necklace.* Journal of the American Chemical Society, 2000. **122**(22): p. 5411–5412
39. Brough, B., et al., *Evaluation of synthetic linear motor-molecule actuation energetics.* Proceedings of the National Academy of Sciences of the United States of America, 2006. **103**(23): p. 8583–8588
40. Lussis, P., et al., *A single synthetic small molecule that generates force against a load.* Nature Nanotechnology, 2011. **6**(9): p. 553–557
41. Voulgarakis, N.K., et al., *Sequencing DNA by dynamic force spectroscopy: Limitations and prospects.* Nano Letters, 2006. **6**(7): p. 1483–1486
42. Qamar, S., P.M. Williams, and S.M. Lindsay, *Can an atomic force microscope sequence DNA using a nanopore?* Biophysical Journal, 2008. **94**(4): p. 1233–1240.
43. Larobina, D. and L. Cipelletti, *Hierarchical cross-linking in physical alginate gels: a rheological and dynamic light scattering investigation.* Soft Matter, 2013. **9**(42): p. 10005–10015
44. Siviello, C., F. Greco, and D. Larobina, *Analysis of linear viscoelastic behaviour of alginate gels: effects of inner relaxation, water diffusion, and syneresis.* Soft Matter, 2015. **11**(30): p. 6045–6054.
45. Mansel, B.W., et al., *Zooming in: Structural Investigations of Rheologically Characterized Hydrogen-Bonded Low-Methoxyl Pectin Networks.* Biomacromolecules, 2015. **16**: p. 3209–3216.
46. Stokke, B.T., et al., *Distribution of Uronate Residues in Alginate Chains in Relation to Alginate Gelling Properties.* Macromolecules, 1991. **24**(16): p. 4637–4645
47. Round, A.N., *et al*, in Preparation
48. S. Ballance et al., *Preparation of high purity monodisperse oligosaccharides derived from mannuronan by size-exclusion chromatography followed by semi-preparative high-performance anion-exchange chromatography with pulsed amperometric detection.* Carbohydrate Research 2009, **344**(2), p. 255–259.
49. Rico, P., et al., *High-Speed Force Spectroscopy Unfolds Titin at the Velocity of Molecular Dynamics Simulations.* Science, 2013. **342**(6159): p. 741–743.
50. Suresh, S.J. and V.M. Naik, *Hydrogen bond thermodynamic properties of water from dielectric constant data.* Journal of Chemical Physics, 2000. **113**(21): p. 9727–9732.
51. Hakem, I.F., et al., *Temperature, pressure, and isotope effects on the structure and properties of liquid water: A lattice approach.* Journal of Chemical Physics, 2007. **127**(22): p. 224106.
52. Borgogna, M., et al., *On the Initial Binding of Alginate by Calcium Ions. The Tilted Egg-Box Hypothesis.* Journal of Physical Chemistry B, 2013. **117**(24): p. 7277–7282
53. Borukhov, I., et al., *Elastically driven linker aggregation between two semiflexible polyelectrolytes.* Physical Review Letters, 2001. **86**(10): p. 2182–2185
54. Corfield, A.P., *Mucins: A biologically relevant glycan barrier in mucosal protection.* Biochimica Et Biophysica Acta-General Subjects, 2015. **1850**(1): p. 236–252
55. Hang, H.C. and C.R. Bertozzi, *The chemistry and biology of mucin-type O-linked glycosylation.* Bioorganic & Medicinal Chemistry, 2005. **13**(17): p. 5021–5034
56. Liu, Z.H., et al., *Polysaccharides-based nanoparticles as drug delivery systems.* Advanced Drug Delivery Reviews, 2008. **60**(15): p. 1650–1662

57. Lai, S.K., Y.Y. Wang, and J. Hanes, *Mucus-penetrating nanoparticles for drug and gene delivery to mucosal tissues.* Advanced Drug Delivery Reviews, 2009. **61**(2): p. 158–171
58. Fuongfuchat, A., et al., *Rheological studies of the interaction of mucins with alginate and polyacrylate.* Carbohydrate Research, 1996. **284**(1): p. 85–99
59. Haugstad, K.E., et al., *Direct Determination of Chitosan-Mucin Interactions Using a Single-Molecule Strategy: Comparison to Alginate-Mucin Interactions.* Polymers, 2015. **7**(2): p. 161–185
60. Menchicchi, B., et al., *Biophysical Analysis of the Molecular Interactions between Polysaccharides and Mucin.* Biomacromolecules, 2015. **16**(3): p. 924–935
61. Taylor, C., et al., *Rheological characterisation of mixed gels of mucin and alginate.* Carbohydrate Polymers, 2005. **59**(2): p. 189–195
62. Nordgard, C.T. and K.I. Draget, *Oligosaccharides As Modulators of Rheology in Complex Mucous Systems.* Biomacromolecules, 2011. **12**(8): p. 3084–3090.
63. Sletmoen, M., et al., *Oligoguluronate induced competitive displacement of mucin-alginate interactions: relevance for mucolytic function.* Soft Matter, 2012. **8**(32): p. 8413–8421
64. Popeski-Dimovski, R., *Work of adhesion between mucin macromolecule and calcium-alginate gels on molecular level.* Carbohydrate Polymers, 2015. **123**: p. 146–149
65. Bucior, I. and M.M. Burger, *Carbohydrate-carbohydrate interactions in cell recognition.* Current Opinion in Structural Biology, 2004. **14**(5): p. 631–637
66. Haugstad, K.E., et al., *Enhanced Self-Association of Mucins Possessing the T and Tn Carbohydrate Cancer Antigens at the Single-Molecule Level.* Biomacromolecules, 2012. **13**(5): p. 1400–1409.
67. Haugstad, K.E., et al., *Single molecule study of heterotypic interactions between mucins possessing the Tn cancer antigen.* Glycobiology, 2015. **25**(5): p. 524–534.
68. Taylor, C., et al., *The gel matrix of gastric mucus is maintained by a complex interplay of transient and nontransient associations.* Biomacromolecules, 2003. **4**(4): p. 922–927
69. Spruijt, E., M.A.C. Stuart, and J. van der Gucht, *Linear Viscoelasticity of Polyelectrolyte Complex Coacervates.* Macromolecules, 2013. **46**(4): p. 1633–1641
70. Thünemann, A.F., et al., *Polyelectrolyte complexes.* Adv. Polym. Sci., 2004. **166**: p. 113–171
71. Bucur, C.B., Z. Sui, and J.B. Schlenoff, *Ideal mixing in polyelectrolyte complexes and multilayers: Entropy driven assembly.* Journal of the American Chemical Society, 2006. **128**(42): p. 13690–13691
72. Arents, G. and E.N. Moudrianakis, *Topography of the histone octamer surface: repeating structural motifs utilized in the docking of nucleosomal DNA.* Proceedings of the National Academy of Sciences, 1993. **90**(22): p. 10489–10493
73. Schiessel, H., *The physics of chromatin.* J. Phys. Condens. Matter, 2003. **15**: p. R699–R774
74. Mangenot, S., et al., *Salt-induced conformation and interaction changes of nucleosome core particles.* Biophysics Journal, 2002. **82**: p. 345–356
75. Mangenot, S., et al., *Interactions between isolated nucleosome core particles.* Eur. Phys. J. E, 2002. **7**: p. 221–231
76. Bertin, A., et al., *Role of histone tails in the conformation and interactions of nucleosome core particles.* Biochemistry, 2004. **43**: p. 4773–4780
77. Müller, M., *Polyelectrolyte Complexes in the Dispersed and Solid State II: Application Aspects.* Advances in Polymer Science. Vol. 256. 2014, Berlin, Heidelberg: Springer Berlin Heidelberg. VII, 264 s. 137 illus., 1 illus. in color. : online resource.
78. Bertin, A., *Polyelectrolyte complexes of DNA and polycations as gene delivery vectors.* Advances in Polymer Science, 2014. **256**: p. 103–196
79. Müller, M., *Sizing, shaping and pharmaceutical applications of polyelectrolyte complex nanoparticles.* Advances in Polymer Science, 2014. **256**: p. 197–260
80. Hud, N.V. and K.H. Downing, *Cryoelectron microscopy of l-phage DNA condensates in vitreous ice: The fine structure of DNA toroids.* Proc. Natl. Acad. Sci. U. S. A., 2001. **98**: p. 14925–14930

81. Maurstad, G. and B.T. Stokke, *Metastable and stable states of xanthan polyelectrolyte complexes studied by atomic force microscopy.* Biopolymers, 2004. **74**: p. 199–213.
82. Priftis, D. and M. Tirrell, *Phase behaviour and complex coacervation of aqueous polypeptide solutions.* Soft Matter, 2012. **8**(36): p. 9396–9405
83. Spruijt, E., et al., *Direct measurement of the strength of single ionic bonds between hydrated charges.* ACS Nano, 2012. **6**(6): p. 5297–5303

Chapter 2
Zero-Shear Viscosities of Polysaccharide Solutions

Takahiro Sato

Abstract Solution viscosities are strongly dependent on the molecular weight, the concentration, and the chain conformation of polymers added. We have formulated the intrinsic viscosity $[\eta]$ at infinite dilution and the zero-shear viscosity η at finite concentrations using the molecular theory based on the wormlike chain and fuzzy cylinder models. In the theory, the solution viscosity at finite concentrations is affected by both hydrodynamic and entanglement interactions. The formulated viscosity equations quantitatively predict $[\eta]$ and η as functions of the wormlike chain parameters and the strength of the hydrodynamic interaction and demonstrated that the relative importance of the hydrodynamic and entanglement interactions in the solution viscosity depends on the chain stiffness. We have compared the formulated viscosity equations with experimental results for solutions of three polysaccharides and two synthetic polymers covering a wide range of the chain stiffness.

Keywords Intrinsic viscosity • Zero-shear viscosity • Hydrodynamic interaction • Entanglement interaction • Wormlike chain model • Fuzzy cylinder model

2.1 Introduction

Various polysaccharides are contained in foods and drinks and provide unique rheological properties. In many cases, foods and drinks need suitable viscosity to taste comfortably, and polysaccharides are utilized as rheology control reagents for foods and drinks. The addition of polysaccharides remarkably changes viscosities of foods and drinks, and it is well known that the viscosities are strongly dependent on the molecular weight, the concentration, and the chain conformation of polysaccharides added [1].

To understand the viscosity behavior of polysaccharide solutions, the present chapter deals with the molecular theory on the solution viscosity of linear-chain

T. Sato (✉)
Department of Macromolecular Science, Osaka University, 1-1 Machikaneyama-cho, Toyonaka, Osaka 560-0043, Japan
e-mail: tsato@chem.sci.osaka-u.ac.jp

I. Kaneda (ed.), *Rheology of Biological Soft Matter*, Soft and Biological Matter,
DOI 10.1007/978-4-431-56080-7_2

polymers. The polymer conformation ranges from the rigid rod to flexible random coil, and the solution concentration covers from the dilute limit to the concentrated regime just below the phase boundary concentration where the liquid crystal phase appears.

Linear-chain polymers are usually viewed as the wormlike chain model [2] and characterized by the persistence length (or the stiffness parameter) q, the contour length h per monomer unit, and the thickness (or the diameter) b of the wormlike chain, as well as the monomer unit molar mass M_0 and the molecular weight M. The intrinsic viscosity $[\eta]$ characterizing the viscosity of infinitely dilute polymer solution is calculated in terms of these wormlike chain parameters. With increasing the polymer concentration, the intermolecular interaction contributes to the solution viscosity. We can divide the intermolecular interaction into the hydrodynamic interaction and the entanglement interaction. The latter interaction can be treated by the fuzzy cylinder model [3], which is explained below. The Huggins coefficient k' consists of the two interaction terms. The relative importance of the two terms depends on the persistence length q and the contour length $L = hM/M_0$. Finally, the zero-shear viscosity η over a wide concentration range is quantitatively predicted by a viscosity equation derived by a molecular theory.

The present chapter consists of seven sections. After briefly mentioning the solution viscosity behavior of a polysaccharide as a typical example (Sect. 3.2), we present an introduction of viscometry (Sect. 2.3) and definition of the viscosity coefficient (Sect. 2.4) as a basic background to explain the molecular theory. Sections 2.5 and 2.6 deal with the molecular theory for polymer solution viscosity at infinite dilution and at finite concentrations, respectively. In Sect. 2.7, we compare the molecular theory explained with viscosity data for five polymer solution systems of which chain stiffness is largely different and discuss how the concentration, molecular weight, and chain stiffness affect the polymer solution viscosity. Section 2.8 summarizes the results and conclusions explained in this chapter.

2.2 Polysaccharides Used as the Viscosity Enhancement Reagent

Polysaccharides are biopolymers abundant in nature, produced by plants, animals, fungus, algae, and microorganisms [4]. Human beings have used the natural polymers for long years, even before chemists did not verify the existence of polymers. Polysaccharides produced by industrial fermentation, say xanthan (xanthan gum), scleroglucan, and succinoglycan, are utilized as viscosity enhancement reagents adding to foods, cosmetics, paints, cements, and so on. Viscosities of aqueous solutions of xanthan and scleroglucan are stable against changes of temperature, added salt, and pH, while the solution viscosity of succinoglycan is known to collapse at a melting temperature, where this polysaccharide undergoes a structural

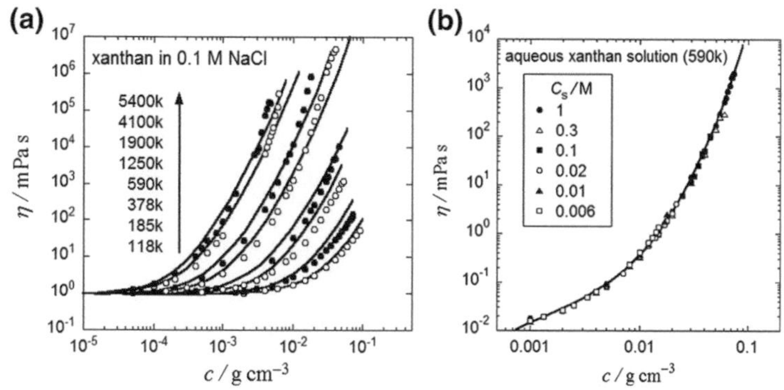

Fig. 2.1 Polymer concentration dependences of the zero-shear viscosity of xanthan in aqueous NaCl solutions; (**a**) data for different molecular weight samples; (**b**) data at different NaCl concentrations [3].

order-disorder transition. The above three polysaccharides are all multistranded helical polymers, and their solution viscosities strongly change by the denaturation of their native helical conformations.

Figure 2.1a shows concentration dependences for 0.1 M aqueous NaCl solutions of xanthan double helices of different molecular weights. In the figure, 5400 k ($=5.4 \times 10^6$) indicates the weight-average molecular weight of the original xanthan sample, and lower molecular weight samples were prepared by sonication of the original sample. For the original xanthan sample, the addition to the aqueous salt solution only by 0.5 wt.% enhances the solution viscosity 10^6 times as high as that of the solvent, but the strong concentration dependence remarkably diminishes with decreasing the molecular weight of the xanthan sample. As mentioned above, the aqueous xanthan solution viscosity is stable by adding salt, in spite of the fact that xanthan is a polyelectrolyte, which is demonstrated in Fig. 2.1b. This is a sharp contrast with the viscosity collapse of aqueous solutions of normal polyelectrolytes by addition of salt. This contrast comes from the stiff double-helical conformation of xanthan.

Because of the strong concentration and molecular weight dependences such as shown in Fig. 2.1a, we can control the viscosity of liquids by adding a small amount of viscosity enhancement reagents. To give a suitable viscosity to liquid products, we have to properly select the kind of viscosity enhancement reagent, as well as its molecular weight and concentration. In what follows, we present a viscosity equation, which gives us a guideline for the selection. Before explaining the viscosity equation, we will review basic knowledge of rheology and molecular theory of polymer solution viscosity.

2.3 Viscometry Using Various Types of Flow

2.3.1 Various Types of Viscometers

There are various types of viscometers, of which principles are illustrated in Fig. 2.2. To determine the viscosity coefficient η, one measures the flow time t for the capillary viscometer (Panel a), falling time t of the ball for falling ball viscometer (Panel b), and the torsional angle Θ of the wire for the coaxial cylinder and corn-and-plate viscometers (Panels c and d). The basic equations to determine η are given by

$$\eta = \begin{cases} \dfrac{\pi a^4 P}{8lQ} = \dfrac{\pi a^4 P}{8lV_t} t & \text{(Panel a)} \\ \dfrac{mg}{6\pi a \upsilon} = \dfrac{mg}{6\pi a L} t & \text{(Panel b)} \\ \dfrac{k}{4\pi l \Omega}\left(\dfrac{1}{r_1{}^2} - \dfrac{1}{r_2{}^2}\right)\Theta & \text{(Panel c)} \\ \dfrac{3k}{2\pi R^3}\dfrac{\phi}{\Omega}\Theta & \text{(Panel d)} \end{cases} \tag{2.1}$$

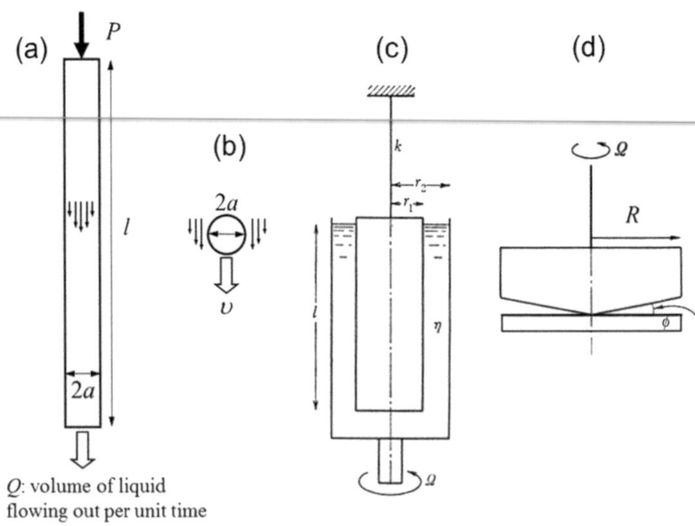

Fig. 2.2 Schematic diagrams of four types of viscometers: (**a**) the capillary viscometer, where l and a are the length and inner radius of the capillary, P is the pressure difference between the inlet and outlet of the capillary, and Q is the volume of the sample liquid flowing out per unit time ($V_t = Qt$ is the volume of the liquid flowing during time t); (**b**) the falling ball viscometer, where a and υ are the radius and the falling velocity of the ball in the sample liquid ($L = \upsilon t$ is the distance for the ball to fall during time t); (**c**) the coaxial cylinder viscometer, where l and r_1 are the length and radius of the inner cylinder and r_2 and Ω are the radius and angular velocity of the outer cylinder; (**d**) the corn-and-plate viscometer, where R and Ω are the radius and angular velocity of the corn and ϕ is the gap angle between the corn and plate

where m is the mass of the ball, g is the gravitational acceleration, k is the torsional elastic constant of the wire suspending the inner cylinder (Panel c) or the corn (Panel d), and other parameters appearing the equations are explained in the caption of Fig. 2.2. In Fig. 2.1, the data of η lower and higher than 1 mPa·s were obtained by a capillary viscometer in Panel a and by a ball viscometer in Panel b, respectively.

The rate of each flow is characterized by the shear rate $\dot{\gamma}$ of which definition will be given in the next subsection. While the shear rate of the flow in Panel d is uniform within the sample liquid, $\dot{\gamma}$ of the flows in Panels a–c are not uniform. The typical shear rate of each viscometer is given below:

$$\dot{\gamma} = \begin{cases} Pa/2\eta l & \text{(Panel a; at the capillary wall)} \\ 2a/3\upsilon & \text{(Panel b; at the equator of the ball)} \\ 2\Omega r_2{}^2/\left(r_2{}^2 - r_1{}^2\right) & \text{(Panel c; at the inner cylinder surface)} \\ \Omega/\phi & \text{(Panel d)} \end{cases} \tag{2.2}$$

As mentioned in the next section, Sect. 2.4, η may depend on $\dot{\gamma}$, and the shear rate dependence is most accurately measured by the corn-and-plate viscometer, where the shear rate is constant within the sample liquid.

2.3.2 Simple Shear Flow

Though we have mentioned various flow types used for rheometry, all the flow can be regarded locally as the simple shear flow, which is defined as the flow in the sliding parallel planes illustrated in Fig. 2.3. In what follows, we consider this simple shear flow. The rate of this flow can be specified by the shear rate $\dot{\gamma}$, the velocity V of the upper plane relative to the bottom place divided by the thickness d of the liquid between the two parallel planes.

Let us define the Cartesian coordinate system where the x and y axes are parallel to the flow direction and vertical to the sliding planes, respectively, and the z axis is

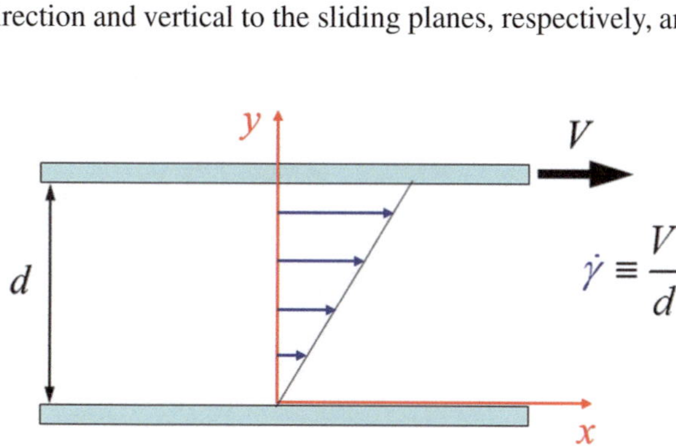

Fig. 2.3 Simple shear flow of the sample liquid sandwiched between the parallel plates

chosen to be vertical to both x and y axes to form the right-hand coordinate system (i.e., the upper side of the paper of Fig. 2.3 is chosen to be the positive z). Then, the velocity vector υ of the liquid can be written as

$$\boldsymbol{\upsilon} = \begin{pmatrix} \dot{\gamma} y \\ 0 \\ 0 \end{pmatrix} = \begin{pmatrix} 0 & \dot{\gamma} & 0 \\ 0 & 0 & 0 \\ 0 & 0 & 0 \end{pmatrix} \begin{pmatrix} x \\ y \\ z \end{pmatrix} = \left[\begin{pmatrix} 0 & \frac{1}{2}\dot{\gamma} & 0 \\ \frac{1}{2}\dot{\gamma} & 0 & 0 \\ 0 & 0 & 0 \end{pmatrix} + \begin{pmatrix} 0 & \frac{1}{2}\dot{\gamma} & 0 \\ -\frac{1}{2}\dot{\gamma} & 0 & 0 \\ 0 & 0 & 0 \end{pmatrix} \right] \begin{pmatrix} x \\ y \\ z \end{pmatrix} \tag{2.3}$$

where the 3×3 matrix in the third side of the above equation is the shear rate matrix, which is divided into the two 3×3 matrices in the fourth side. As shown in Fig. 2.4, the two matrices in the fourth side represent the extension and rotation, respectively. Oppositely speaking, the simple shear flow consists of the vortex and rotation. We will discuss the effect of the rotation matrix on the polymer solution viscosity in the following.

To slide the upper plane by the velocity V, we need to apply the force F to the plane, and we define the shear stress σ by F/A where A is the area of the plane. The force is proportional to V, so that we can write

$$\sigma \equiv \frac{F}{A} = \eta \frac{V}{d} = \eta \dot{\gamma} \tag{2.4}$$

where η is the proportional constant and we have used the definition of $\dot{\gamma}$ in the last equation. This is the definition of the viscosity coefficient η. Since we apply the force F to the plane and the plane moves the distance V per unit time, we make the work FV per unit time to the flowing liquid. Using σ and $\dot{\gamma}$, we can write the work

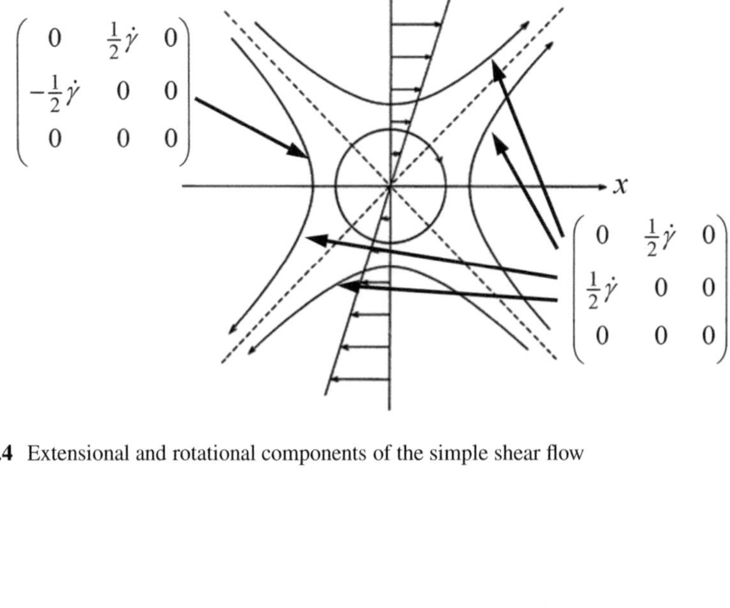

Fig. 2.4 Extensional and rotational components of the simple shear flow

w per unit time and unit volume in the form

$$w = \frac{FV}{Ad} = \sigma\dot{\gamma} = \eta\dot{\gamma}^2 \tag{2.5}$$

The last equation was obtained from Eq. 2.4.

2.4 Viscosity Coefficient

The viscosity coefficient η for a given polymer-solvent system depends on not only the shear rate $\dot{\gamma}$ of the flow, the temperature, and the pressure but also the polymer concentration c (the mass concentration with the units of g/cm^3) and the molecular weight M. Although we mostly deal with the c and M dependences in this chapter, we briefly mention the dependences on the shear rate and temperature.

If the shear rate of the solution flow is low enough, η is independent of $\dot{\gamma}$. This flow behavior is referred to as the Newtonian, and the viscosity coefficient at the low shear rate limit is called as the zero-shear viscosity. When the shear rate is increased in the rodlike and flexible polymer solutions, the orientation of rodlike polymer becomes anisotropic, and the conformation of the flexible polymer chain deforms from the thermally equilibrium one, respectively, in the solutions, which reduces the viscosity coefficient (the shear thinning effect). For aggregating polymer solutions, the shear flow may dissociate the aggregates, which also brings about the shear thinning effect. Although the shear thinning effect is important in some cases, we consider only the zero-shear viscosity in what follows.

The solvent viscosity strongly depends on the temperature. According to the hole theory of Cohen and Turnbull [5], the flow of the solvent composed of spherical particles occurs by the jump of the particle into a hole nearby, which is an activation process of a relatively high activation energy, leading to the strong temperature dependence. The conformation of the polymer chain is also dependent on the temperature, but its contribution to the temperature dependence of the solution viscosity is usually minor. If the heating dissociates polymer aggregates or thermal denaturation, the solution viscosity can reduce remarkably.

It is well known that the addition of the polymer to the solvent can increase remarkably the solution viscosity. For dilute polymer solutions, the c dependence of the zero-shear viscosity η is usually written as

$$\eta = \eta_s\left(1 + [\eta]\,c + k'[\eta]^2c^2 + \cdots\right) \tag{2.6}$$

where η_s is the solvent zero-shear viscosity, $[\eta]$ is the intrinsic viscosity, and k' is the Huggins coefficient. The intrinsic viscosity $[\eta]$ is a measure of the ability of the viscosity enhancement of the polymer species, which we will consider theoretically later. The concentration dependence for the polymer solution viscosity is usually so strong that terms with higher order than c^2 in Eq. 2.6 become important with increasing c.

The zero-shear viscosity also strongly depends on the molecular weight M of the solute polymer. Historically, the M dependence of the intrinsic viscosity $[\eta]$ has been important because one can determine M from $[\eta]$. The famous Mark-Houwink-Sakurada equation given by

$$[\eta] = KM^a \tag{2.7}$$

with two constants specific to the polymer-solvent system was proposed for this purpose. However, it should be noticed that this equation was originally proposed for flexible polymers and does not hold for stiff polymers over a wide M range. On the basis of the scaling concept [6, 7], many polymer solution properties in the semi-dilute regime can be expressed as a universal function of c/c^*, where c^* is the overlap concentration. Since c^* may be equated to the reciprocal of $[\eta]$, η_0 becomes a universal function of $c[\eta]$. Using Eq. 2.6, we can say that both c and M dependences of η_0 can be expressed as a function of $c[\eta]$. We should note again that the scaling concept was proposed for flexible polymer solutions, and it is questionable to apply the above argument to stiff polymer solutions.

2.5 Intrinsic Viscosity [2, 8]

2.5.1 Friction Between Polymer and Solvent

Let us calculate the intrinsic viscosity of a rodlike and flexible polymer chains composed by N beads of a diameter b suspended in the simple shear flow of the solvent (cf. Fig. 2.5). As mentioned in the previous section, the simple shear flow can be divided into the extension and rotation. At a sufficiently low shear rate, the former flow component does not affect the motion of the polymer chain, but the latter flow component rotates the polymer chain around the center of mass of the polymer chain with the angular velocity of $\dot{\gamma}/2$. (The center of mass itself moves along the x direction, but this motion does not contribute to η. We choose the center of mass as the origin of the Cartesian coordinate system and consider the relative motion of each bead to the center of mass.)

The simple shear flow behavior introduced in Sect. 3 does not depend on the z coordinate. Therefore, we omit the z component of vectors and matrices appearing in what follows, just for simplicity. The (relative) velocity υ_i of the i-th bead located at $\mathbf{r}_i = (x_i, y_i)$ and the (unperturbed) solvent flow velocity $\upsilon^\circ{}_s$ at the same position are given by

$$\boldsymbol{\upsilon}_i = \begin{pmatrix} 0 & \frac{1}{2}\dot{\gamma} \\ -\frac{1}{2}\dot{\gamma} & 0 \end{pmatrix} \mathbf{r}_i, \ \boldsymbol{\upsilon}^\circ{}_\mathrm{s} = \begin{pmatrix} 0 & \dot{\gamma} \\ 0 & 0 \end{pmatrix} \mathbf{r}_i \tag{2.8}$$

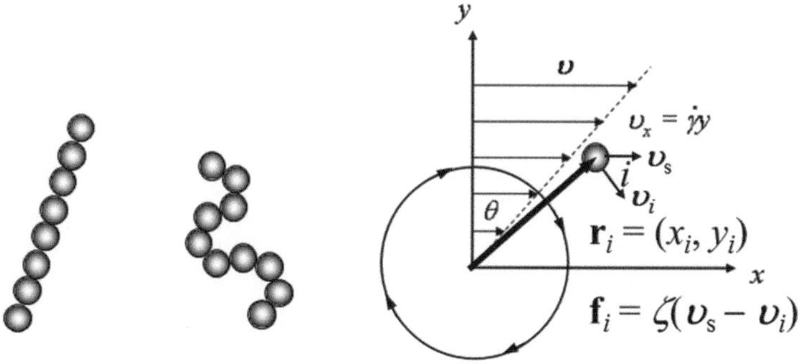

Fig. 2.5 Bead models for the rodlike polymer and the Gaussian chain and a bead under the simple shear flow

The difference between υ_i and $\upsilon^\circ{}_s$ produces the frictional force $\mathbf{f}_i = \zeta(\upsilon^\circ{}_s - \upsilon_i)$ $\mathbf{r}_i$, where ζ $(=3\pi\eta_s b)$ is the frictional coefficient of the bead. The energy dissipation by this frictional force per unit time is given by $\mathbf{f}_i \cdot \upsilon^\circ{}_s$, and their sum over beads in a unit volume of the solution should be equal to w in Eq. 2.5 from the energy conservation rule, and we have the following relation:

$$w = \frac{cN_A}{M}\sum_{i=1}^{N} \mathbf{f}_i \cdot \boldsymbol{\upsilon}^\circ{}_s = \frac{3\pi\eta_s bcN_A}{2M}\dot{\gamma}^2 \sum_{i=1}^{N} y_i^2 \tag{2.9}$$

where N_A is the Avogadro constant and the prefactor cN_A/M is the number of polymer chains per unit volume of the solution. Combining this equation with Eq. 2.5, we obtain the molecular expression for the zero-shear viscosity induced by the friction between polymer and solvent (the viscous component of η) in the form

$$\eta_V = \frac{3\pi\eta_s bcN_A}{2M}\sum_{i=1}^{N} y_i^2 \tag{2.10}$$

In the above calculation of w, we did not consider the interchain hydrodynamic interaction, so that Eq. 2.10 should be applied to dilute solution where c^2 and the higher-order terms are neglected. Thus, the comparison between Eqs. 3.4.1 and 3.5.3 gives us the equation:

$$[\eta] = \frac{3\pi bN_A}{2M}\sum_{i=1}^{N} y_i^2 \tag{2.11}$$

Applying this equation to the rodlike polymer and the Gaussian chains, we obtain the following expressions:

$$[\eta] = \begin{cases} \frac{\pi N_A (bN)^3}{8M} & \text{(rodlike polymer)} \\ \frac{\pi N_A b^3 N^2}{12M} & \text{(Gaussian chain)} \end{cases} \tag{2.12}$$

From the latter equation, it turns out that the exponent a in Eq. 2.7 is 1 for the Gaussian chain, which is the famous Staudinger equation for $[\eta]$. Ironically, Staudinger himself believed for a long time that all polymer chains are rodlike, but a for the rodlike polymer is 2 in Eq. 2.12, instead of his equation. Moreover, experimental M dependences of $[\eta]$ reported for many flexible polymer-solvent systems did not obey the Eq. 2.7 for the Gaussian chain.

2.5.2 Intramolecular Hydrodynamic Interaction

The above calculation, however, has not taken into account the perturbation of the solvent flow by motions of beads surrounding each bead. But actually this perturbation is important, and we have to modify Eq. 2.12 considering this effect. Figure 2.6 illustrates the solvent flow around the bead moving at the velocity υ by

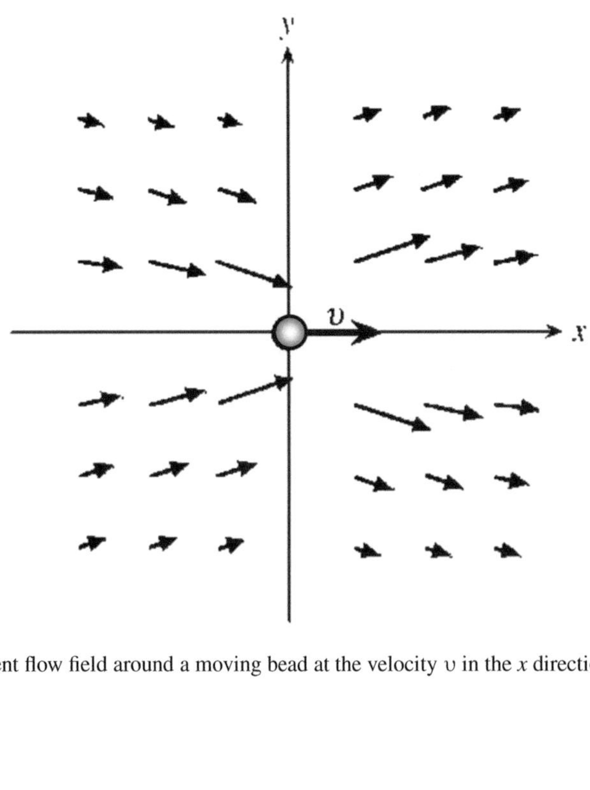

Fig. 2.6 Solvent flow field around a moving bead at the velocity υ in the x direction

the external force $\mathbf{F} = 3\pi\eta_s \upsilon$. Mathematically, the perturbation of the solvent flow υ'_s at the position $\mathbf{r}$ from the moving bead can be expressed by

$$\boldsymbol{\upsilon}'_s = \mathbf{TF}, \ \mathbf{T} \equiv \frac{1}{8\pi\eta_s r}\left(\mathbf{I} + \frac{\mathbf{rr}}{r^2}\right) \tag{2.13}$$

where $r \equiv |\mathbf{r}|$, $\mathbf{I}$ is the unit tensor, $\mathbf{rr}$ is the dyadic $[(\mathbf{rr})_{\alpha\beta} = \alpha\beta;\ \alpha,\ \beta = x,\ y,\ z]$, and $\mathbf{T}$ is called as the Oseen tensor. By this perturbation, the solvent flow velocity υ_s at the position of the i-th bead must be written as

$$\boldsymbol{\upsilon}_s = \boldsymbol{\upsilon}^{\circ}{}_s + \sum_{j \neq i} \boldsymbol{\upsilon}'_{s,j} \tag{2.14}$$

instead of Eq. 2.8, where $\upsilon'_{s,j}$ is the perturbation by the j-th bead calculated by Eq. 2.13 with $\mathbf{r} = \mathbf{r}_i - \mathbf{r}_j$ ($\mathbf{r}_j$: the position vector of the j-th bead).

Using this equation for υ_s, Kirkwood and Riseman calculated $[\eta]$ for the rodlike polymer and the Gaussian chain both with sufficiently large N. Their results are written as

$$[\eta] = \begin{cases} \dfrac{\pi N_A L^3}{90M\,\ln(L/b)} & \text{(rodlike polymer)} \\ 6^{3/2}\Phi\dfrac{\langle S^2\rangle^{3/2}}{M} & \text{(Gaussian chain)} \end{cases} \tag{2.15}$$

where L (= bN) is the length of the rod, $\langle S^2\rangle$ (= $b^2N/6$) is the square radius of gyration of the Gaussian chain, and Φ is the Flory viscosity constant (=2.87×10^{23}). The Mark-Houwink-Sakurada exponent a is 0.5 for the Gaussian chain, which agrees with experimental results. Compared with the Einstein viscosity equation for a hard sphere, it can be seen that the Gaussian chain with the radius of gyration of $\langle S^2\rangle^{1/2}$ is equivalent to the sphere with the radius of $0.87\langle S^2\rangle^{1/2}$, giving the same $[\eta]$. Because the solvent inside the Gaussian chain rotates together with the chain by the intramolecular hydrodynamic interaction, the Gaussian chain behaves like a sphere. The double logarithmic plot of $[\eta]$ against M for rodlike polymer follows a concave curve due to the factor $1/\ln(L/b)$ in Eq. 2.15, and the slope approaches ca. 1.7 at high M, which is also consistent with experimental results, as mentioned below. Thus, for the rodlike polymer, $[\eta]$ does not obey the Mark-Houwink-Sakurada equation.

2.5.3 *Contribution of the Elastic Stress*

It is well known that polymer molecules move and rotate randomly by thermal agitation in solution, but so far we have not taken into account this Brownian motion, which may contribute to the solution viscosity. When a stepwise shear deformation is exerted on the rodlike polymer solution, the extension component (cf. Eq. 2.2) of the very fast shear flow orients rods to the direction along the line $y = x$. After the deformation, the shear flow is stopped and the orientation of the rods is randomized by the rotational diffusion. Thus, the stepwise deformation provides the transient orientational entropy loss. According to the theory of rubber elasticity, the entropy loss induces the force (or the stress) to recover the system to the original state before the deformation (the Le Chatelier principle), and the transient stress after the stepwise shear deformation is written as

$$\sigma = \frac{k_B T}{5} \frac{c N_A}{M} \exp(-2D_r t) \tag{2.16}$$

for thin rods ($L >> b$), where $k_B T$ is the Boltzmann constant multiplied by the absolute temperature, t is the time elapsing after the deformation, and D_r is the rotational diffusion coefficient of the rodlike polymer in the solution.

This stress relaxation followed by the stepwise deformation can be expressed by the Maxwell model consisting of an elastic spring with the spring constant k and a dashpot including a fluid of the viscosity coefficient η_d in a series, which writes the transient stress as

$$\sigma = k \exp(-t/\tau) \tag{2.17}$$

where τ is the relaxation time given by η_d/k. Comparing Eqs. 2.16 and 2.17, we have the relations $k = (k_B T/5)(cN_A/M)$ and $\tau = 1/2D_r$. On the other hand, for the steady shear flow, the Maxwell model gives the equation for the shear stress as

$$\sigma = k\tau\dot{\gamma} \tag{2.18}$$

Compared this equation with Eq. 2.3, the elastic component of the zero-shear viscosity η_E is related to D_r by

$$\eta_E = \frac{RTc}{10MD_r} \tag{2.19}$$

with the gas constant R ($=k_B N_A$).

The hydrodynamic calculation gives us the rotational diffusion coefficient $D^\circ_{r,0}$ at infinite dilution in the form

$$D^\circ_{r,0} = \frac{3k_B T \ln(L/b)}{\pi \eta_s L^3} \tag{2.20}$$

in the condition of $L >> b$. By adding the elastic component to Eq. 2.15 of $[\eta]$ for the rod with $L >> b$, we obtain the total intrinsic viscosity as

$$[\eta] = \frac{2\pi N_A L^3}{45M \ln(L/b)} \tag{2.21}$$

The elastic component enlarges $[\eta]$ by the factor 4, but the M dependence does not change.

The Gaussian chain has a lot of internal degrees of freedom, and the micro-Brownian motion with respect to the degrees of freedom can be treated by using the bead-spring model where N beads are connected by $N-1$ elastic springs. Zimm calculated $[\eta]$ for the bead-spring model including the effect of the intramolecular hydrodynamic interaction and micro-Brownian motion. The result is given in the same form as the second line of Eq. 2.15, where Φ is written as

$$\Phi = \frac{\pi}{4}\sqrt{\frac{\pi}{3}}N_{\mathrm{A}}\sum_{j=1}^{N}\frac{1}{\lambda'_j} \tag{2.22}$$

where λ'_j is the eigenvalue of the transformation matrix into the normal coordinates. The numerical calculation of λ'_j gives us Φ to be 2.84×10^{23}, which is in a good agreement with Φ obtained by Kirkwood and Riseman (Eq. 2.15), but this agreement may be just by accident, because the two theories used the different polymer models. For the bead-spring model in solution, η cannot be divided into the viscous and elastic components. We can say from Eq. 2.22 that the normal mode $j=1$, which approximately corresponds to the end-over-end rotation of the chain, contributes to η ca. 42 %.

2.5.4 Wormlike Chain Model and Excluded Volume Effect

Cellulose and its derivatives are known to be semiflexible polymers. The double-helical polysaccharide xanthan, triple-helical polysaccharide schizophyllan, and so on are typical stiff polymers. Their molecular conformation can be represented as the wormlike chain model, which interpolates between the Gaussian chain and rodlike polymer. This model is characterized by the contour length L and the persistence length q. The latter quantity is larger for the stiffer polymer. When this model is applied to a polysaccharide with the molecular weight M, we can calculate L by hM/M_0, where h and M_0 are the (average) contour length and molar mass per monosaccharide unit of the polysaccharide main chain, respectively. This model becomes identical with the Gaussian chain, if M tends to infinity. The parameters b and N of the Gaussian chain correspond to $2q$ and $N_{\mathrm{K}}\equiv L/2q$ of the wormlike chain, where $2q$ and N_{K} are sometimes referred to as the length and number of Kuhn's statistical segments, respectively. The radius of gyration of the wormlike chain is written as

$$\langle S^2\rangle = (2q)^2\left\{\frac{1}{6}N_{\mathrm{K}} - \frac{1}{4} + \frac{1}{4N_{\mathrm{K}}}\left[1 - \frac{1}{2N_{\mathrm{K}}}\left(1-\mathrm{e}^{-2N_{\mathrm{K}}}\right)\right]\right\} \tag{2.23}$$

At low and high M limits, the above equation tends to

$$\langle S^2 \rangle = \begin{cases} \frac{1}{12}L^2 \ (M \to 0) \\ \frac{1}{6}(2q)^2 N_K \ (M \to \infty) \end{cases} \tag{2.24}$$

that is, the rod and Gaussian chain limits, respectively.

The intrinsic viscosity for the wormlike chain model (the wormlike cylinder or wormlike touched-bead model) was formulated by Yamakawa et al. The intrinsic viscosity equation is not given in a simple analytic equation but in a complex form of the empirical interpolation formulas. The molecular weight dependence of $[\eta]$ for the wormlike chain model obeys that for the rodlike polymer in the low M region and the Gaussian $M^{1/2}$ relation in the high M region. The double logarithmic plot of $[\eta]$ vs. M thus does not follow the Mark-Houwink-Sakurada equation.

Parts of a polymer chain cannot overlap, but the above equations for $\langle S^2 \rangle$ and $[\eta]$ of the wormlike chain do not consider this excluded volume effect. While this effect disappears at the rod limit because of no possibility of the self-interaction, it becomes important with increasing the chain flexibility. The excluded volume effect for the wormlike chain was considered also by Yamakawa et al. By this effect, the polymer chain expands, and the degree of the expansion is expressed in terms of the expansion factor α_S defined by the radius of gyration divided by the unperturbed one or α_η defined by $[\eta]^{1/3}$ divided by the unperturbed one. The strength of the excluded volume effect is characterized by the excluded volume strength B, and the expansion factors α_S and α_η are given as functions of $B/2q$ and N_K. When $B/2q$ or N_K increases from zero, the excluded volume effect becomes important, and α_S and α_η increase from unity. At the Gaussian chain limit ($N_K \to \infty$), α_S and α_η are proportional to ${N_K}^{1/10}$ in a good solvent, where the chain has a positive value of B, and $\langle S^2 \rangle^{1/2}$ and $[\eta]$ are proportional to $M^{0.6}$ and $M^{0.8}$, respectively.

2.6 Zero-Shear Viscosity at Finite Concentrations [3, 7]

2.6.1 Rodlike Polymer Solutions

As already mentioned above, the zero-shear viscosity of rodlike polymer solutions consists of the solvent, viscous, and elastic components:

$$\eta = \eta_s + \eta_V + \eta_E \tag{2.25}$$

The viscous component η_V is related to the rotational diffusion coefficient $D_{r,0}$ by

$$\eta_V = \frac{RTc}{30MD_{r,0}} \tag{2.26}$$

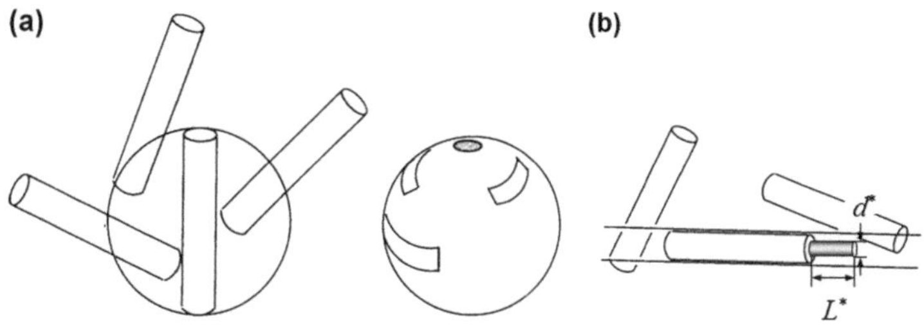

Fig. 2.7 Rods in a concentrated solution, where a rotating rod (**a**) or a longitudinally moving rod (**b**) collides with surrounding rods

From Eqs. 2.15 to 2.20, $D_{r,0} = D^{\circ}{}_{r,0}$ at the infinite dilution. With increasing the polymer concentration, the motion of surrounding rods perturbs the solvent flow at the rod taking into consideration, which affects the rotational diffusion coefficient, and then η_V. Although there are many hydrodynamic calculations, the long-range nature of the intermolecular hydrodynamic interaction makes mathematical calculations difficult, and rigorous calculation has not been performed yet. Here, the intermolecular hydrodynamic interaction is empirically expressed using an adjustable parameter k'_{HI}, and $D_{r,0}$ is written as

$$D_{r,0}{}^{-1} = D^{\circ}_{r,0}{}^{-1}\left(1 + k'_{HI}\,[\eta]\,c\right) \tag{2.27}$$

For the elastic component η_E, Eq. 2.19 holds even at finite polymer concentrations. The rotational motion of each rod is affected by surrounding rods at finite concentration through not only the intermolecular hydrodynamic interaction but also the direct collision among rods (cf. Fig. 2.7a). Because η_V arises from the friction between the polymer and solvent, we do not consider the effect of the direct collision, but we have to consider both effects on η_E through D_r.

2.6.2 *Rotational Diffusion Coefficient and Viscosity Equation*

As illustrated in Fig. 2.7a, the rotational diffusion motion of a rodlike polymer (enclosed by the circle in the figure) is hindered by the collision with surrounding rods in solution. This hindered rotational diffusion motion can be treated as the two-dimensional translational diffusion motion of a circular object on the spherical surface, where ribbonlike objects (the projection of the hindering rod onto the spherical surface) hinder the translational diffusion motion.

The collision effect on D_r, sometimes called as the entanglement effect, can be treated by the mean-field Green function method. Originally, this method was applied by Edwards and Evans to the longitudinal diffusion motion in the concentrated rodlike polymer solution and then extended by Teraoka and Hayakawa to the transverse and rotational diffusion motions.

Although the rotational diffusion motion is two-dimensional, we briefly explain the mean-field Green function method for the one-dimensional diffusion motion of a particle. The extension to the rotational diffusion motion of a rod is straightforward, but mathematically more complex, which is the reason why we explain the method in the one dimension.

The diffusion motion is stochastic, so that we can discuss only the probability of the motion. Let x be the coordinate of the diffusing particle in its one-dimensional path. If no barriers are present in the path, the *unperturbed* conditional probability $G_0(x, x'; t, t')$ that the particle at the position x' at time t' moves to the position x at time t is given by

$$G_0\left(x, x'; t, t'\right) = \frac{1}{\sqrt{4\pi D_0\left(t - t'\right)}} \exp\left[-\frac{\left(x - x'\right)^2}{4D_0\left(t - t'\right)}\right] \tag{2.28}$$

where D_0 is the unperturbed diffusion coefficient of the particle. Mathematically, this conditional probability is referred to as the Green function. Figure 2.8 illustrates functional forms of $G_0(x, x'; t, t')$ given by Eq. 2.28 at three different $t - t'$. If one drops ink in water at x', ink diffuses and the concentration distribution of ink changes with time according to the functional forms in the figure.

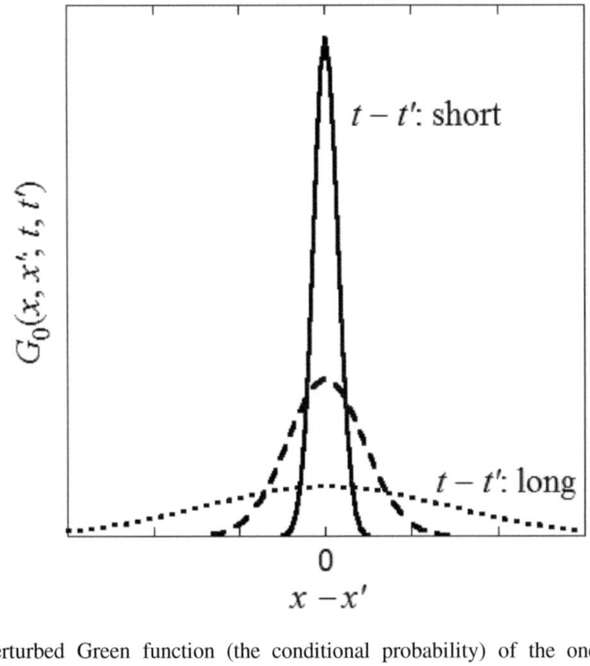

Fig. 2.8 Unperturbed Green function (the conditional probability) of the one-dimensionally diffusing particle

Suppose now a single reflecting point barrier is placed at position R in the path at time t_R and removed at time t_Q. Following Edwards and Evans, the Green function perturbed by this barrier can be expressed by

$$G\left(x, x'; t, t'\right) = \int_{-\infty}^{\infty} \mathrm{d}x_1 \int_{-\infty}^{\infty} \mathrm{d}x_2 G_0\left(x, x_2; t, t_\mathrm{Q}\right) Q'\left(x_2, x_1; t_\mathrm{Q}, t_\mathrm{R}\right) G_0\left(x_1, x'; t_\mathrm{R}, t'\right) \tag{2.29}$$

where $Q'(x_2, x_1; t_\mathrm{Q}, t_\mathrm{R})$ is the transition probability that the particle at the position x_1 at time t_R moves to the position x_2 at time t_Q. After averaging the Fourier transform of Eq. 2.29 with respect to the random variables R, t_R, and t_Q, and making some mathematical manipulations, we obtain

$$\left\langle \widehat{G}\left(k; t, t'\right)\right\rangle \equiv \int_{-\infty}^{\infty} G\left(x, x'; t, t'\right) \mathrm{e}^{\mathrm{i}k(x-x')} \mathrm{d}\left(x - x'\right) = \mathrm{e}^{-D_0(t-t')k^2} + \left\langle \Delta\widehat{G}\left(k; t, t'\right)\right\rangle \tag{2.30}$$

with the perturbed part $\langle \Delta\hat{G}(k; t, t')\rangle$ given by

$$\left\langle \Delta\widehat{G}\left(k; t, t'\right)\right\rangle = \frac{8}{3\sqrt{\pi}l}(D_0\tau)^{3/2}k^2 + O\left(k^4\right) \tag{2.31}$$

for a sufficiently long time interval $t - t'$. Here, l is a path length comparable to the maximum distance that the particle can diffuse during $t - t'$, and τ is the mean lifetime of the barrier. The brackets $\langle \cdots \rangle$ represent the average with respect to R, t_R, and t_Q. The effective diffusion coefficient D can be calculated from

$$D = -\lim_{t-t'\to\infty} \frac{1}{2\left(t - t'\right)} \left.\frac{\partial^2 \left\langle \widehat{G}\left(k; t, t'\right)\right\rangle}{\partial k^2}\right|_{k=0} \tag{2.32}$$

From Eqs. 2.31 to 2.32, we get D in the presence of a single reflecting barrier as

$$D = D_0 - \frac{8(D_0\tau)^{3/2}}{3\sqrt{\pi}l\left(t - t'\right)} \tag{2.33}$$

Next, we consider the case where more than one barrier appears in the diffusion path, and particle receives multiple perturbations. Teraoka and Hayakawa assumed that the Green function for this case satisfies the following Dyson equation:

$$\langle G\rangle = G_0 + G_0\left\langle \sum_{i=1}^{M} Q_i\right\rangle \langle G\rangle \tag{2.34}$$

where Q_i is the i-th perturbation element and M is the number of elements appearing on the path length l between t and t'. In Eq. 2.34, we have omitted the arguments of the Green functions and Q_i as well as the integral symbol; the second term on the right-hand side of Eq. 2.34 involves the same integration as in Eq. 2.29. When the whole right-hand side is inserted into $\langle G \rangle$ on the right- hand side, it turns out that Eq. 2.34 gives a Green function perturbed by any number of sequential independent elements. Eq. 2.34 can be solved recursively, and the Fourier transform of the solved $\langle G \rangle$ is inserted into Eq. 2.32 to obtain

$$D = D_0 \left(1 + \frac{4}{3\sqrt{\pi}} \rho_{\mathrm{B}} \tau^{3/2} {D_0}^{1/2} \right)^{-2} \tag{2.35}$$

where ρ_{B} [$\equiv M/l(t - t')$] is the number of the barriers appearing per unit path length and unit time. If only a barrier appears for a sufficiently long time interval $t - t'$, Eq. 2.35 reduces to Eq. 2.33 for the first-order perturbation.

We can extend the above theory to the two-dimensional rotational diffusion motion. The perturbed rotational diffusion coefficient D_{r} can be written as

$$D_{\mathrm{r}} = D_{\mathrm{r},0} \left[1 + \sqrt{\frac{6}{\beta} \frac{N_{\mathrm{A}} c L^3}{M}} \left(\frac{D_{||,0}}{D_{||}} \right)^{1/2} \right]^{-2} \tag{2.36}$$

where β is a constant and $D_{||}$ is the perturbed longitudinal diffusion coefficients of the rod. To obtain this equation, we have assumed $\tau \propto L^2/D_{||}$, because the hindrance of the rotational motion by surrounding rods is released mainly through the longitudinal motion of the hindering or rotating rod. Teraoka calculated β to be 1350 by simulation for entangled rods.

We need an equation for $D_{||}$ to calculate D_{r} from Eq. 2.36. Edwards and Evans formulated $D_{||}$ by the mean-field Green function method, but their equation can be only applied to very concentrated solutions of rodlike polymers, where the transverse motion of each rod is almost completely prohibited. We want to formulate the viscosity equation for rodlike polymers in less concentrated solutions. Sato and Teramoto proposed to formulate $D_{||}$ by the hole theory. This method is similar to that used by Cohen and Turnbull to obtain an expression of the self-diffusion coefficient in liquids of small molecules and metals. The basic assumption in our hole theory is that the longitudinal diffusion of a test chain occurs only when a "hole" that is larger than a critical hole exists in front of the test chain. Here the hole means a region that contains no chain segments in the solution (not true vacuum region), and then one should not confuse that with the free volume that often appears in the literature on polymer melts and solutions.

From Cohen and Turnbull's argument [5], it turns out that $D_{||}$ is proportional to the total probability P_{h} of finding a hole that is larger than the critical hole. The probability P_{h} is written in terms of the average mutual excluded volume ${V_{\mathrm{ex}}}^*$ between the critical hole and the solute molecule as

$$P_{\mathrm{h}} = \exp\left(-{V_{\mathrm{ex}}}^* c'\right) \tag{2.37}$$

and thus $D_{||}$ is given by

$$D_{||} = D_{||,0} \exp\left(-V_{ex}{}^{*}c'\right) \tag{2.38}$$

When the critical hole for the rodlike polymer with the length L and diameter d is assumed to be a cylinder of the length $L^{*} = \lambda^{*}L$ and diameter $d^{*} = \lambda^{*}d$, where λ^{*} is the constant similitude ratio (cf. Fig. 2.7b), $V_{ex}{}^{*}$ is written as

$$V_{ex}{}^{*} = \frac{\pi}{4}\Big[LL^{*}\left(d+d^{*}\right) + Ld^{2} + L^{*}d^{*2} + \frac{1}{2}\left(Ld^{*2} + L^{*}d^{2}\right) + \frac{\pi}{2}\left(L+L^{*}\right)dd^{*} + \frac{\pi}{4}\left(d+d^{*}\right)dd^{*}\Big] \tag{2.39}$$

Combining all the results (Eqs. 2.25–2.27, 2.19, 2.36, and 2.38), we obtain the viscosity equation for rodlike polymer solutions:

$$\frac{\eta-\eta_s}{\eta_s} = [\eta]c\left(1+k'_{HI}[\eta]c\right)\left\{\frac{1}{4}+\frac{3}{4}\left[1+\sqrt{\frac{6}{\beta}\frac{N_A cL^3}{M}}\exp\left(\frac{1}{2}V_{ex}{}^{*}c'\right)\right]^2\right\} \tag{2.40}$$

Expanding the above equation in the power series of c and comparing the second-order term with Eq. 2.6, we can express the Huggins coefficient k' in the form

$$k' = k'_{HI} + k'_{EI} \tag{2.41}$$

where k'_{EI} is the contribution of the entanglement interaction to k', given by

$$k'_{EI} = \frac{135}{4\pi\sqrt{\beta}}\ln\left(L/b\right) \tag{2.42}$$

On the other hand, the contribution k'_{HI} of the hydrodynamic interaction for rodlike polymer solutions has formulated by Riseman and Ullman. However, their calculation contains an improper integration, and thus the results are not conclusive.

When both hydrodynamic interaction and the concentration dependence of $D_{||}$ are neglected in Eq. 2.40, η of the rodlike polymer solution at high concentrations depends on the molecular weight and concentration as $\eta \propto c^3M^6/\ln(L/b)$. These dependences were predicted by Doi and Edwards [7]. However, these M and c dependences are modified by the effects of the hydrodynamic interaction and the concentration dependence of $D_{||}$. If the effect of the hydrodynamic interaction is taken into account, η depends on M and c as $\eta \propto c^4M^8/\ln^2(L/b)$ at high concentrations.

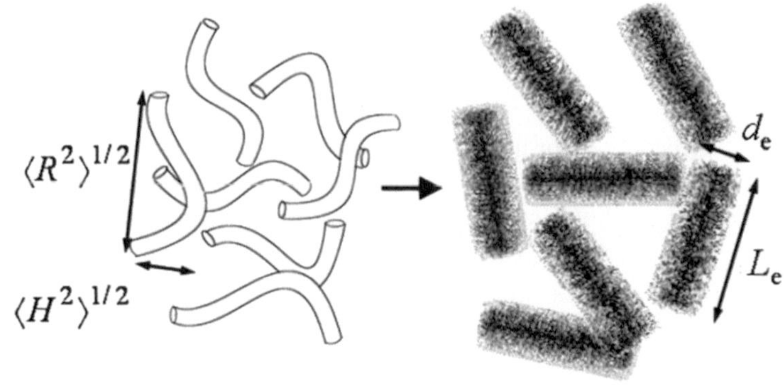

Fig. 2.9 Fuzzy cylinder model for semiflexible polymers in a concentrated solution

2.6.3 Flexibility and Finite Thickness Effects

The left illustration of Fig. 2.9 depicts semiflexible polymers in an isotropic concentrated solution. Each polymer chain rotates, translates, and changes its conformation in the solution. Among those molecular motions, the conformational change is rapidest, which smears the segment distribution as shown in the right illustration of Fig. 2.9. To formulate D_r of the semiflexible polymer in a concentrated solution, we view the polymer chain as this smeared cylindrical segment distribution, called as the fuzzy cylinder. The effective length L_e and diameter d_e of the fuzzy cylinder are defined using the mean square end-to-end distance $\langle R^2\rangle$ and mean square distance $\langle H^2\rangle$ between the midpoint and the end-to-end axis of the wormlike chain by

$$L_e = \langle R^2\rangle^{1/2},\ d_e = \left(\langle H^2\rangle + b^2\right)^{1/2} \tag{2.43}$$

where b is the thickness of the wormlike chain polymer. Both $\langle R^2\rangle$ and $\langle H^2\rangle$ can be calculated from q and N_K of the wormlike chain; $L_e = L$ and $d_e = b$ in the rod limit, and $L_e = 6^{1/2}d_e$ in the Gaussian coil limit. (Strictly speaking, $\langle H^2\rangle$ has not been formulated for the wormlike chain, but for the Tagami model, allowing the chain contour to stretch in contrast to the wormlike chain. However, the Tagami model provides the identical expressions of $\langle R^2\rangle$ and $\langle S^2\rangle$, so that we can expect its $\langle H^2\rangle$ is a good approximation of that for the wormlike chain.)

Using the fuzzy cylinder model, we may extend the viscosity equation for the rodlike polymer solution in the previous subsection to semiflexible one by replacing L with L_e. However, we have to consider the finite thickness effect on the viscosity equation, because the axial ratio L_e/d_e of the fuzzy cylinder decreases with increasing the chain flexibility. The extension of Eq. 2.36 for D_r to the fuzzy cylinder system can be written as

$$\frac{D_r}{D_{r0}} = \left[1 + \sqrt{\frac{6}{\beta}\frac{N_A c L_e{}^4}{ML}} F \left(\frac{D_{||,0}}{D_{||}}\right)^{1/2}\right]^{-2} \tag{2.44}$$

where F is a factor relating to the finite thickness effect, empirically given by

$$F \equiv \left(1 + C_r \frac{d_e}{L_e}\right)^3 \left(1 - C_r \frac{d_e}{5L_e}\right), \; C_r = \frac{1}{2}\left[\tanh\left(\frac{1}{4}N_K - 1\right) + 1\right] \tag{2.45}$$

The final viscosity equation for semiflexible polymer solutions at finite concentration is given by

$$\frac{\eta - \eta_s}{\eta_s} = [\eta]\, c\left(1 + k'_{HI}\,[\eta]\, c\right)\left\{\frac{1}{4} + \frac{3}{4}\left[1 + \sqrt{\frac{6}{\beta}\frac{N_A c L_e{}^4}{ML}} F \exp\left(\frac{1}{2}V_{ex}{}^* c'\right)\right]^2\right\} \tag{2.46}$$

instead of Eq. 2.40. Here, $V_{ex}{}^*$ is calculated by Eq. 2.39, using $L^* = \lambda^* L_e$ and diameter $d^* = \lambda^* d_e$. (Sato et al. further extended the viscosity equation, including the finite thickness effect of the polymer chain on frictional properties, but this effect is not important if $L >> b$.)

Expanding the above equation in the power series of c, we have the expression for the Huggins coefficient k' as the sum of the contributions of the entanglement and hydrodynamic interactions as in Eq. 2.41. The contribution k'_{EI} of the entanglement interaction to k' is given by

$$k'_{EI} = \frac{3}{2}\sqrt{\frac{6}{\beta}\frac{N_A L_e{}^4}{[\eta]\, ML}} F \tag{2.47}$$

instead of Eq. 2.42.

When the concentration dependence of $D_{||}$ is neglected in Eq. 2.46, the polymer concentration dependence η at high concentrations is not affected by the chain flexibility, i.e., $\eta \propto c^3$ and c^4, when the hydrodynamic interaction is neglected and not neglected, respectively. The molecular weight dependence of η for semiflexible polymer solutions does not obey any power law, because the polymer chain conformation changes from the rodlike to random coil with increasing M. In the coil limit, both L_e and $[\eta]$ are proportional to $M^{1/2}$ (without considering the excluded volume effect); Eq. 2.46 at $V_{ex}{}^* = 0$ gives us the relation $\eta \propto M^{1/2}$ and M in the dilute and concentrated limits, respectively. The dependence in the concentrated limit is very much different from the M dependence of η for the rodlike polymer solution, mentioned in the previous subsection [$\eta \propto M^8/\ln^2(L/b)$]. It is noted however that when the polymer concentration is high enough, the fuzzy cylinder model becomes inappropriate, and we have to consider the reptation-like motion of the flexible chain in the entangled polymer network. We expect the well-known relation, $\eta \propto M^{3.4}$, for very concentrated flexible polymer solutions.

2.7 Comparison with Experimental Polymer Solution Viscosities [9]

2.7.1 *Polymers with Different Chain Stiffness*

While most of synthetic polymers are flexible, except for conjugated polymers, e.g., polyacetylenes and aromatic polyamides, biopolymers have a variety of the chain stiffness. Here, since we are mostly interested in the chain stiffness effect on the polymer solution viscosity, we show experimental solution viscosity data of different chain stiffness polymers shown in Fig. 2.10. In the figure, schizophyllan (SPG) is an extracellular polysaccharide produced by a fungus *Schizophyllum commune*. This polysaccharide forms a rigid triple helix in aqueous solution, which is one of the stiffest polymers we know so far. Xanthan is a bacterial polysaccharide produced by the *Xanthomonas campestris*, which forms a stiff double helix in aqueous salt solution. Cellulose is known to be a semiflexible polymer, and cellulose

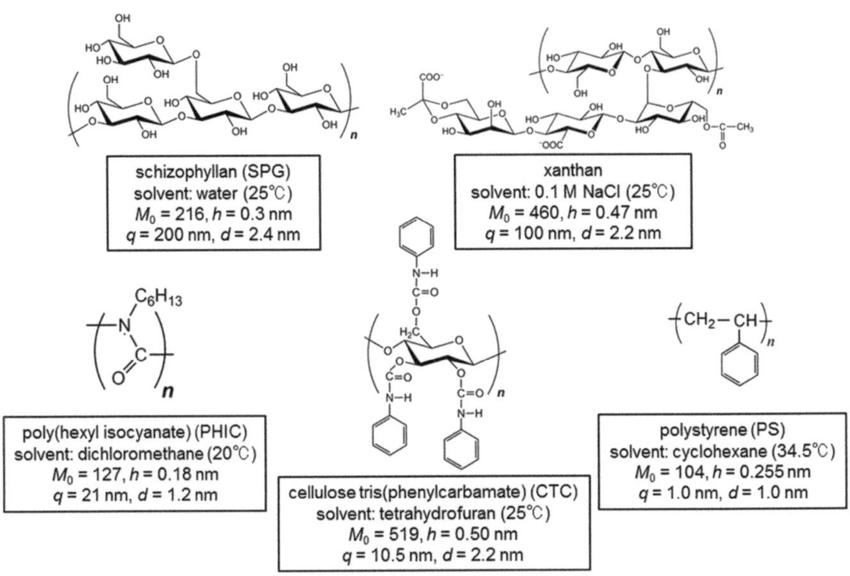

Fig. 2.10 Chemical structures and molecular characteristics of polymers with different chain stiffness

tri(phenylcarbamate) (CTC) is its derivative dissolved in organic solvents, say, tetrahydrofuran (THF).

The chain stiffness of those polysaccharides (and the derivative) is represented in terms of the persistence length q, which is listed in Fig. 2.10, as well as other molecular parameters, the molar mass M_0 and contour length h per glucose residue, and the diameter or thickness of the polymer chain d. While the values of M_0 and h were determined from the chemical structure, q was estimated from the molecular weight dependence of the radius of gyration or of the hydrodynamic properties (the hydrodynamic radius, sedimentation coefficient, or intrinsic viscosity). The diameter d may be calculated from both the partial specific volume of the polymer and the hydrodynamic properties. The d values estimated from the latter are sometimes slightly larger than those from the former. To fill up the gap between q of xanthan and CTC, I add a synthetic polymer, poly(hexyl isocyanate) (PHIC), which is a main-chain conjugated polymer with $q = 21$ nm in dichloromethane (DCM). Furthermore, I add also polystyrene (PS) as the typical flexible polymer, to examine the chain stiffness effect on the polymer solution viscosity. The molecular parameters were determined in the same way.

The intramolecular excluded volume effect for PS does not play any role in the theta solvent cyclohexane (CH). Furthermore, the same effect is also not important for stiff polymers because of the scarce probability of the self-interaction. Only the excluded volume effect of CTC in the good solvent THF may be appreciable at high molecular weights.

2.7.2 Dilute Region

As already shown in Fig. 2.1 for aqueous NaCl solutions of xanthan, the polymer concentration dependence of the zero-shear viscosity η strongly depends on the polymer molecular weight. Figure 2.11 illustrates the concentration dependences in a dilute region in the form of the linear plot.

From the initial slope, one obtains the intrinsic viscosity $[\eta]$ (cf. Eq. 2.6), but $[\eta]$ are usually determined by the use of the Huggins plot, $[(\eta/\eta_s) - 1]/c$ vs. c, or the Mead-Fuoss plot, $[\ln(\eta/\eta_s) - 1]/c$ vs. c, although those plots are not shown in Fig. 2.11. The molecular weight dependence of $[\eta]$ for xanthan in 0.1 M aqueous NaCl is shown in Fig. 2.12a, along with the results of the other four polymers in dilute solution. In the abscissa, M_w is the weight-average molecular weight. Data points for PS in cyclohexane obey the straight line with the slope of 0.5 in the double logarithmic plot, which agrees with the prediction of Eqs. 2.15 and 2.24 for the Gaussian chain. On the other hand, data points for the other stiffer polymers follow convex curves, not obeying the Mark-Houwink-Sakurada equation (Eq. 2.7) with constant K and a over the whole M_w ranges investigated. The curves for xanthan and SPG at $M_w < 5 \times 10^5$ have a slope of 1.6, which indicates that the xanthan double helix and SPG triple helix exist as rigid rods (cf. Eq. 2.15) in the M_w region. However, data points at the higher M_w follow the convex curves with smaller slopes,

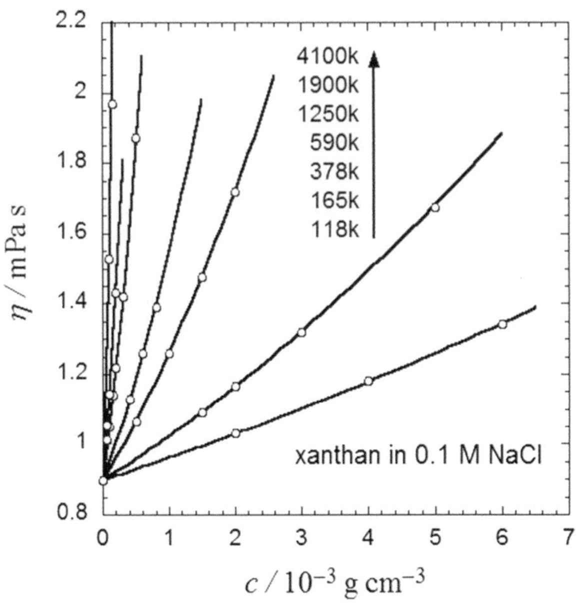

Fig. 2.11 Linear plots of the concentration c dependence of η for 0.1 M aqueous NaCl solutions of xanthan double helices with different molecular weights

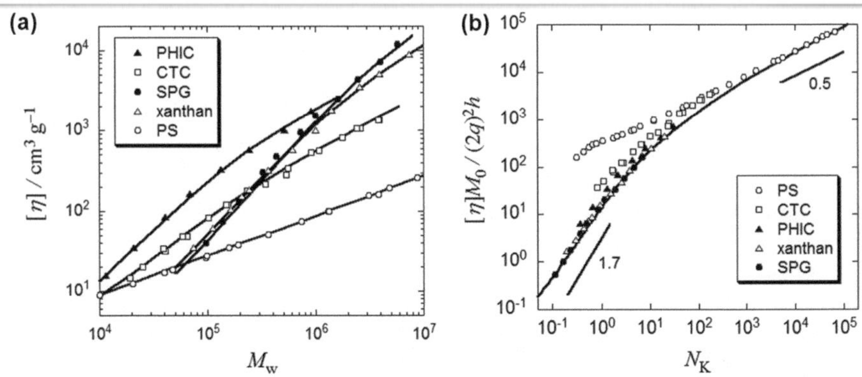

Fig. 2.12 Molecular weight dependence of $[\eta]$ for the five polymers; (**a**) plots of $[\eta]$ against M_w with theoretical curves (*solid curves*) calculated by the wormlike chain model using the parameters listed in Fig. 2.10; (**b**) universal plots of $[\eta]M_0/(2q)^2h$ against $N_K \equiv M_w/[2q(M_0/h)]$ with theoretical curve (*solid curve*) calculated by the wormlike chain model for SPG

which reflect the flexibility of the helices. The curves for CTC and PHIC have slopes of 1.0 and 1.3 at $M_w < 10^5$, respectively, slightly smaller than that for xanthan and SPG, and the slopes decrease with increasing M_w. The slopes at the highest M_w are 0.85 (PHIC) and 0.7 (CTC), which are close to the slope (0.5) for PS in cyclohexane.

Figure 2.12b replots the $[\eta]$ data for the five polymers in the form of $[\eta]M_0/(2q)^2h$ vs. $N_K \equiv M_w/[2q(M_0/h)]$ (the number of Kuhn's statistical segments). All the data points except for PS at low M_w almost obey a single composite curve (the solid curve) with slopes of 1.7 at low N_K (for the rod limit) and 0.5 at higher N_K (for the Gaussian limit). Deviations of data points for PS and CTC in low M_w regions from the composite curve come from the finite thickness effect on $[\eta]$. Because this effect is not important at sufficiently large axial ratio L/b, all the data points (except for CTC) approach to the composite curve in the high N_K region. Deviations of data points for CTC in high M_w region are due to the excluded volume effect.

From the curvature in the plot of η vs. c as shown in Fig. 2.11 or from the initial slope of the Huggins or Mead-Fuoss plots, we can determine the Huggins coefficient k'. The results for the five polymers are shown in Fig. 2.13a. Most of the data of k' are in between 0.3 and 0.5, irrespective of M_w and the chain stiffness. Therefore, k' is sometimes called as the Huggins constant, although it is not a constant, strictly speaking. Data of k' for PS at low M_w exceed 0.5 slightly.

According to Eq. 2.41, k' can be divided into the contributions of the entanglement and hydrodynamic interactions. The latter contribution k'_{EI} is calculated by Eq. 2.47, but the former contribution k'_{HI} cannot be calculated theoretically. Figure 2.13b shows k'_{EI} calculated by Eq. 2.47 for the five polymers by solid curves, and k'_{HI} estimated by subtracting k'_{HI} from the experimental k' in Panel a by symbols, as functions of N_K. (The theoretical values of k'_{EI} include the finite thickness effect of the polymer chain, which is neglected in Eq. 2.47.)

For stiff polymers, k'_{EI} exhibits a broad peak around $N_K = 1 - 10$. The peak shifts to larger N_K, and the peak height decreases with increasing the flexibility. On the other hand, k'_{HI}, except for PS, shows a broad minimum in the same N_K region and increases monotonically with N_K at large N_K, obeying a common asymptotic line. The deviation from the common line starts at larger N_K for more flexible polymers. When k'_{EI} and k'_{EI} for the same polymer are compared, k'_{EI} is superior to k'_{HI} for

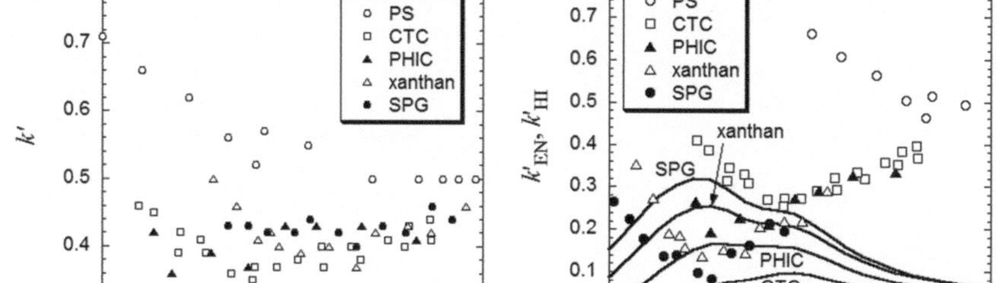

Fig. 2.13 (**a**) Molecular weight dependence of the Huggins coefficient k' and (**b**) contributions to the Huggins coefficient of the entanglement interaction k'_{EI} (*solid curves*) and of the hydrodynamic interaction k'_{HI} (*symbols*) for the five polymers

SPG and xanthan, they are comparable for PHIC, and k'_{HI} is predominant for CTC in the peak region. The hydrodynamic interaction term is much larger for PS over the entire N_K region.

2.7.3 Zero-Shear Viscosity Over a Wide Concentration Range

Figure 2.14a shows the concentration dependence of η for aqueous solution of seven schizophyllan (SPG) samples with different molecular weights, from 128 k ($=1.28 \times 10^5$) to 4300 k ($=4.3 \times 10^6$), over a wide concentration range. (Numbers in the parentheses in Fig. 2.14a indicate N_K values of the samples.) Solid curves in the figure indicate theoretical values calculated by Eq. 2.46 derived for the fuzzy cylinder model. In the equation, parameters L_e, d_e, L, and $[\eta]$ were estimated from the wormlike chain parameters listed in Fig. 2.10, and strengths of the hydrodynamic interaction k'_{HI} were estimated by subtracting k'_{HI} calculated by Eq. 2.47 (cf. Fig. 2.13b) from the experimental k'. Thus, the unknown parameter in Eq. 2.46 is only λ^* in V_{ex}^* (cf. Eq. 2.39). When λ^* is chosen to be 0.04, the theoretical curves nicely fit the experimental data points.

The contribution of V_{ex}^* or the effect of the concentration dependence of $D_{||}$ is not however so important, as demonstrated by dotted curves in Fig. 2.14b. Because this contribution becomes important exponentially with c, we might expect the importance at higher c. But the SPG solutions form a liquid crystal phase at high c, of which viscosity cannot be described by Eq. 2.46. We can say that the effect of the

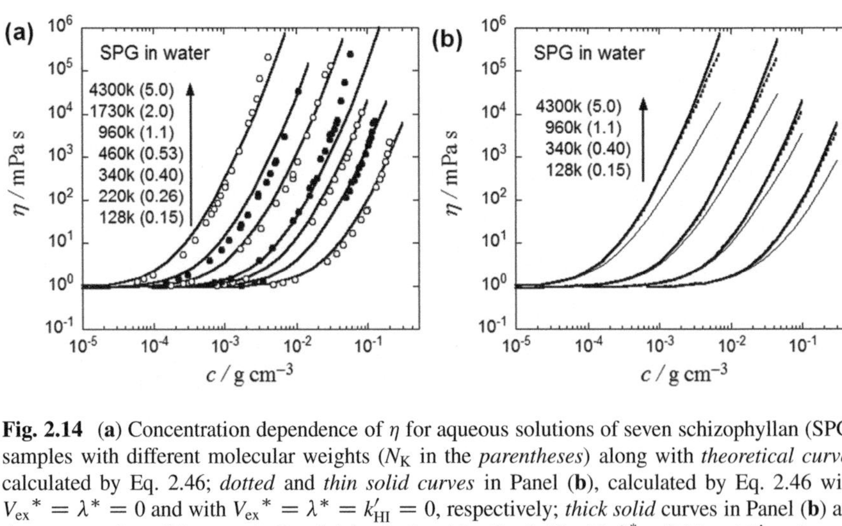

Fig. 2.14 (**a**) Concentration dependence of η for aqueous solutions of seven schizophyllan (SPG) samples with different molecular weights (N_K in the *parentheses*) along with *theoretical curves* calculated by Eq. 2.46; *dotted* and *thin solid curves* in Panel (**b**), calculated by Eq. 2.46 with $V_{ex}^* = \lambda^* = 0$ and with $V_{ex}^* = \lambda^* = k'_{HI} = 0$, respectively; *thick solid* curves in Panel (**b**) are the same as the solid curves in Panel (**a**) calculated by Eq. 2.46 with $\lambda^* = 0.04$ and k'_{HI} given in Fig. 2.13b

concentration dependence of $D_{||}$ (the so-called jamming effect) is not important in η for SPG isotropic solutions.

Figure 2.14b also displays thin solid curves calculated by Eq. 2.46 with $V_{ex}^* = \lambda^* = k'_{HI} = 0$. These curves deviate remarkably downward, indicating that the effect of the hydrodynamic interaction plays an important role in η at high c. Equation 2.46 predicts c^4 and c^3 dependences of η for the dotted and thin solid curves, respectively, in the high c region. The actual concentration dependence is slightly stronger than the c^4 dependence, because of the jamming effect.

Comparisons between theory and experiment for the solution viscosity of the remaining four polymers are shown in Fig. 2.15. The solid curves in each Panel are the theoretical curves calculated by Eq. 2.46 in the same way as in Fig. 2.14a. It is noted that there are no adjustable parameters in Eq. 2.46, if we select the same value for λ^* as in the case of SPG (=0.04). The agreement between the theory and experiment is good except for the highest M sample of PHIC ($N_K \sim 100$), the lowest M sample of CTC (L/b <10), as well as all the samples of PS (N_K >50). We can say

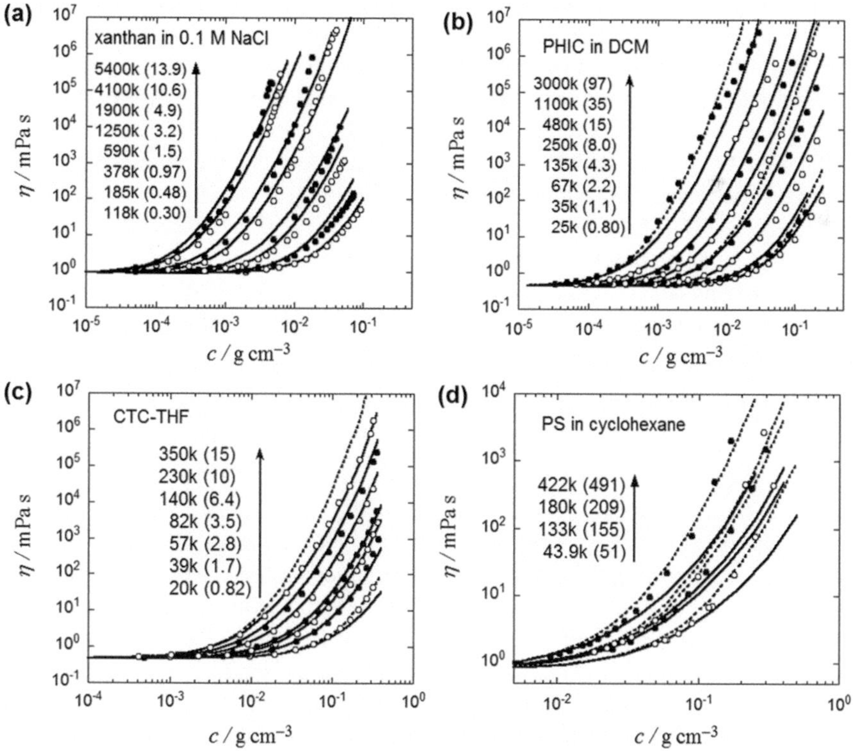

Fig. 2.15 Comparisons between theory and experiment for η of xanthan, PHIC, CTC, and PS solutions with different polymer molecular weights over wide concentration ranges; *solid curves* in each Panel, calculated by Eq. 2.46 with $\lambda^* = 0.04$ and k'_{HI} given in Fig. 2.13b; *dotted curves*, calculated by Eq. 2.48

that the viscosity equation (Eq. 2.46) predicts quantitatively the polymer solution viscosity if N_K <50 and L/b >10.

As indicated in Fig. 2.13b, the hydrodynamic interaction is more important in the Huggins coefficient than the entanglement interaction at N_K >50. In Eq. 2.46, the effect of the hydrodynamic interaction is considered only up to the linear order of c (cf. Eq. 2.27), but the higher-order term may be important at N_K >50. Beenakker [10] calculated η for spherical particle solutions up to the order of c^3. Applying his result to consider the higher-order effect of the hydrodynamic interaction, we modify Eq. 2.46 by

$$\frac{\eta-\eta_s}{\eta_s} = [\eta]c\left(1 + k'_{HI}[\eta]c + \left(k'_{HI}[\eta]c\right)^2\right) \times\left\{\frac{1}{4} + \frac{3}{4}\left[1 + \sqrt{\frac{6}{\beta}}\frac{N_A c L_e{}^4}{ML}F\exp\left(\frac{1}{2}V_{ex}{}^*c'\right)\right]^2\right\} \tag{2.48}$$

The dotted curves in Fig. 2.15 (Panels b–d) indicate theoretical values of η calculated by Eq. 2.48. The agreement between theory and experiment is improved by the modification for PHIC with $N_K = 97$ at intermediate c (Panel b) and for PS with $N_K > 50$ (Panel d). However, data points for PHIC with $N_K = 97$ at high c and with $N_K = 4.3$ and 0.80 in the whole c range (Panel b) are closer to the solid curve, indicating that the original Eq. 2.46 is better. In Panel c, data points for CTC with $N_K = 15$ are fitted to the solid curve, but data points with $N_K = 0.82$ and $L/b = 7.8$ to the dotted curve. From these comparison, we can say that the modified Eq. 2.48 improves the agreement with experimental data with N_K >50 and L/b <10, but not otherwise. The hydrodynamic interaction affects the polymer solution viscosity delicately, depending on the stiffness and the axial ratio of the polymer chain.

The viscosity equation Eq. 2.46 or 2.48 possesses two factors originated from the hydrodynamic interaction Λ_{HI} and the entanglement interaction Λ_{EI}, defined by

$$\begin{cases} \Lambda_{HI} \equiv 1 + k'_{HI}[\eta]c \text{ or } 1 + k'_{HI}[\eta]c + \left(k'_{HI}[\eta]c\right)^2 \\ \Lambda_{EI} \equiv \frac{1}{4} + \frac{3}{4}\left[1 + \sqrt{\frac{6}{\beta}}\frac{N_A c L_e{}^4}{ML}F\exp\left(\frac{1}{2}V_{ex}{}^*c'\right)\right]^2 \end{cases} \tag{2.49}$$

Figure 2.16 shows the ratio $\Lambda_{HI}/\Lambda_{EI}$ for SPG (solid curves), PHIC (dot-dash curve), CTC (dotted curves), and PS (dashed curves). For stiff polymers, SPG and PHIC, the factor Λ_{EI} is superior to the factor Λ_{HI} at high concentrations. Although not shown in the figure, Λ_{EI} is predominant also for PHIC with higher M as well as for xanthan over the entire M. On the other hand, for the flexible polymer PS, the factor Λ_{HI} is superior to the factor Λ_{EI} at high concentrations. For the semiflexible polymer CTC with an intermediate chain stiffness, the importance of the two interactions depends on the molecular weight; Λ_{HI} is more important at low M but Λ_{EI} is more important at higher M. In Fig. 2.13b, we can see that $k'_{HI} > k'_{EI}$ for PHIC, CTC, and PS over the entire N_K (or M). Therefore, the predominance of the entanglement interaction is enhanced in solution viscosity at higher concentrations.

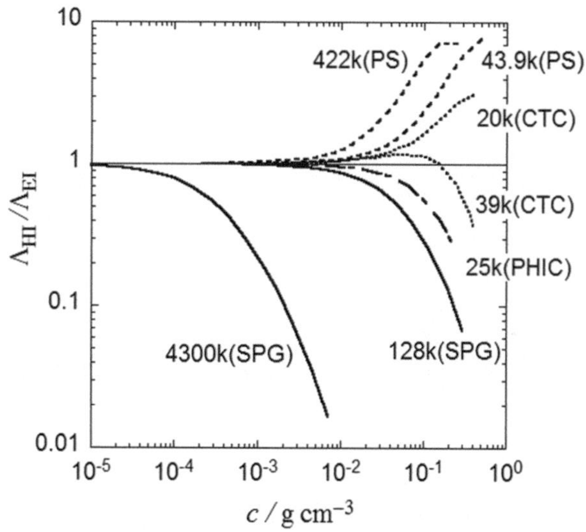

Fig. 2.16 Ratio of the factor Λ_{HI} originated from the hydrodynamic interaction to the factor Λ_{EI} from the entanglement interaction for SPG (*solid curves*), PHIC (*dot-dash curve*), CTC (*dotted curves*), and PS (*dashed curves*) in solution. Λ_{HI} for PS and CTC at 20 k were calculated by Eq. 2.49 including c^2 term

2.8 Conclusions

Naturally occurring polysaccharides have a variety of the chain conformation. Schizophyllan (or scleroglucan with the same chemical structure but different origin), xanthan (xanthan gum), and succinoglycan are rigid helical polymers with large persistence length q, On the other hand, amylose and pullulan are flexible polymers, of which persistence length is as small as that of PS. Cellulose and its derivatives (including CTC), as well as hyaluronic acid, are known as semiflexible polymers with intermediate q values.

The present chapter introduced molecular theories to formulate the intrinsic viscosity $[\eta]$, the Huggins coefficient k', and the zero-shear viscosity η at finite concentrations. According to the theories, $[\eta]$ for the rigid helical polymers are more strongly dependent on the molecular weight than those of more flexible polysaccharides. Therefore, the stiffer polysaccharides with high molecular weights are more suitable as the viscosity enhancement reagent.

Polysaccharide solution viscosities exhibit strong molecular weight and polymer concentration dependences. The molecular theory explained in the present chapter teaches us that those strong dependences arise from both entanglement and hydrodynamic interactions among polymer chains in the solution. The contribution of the entanglement interaction to the solution viscosity is more important than the contribution of the hydrodynamic interaction for stiff polymer solutions, but opposite for flexible polymer solutions. We can anticipate the effectiveness of newly found polysaccharides as the viscosity enhancement reagent, if we know the molecular characteristics of the polysaccharides.

References

1. Brant D. A., ed, "*Solution Properties of Polysaccharides*," ACS Symp. Ser., No. 150, Am. Chem. Soc., Washington, D.C., 1981.
2. Yamakawa, H., "*Helical Wormlike Chains in Polymer Solutions*," Springer, Berlin & Heidelberg, 1997.
3. Sato T., Teramoto, A., *Adv. Polym. Sci.*, 1996, 126: 85.
4. Ree, D. A., "*Polysaccharide Shapes*," Outline Studies in Biology, Chapman & Hall, London, 1977.
5. Cohen, M. H., Turnbull, D., *J. Chem. Phys.*, 1959, 31: 1164.
6. De Gennes, P.-G., "*Scaling Concepts in Polymer Physics*," Cornell University Press, Ithaca, 1979.
7. Doi, M., Edwards, S. F., "*The Theory of Polymer Dynamics*," Oxford University Press, Oxford, 1986.
8. Yamakawa, H. "*Modern Theory of Polymer Solutions*," Harper & Row Publishers, New York, 1971.
9. Sato, T., *Kobunshi Ronbunshu*, 2012, 69: 613, and the references therein.
10. Beenakker, C. W. J., *Physica A*, 1984, 128: 48.

Chapter 3
Gel-Solvent Friction

Masayuki Tokita

Abstract The science of gel draws much attention after the discovery of the volume phase transition of gel. Among others, the information on the dynamics of gel is of importance to understand the kinetic behaviors of the volume phase transition of the gel. It is well established that the dynamics of gel is governed mainly by the collective diffusion of the polymer network of gel and the collective diffusion itself is determined by the balance between two forces. One is the elastic force due to the deformation of three-dimensional polymer network of gel, and the other is the frictional drag force between the polymer network of gel and the gel fluid. In early stage of the study in gels, the elastic properties of gel attract much attention, and, hence, considerable effort has been devoted to clarify the elastic behaviors of gel. The elastic properties of various gels under various experimental conditions are gained and reported so far. In contrast, much attention has not been paid to the frictional properties of gel since the experimental method in obtaining the reliable values of the friction coefficient has not been established until recently. The systematic studies on the frictional properties of gel begun only recently. Here, we would like to overview the earlier studies on the frictional property of gels including why the frictional study of gel is difficult and how we can solve the difficulty to obtain the reliable values of the friction coefficient of gel. The recent advancement on the frictional study of colloid gel will also be reviewed.

Keywords Friction • Collective diffusion • Colloid gel • Network gel

3.1 Introduction

Gel is a state of matter that is familiar even in our daily life. The unique physical properties of a gel arise from its structure, namely, a three-dimensional cross-linked polymer network immersed in a fluid. Because of this structure, gels play important roles in a wide variety of biological and chemical systems. In earlier studies of

M. Tokita (✉)
Department of Physics, Faculty of Science, Kyushu University, 744 Motooka, 819-0395, Fukuoka, Japan
e-mail: tokita@phys.kyushu-u.ac.jp

I. Kaneda (ed.), *Rheology of Biological Soft Matter*, Soft and Biological Matter,
DOI 10.1007/978-4-431-56080-7_3

gel, the elastic property of gel draws much attention because the formation of a gel is similar phenomenon with the solidification, and, hence, the drastic change in the elastic properties appears with the onset of the formation of three-dimensional polymer network. The phenomenon is simply called as the gelation. In addition to this, many experimental techniques are developed since the elasticity measurement is rather easy. As a result, many experimental studies are made on various gels having various chemical structures. The scientific knowledge on the gel and gelling processes are gained and accumulated so far.[1]

A new insight into the viscoelasticity of gel is developed theoretically, later which is called as THB theory, with pioneering experimental studies [1]. In THB theory, the viscoelastic property of gel is explained in terms of the balance between two forces, namely, the elastic force due to the deformation of three-dimensional polymer network of gel and the frictional drag force between the polymer network and the gel fluid. In the 1970s to 1980s, the volume phase transition of gel is discovered and the studies on the kinetics of the phase transition are accelerated for the better understanding of the volume phase transition of gel and also for the practical use of gels [2–4]. It becomes eventually clear that the frictional property of gel also plays the essential roles in the kinetic behaviors of gel. Although the important roles played by the friction in the dynamics of gel becomes known, no one succeeded to obtain the reliable values of the friction coefficient of gel. Hence, the study on the frictional property of gel is left as it is. The reason is simple. The measurement of the friction coefficient of gel is considerably difficult, and it also requires patience to the researcher. In a metaphorical sense, the situation of the gel science until the 1990s is said as follows: elasticity is the queen of the gel world and friction is her shadow. However, the situation is changed drastically after the discovery of the critical behavior of the friction near the volume phase transition point of gel in 1991, which opens a new insight into the critical transport phenomena in the gel world. Then, the frictional property of gel eventually draws much attention. Today, we realize that the friction of the gel is the most characteristic physical property of the gel. The elasticity is, of course, a characteristic property of gel; however, it is not a characteristic property that is restricted to the gel since we know that the polymer solutions also show the viscoelastic properties especially in their concentrated states. In contrast, the friction is only defined for the system with the polymer network and fluid, that is, the gel.[2]

The viscoelastic properties of gels are uniquely described by three parameters: the osmotic bulk modulus K, the shear modulus μ, and the friction coefficient f between the polymer network and the gel fluid [1]. It has been established by

[1]We do not cite the reports of the rheological studies on the gel and gelling processes here because there are actually so many studies to be cited. Readers who are interested in the rheological studies of gels, please refer to well-known books of rheology such as *Viscoelastic Properties of Polymers* written by J. D. Ferry. Many references, including the experimental method of rheology, are cited therein.

[2]See the definition of the friction coefficient, Eq. (3.4). The friction we will discuss here is a property that is defined to a system that consists of the *porous solid* and the fluid.

THB theory that the viscoelastic properties of a gel can be determined by dynamic light scattering from the collective mode of the density fluctuations of the polymer network of the gel. Light is scattered by the collective mode, or the longitudinal density fluctuations, that decays with a characteristic time τ_c as follows:

$$\tau_c = \frac{\lambda^2}{4\pi D_c} \tag{3.1}$$

$$D_c = \frac{E}{f} = \frac{K + \frac{4}{3}\mu}{f} \tag{3.2}$$

Here, D_c is so-called collective diffusion coefficient, λ is the scattering wavelength of light in the gel, and E the longitudinal modulus only of the polymer network. The amplitude of the scattered light intensity I is given as follows [5]:

$$I \propto \frac{kT}{E} \tag{3.3}$$

Equations (3.1) to (3.3) suggest that it is possible to determine both the longitudinal modulus and the collective diffusion coefficient from the intensity and the decay time of fluctuations of light, which were experimentally proven by many studies on various gels [5–8]. Thus, by combining with a shear modulus, which can be determined by a simple mechanical measurement, all three parameters can be obtained using the dynamic light scattering spectroscopy.

The mechanical studies of gel have been made extensively so far. Young's modulus and the shear modulus of various gels have been measured under various experimental conditions. The experimental results on the concentration dependence of the elastic modulus are in good agreement with the prediction of the scaling theory [9]. In contrast, there have been not much systematic experimental and theoretical studies on the frictional property of gels. The concept of the friction is simple, but the actual measurements are not trivial and require a careful design and treatment since gels are not solid but fragile object.

The principle of the mechanical measurement of the friction coefficient of a gel is schematically illustrated in Fig. 3.1. A thin slab of a gel of thickness L is fixed in a space with the rim attached to the wall of a cell. The water is pushed by a small pressure P. The velocity of the water flow through the gel v_s is determined by measuring the rate at which water flows out of the gel in a steady state. Then, the friction coefficient of the gel f is defined as follows:

$$f = \frac{P}{v_s L} \tag{3.4}$$

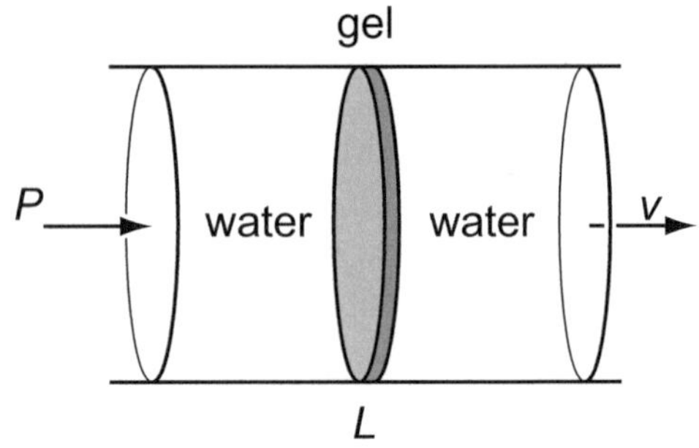

Fig. 3.1 Schematic illustration of the measurements of the friction coefficient *f* between a gel and water

This equation indicates that the steady-state velocity of flowing water in gel is a linear function of the pressure.[3]

The frictional property of gels, however, has been studied in not ideal experimental conditions so far [1, 10–13]. In the previous studies, relatively large gels were used to determine the friction coefficient. If the dimension of the gel became larger, the relaxation time for the mechanical deformation by the applied pressure became longer. The relaxation time of a typical gel having the characteristic size of about 1 cm is of the order of 10 days, indicating that it takes more than 10 days to determine the friction coefficient of the gel at a stationary state [4]. However, such long time measurements have not been made so far. Furthermore, the apparent velocity of water flow is slow for this thickness, and a relatively high pressure is necessary to secure the accurate determination of the velocity. When the applied pressure is high, however, the other effects become pronounced, for instance, the destruction of the gel structure, the nonlinear relationship between the velocity of the fluid and the applied pressure, and the shrinkage of the gel due to the frictional force. Solving of such problems seems to be required for reliable and accurate determination of the friction coefficient of gels.

First of all, we describe the details of the apparatus that is critical for the accurate determination of the friction coefficient of gel [14]. There are actually some difficulties in obtaining a reliable value of the friction coefficient of gel as described above. We shall also describe such difficulties in detail and the way they are solved.

[3]Equation (3.4) is sometime called as Darcy's law. So far, the equation is applied to the studies of the fluid flow in the porous hard materials such as the fluid flow in the sintered glass filters and the subterranean water flow in the soil.

3.2 Experimental

3.2.1 Apparatus

Since the gels we study are typical fragile object, the pressure that is exerted on the gels should be small enough not to break the gel. Besides, the friction should be small enough not to impose a substantial frictional resistance that induces a shrinkage of the gel. Since the osmotic compressibility of the gel is very small, typically 10^3–$10^4\,\mathrm{N/m^2}$, the applied pressure should be comparable or less than these values to assure no shrinkage of the gel. In addition, we designed to fix the gel volume by choosing the condition at which the gel has a positive osmotic pressure and also by mechanically prohibiting the gel from swelling using filters.

A simple calculation using the previously available results of the dynamic light scattering and the mechanical measurements suggests that the friction coefficient of typical poly(acrylamide) gel is expected to be of the order of 10^{14}–$10^{15}\,\mathrm{N\,s/m^4}$ [5]. The velocity of the flow of fluid inside the a gel having $L = 1\,\mathrm{cm}$ becomes, therefore, of the order of 10^{-9}–$10^{-10}\,\mathrm{m/s}$ at a pressure of $10^3\,\mathrm{N/m^2}$ according to Eq. (3.4). In order to determine such slow velocity of water in the gel, a capillary with a uniform and well-calibrated diameter is used to amplify the velocity of flowing water in gel.

Since the water flow is slow, it is also important to avoid any leakage of water around the gel sample. This is achieved by using a plastic film with a special surface treatment to which the gel is chemically attached. The chemical bonding of the gel to the film secures no leakage of water around the gel sample.

The apparent velocity of water in the capillary is fast at first, slows down, and reaches a constant value in a steady state. This apparent fast initial velocity seems to be due to the bending deformation of the gel and the gel holder and the gel shrinking due to the frictional resistance. Although the deformation of gel due to the frictional pressure of the flowing water is small, it contributes to the apparent flow of water out of the capillary. Thus, the velocity of the water inside the gel is time dependent. The order of the relaxation time at which the flow of fluid reaches a stationary state due to the gel shrinking can be estimated using the kinetic theory of the swelling of gel. It has been shown that the relaxation time of the swelling and shrinking of a gel is proportional to the square of a typical linear size of the gel [4]. In the present case, it is the thickness of the gel L and relaxation time $= L^2/\pi^2 D_c$. For example, the characteristic time of the swelling and shrinking of a gel having a thickness of $L = 10^{-3}\,\mathrm{m}$ and the typical collective diffusion coefficient of $D_c = 10^{-11}\,\mathrm{m^2/s}$ is the order of $10^4\,\mathrm{s} \sim 3\,\mathrm{h}$. The thinner the thickness of the gel, the shorter the relaxation time is. It is of further advantage to have a thinner gel, that is, the velocity of the fluid becomes larger if the gel becomes thinner making it possible to have more accurate determination of the velocity of the fluid.

The characteristic time needed for bending deformation of the gel and the gel holder is approximately 30 min as estimated by the control experiments, where the gel is molded between two bonding films with no openings so that no water flow is

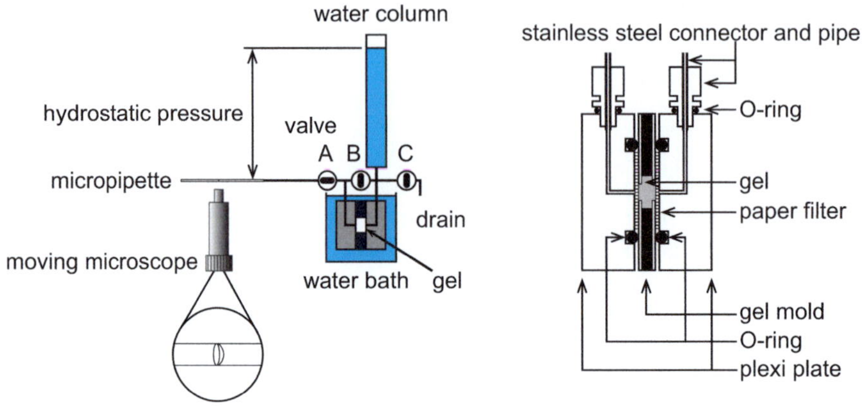

Fig. 3.2 Schematic illustration of apparatus is shown in the *left figure*. The apparatus is set on a vibration-free optical table to avoid the mechanical disturbances. The cross section of the sample cell is illustrated in the *right figure*. Both plexi plates are tightly fixed to avoid the swelling of gel

allowed. Thus, the movement of water in the capillary is only due to the bending of the gel-film system. It is also important to use a stainless steel gel holder rather than a plastic one to reduce the amplitude of the bending deformation. The former has ten times less deformation than the latter. The steady-state flow of water is observed after all these relaxation phenomena are diminished.

We construct the apparatus in which all the aspects described above are taking into account [14]. The outline of the apparatus is schematically shown in Fig. 3.2. The cell is made of plexiglass. The connectors, pipes, and valves are made of stainless steel to avoid the expansion due to the applied hydrostatic pressure. The chromatography column of 50 cm length is used as a reservoir for the water to generate the hydrostatic pressure. The range of the height of the water column can be changed from about 20–60 cm, which corresponds to the pressure from about 2×10^3 to $6 \times 10^3\ \mathrm{N/m^2}$. The temperature of the cell is controlled within an accuracy better than 0.1 degree using a water circulating system. The apparatus is set on the vibration-free optical table to avoid any external mechanical disturbances.

The structure of the gel holder is shown in Fig. 3.3. The gel mold is made of stainless steel. The gel bonding films, which have chemically active double bonds on the surface, are stuck on both sides of the mold using silicone glue. The final thickness of the gel mold including gel bonding film is about 1 or 2 mm. The reaction between the double bonds on the gel bonding film and the monomer and/or the cross-linker in the pre-gel solution occurs during the gelation reaction. Because the gel is chemically bonded to the gel bonding films, the leakage of water around the gel is completely avoided.

After the gel has been cast, the gel mold is taken out of the reaction bath, and glass plates are removed carefully under the water. Then the gel mold, which contains the sample gel, is tightly fixed between two cells made of plexiglass of about 1 cm in thickness. The O-rings are almost completely embedded in the

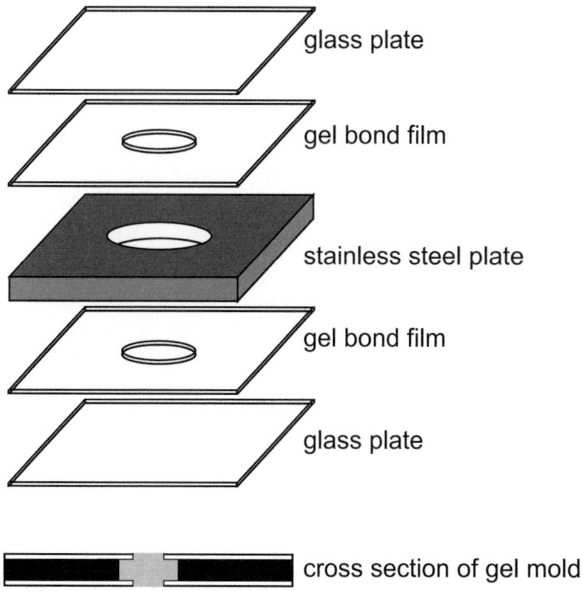

Fig. 3.3 Structure of gel holder. Two gel bond films are glued on the stainless steel plate. Then the gel mold is soaked into the pre-gel solution with both sides are covered by glass plates. After the reaction is finished, both glass plates are removed carefully under water. Finally, the gel mold is fixed in the sample cell which is shown in Fig. 3.2

plexiglass plate cells to minimize the bending of the gel mold. At the end of outflow stainless steel pipe, a calibrated micropipette of inner diameter ranging from 0.34 to 0.50 mm is attached to measure the flow rate. Different micropipettes are used depending on the applied pressure and the friction coefficient of gel. The micropipettes are commercially available and indicated as 5 and 10 $\mu\ell$.

The gel is left for overnight without pressure to reach equilibrium in water at a fixed temperature. The valve A in Fig. 3.2 is closed and the valve B is left open so that the no pressure is exerted upon the gel. At the time of measurement, the valve B is closed and A is opened immediately to apply the hydrostatic pressure to the gel. The position of the meniscus in the micropipette is measured as a function of time using a microscope staged on a micro-translator. After the measurement is done, the valve A is closed and the valve B is opened to release the pressure, which allows the gel to recover its initial state.

The friction coefficient of gel is huge, say the order of about $10^{14}\,\mathrm{N\,s/m^4}$ as described above, if the polymer network of gel is uniform in space. The friction of gel, however, becomes very small when the structure of the polymer network is changed. The structural change of the polymer network is caused by the dynamic fluctuations as well as the static fluctuations. In the former case, it has been shown that the friction coefficient of gel becomes smaller more than three orders of magnitudes and appears to diminish as the gel approaches a certain temperature

Fig. 3.4 Friction measurement in tenuous gels. The gel is prepared in one end of a micropipette at a thickness of L. Then the micropipette is soaked into water at a desired depth h_0 to generate the pressure P. The elution height, $h(t)$, is then measured as a function of time. The results are analyzed using Eqs. (3.5) and (3.6)

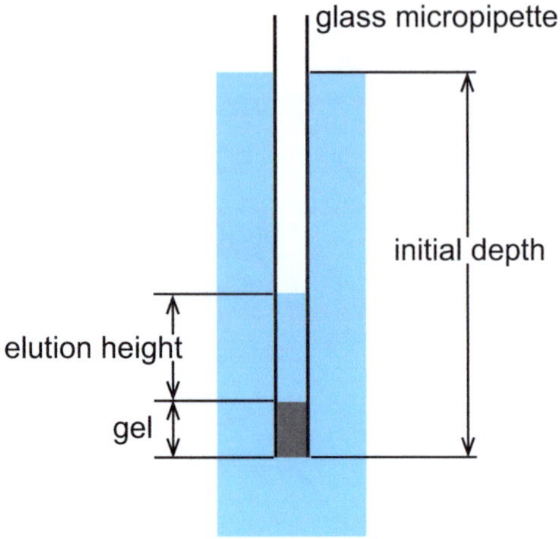

at constant volume of the gel. The results, which are observed in the vicinity of the volume phase transition point of the thermosensitive gel, are discussed in terms of the critical density fluctuations in the polymer network of gel. On the other hand, if the nonuniform structure, which is caused by the dynamic fluctuations, is frozen into the polymer network of the gel during the gelation reaction, the structure of the polymer network is totally altered from the uniform gel. The nonuniform structure thus frozen into the gel is static because it is chemically fixed in the space of polymer network of the gel. As a result of frozen-in density fluctuations, the gel becomes translucent and/or opaque. In such a nonuniform gel, we find that the friction of the gel decreases more than four orders of magnitude. Therefore, the friction coefficient of gel can be determined by much simpler method in the case of opaque gels. In Fig. 3.4, we show the example. In this case, the gel is prepared in the micropipette of known inner diameter. Then the micropipette is soaked into water at a desired depth, h_0. The water eventually passing through the gel due to the pressure gradient and the height of the water column in the micropipette becomes higher. If the height of water column is measured as a function of time, $h(t)$, then the friction coefficient of gel f is determined from the characteristic time τ_e of the elution time course.

$$\frac{h_0 - h(t)}{h_0} = \exp\left(-\frac{t}{\tau_e}\right) \tag{3.5}$$

$$\tau_e = \frac{Lf}{\rho g} \tag{3.6}$$

Here, L, ρ, and g are the thickness of the sample gel, the density of fluid, and the acceleration of gravity. The friction coefficient less than $10^{13}\,\mathrm{N\,s/m^4}$ can be determined by using this simple apparatus.

3.2.2 *Sample*

The sample poly(acrylamide) gels are prepared by a standard-free radical polymerization. All the reagents used here are of the electrophoresis grade (BioRad) and used without further purification. The predetermined amount of acrylamide (main chain component), N,N'-methylenebisacrylamide (cross-linking agent) are dissolved into a distilled and de-ionized water prepared by a Milli-Q system. The predetermined amount of the accelerator, N,N,N',N'-tetramethylethylenediamine, is added. Then, the pre-gel solution is de-gassed for about 30–40 min. The initiator solution is added to the pre-gel solution to initiate the gelation reaction. The solution is kept at the desired temperature undisturbed until the reaction is completed. In the case of poly(N-isoprppylacrylamide) gel (thermosensitive gel), the monomer is re-crystallized before polymerization [15]. The polymerization is made mostly in the same manner except the temperature control since the gel becomes opaque by the reaction heat if the temperature is not controlled. The gelation reaction is made usually below 20 °C.

3.3 Results

3.3.1 *Calibration of Apparatus*

Typical experimental results of the elution time course for the poly(acrylamide) gels are shown in Fig. 3.5. Poly(acrylamide) gel at a total concentration of about 8 wt% with a cross-linking density of about 1 mol% is employed for this measurement. The position of the meniscus of water in the micropipette of 10 $\mu\ell$ is plotted as a function of time in this figure. Different curves denote the measurement at different applied

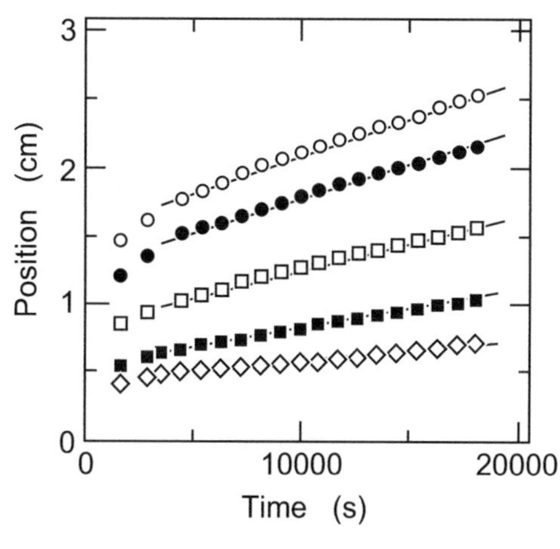

Fig. 3.5 The time course of the position of meniscus in the micropipette of 10 $\mu\ell$. The applied pressures are 5.88, 4.90, 3.92, 2.94, and 1.96 (in 10^3 N/m^2) from top to bottom. The thickness of the gel is 1.95 mm. In this figure, only the results near the steady-state regions are shown. The rapid movement of meniscus in the very initial state is not sown for the simplicity. *Lines* are drawn only guide to the eye

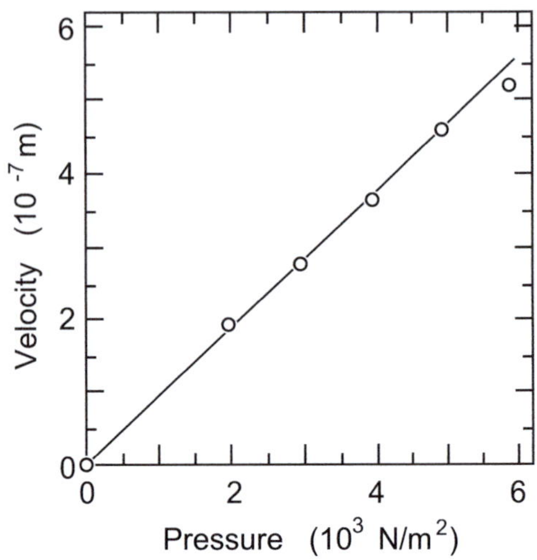

Fig. 3.6 The pressure dependence of the velocity of the fluid in the micropipette at the stationary state. The *straight line* in this figure is obtained by the method of least squares

pressures. We find that the position of the meniscus rapidly moves as soon as the pressure is applied on the gel. The velocity of the meniscus decreases with time and approaches to an asymptote in the time region of about 1×10^4 s. The velocity of the fluid in the micropipette is at a stationary state v_{sc} is obtained from the slope in this steady-state region. The rapid movement of the meniscus in the very initial stage of the measurement can be considerably decreased if the friction cell is improved. For instance, most parts of the friction cell is made of stainless steel in the improved cell to avoid the bending of the gel mold. The friction measurement using such an improved cell is reported recently [16].[4]

The velocity of fluid at the stationary state in the micropipette is plotted as a function of the applied pressure in Fig. 3.6. The relationship between the applied pressure and the velocity is linear as predicted from Eq. (3.4). The friction coefficient, $f = 1.0 \times 10^{15}\ \mathrm{N\,s/m^4}$, is obtained from the straight line in Fig. 3.6 using the following equation.

$$f = \left(\frac{dv_{sc}}{dP}\right)^{-1} \frac{1}{L} \left(\frac{R}{r}\right)^2 \tag{3.7}$$

Here, $(dv_{sc}/dP) = 8.95 \times 10^{-11}\ \mathrm{m^3/N\,s}$ is determine from the slope of the straight line in Fig. 3.6. The thickness of the gel $L = 1.9 \times 10^{-3}$ m. The factor $(R/r)^2$ is the ratio of the area of the cross section of the micropipette of radius $r = 2.5 \times 10^{-4}$ m and the area of the hole on the gel bonding film with the radius $R = 3.3 \times 10^{-3}$ m, which is necessary to convert v_{sc} to the velocity of the fluid in the gel at a stationary state v_s. The value of the friction coefficient obtained here is in a reasonable

[4]The result of the elution time course from gel can be fitted to the theoretical one. We find that the results are in good agreement with the independent elasticity measurement.

agreement with the one obtained from the dynamic light scattering studies. These results indicate that it is possible to accurately determine the friction coefficient of a gel using this apparatus.

3.3.2 *Uniform Gel*

We have constructed an apparatus which work well for the friction measurement of gel. Thus, we study all the aspect of the frictional properties of uniform gel. Among others, poly(acrylamide) gel is employed as a sample gel because this gel is most commonly used in the chemical industry and biotechnology as a medium for the electrophoresis as well as the gel permeation chromatography. It is well known that uniform and transparent gel is prepared and used in the poly(acrylamide) gel electrophoresis. Therefore, our first aim is set to understand all frictional properties of uniform poly(acrylamide) gel.

3.3.2.1 Temperature Dependence of the Friction Coefficient

Firstly, the temperature dependence of the friction coefficient of poly(acrylamide) gel is studied. The measurements are made at two concentrations of the gel, namely, 5 wt% (main constituent : cross-linker = 693 mM : 7 mM) and 8 wt% (main constituent : cross-linker = 1240 mM : 22.4 mM). The composition of former gel is almost the same with the composition typically recommended for the acrylamide gel electrophoresis. The applied pressure is fixed at 5.88×10^3 N/m^2 and the temperature is varied from 0 to 60 °C.

The results are shown in Fig. 3.7. The open symbols in this figure indicate the results obtained upon lowering the temperature and the solid symbols represent the ones obtained while the system is heated step by step. The hysteresis is not found between the heating and the cooling processes. The result indicates that the friction coefficient of these gels decreases slightly but monotonically with the temperature.

The friction coefficient of the polymer network of a gel and a fluid should be proportional to the viscosity of the fluid passing through the gel and also determined by the average mesh size of the polymer network. The average mesh size of the polymer network of a gel is reasonably assumed to be proportional to the correlation length ξ of the gel, which, in a uniform gel in a good solvent, represents the average distance between the neighboring contact points of the polymers. The friction coefficient of a uniform gel for a good solvent is inversely proportional to the square of the average mesh size of the polymer network, that is, the void area of the mesh. Hence, we expect following simple relationship.

$$f \propto \frac{\eta}{\xi^2} \tag{3.8}$$

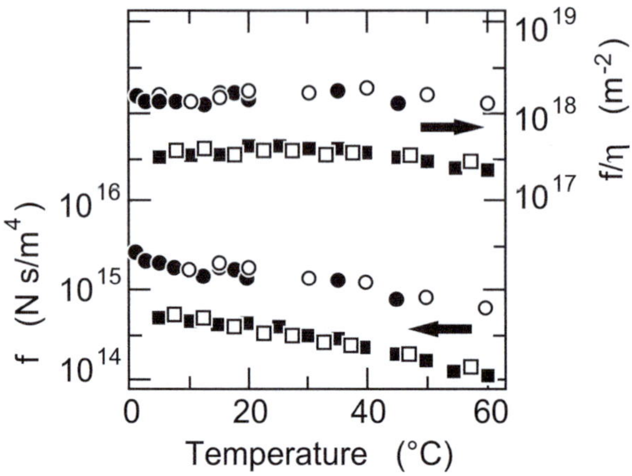

Fig. 3.7 The temperature dependence of the friction coefficient of the poly(acrylamide) gels: bottom portion. The total concentrations of acrylamide are 1.24 M (*circles*) and 693 mM (*squares*). The friction coefficient of gel normalized by the viscosity of water, $f(T)/\eta(T)$, are also plotted: top portion. Open symbols are upon lowering the temperature and closed symbols represent the data taken upon increasing the temperature

Therefore, in order to discuss the temperature dependence of the correlation length of the gel, it is convenient to use $f(T)/\eta(T)$ rather than the law value of the friction coefficient $f(T)$. This ratio directly represents the effective pore sizes and their distribution. We, thus, also show the temperature dependence of the friction coefficient of gel, which is normalized by the viscosity of water, in Fig. 3.7. It is found from this figure that the normalized friction coefficient of poly(acrylamide) gel is independent of the temperature. It indicates that the pore size of the poly(acrylamide) gel is stable in the temperature region studied. Such a property is required and also suitable for the stable separation in the electrophoresis systems.

3.3.2.2 Total Polymer Concentration Dependence

The friction coefficient of poly(acrylamide) gel is measured at various total concentration of polymer gels. In this series of experiments, the total concentration of polymer is changed at a constant molar ratio of cross-linker of 1 mol%. The total concentration is changed from 3 to 20 wt% and the friction measurements are made at a temperature of 20.0 ± 0.1 °C.

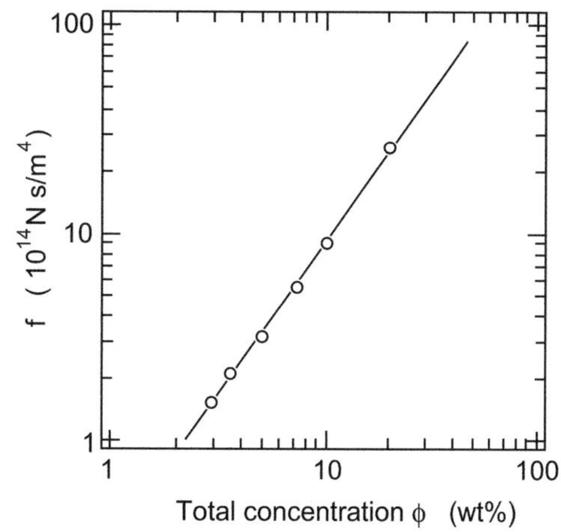

Fig. 3.8 The double logarithmic plot of the friction coefficient as a function of the total concentration of the gel. The *straight line* in this figure indicates the result of the least squares analysis. The slope of this straight line is 1.5

The results are shown in Fig. 3.8 where the friction coefficient f is plotted as a function of the polymer concentration ϕ in the double logarithmic scale. The data are well represented by a straight line with a slope of 1.5, which is determined by the least-square fitting. The results indicate that the concentration dependence of the friction coefficient is expressed by a power law relationship.

$$f \propto \phi^{1.5} \tag{3.9}$$

The total concentration dependence of the friction coefficient is well expressed by the power law relationship. Since the cross-linking density is low enough, we assume that the correlation length should depend on the concentration in the following power law relationship [9].

$$\xi \propto \phi^{-3/4} \tag{3.10}$$

Substitution of Eq. (3.10) into Eq. (3.8) gives the concentration dependence of the friction coefficient as follows.

$$f \propto \eta\phi^{3/2} \tag{3.11}$$

The power 1.5, which is obtained experimentally, is, therefore, in a god agreement with the scaling theory that predicts the exponent of 3/2. The results indicate that the frictional property of a gel is mainly governed by the pore size equal to the correlation length of the gel for a good solvent.

The experimental result Eq. (3.9) is also consistent with the previous experimental results of the light scattering and the mechanical studies. It has been reported that the concentration dependence of the collective diffusion coefficient of the gel and that of the elastic modulus are well represented by the scaling form as follows.

$$D_c \propto \phi^{3/4} \tag{3.12}$$

$$E \propto \phi^{9/4} \tag{3.13}$$

The experimental results Eqs. (3.12) and (3.13) together with the results Eq. (3.9) confirm that the relationship between the collective diffusion coefficient, the elastic modulus, and the friction coefficient should be given by Eq. (3.2).

3.3.2.3 Cross-Linker Concentration Dependence

The friction coefficient of the poly(acrylamide) gel is measured at various cross-linker concentrations at a temperature of 20.0 ± 0.1 °C. The cross-linker concentration is changed from 1.4 to 70 mM at a constant total concentration of 700 mM. The results are shown in Fig. 3.9. The friction coefficient of gel decreases with an increase of the cross-linker concentration in most of the concentration region, except that it sharply decreases at 0.2 mol% of the cross-linker concentration. The measurements cannot be made below this concentration because the gel becomes too soft to be used in the present experimental setup. It suggests that system is close to the gelation threshold. The sharp decrease of the friction observed in the 0.2 mol% gel may be due to the effect of the sol-gel transition. It is also found from the naked eye inspection that the gels become opaque above 3 mol% cross-linker concentration.

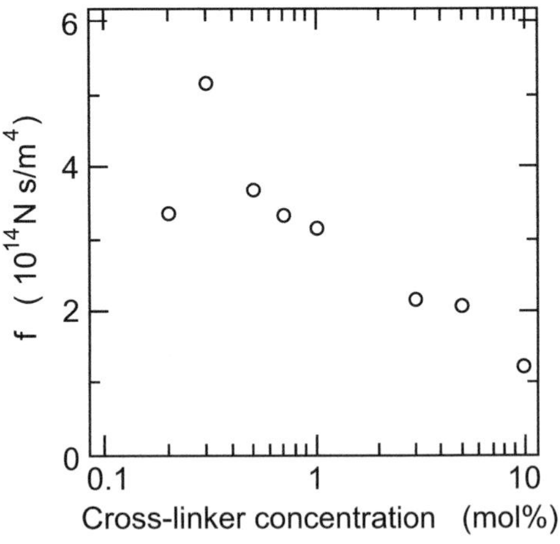

Fig. 3.9 The cross-linker concentration dependence of the friction coefficient of the poly(acrylamide) gel. The total concentration of monomer and cross-linker is fixed at 700 mM

It may be natural to expect from Eq. (3.8) that the friction coefficient of the gel should increase with an increase of the cross-linker concentration. Since the length of the polymer chain that connects two cross-linking points decreases with the cross-linking density, and thus the pore size is expected to be smaller. The results obtained here, however, are entirely different from our expectation, namely, the friction coefficient decreases upon increasing the cross-linking density of gel. Besides, it is observed that the gel becomes opaque above 3 mol% of the cross-linker concentration. Although the quantitative studies have not been made so far, similar results of the opacity change were reported. The appearance of the opacity of the gel suggests that the structure of the polymer network becomes nonuniform with the concentration of the cross-linker. The cross-linking agent, when polymerized, is not very soluble in water. Upon gelation such insoluble clusters are formed and frozen into the polymer network of gel to form a nonuniform polymer network. Some regions are more dilute and the other regions are denser. These nonuniform structures create the nonuniform refractive index within the gel and scatter light. This is the reason for the opacity of the gel prepared with a larger concentration of the cross-linker.

The friction is expected to be small for such a nonuniform gel, since water flows choosing mainly the dilute regions, that serve as a more open and free path, while avoiding the denser regions, that block the flow. The opaque gel has, therefore, a much smaller friction coefficient than a transparent gel. The friction and the structure of opaque gel will be discussed in detail in the following section. It should, however, be noted here that the effect of inhomogeneities on the friction coefficient appears much earlier than the visual opacification appears in the gel. Indeed the friction coefficient decreases with the increasing cross-linker concentration even in the lower concentration regions of the cross-linker where the gel looks completely transparent. The results suggest that the seeds of the concentration fluctuations are already formed in the lower cross-linker concentration region, which have not been detected by optical measurements. Thus the friction coefficient of a gel reveals, in an extremely sensitive way, the spatial nonuniform structure of the gel.

3.3.3 Nonuniform Gel

The friction coefficient is extremely sensitive to the spatial nonuniform structure of the gel, namely, the structure of the polymer network, as described in the previous section. It may be natural to expect that one can clarify the relationship between the structure and the property of gel by combining the structural and the frictional studies of gels. Although such a study is difficult, in some cases it is possible to study the structure-property relationship of the gel by measuring the friction of gel. We present two examples of such studies in the following sections:

3.3.3.1 Friction and Critical Density Fluctuation of Gel

When a gel is brought near its critical point, the density fluctuations become dominant and the gel becomes opaque as a result of light scattering from the increasing density fluctuations. The phenomenon is known as the critical opalescence, which, in binary mixture of simple liquids, appears and diverges to infinity at the critical point at a constant volume of the binary mixture. Similar phenomenon is known to be observed in the vicinity of the volume phase transition temperature of the thermosensitive gel of poly(N-isopropylacrylamide) gel [15]. Therefore, we study the critical behavior of friction in the vicinity of the volume phase transition point of poly(N-isopropylacrylamide) gel.

The sample gel is prepared in the similar manner with poly(acrylamide) gel except the temperature of the reaction system is controlled at 20.0 °C to avoid the emergence of frozen-in density fluctuations. The resultant gel is therefore free from the density fluctuations, and the gel looks completely transparent at least by the naked eye inspection. The gel undergoes a very small discontinuous transition at 33.6 °C if the gel is allowed to swell freely in water as shown in the swelling curve in Fig. 3.10. The swelling curve thus obtained corresponds to the isobar at zero osmotic pressure. In contrast the friction experiment is, in the case of our apparatus, carried out under the constant volume conditions of $V/V_0 = 1$. Such an experimental path is called as the isochore conditions and is also shown in the swelling curve of the gel in Fig. 3.10. The gel has a lower critical point because of hydrophobic interaction.

When the temperature is raised along the isochore, the gel presumably goes into the negative osmotic pressure region then into the coexistence regime. Temperature dependence of the swelling and shrinking behavior of the gel is checked on a sample prepared in the same mold. Without paper filter, the gel slightly swells in the open portion at lower temperatures. No visible shrinkage is observed up to the highest

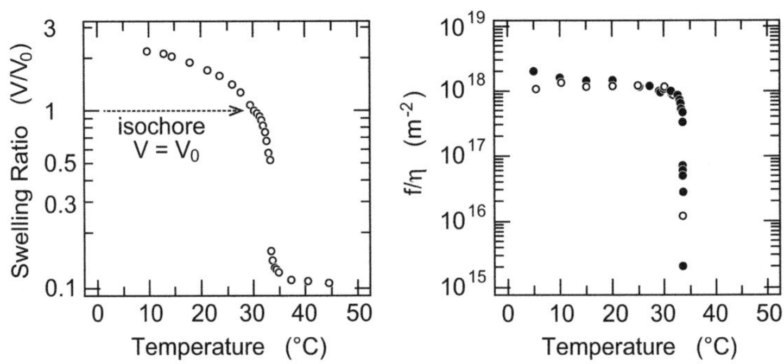

Fig. 3.10 The swelling curve of poly(N-isopropylacrylamide)gel (*left*) and the temperature dependence of the normalized friction coefficient of poly(N-isopropylacrylamide)gel, $f(T)/\eta(T)$ (*right*). *Solid symbols* in the right figure indicate that data are taken with increasing temperature and open symbols are the results of decreasing temperature. The reversible decrease of the friction occurs near the volume phase transition temperature

temperature of the friction experiment. The attachment to the gel bonding film remained intact in the entire temperature range of the friction experiment.

During the friction experiment, a test tube containing a gel without a mold is placed in the same water bath, and the appearance of the gel is continuously monitored. At the highest temperature of our experiment, the decrease in the free gel diameter from the isochore diameter is less than 10 %. These observations confirm that the gel swelling or shrinking as a whole should not have significant effect on the frictional measurement. A slight opacity developed in the gel near the temperature at which the friction diminishes, presumably due to the dynamic density fluctuations. As we shall see later, the fluctuations seem to be responsible for the diminishing of the gel-solvent friction.

The friction coefficient of gel is measured as a function of temperature, $f(T)$. Then, the results are normalized by the viscosity of water, and the temperature dependence of the normalized friction of poly(N-isopropylacrylamide) gel, $f(T)/\eta(T)$, thus obtained is plotted in Fig. 3.10 [17]. The viscosity of water is taken from a table. It is clear from Fig. 3.10 that the friction of poly(N-isopropylacrylamide) gel is independent of the temperature in the lower temperature regions, while it rapidly decreases three orders of magnitude as the temperature approaches 33.6 °C. Besides, the drastic decrease of the friction of poly(N-isopropylacrylamide) gel is entirely reversible.

The phenomenon of the reversible decrease of friction may be intuitively explained as follows. The gel-solvent friction is primarily determined by the pore size of the polymer network and the viscosity of the solvent as given in Eq. (3.8). The latter should not show any anomaly in the present temperature region of experiments. When the network is uniform at the lower temperature region, the pore size should be given by the average distance between the nearest polymer-polymer contacts because the values of friction of poly(N-isopropylacrylamide) gel, $f(T)/\eta(T)$, is almost the same with that of poly(acrylamide) gel shown in Fig. 3.7. Under certain conditions, however, the polymer network undergoes substantial density fluctuations in space and time. Some portion of the gel swells, while the other portions shrink maintaining the total gel volume constant. The effective pore size is then given by the distance over which the network density fluctuations are correlated. The water passes through the swollen open space avoiding the shrunken regions.

The pore size practically diverges as the gel passes the coexistence curve into the two-phase region, probably for one of the following reasons:

- The gel may remain in the metastable single phase as a superheated gel. The metastable state has been observed in the hysteresis of the swelling curves of various gels including poly(N-isopropylacrylamide) gel. The temperature gap at a hysteresis can be several degrees. In this case, the fluctuations are dynamic and should diverge on the spinodal line.
- The gel may undergo phase separation creating domains of swollen and shrunken phases. The density fluctuations are static and would diverge on or near the coexistence curve.

In both cases the effective pore size diverges, making the friction diminish. The pore structure is not permanent but is reversibly enlarged or reduced with temperature in the vicinity of the phase transition point. Hence, the friction of the gel changes reversibly.

By choosing an optimal combination of solvent and temperature, the phenomenon should be universally observed in any gel. A drastic and reversible change in the friction may have applications in separation technology and may be relevant to some biological transport phenomena.

3.3.3.2 Friction of Opaque Gel

We have already seen in Fig. 3.9 that the friction of poly(acrylamide) gel decreases upon increasing the mole fraction of the cross-linker in gel. The decrease in friction is not very large but is systematic in the concentration range studied. It is also noted that the gel becomes opaque with the concentration of cross-linker. The phenomenon observed here attracts our attention, namely, what happens to the gel if we increase the cross-linker concentration further and how the friction of such gels can be explained. A pioneering study on the opacification of poly(acrylamide) gel has been made more than 40 years ago, and the results are summarized as a *phase diagram of reaction system* [18]. It has been reported that poly(acrylamide) gel becomes completely opaque upon increasing the concentration of the cross-linker.

In Fig. 3.11, we show a photograph of poly(acrylamide) gels that are prepared at various concentrations of the cross-linker under constant total concentration of the gel. It is clear from the image that the opacity of the gel increases with the concentration of the cross-linker. Firstly, the gel becomes translucent when the concentration of cross-linker becomes about 3 mol%, then the gel gets completely opaque at the mole fraction of the cross-linker becomes 10 % at a total concentration of 700 mM [19]. The naked eye inspection of opaque gel indicates that the opacity of gel is independent of time suggesting the presence of frozen-in density fluctuations within the polymer network of gel. It further suggests the

Fig. 3.11 The photograph of opaque gels. The gels are polymerized in glass tubes and the photograph is taken. The mole fraction of the cross-linker is changed as 7, 21, 35, and 70 mM at a constant total concentration of 700 mM which corresponds to 1, 3, 5, and 10 mol%, respectively

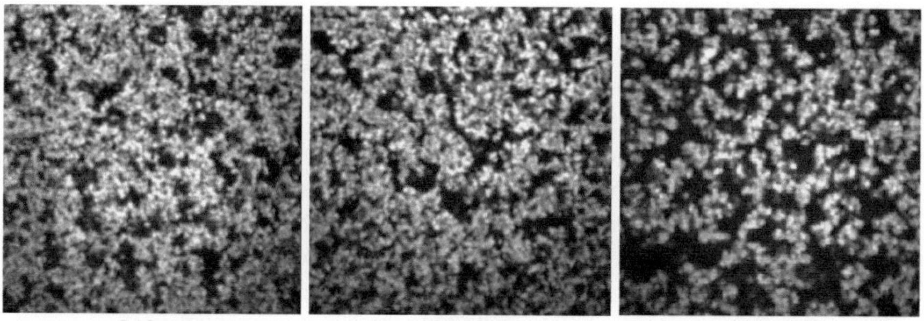

Fig. 3.12 The confocal laser scanning microscope images of opaque gels. The total concentration of the gel is fixed at 700 mM, while the mole fraction of the cross-linker is changed from 30 % (210 mM), 40 % (280 mM), and 50 % (350 mM) from *left* to *right*. Gels ate prepared at a temperature of 20.0 °C. The magnification of the objective lens is 100× with N.A. = 1.4, and the sides of the image corresponds to 28 μm

possibility of studying the structure of gel by recent advanced optical microscope technique, that is, the confocal laser scanning microscope. The real space structure of opaque poly(acrylamide) gel has been studied in detail by using a confocal laser scanning microscope [20]. In Fig. 3.12, we show the typical confocal laser scanning microscope images of opaque poly(acrylamide) gels. In this study, the polymer network of the gel is labeled by a fluorescent dye, and then the fluorescence image of the gel is gained. Therefore, the brighter regions of the image correspond to the denser region of polymer network, while the density of polymer network is lower, or practically zero, in darker regions of the image. In addition to this, many particles can be seen in the brighter region of the confocal laser scanning microscope images. The confocal laser scanning microscope images that are shown in Fig. 3.12, thus, strongly suggest that the opaque poly(acrylamide) gel consists of the bicontinuous structure of the aggregate of spherical particles and the void region. The structural parameters that characterize the structure in Fig. 3.12 are the diameter of the particle, the occupied area by the particle, and the spatial distribution of the particles. These parameters can be easily determined by the image analysis of the confocal laser scanning microscope images and the following points are clarified for the structure of opaque poly(acrylamide) gels:

- The opaque gels do not swell under water that is found by independent measurement of the swelling ratio of the gel. Their swelling ratio is almost unity, $V/V_0 \simeq 1$. In other words, the gels maintain the original volume and hence the original structure at the preparation.
- The opaque poly(acrylamide) gel consists of the fractal aggregates of the spherical particles. The fractal dimension of $D_F \sim 1.8$ is obtained for the confocal laser scanning microscope images. The results are consistent with the one obtained from the small angle neutron scattering results [21].

- There are also darker regions, where the density of the polymer network is lower, spreading in all portion of the image.
- The radius of the spherical particle is of the order of 1 μm or more which depends on the composition of the gel.
- The density of the particle is found to be of the order of $10^3\,\mathrm{kg/m^3}$ under the complete reaction conditions. The density thus obtained is close to that of acrylamide and/or N,N'-methylenebisacrylamide in the solid state.
- The particle is, therefore, regarded as a hard sphere in which acrylamide and N,N'-methylenebisacrylamide are closely polymerized.
- The opaque gel do not contain long flexible chains.
- The opaque gel is regarded as a colloid gel rather than a network gel.

It is well known that the opaque poly(acrylamide) gel losses high elasticity and it becomes considerably brittle. The changes in the elastic properties are reasonably understood intuitively from the structure of opaque gel since they are made of the aggregates of colloidal particles as shown in Fig. 3.12. These results further suggest that all chemicals in feed are polymerized into the colloidal particles and there are no long flexible chains in the opaque gels because the swelling ratio of the gel under water is a constant of about $V/V_0 = 1$. As far as our knowledge, the frictional property of opaque gel has yet to be studied. It is, however, easily expected from the confocal laser scanning microscope images in Fig. 3.12 that the friction of opaque gel should be small since the darker region, which spreads all portion of the image, serves as the path for the water flow. Actually, the friction of opaque gel is found to be very small by the preliminary friction experiment, and the simple method given in Fig. 3.4 can be used to determine the friction coefficient of opaque gels. Typical results are given in Fig. 3.13. It is clear from Fig. 3.13 that the normalized elution time course is well explained by Eq. (3.5). The friction coefficient of opaque gel is obtained from these results by using Eq. (3.6). It should be worth noting here that the normalized elution time course of opaque gels given in Fig. 3.13 does not show any deviation from the straight line that is expected from Eq. (3.5) in the time region measured here. The results indicate that the time course of water flow in opaque gels is governed only by a characteristic time. Besides, the friction coefficient of opaque gel can be determined about 1 h though much thicker sample gels are used in these experiments.[5]

In Fig. 3.14, the friction coefficient of poly(acrylamide) gel is plotted as a function of the cross-linker concentration [22]. The results shown in Fig. 3.9 are also plotted in this figure. It is clear from this figure that the friction of gel is still enormous even the cross-linker concentration is increased up to 10 mol% where the gel looks almost completely opaque as shown in Fig. 3.11. Although the opacity grows at this cross-linker concentration, the polymer network is still a mesh of molecular size in terms of the frictional property since the friction coefficient of the gels are still in the order of $f \sim 10^{14}\,\mathrm{N\,s/m^4}$ which is typical value of the

[5] In these measurements the thickness of the sample gels are about 5–7 mm or more.

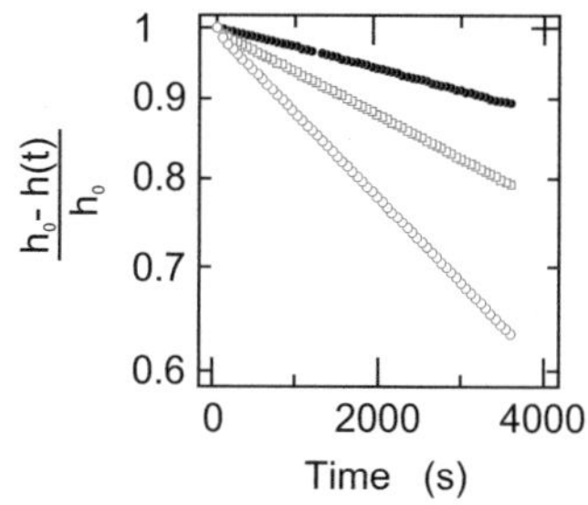

Fig. 3.13 The logarithm of the normalized elution time course, Eq. (3.5), is plotted as a function of time. The mole fraction of the cross-linker in the total concentration of 700 mM gel are 0.3 (210 mM), 0.4 (280 mM), and 0.5 (350 mM) from top to bottom

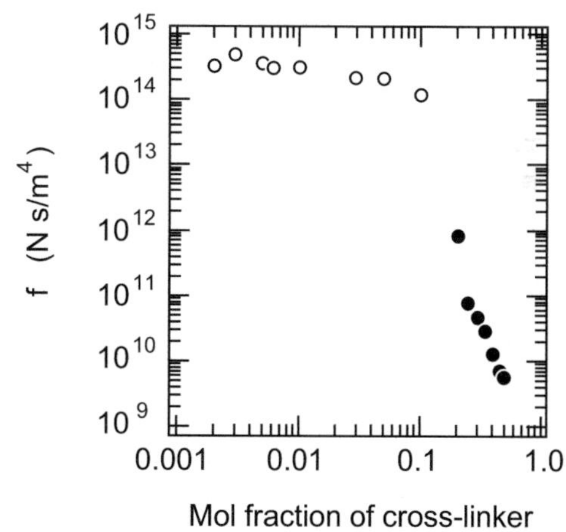

Fig. 3.14 The cross-linker concentration dependence of friction coefficient. *Open symbols* are the results already shown in Fig. 3.9. *Closed symbols* are the results for the gels that are prepared at higher concentration region of cross-linker. The structure of these gels are observed by the confocal laser scanning microscope after friction measurements

friction coefficient for uniform (transparent) gels. The friction coefficient of gel, however, decreases almost four orders of magnitude when the concentration of cross-linker is increased more than 10 mol%. The drastic decrease of the friction is due to the formation of the bicontinuous structure of the aggregated colloidal particles and the void region as shown in the confocal laser scanning microscope images given in Fig. 3.12. In other words, the structure of the polymer network of gel transforms from the molecular network to the aggregated structure of colloidal particles upon increasing the concentration of the cross-linker. The change of the structure of polymer network is, thus, regarded as a structural transition from the network gel to the colloid gel.

It is clear from the real space structure analysis of the opaque gel that the substances in feed are polymerized into the brighter regions to form the spherical particles while darker region, where the density of chemicals is practically zero, is also spreading in the whole region of the space. This is the reason why the opaque poly(acrylamide) gel shows such a lower friction coefficient. The water can pass through the darker region easily which serves as an open space for the water flow. Since the long polymer chains do not exist in the opaque gel, the colloidal particle, which is regarded as a hard sphere, simply behaves as the obstacle for the water flow. In the case of opaque gels, therefore, water flows around the aggregates of *N* spherical particle of radius R. We thus expect that the friction of opaque gel can be written by following Stokes relationship.

$$f = N6\pi\eta R \tag{3.14}$$

Since the viscosity of water η is known, the friction coefficient of gel f can be calculated if the number of colloidal particle in a unit volume of the gel N and the radius of the colloidal particle R are determined from the confocal laser scanning microscope images. The radius of the particle can be determined directly from the confocal laser scanning microscope images. On the other hand, the number of the particle is proportional to the area occupied by the colloidal particles, σ, in the confocal laser scanning microscope image. The number of particles in a unit volume of the gel is then expressed as $\sigma/\pi R^2 V_{\mathrm{obs}}$ where the observation volume of the confocal laser scanning microscope image is written as $V_{\mathrm{obs}} = \sigma_0\delta$. Here, δ and σ_0 are the thickness of the focal plane and the total area of the image, and hence these parameters are known constants that are determined only by the experimental setup.[6] The friction coefficient of opaque gel is finally expressed as follows:

$$f = \frac{6\eta}{\delta\sigma_0}\frac{\sigma}{R} \tag{3.15}$$

The structural parameters of gel, R and σ/σ_0, are plotted in Fig. 3.15 as a function of the mole fraction of cross-linker. The friction coefficient of gel is then calculated by using Eq. (3.15) and is plotted in Fig. 3.16 in double logarithmic manner. The calculated values of the friction coefficient are almost one half of the experimentally measured values. Although the thickness of the focal plane of the confocal laser scanning microscope image is thin, it has a finite thickness. In addition to this, the diameter of the particle is the same order of magnitude with the thickness of the focal plane. The smaller particles thus can easily be hidden behind the larger particles. Thus, the number of the particle that is determined from the two-dimensional confocal laser scanning microscope image is underestimated by the overlapping of the particles. If the three-dimensional image of the bicontinuous structure of opaque gel is reconstructed from the slices of confocal laser scanning microscope images, much reliable values of the friction coefficient are obtained. We, therefore, can conclude that the frictional property of the opaque gel is well explained in terms of the structure of the opaque gel that is revealed by the confocal laser scanning microscope imaging of the gel.

Some aspects of the polymer network dynamics that is related to the water flow through the gel are worth noting here. It has been shown in the early studies of friction of the gel that the velocity of water flow in the gel is fast in the beginning

[6] In the case of present confocal laser scanning microscope, $\delta \simeq 1\,\mu\mathrm{m}$ when the objective lens with the magnification of 100× and N.A. = 1.4 is employed. The area σ_0 depends on the setting of the image analysis software. The one side of the image is usually fixed from 20 to 30 μm.

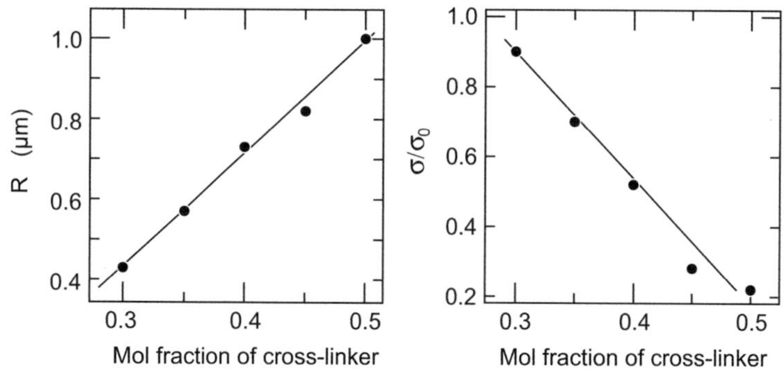

Fig. 3.15 The cross-linker concentration dependence of the structure factor of opaque gels, the radius of the particle R (*left*), and the relative occupied area by the aggregates of particles σ/σ_0 (*right*). *Lines* are drawn only guide to the eye

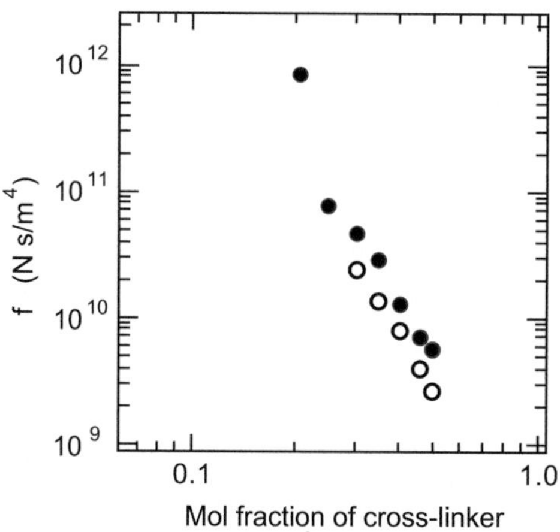

Fig. 3.16 The calculated values of friction coefficient of opaque gel according to Eq. (3.15) with the data shown in Fig. 3.14. Open symbols represent the calculated values of friction coefficient of the gel, and closed symbols are the measured values of friction coefficient of opaque gels that is shown in Fig. 3.14

and then decreases with time to a steady-state value. In the case of the uniform gels, it requires few hours to achieve a steady state even in the case of about 1 mm thick gel. The polymer network of the gel easily deforms from the equilibrium state due to the drag force of the flowing water when the polymer network of gel consists of long flexible chains. Therefore, the restoring force due to the rubber elasticity arises and it competes with the drag force. The balance between these forces determines the characteristic time interval that is required to attain the steady-state flow of water as given by the collective diffusion coefficient of the gel as shown in Eq. (3.2). This characteristic time is known to be proportional to the inverse of the square of the typical size of the gel. In addition to this, the flow time course is expressed by the superposition of the exponential functions in the case of the transparent gels. These are the reason for such a long equilibrium time for the stationary state in the uniform gel [4, 16]. In contrast, the flow time course of the opaque gel is expressed by a single exponential function as described by Eq. (3.5) [19, 22]. Besides, the velocity of the flowing water is almost constant throughout the measurement even in the thicker gels of about 5 mm or more. Therefore, the polymer network dynamics of the opaque gel is totally different from that of the transparent gel. The differences in the network dynamics may also be attributed to the structure of opaque gels which strongly suggest that the opaque gels do not contain long flexible chains that can deform under the flow of water and exhibit the rubber elasticity. These results are also confirmed by the fact that the friction of opaque gel is simply expressed by summation of the hydrodynamic friction of hard sphere that is described by Stokes' law.

3.4 Concluding Remarks

The studies on the frictional properties of gels are reviewed. The frictional property of gel is one of the important properties of gel that determines the dynamics of the swelling and shrinking of gel. The frictional property of gel also plays the essential roles in the pattern formation of gel, the extensive compression of the gel, and so forth [23–27]. Thus, the information on the frictional property of gel is of useful in many applications such as medicine, food, chemical engineering, and biotechnology.

The knowledge gained so far is, however, still limited. For instance, we do not have any information on the numerical constant of Eq. (3.8). In the case of well-defined flow, the equation can be written definitively including the numerical constant. The Hagen-Poiseuille law of capillary flow is such a case. In many literatures of the hydrodynamics, the Hagen-Poiseuille law is expressed as follows:

$$P = 8\eta L \frac{v}{r^2} \tag{3.16}$$

Here, P, L, r, η, and v are the pressure, the length of the capillary, the radius of the capillary, the viscosity of the fluid, and the velocity of the flowing fluid in the capillary. Equation (3.16) can be rewritten as follows:

$$\frac{P}{L} = 8\frac{\eta}{r^2}v = fv \tag{3.17}$$

The friction f of a capillary of radius r is, therefore, $8\eta/r^2$. Since the typical frictional pore size for a capillary is the radius, result agrees with Eq. (3.8), namely, $f \propto \eta/\xi^2$. Therefore, the proportional constant of Eq. (3.8) for the friction of a capillary with well-defined radius of r is a simple numerical value of 8 [28].

Therefore, Eq. (3.8) should generally be written as follows:

$$f = Q\frac{\eta}{\xi^2} \tag{3.18}$$

The constant Q reflects the structure of the pore, and hence it may be regarded as the structure factor of porous material. If we can determine the structure factor of the porous material experimentally and/or theoretically, it will be possible to design a material with the desired frictional properties. Such information will be of useful in designing the purification and separation systems. Further detailed study is, therefore, clearly required not only for the applications but also for the better understanding of the physics of gel since gel is a universal state of matter.

References

1. T. Tanaka, L. O. Hocker, and G. B. Benedek, J. Chem. Phys., **59**, 5151 (1973).
2. T. Tanaka, Phys. Rev. Lett., **40**, 820 (1978).
3. T. Tanaka, D. J. Fillmore, S. -T. Sun, I. Nishio, G. Swislow, and A. Shah, Phys. Rev. Lett., **45**, 1636 (1980).
4. T. Tanaka and D. J. Fillmore, J. Chem. Phys., **70**, 1214 (1979).
5. T. Tanaka, S. Ishiwata, and C. Ishimoto, Phys. Rev. Lett., **38**, 771 (1977).
6. J. P. Munch, S. Candau, J. Herz, and G. Hild, J. Phys. (Paris) **38**, 971 (1977).
7. J. P. Munch, P. Lemarechal, and S. Candau, J. Phys. (Paris) **38**, 1499 (1977).
8. T. Takebe, K. Nawa, S. Suehiro and T. Hashimoto, J. Chem. Phys., **59**, 4360 (1989).
9. P. G. de Gennes, *Scaling Concepts in Polymer Physics*, (Cornell University Press, Ithaca, 1979).
10. N. Weiss and A. Silberberg, Polym. Prepr. Am. Chem. Soc., Div. Polym. Chem., **16**, 289 (1979).
11. N. Weiss, T. van Vilet, and A. Silberberg, J. Polym. Sci. Polym. Phys. Eds., **17**, 2229 (1979).
12. A. M. Hecht and E. Geissler, J. Chem. Phys., **73**, 4077 (1980).
13. E. Geissler and A. M. Hecht, J. Chem. Phys., **77**, 1548 (1982).
14. M. Tokita and T. Tanaka, J. Chem. Phys., **95**, 4613 (1991).
15. Y. Hirokawa and T. Tanaka, J. Chem. Phys., **81**, 6379 (1984).
16. Y. Y. Suzuki, M. Tokita, and S. Mukai, Euro. Phys. J. E, **29**, 415 (2009).
17. M. Tokita and T. Tanaka, Science, **253**, 1121 (1991).
18. E. G. Richards and C. J. Temple, Nature (Phys. Sci.), **230**, 92 (1971).
19. M. Tokita, Polymers, **6**, 651 (2014).
20. Y. Doi and M. Tokita, Langmuir, **21**, 5285 (2005).
21. S. Mukai, H. Miki, V. Garamus, R. Willmeit, and M. Tokiya, Progre. Colloid Polymer Sci., **136**, 95 (2009).
22. Y. Doi and M. Tokita, Langmuir, **21**, 9420 (2005).
23. E. S. Matsuo and T. Tanaka, Nature, **358**, 482 (1992).
24. J. Maskawa, T. Takeuchi, K. Maki, K. Tsujii, and T. Tanaka, J. Chem. Phys., **110**, 10993 (1999).
25. M. Tokita, S. Suzuki, K. Miyamoto, and T. Komai, J. Phys. Soc. Jpn., **68**, 330 (1999).
26. M. Tokita, K. Miyamoto, and T. Komai, J. Chem. Phys., **113**, 1647 (2000).
27. K. Nakamura, E. Shinoda, and M. Tokita, Food Hydrocolloids, **15**, 247 (2001).
28. M. Tokita, Adv. Polym. Sci., **110**, 27 (1993).

Part II
Applications: Foods

Chapter 4
Swallowing and the Rheological Properties of Soft Drink and Agar Gel

Hatsue Moritaka

Abstract Mastication and swallowing are essential steps in food ingestion and nutrient absorption in humans. Mastication is the process in which food taken from the mouth is crushed and mixed with saliva in the oral cavity to create food boluses that are optimal for swallowing. The use of facial and jaw muscles, in addition to teeth, is necessary for mastication. Foods with gel-like texture are crushed by mastication, and the resulting boluses are transferred from the oral cavity to the pharynx with assistance from tongue pressure and then to the esophagus by suction generated upon elevation of the larynx. Bolus formation by mastication is closely associated with the swallowing process. Meanwhile, liquid foods are transferred from the oral cavity to the pharynx without mastication. However, the swallowing method of a drink changes by the additions included in the drink. It is important to know the relationship between physical property of drink and the linguapalatal swallowing pressure, the relationship between the preprandial food size and the nature of food bolus, and then the relationship between the texture property of the food bolus and the swallowing.

Keywords Soft drink • Ager gel • Rheological property • Mastication • Linguapalatal swallowing pressure

Mastication and swallowing are essential steps in food ingestion and nutrient absorption in humans. Mastication is the process in which food taken from the mouth is crushed and mixed with saliva in the oral cavity to create food boluses that are optimal for swallowing. The use of facial and jaw muscles, in addition to teeth, is necessary for mastication. Foods with gel-like texture are crushed by mastication, and the resulting boluses are transferred from the oral cavity to the pharynx with assistance from tongue pressure and then to the esophagus by suction generated upon elevation of the larynx. Bolus formation by mastication is closely associated

H. Moritaka (✉)
Division of Food Science and Nutrition, Showa Women's University, Graduate School of Human Life Science, Tokyo, Japan
e-mail: hatsue.2110.moritaka@gmail.com

I. Kaneda (ed.), *Rheology of Biological Soft Matter*, Soft and Biological Matter,
DOI 10.1007/978-4-431-56080-7_4

with the swallowing process. Meanwhile, liquid foods are transferred from the oral cavity to the pharynx without mastication. However, the swallowing method of a drink changes by the additions included in the drink. It is important to know the relationship between physical property of drink and the linguapalatal swallowing pressure, the relationship between the preprandial food size and the nature of food bolus, and then the relationship between the texture property of the food bolus and the swallowing.

4.1 Soft Drink

4.1.1 Vegetable Juices with Carrot Puree

It is recognized that dietary fiber is an important food component because it has beneficial effects; reducing blood cholesterol, reducing the risk of diabetes, and improving laxation [1, 2], but the majority of Japanese and North Americans do not reach the recommended daily intake of dietary fiber [3]. A convenient way to add functional components, such as dietary fiber, into the daily diet is to add them to processed foods [4], and in fact, fiber-rich juice and other food products are on the market. On the other hand, the soluble and insoluble dietary fibers influence the rheological and textual properties of processed food. The rheological properties of alimentary bolus greatly influence the safety of its transport from the oral cavity to the stomach, and this is a key concern in the management of dysphagia [5–9]. Here, the effects of substitution from carrot juice by carrot puree on rheological properties and swallowing characteristic of mixed carrot and apple juice are mentioned.

4.1.1.1 Rheological Properties of Vegetable Juices with Carrot Puree [10]

Vegetable juice is prepared by mixing equal amounts of apple juice and carrot juice. The median diameter of carrot puree (CP) is 250 μm.

The apparent viscosity of the vegetable juices at high CP concentrations is greater than those at low CP concentrations, and the apparent viscosity decreases with increasing shear rates (Fig. 4.1). The reduction in the apparent viscosity is greater for vegetable juices at high CP concentrations than those at low concentrations. In the vegetable juices at high CP concentrations, the pieces of crushed carrot tissue flow as an aggregate, resulting in a reduction in the resistance of the juice. This may be to reduce shear stress in vegetable juices at high CP concentrations at high shear rates. The apparent viscosity of the vegetable juice even at 0 % CP decreases to a shear rate of about 3 s^{-1}. This might be due to soluble pectin and hemicellulose in vegetable juice at 0 % CP.

Figure 4.2 shows the relative viscosity of vegetable juice at a shear rate of 10 s^{-1} as representative. The relative viscosities of vegetable juices at 10.0–30.0 %

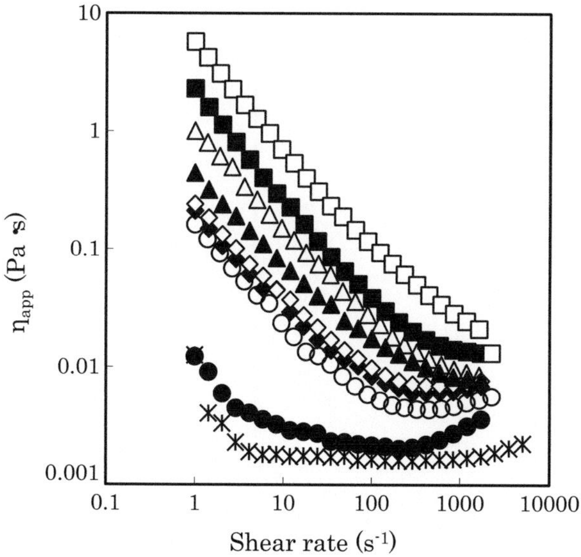

Fig. 4.1 Apparent viscosity of vegetable juice with carrot puree [10]. Concentration of carrot puree: * 0 %, ● 5.0 %, ○ 10.0 %, ♦12.5 %, ◇ 15.0 %, ▲17.5 %, △20.0 %, ■ 25.0 %, □ 30.0 %

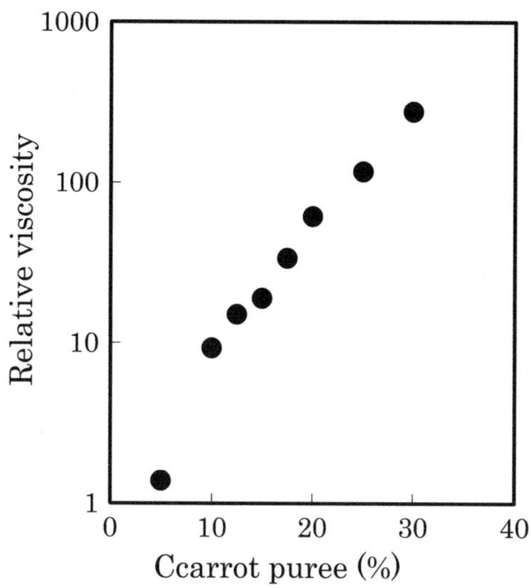

Fig. 4.2 Relative viscosity of vegetable juice with carrot puree [10]

CP show linear relationships with the CP concentration. On the other hand, the relationship between the relative viscosity and the CP concentration of vegetable juice at 5.0 % CP deviates from the linear relationships demonstrated by the vegetable juices at higher CP concentrations. The vegetable juice at 5.0 % CP may

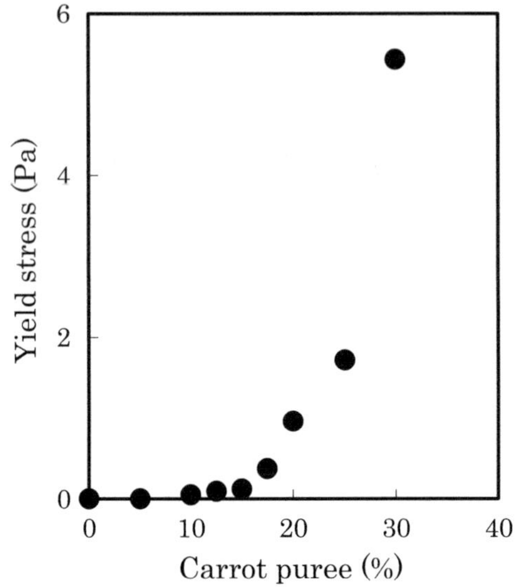

Fig. 4.3 Yield stress of vegetable juice with carrot puree [10]

be considered a dilute solution in which the flecks of carrot tissue are dispersed so as to be independent of each other and can move freely. On the other hand, in the vegetable juice at a concentration >10.0 % CP, the flecks of carrot tissue begin to aggregate. This appears to be why the long-range force increases as the CP concentration increases, turning the vegetable juice into a semi-thick solution. Furthermore, the soluble pectin, cellulose, and hemicellulose in the CP are added, making the pectin, cellulose, and hemicellulose concentrations of the juice higher.

Yield stress is not detected in the vegetable juice at 0 or 5.0 % CP, but it increases gradually in the 10.0–15.0 % CP juice (Fig. 4.3). The increase in yield stress is concentration-dependent in the vegetable juice at >17.5 % CP, and it shows a significant increase in the 30.0 % CP juice.

It has been reported that the feeling of initial contact in the mouth is correlated with the physical property in small deformation. The storage modulus (G′) is larger in the vegetable juices at high CP concentrations (Fig. 4.4). The relationship between G′ and the loss modulus (G″) in all the juices is G′ > G″ at all frequencies. This is thought to be because the time needed for the pieces of carrot tissue to rearrange takes longer than the time scales of the deformation velocity of the pieces, resulting in elastic transformation of the carrot pieces.

4.1.1.2 Sensory Evaluation [10]

For the item swallowing smoothness, the vegetable juices at 0–12.5 % CP and those at >17.5 % CP are placed in different same-quality subgroups, and the evaluation of vegetable juices at >17.5 % CP is low (Fig. 4.5a). The subjects evaluate the sizes of

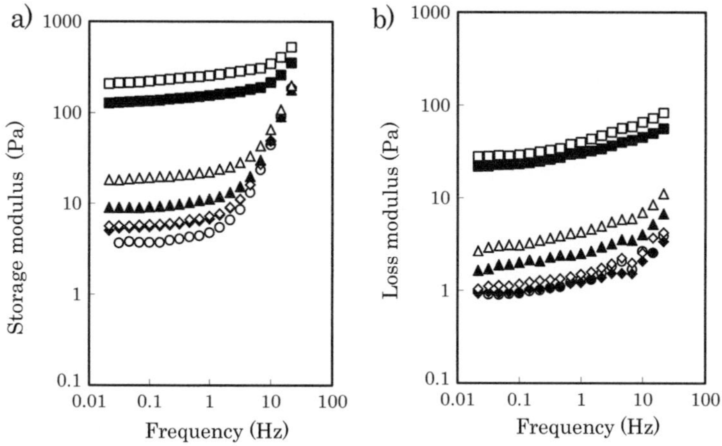

Fig. 4.4 Storage modulus and loss modulus of vegetable juice with carrot puree [10]. (**a**) Storage modulus, (**b**) loss modulus. Concentration of carrot puree: ●5.0 %, ○10.0 %, ♦12.5 %, ◇ 15.0 %, ▲17.5 %, △20.0 %, ■25.0 %, □30.0 %

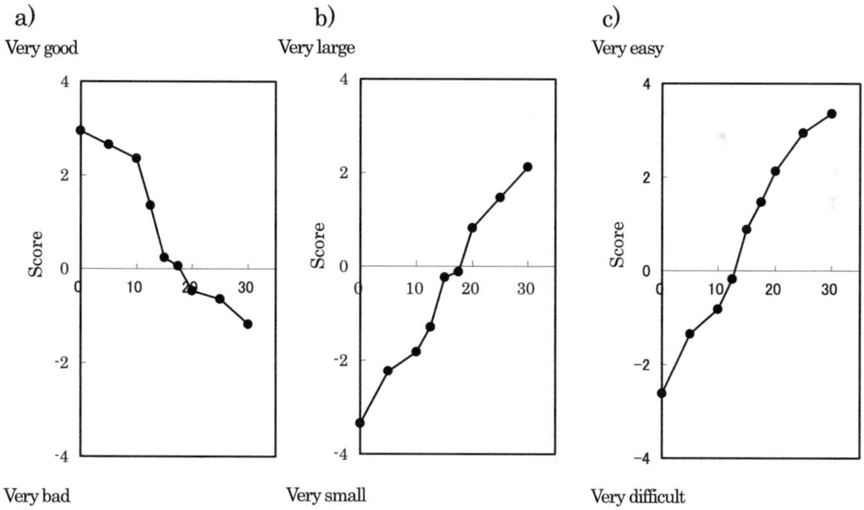

Fig. 4.5 Sensory evaluation of vegetable juice with carrot puree [10]. (**a**) Swallowing smoothness, (**b**) size of carrot pieces, (**c**) easy of collecting

the pieces of carrot tissue in the vegetable juice at 15.0–30.0 % CP to be larger than that in the juices at 0–12.5 % CP (Fig. 4.5b).

Although the size of these pieces is the same regardless of CP concentration differences, it is perceived to become larger with increasing concentrations of CP. This is thought to be because flecks of carrot tissue aggregate more as the CP concentration increases, causing subjects to perceive the size to be larger. The easy collecting of the vegetable juices at 0–10.0 % CP is evaluated to be more difficult than that at 17.5–30.0 % CP. The vegetable juices at 17.5–30.0 % CP are perceived to agglomerate collectively (Fig. 4.5c).

4.1.1.3 Transport Characteristics of Juice in the Pharyngeal Region [10]

The maximum velocity of the vegetable juice with CP in the pharyngeal region decreases with increasing CP concentration. The maximum velocity of the vegetable juice classify into three same-quality subgroups: vegetable juice at 0–12.5 % CP, vegetable juice at 10.0–25.0 % CP, and vegetable juice at 17.5–30.0 % CP (Fig. 4.6).

The maximum velocity of vegetable juice passing through the pharyngeal region is the fastest moving part of juice bolus. Average maximum velocity is a mean value obtained at the tip of the flow velocity spectrum, and average velocity is a mean value calculated from the entire flow velocity spectrum of the juice bolus.

Figure 4.7 shows the apparent shear rate and apparent shear stress calculated from the diameter of the pharyngeal region, the maximum velocity, the average maximum velocity, and the average velocity. The average apparent shear rate calculated from the maximum velocity is 63 s^{-1}, that calculated from the average maximum velocity is 36 s^{-1}, and then that calculated from the average maximum velocity is 12 s^{-1}. Figure 4.8 shows the relations between the maximum velocity and apparent viscosity at the apparent shear rates of 65 s^{-1}, 35 s^{-1}, and 10 s^{-1} in a log-log graph. The coefficients of determination calculated from these relations are the high values and are comparable. The results obtained in this research do not differ greatly from preceding reports [11–13].

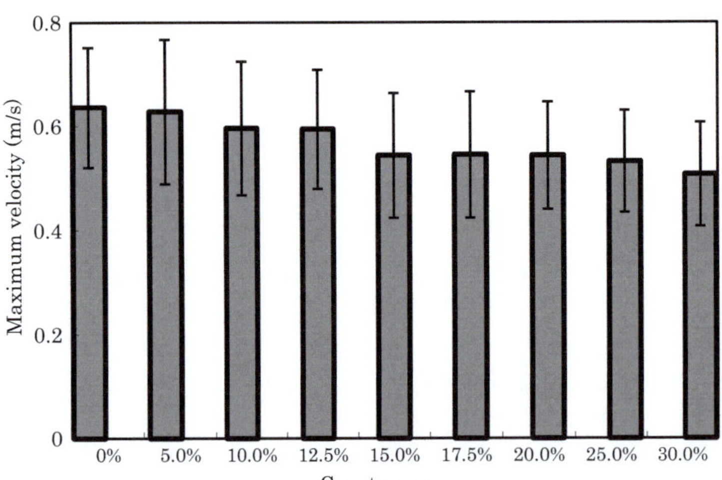

Fig. 4.6 Maximum velocity of vegetable juice with carrot puree in the pharyngeal region [10]

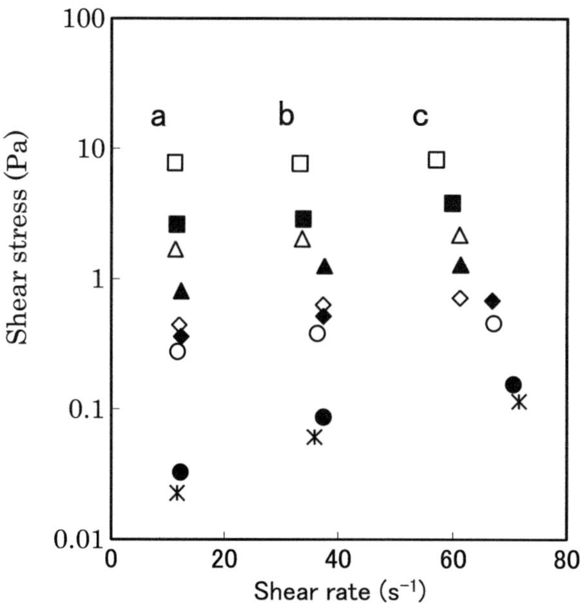

Fig. 4.7 Shear stress and shear rate of vegetable juice with carrot puree in pharyngeal region [10]. (**a**) Shear stress and shear rate calculated from average velocity, (**b**) shear stress and shear rate calculated from average maximum velocity, (**c**) shear stress and shear rate calculated from the maximum velocity. Concentration of carrot puree: * 0 %, ●5.0 %, ○ 10.0 %, ♦12.5 %, ◊ 15.0 %, ▲17.5 %, Δ20.0 %, ■ 25.0 %, □ 30.0 %

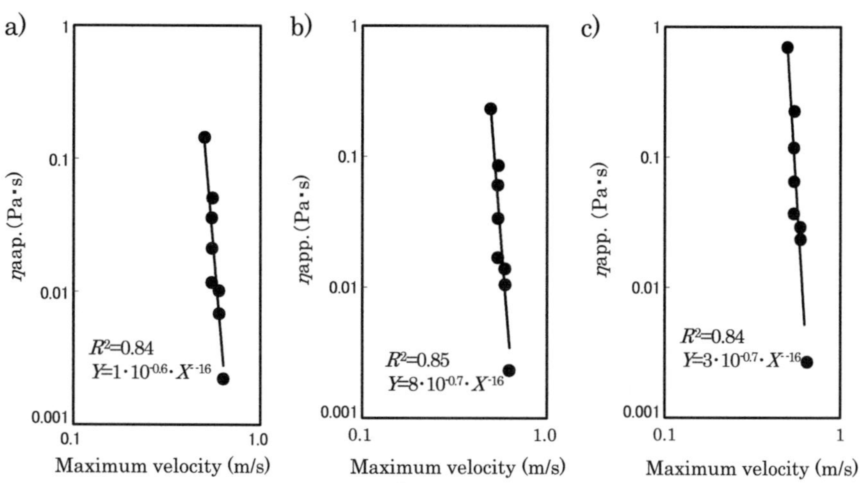

Fig. 4.8 Relationship between maximum velocity in the pharyngeal region and apparent shear rate of vegetable juice with carrot puree [10]. (**a**) Apparent shear rate of 65 s^{-1}, (**b**) apparent shear rate of 35 s^{-1}, (**c**) apparent shear rate of 10 s^{-1}

4.1.2 Carbonated Drink

The tongue plays an important role in the swallowing process. The tongue responds to chemical or physical stimulation, particularly the chemical stimulation caused by taste substances and pungent condiments [14]. These receptor cells are distributed throughout the oral epithelium and are innervated by the trigeminal and glossopharyngeal nerves.

Various evaluation methods for linguapalatal swallowing pressure are available, but quantitative evaluation has proven difficult and is associated with many problems.

Author reports the influence of carbon dioxide content dissolved in carbonated drinks on linguapalatal swallowing pressure, maximum velocity of a bolus through the pharynx, and intraoral carbonation perception.

4.1.2.1 Number and Area of Air Bubbles [15]

Three different carbonated drinks with gas volumes (GV) of 0, 1.5, and 2.7 (hereinafter referred to as the GV0, GV1.5, and GV2.7 drinks, respectively) are used as sample.

The number of bubbles per cm^2 for the GV2.7 drink rapidly decreases after being poured into the laboratory dish and is stable (Fig. 4.9a). The bubbles in the GV1.5 and GV2.7 drinks show repeated disappearance and formation. The number of bubbles in the GV1.5 drink is low immediately after being poured into the laboratory dish and decreases after reaching a maximum. The average area per

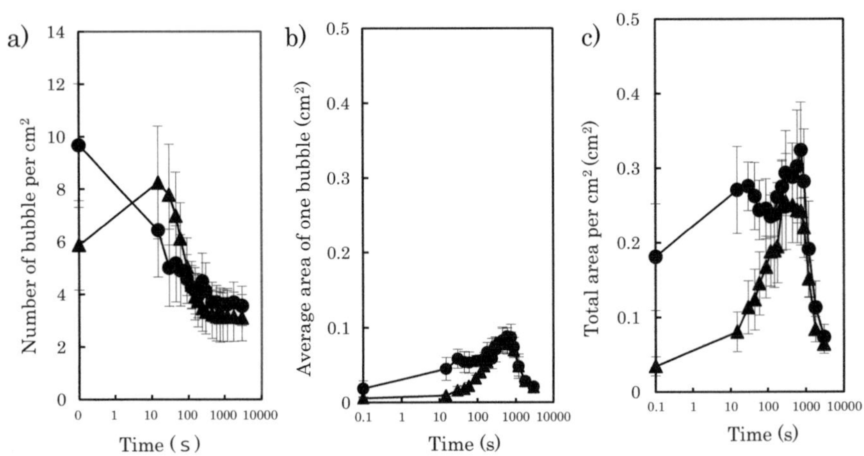

Fig. 4.9 Number and area of air bubbles of carbonated drink with different GV [15]. (**a**) Number of bubble per cm^2, (**b**) average area of one bubble, (**c**) total area of bubble per cm^2. ▲ GV1.5 carbonated drink, ● GV2.7 carbonated drink. The *line* in a figure expresses the standard error

bubble just after being poured is larger for the GV2.7 drink than for the GV1.5 drink (Fig. 4.9b). The average area per bubble for the GV2.7 drink decreases and then becomes comparable to that of the GV1.5 drink. Although the average area per bubble in the GV1.5 drink is small immediately after being poured, it increases and then becomes comparable to that of the GV2.7 drink. The total area of carbon dioxide bubbles per cm^2 tends to be larger for the GV2.7 drink than for the GV1.5 drink (Fig. 4.9c).

4.1.2.2 Sensory Evaluation [15]

Except for the sensory evaluation item palatability, significant differences for other items are recognized between GV0, GV1.5, and GV2.7 drinks (Fig. 4.10). For GV1.5 and GV2.7 drinks, the sensory evaluation items' stinging sensation during swallowing, strength of swallowing, stinging sensation after swallowing, and ease of swallowing are estimated to be higher than those of GV0 drink.

4.1.2.3 Linguapalatal Swallowing Pressure [15]

The tactile sensor system, swallow scan with a special sensor sheet for measuring linguapalatal swallowing pressure [16], is used (Figs. 4.11 and 4.12). This sensor sheet, composed of 0.05-mm resin, has five measuring points.

For all three drinks, onset time in the anterior median area (Chap. 1) appears earlier than that in the posterior median area (Chap. 3) and posterior circumferential areas (Chaps. 4 and 5) (Fig. 4.13a). These results indicate the contact pattern between the tongue and hard palate when swallowing carbonated drinks is similar to that when swallowing water. There are no significant differences in onset time or peak time among the three drinks in any channel (Fig. 4.13a, b).

For each of the three drinks, duration of linguapalatal swallowing pressure at the posterior median area (Chap. 3) is shorter than that of all other hard palate areas (Chaps. 1, 2, 4, and 5) (Fig. 4.13c). The duration at the anterior median and the mid-median areas (Chaps. 1 and 2) of GV0 and GV2.7 drinks are longer than those at the posterior circumferential areas (Chaps. 4 and 5). At all channels (Chaps. 1, 2, 3, 4, and 5), linguapalatal swallowing pressure is significantly shorter for the GV0 drink than for the GV1.5 and GV2.7 drinks except at the mid-median area (Chap. 2).

At all hard palate areas (Chaps. 1, 2, 3, 4 and 5), peak magnitude is significantly smaller with the GV0 drink than with the GV2.7 drink (Fig. 4.13d). At the posterior circumferential areas (Chaps. 4 and 5), peak magnitude of the GV1.5 drink is larger than that of the GV0 drink.

In each drink, the integrated value of peak linguapalatal swallowing pressure is larger at the anterior median area (Chap. 1) and mid-median area (Chap. 2) than at the other channels (Chaps. 3, 4, and 5) (Fig. 4.13e). At all channels (Chaps. 1, 2, 3, 4, and 5), the integrated value of the GV0 drink is smaller than that of the GV1.5 and GV2.7 drinks. At the mid-median area (Chap. 2) and posterior circumferential

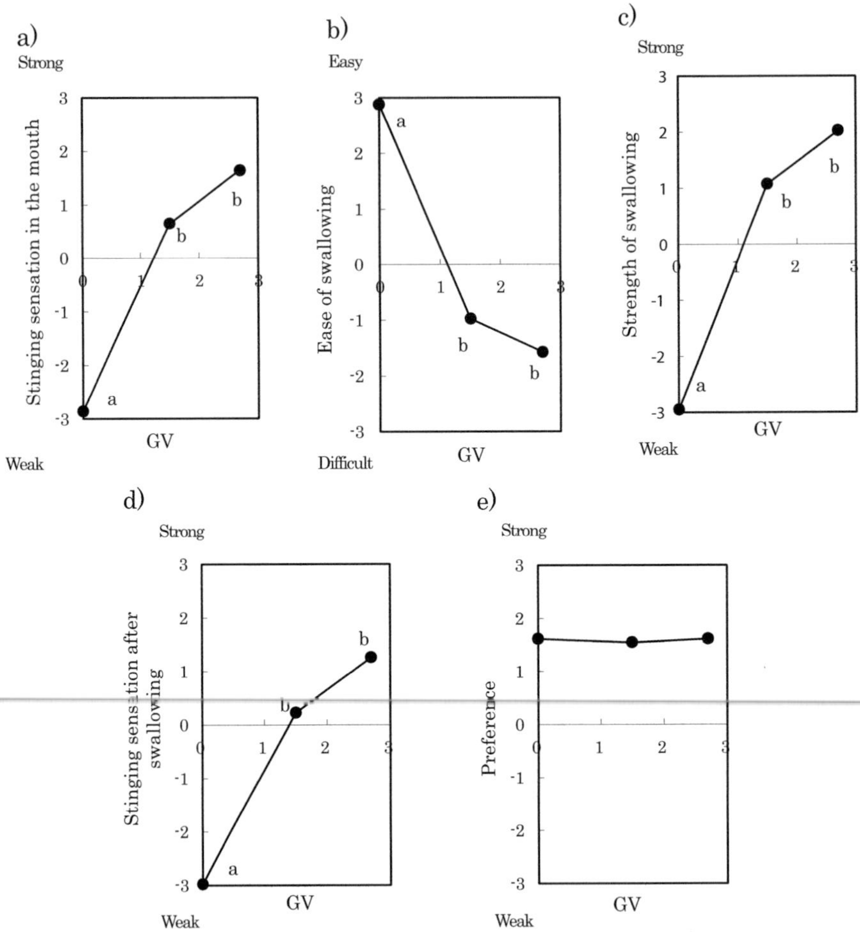

Fig. 4.10 Distance scale of the sensory evaluation of carbonated drink with different GV [15]. (**a**) Stinging sensation in the mouth, (**b**) ease of swallowing, (**c**) strength of swallowing, (**d**) stinging sensation after swallowing, (**e**) preference

area (Chap. 4), the integrated value is smallest for the GV0 drink, followed by the GV1.5 and then the GV2.7 drinks.

The total integrated value of the GV0 drink is smaller than that of the GV1.5 and GV2.7 drinks (Fig. 4.13f). However, there is no significant difference in total integrated value between the GV1.5 and GV2.7 drinks.

Above results suggest for all areas of the hard palate that duration, maximal magnitude, and integrated value of linguapalatal swallowing pressure (Fig. 4.13c–f) are affected by the carbon dioxide content of the drink.

Carbonation irritates oral mucosal nociceptors when the carbon dioxide dissolved in the drink reacts with the salivary enzyme carbonic anhydrase 4 to form H_2CO_3

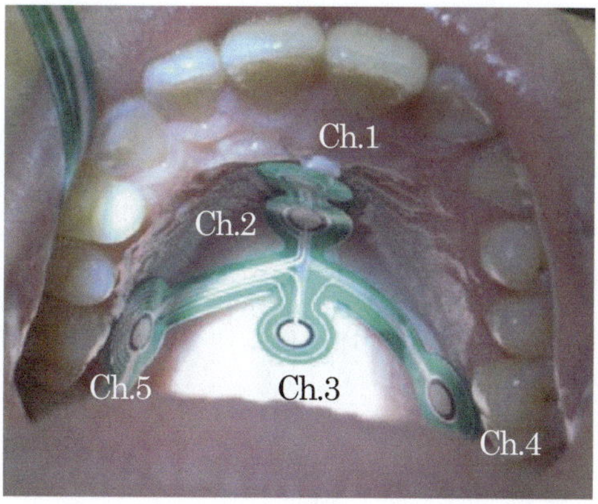

Fig. 4.11 The sensor sheet for measuring linguapalatal swallowing pressure at five measuring points (Chaps. 1, 2, 3, 4 and 5) attached to the palatal mucosa with a sheet-type denture adhesive [15]

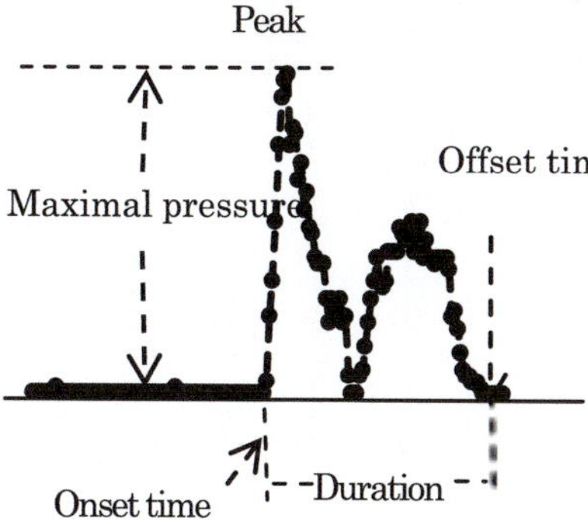

Fig. 4.12 Items for evaluating the state of linguapalatal swallowing pressure production at Chap. 1 [15]

[17]. When the H_2CO_3 separates into bicarbonate ions and free protons in the oral cavity, protons stimulate sour-sensitive taste receptor cells innervated by the facial nerve, yielding a perception of sourness:

$$CO_2 + H_2O \leftrightarrow H_2CO_3 \leftrightarrow HCO^{3-} + H^+$$

Above results add to the evidence that orally chemesthetic beverages stimulate more neuromotor activity compared to water during the oral stage of swallowing.

4.1.2.4 Transit Velocity Through the Pharynx [15]

The maximum velocity of the drinks moving through the pharynx is measured by the pulse Doppler method using ultrasonic diagnostic imaging equipment.

The maximum transit velocity through the pharynx of the GV2.7 drink is faster than that of the GV0 drink (Fig. 4.14). However, there is no significant difference in maximum transit velocity between the GV1.5 and GV2.7 drinks.

The flow of carbonated drinks is hampered by the presence of carbon dioxide bubbles. These bubbles underwent repeated disappearance and generation over time

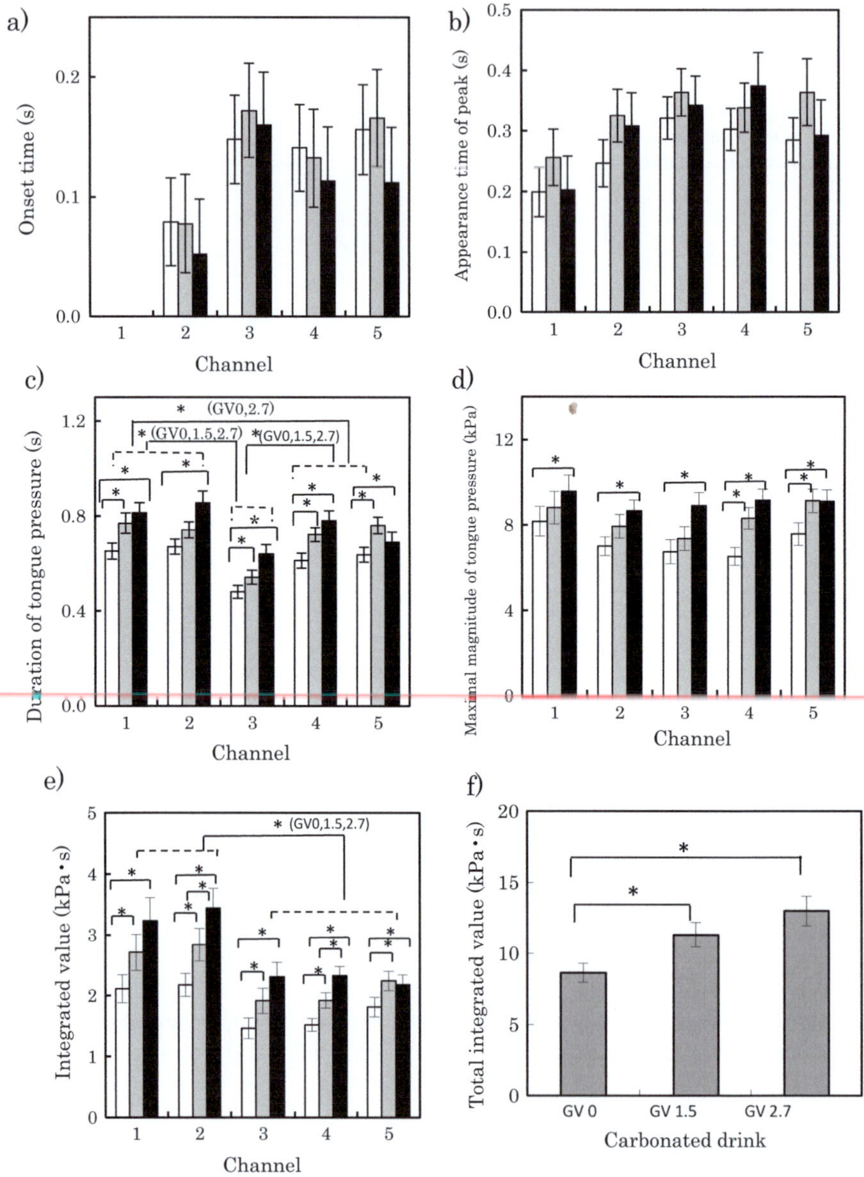

Fig. 4.13 Linguapalatal swallowing pressure of carbonated drink with different GV [15]. (**a**) Onset time of linguapalatal swallowing pressure, (**b**) peak time of linguapalatal swallowing pressure, (**c**) duration of linguapalatal swallowing pressure, (**d**) maximal magnitude of linguapalatal swallowing pressure, (**e**) integrated value of peak, (**f**) total integrated value of all channels. *Open square*: GV 0 carbonated drink, closed square: GV1.5 carbonated drink, *closed square* (*black*): GV2.7 carbonated drink. The *line* in a figure expresses the standard error. *$P < 0.05$

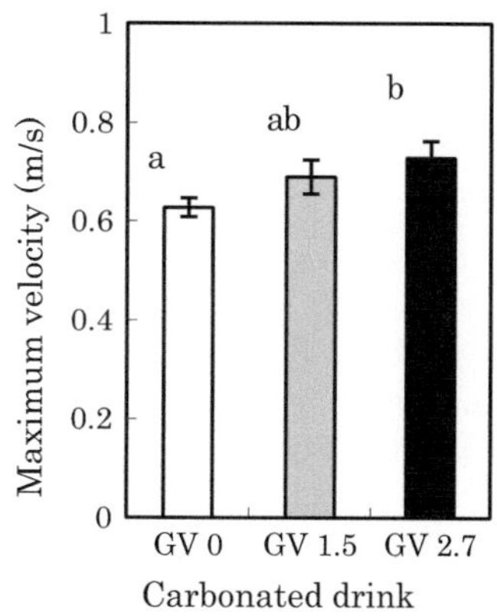

Fig. 4.14 Maximal transit velocity through the pharynx of carbonated drink with different GV [15]. The *line* in a figure expresses the standard error

and then decreased. As measuring time became longer, the apparent viscosity of the GV2.7 and GV1.5 drinks decreased, and when they were vigorously stirred with a glass stick, their apparent viscosity becomes the same as that of the noncarbonated GV0 drink. When strong positive pressure acts on a carbonated drink in the oral cavity, the disappearance of air bubbles is promoted, and the number of bubbles disturbing the flow of the carbonated drink decreases. The carbonated GV2.7 drink may pass more quickly through the pharyngeal region than the noncarbonated GV0 drink due to a decrease in the number of bubbles as a result of strong positive tongue pressure (Fig. 4.13d–f) and strong forceful movement through the pharynx.

The results of this study clearly show that stimulation by carbon dioxide can affect tongue movement, which produces the power needed to move a drink bolus through the oral cavity and that a high concentration of carbon dioxide causes greater impact between the hard palate and the tongue than occurs with a low concentration. These findings provide clinically valuable basic data on the effect of a stimulant in the oral stage of swallowing.

4.2 Agar Gel

In Japan, the ratio of the elderly person increases and exceeds 25 % in 2014. The physical function of the elderly person falls with increasing age, and the swallowing and chewing function also is reduced. The meal chopped finely and the meal with thickener are prepared in order to easily chew and safely swallow for the elderly person. However, the relationships between natures of chewing or swallowing and the physical properties of the food or food bolus have not been studied sufficiently. It is important to clarify these relationships. Here, it is described that the size of agar gel effects on the size distribution of the food fragment generated by chewing.

4.2.1 Pre-mastication Size of Foods and Size Distribution of Food Pieces in Boluses [18, 19]

The number of food pieces generated from mastication of 4 g of agar gel cubes with side length of 3.5 mm (hereinafter, 3.5-mm gel samples) and 4 g of agar gel cubes with side length 15 mm (hereinafter, 15-mm gel samples) increases almost proportionally to the number of chewing strokes (0–10). Although 3.5-mm gel samples results in more pieces than 15-mm gel samples after the same number of chewing strokes, the rate of increase in the number of pieces is higher for 15-mm gel samples than for 3.5-mm gel samples.

The ratios of the number of food pieces of certain sizes (n_s) to the total number of food pieces in boluses (N_T) are calculated using the formula $(n_s/N_T) \times 100$. It is found that the size distributions of food pieces after applying one to ten chewing strokes are different between 4-g gel samples that have distinct pre-mastication sizes. When the number of chewing strokes is low, the size distribution of pieces of 15-mm gel samples is broad and some of the pieces fall into the small-size region, while pieces of 3.5-mm gel samples show a near-normal distribution indicative of monodisperse particles (Fig. 4.15a, b). As the number of chewing strokes increases, the size distribution of pieces of 3.5-mm gel samples becomes similar to that of 15-mm gel samples (Fig. 4.15c). Further, after applying 30 chewing strokes, the size distributions of pieces of both 3.5-mm and 15-mm gel samples become similar to the distribution of pieces of pre-crushed gel samples (Fig. 4.15d). However, 3.5-mm gel samples give more medium-sized pieces than those other types of gel samples. Also, the number of small food pieces generated decreases in the following order: pre-crushed >15-mm >3.5-mm gel samples. This suggests that chewing stroke number

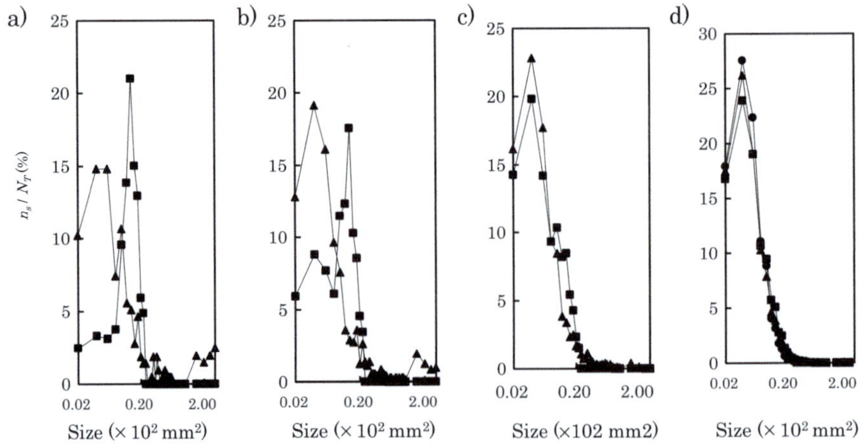

Fig. 4.15 Size distributions of food pieces in boluses after different numbers of chewing strokes [18, 19]. (**a**) 1 stroke, (**b**) 3 strokes, (**c**) 10 strokes, (**d**) 30 strokes. ■3.5-mm gel, ▲15-mm gel, ●pre-crushed gel

has a smaller influence on post-mastication size for 3.5-mm gel samples compared with 15-mm gel samples.

4.2.2 Mean Sizes of Ten Groups (Grouped by Sizes of Gel Pieces) [19]

Figure 4.16 shows the mean size of each of the ten groups, divided by the post-mastication sizes (Gp1–Gp10, from largest to smallest) of 4 g of 3.5-mm and 15-mm gel samples. The mean size of pieces generated from 15-mm gel samples exponentially decreases as the chewing stroke number increases, but the pattern of exponential decay changes after four to five chewing strokes (crossover structure) in every group (Fig. 4.16a). During mastication of 15-mm gel samples, the absolute value of the slope of the regression line (indicating an exponential decrement) is larger in the range from one to four or five chewing strokes than in the range of higher numbers of chewing strokes. This indicates more effective food size reduction in the early stage than in the late stage of mastication. The exponential decrease in the mean size of food pieces at the early stage of mastication is largest in Gp1 (the group with the largest mean size), suggesting a strongly destructive effect of mastication in this group compared with the others for the 15-mm gel samples. On the other hand, the effect of mastication is not so prominent at the later stage in groups with a small mean size.

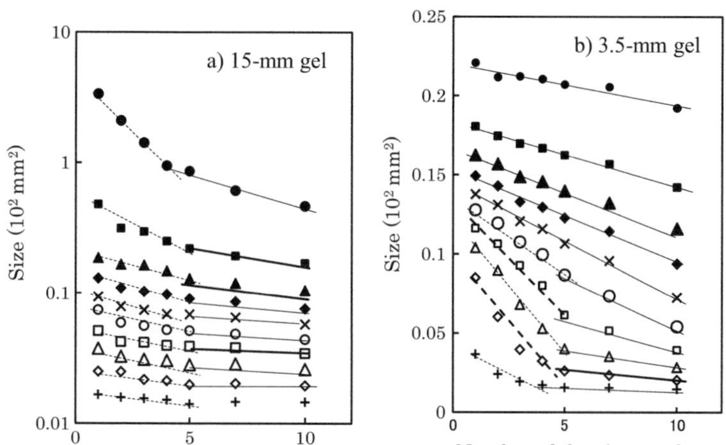

Fig. 4.16 Mean sizes of food pieces in size groups Gp1–Gp10 plotted against the number of chewing strokes [19]. (**a**) 15-mm gel samples, (**b**) 3.5-mm gel samples. ●Gp1, ■Gp2, ▲Gp3, ♦Gp4, ×Gp5, ○Gp6, □Gp7, △Gp8, ◇Gp9, +GP10

Meanwhile, for 3.5-mm gel samples, only the groups with smaller sizes of food pieces (Gp6–Gp10) show a crossover structure (Fig. 4.16b). At low numbers of chewing strokes, the absolute value of the slope of the regression line becomes larger with increasing group number (excluding Gp10), indicating a completely different trend from that observed with 15-mm gel samples. At high numbers of chewing strokes, the decrease in the size of food pieces becomes smaller with increasing group number, similar to the phenomena seen in mastication of 15-mm gel samples. This indicates a narrow distribution of food pieces in Gp1–Gp5, while there is a broad distribution in Gp6–Gp10 for 3.5-mm gel samples.

4.2.3 Model of the Mastication Process [18]

If the results of the preceding breakdown steps influence the following breakdown steps, a lognormal distribution would fit the data from the process. On the other hand, if individual breakdown steps constitute a series of completely random phenomena, a cumulative exponential distribution would fit the data [20–22]. The following equation is used for lognormal distribution fitting [20, 21]:

$$N(s) = (N_T/2)\left[1 - \mathrm{erf}\left\{(\log (s/s_a)) / \left(\sqrt{2\sigma}\right)\right\}\right] \tag{4.1}$$

$$\mathrm{erf}(x) \equiv \left(2/\pi^{1/2}\right) \int_0^{\mathrm{x}} \exp\left(-\mathrm{y}^2\right) \mathrm{dy} \tag{4.2}$$

Here, $N(s)$ is the cumulative size distribution function, N_T is the total number of food pieces, σ is a related parameter, and erf(x) is an error function:

The size distribution of pieces generated from 15-mm gel samples after five chewing strokes shows a greater outward curvature (misfit region) compared to the standard lognormal distribution in the large-size region but that after 30 chewing strokes shows a good fit (Fig. 4.17). A cumulative exponential distribution fits the misfit region. This suggests that the crushing process of 15-mm gel samples by mastication includes two phases: one is when food pieces are small, during which the crushing of gel pieces is influenced by the result of the preceding chewing strokes, and the other is when food pieces are large, during which crushing of food pieces is completely random at each step. Meanwhile, there are theoretical differences in the mastication process between 3.5-mm gel samples and 15-mm gel samples. Each piece generated from 3.5-mm gel sample by mastication is crushed independently, and thus, a lognormal distribution model cannot be used.

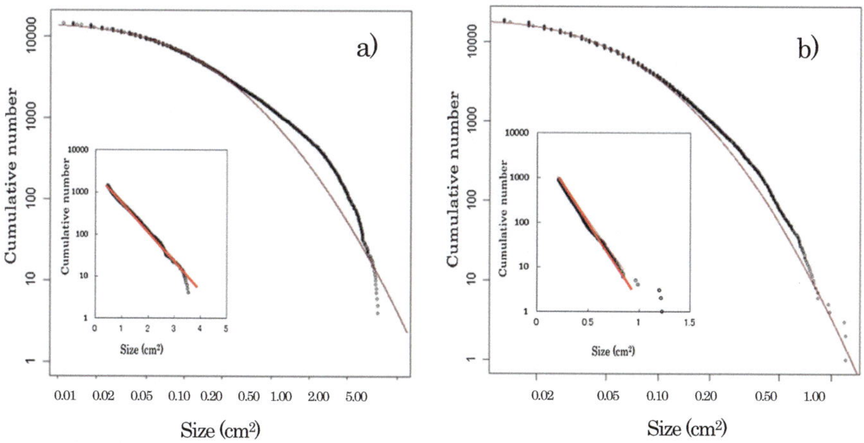

Fig. 4.17 The standard lognormal distribution and the accumulation size of pieces generated from 15-mm gel samples [18]. (**a**) After 5 chewing strokes, (**b**) after 30 chewing strokes

4.2.4 Masseter Activity During Mastication [23]

Peak masseter activity, which is indicative of chewing intensity, decreases as the number of chewing strokes increased when masticating 3.5-mm, 15-mm, and pre-crushed gel samples (Fig. 4.18a). The peak masseter activity during 5–30 chewing strokes is highest for 3.5-mm gel samples, followed by that for 15-mm gel samples and then for pre-crushed gel samples. As the mastication process proceeded to 50 chewing strokes, the peak masseter activity for 3.5-mm gel samples and that for 15-mm gel samples becomes comparable. The activity is smaller for mastication of pre-crushed gel samples, indicating that the intensity of the masseter activity varies depending on the gel sample form. The peak value here is obtained by dividing the sum of means of right and left masseter activity measurements by the number of chewing strokes.

To apply five chewing strokes, 15-mm gel samples require a longer duration of masseter activation than other types of gel samples (Fig. 4.18b). The duration of masseter activation becomes shorter as the number of chewing stroke increases (Fig. 4.18b). The cross-sectional area of the masseter muscle curve becomes smaller as the number of chewing strokes increases (Fig. 4.18c). The cross-sectional area after five chewing strokes decreases in the following order: 15-mm gel samples >3.5-mm gel samples >pre-crushed gel samples. However, after 20 chewing strokes, the cross-sectional area becomes larger when chewing 3.5-mm gel samples than when chewing 15-mm gel samples: after 30 chewing strokes, the cross-sectional areas of the masseter muscle are not significantly different, regardless of types of gel samples. The decreases in the cross-sectional area with increasing chewing stroke number are moderate for 3.5-mm gel samples, while it is more prominent for 15-mm gel samples.

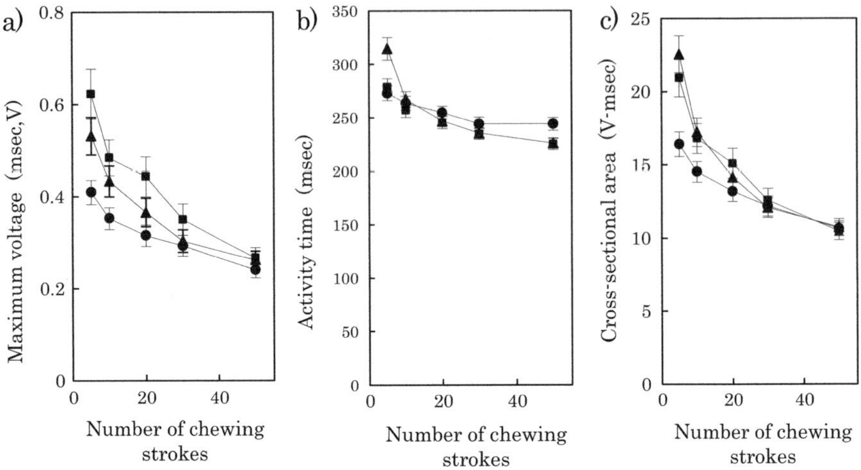

Fig. 4.18 Characteristics of masseter activity plotted against the number of chewing strokes [23]. (**a**) Maximum voltage, (**b**) activity time, (**c**) cross-sectional area. ■3.5-mm gel, ▲15-mm gel, ●pre-crushed gel

4.2.5 Time Required for Swallowing After the End of Mastication [23]

The times required for swallowing after the end of mastication (hereinafter, the swallowing time) are comparable between 3.5-mm and 15-mm gel samples after applying five chewing strokes (Fig. 4.19). The swallowing time of 15-mm gel samples declines markedly after ten chewing strokes, while that of 3.5-mm gel samples is higher than that of other types of gel samples until 20 chewing strokes. This suggests that 15-mm gel samples are effectively crushed to produce favorable boluses by mastication, while 3.5-mm gel samples are less influenced by mastication, and thus, favorable boluses are not produced.

4.2.6 Texture Features [18]

The increases in the number of chewing strokes have no significant influence on the hardness of boluses of pre-crushed gel samples. On the other hand, the hardness of boluses of both 3.5-mm and 15-mm gel samples decreases with the increase in the number of chewing strokes: the bolus hardness of 15-mm gel samples drops markedly after applying ten chewing strokes (Fig. 4.20a). After 30 chewing strokes, the bolus hardness becomes comparable among all three types of gel samples. Adhesiveness increases with mastication irrespective of the type of gel samples, but that of 3.5-mm gel samples becomes lower than that of 15-mm gel samples

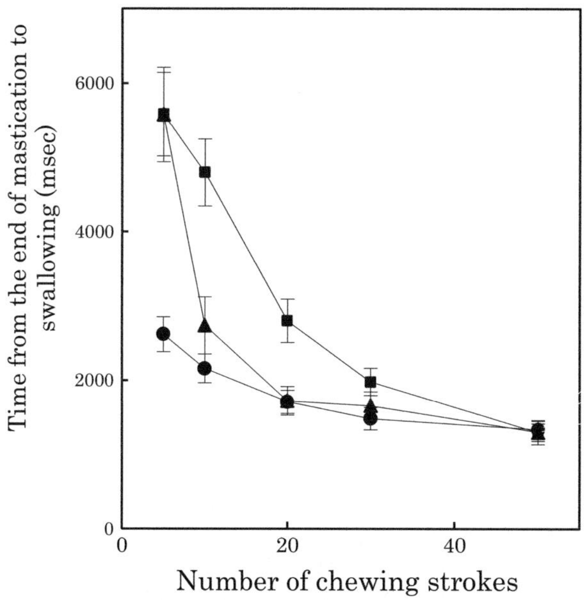

Fig. 4.19 Time required for swallowing from the end of mastication, plotted against the number of chewing strokes [23]. ■3.5-mm gel, ▲15-mm gel, ●pre-crushed gel

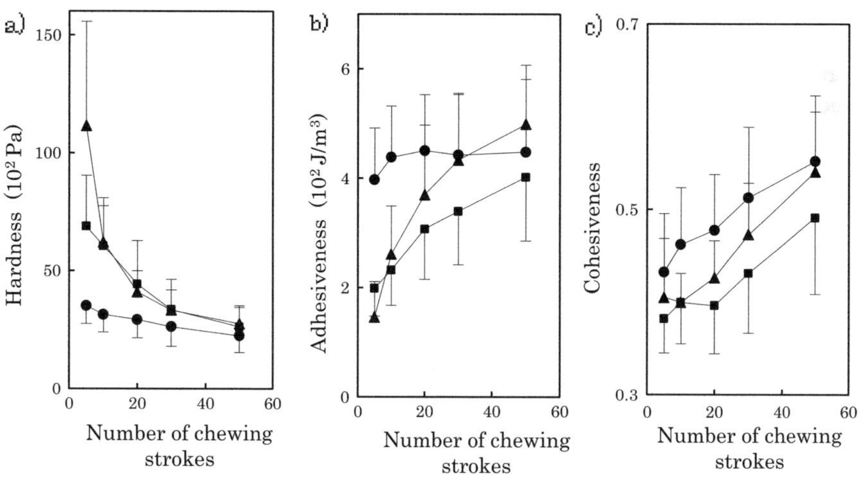

Fig. 4.20 Texture feature levels plotted against the number of chewing strokes [18]. (**a**) Hardness, (**b**) adhesiveness, (**c**) cohesiveness. ■3.5-mm gel, ▲15-mm gel, ●pre-crushed gel

after exceeding 20 chewing strokes (Fig. 4.20b). The cohesiveness increases as the number of chewing strokes increases, but the cohesiveness of 3.5-mm gel samples is lower than that of 15-mm gel samples after exceeding 20 chewing strokes

(Fig. 4.20c). Boluses generated from 3.5-mm gel samples are less adhesive and less cohesive than those from 15-mm gel samples and pre-crushed samples, and their size distributions after mastication are involved in such differences.

4.2.7 Mobility Features of Boluses in the Pharynx [18]

The maximum velocity and the spectrum area at the pharynx decrease during the course of the mastication process (Fig. 4.21a, b). This is because food pieces become smaller, more adhesive, and more cohesive, resulting in the formation of favorable boluses that require less swallowing force. Also, the food transit time at the pharynx after five chewing strokes is significantly shorter for 15-mm gel samples than for other types of gel samples (Fig. 4.21c), probably because of the presence of residual large pieces in the boluses.

Taken together, the size distribution of pieces generated from 3.5-mm gel samples is a near-normal distribution when the number of chewing strokes is low, while a broad distribution, distinct from a normal distribution, can be seen for 15-mm gel samples at the same mastication stage. Specifically, the size distribution of pre-crushed gel samples indicates the presence of small-sized gel pieces even at the same stage. It is known that the space density is higher when a certain space is filled with various sizes of particles, than with monodisperse particles [24]. The higher space density means higher contact area among particles, which is associated with higher cohesiveness. Meanwhile, large-sized pieces remain after applying a small number of chewing strokes on 15-mm gel samples, and thus, the surface

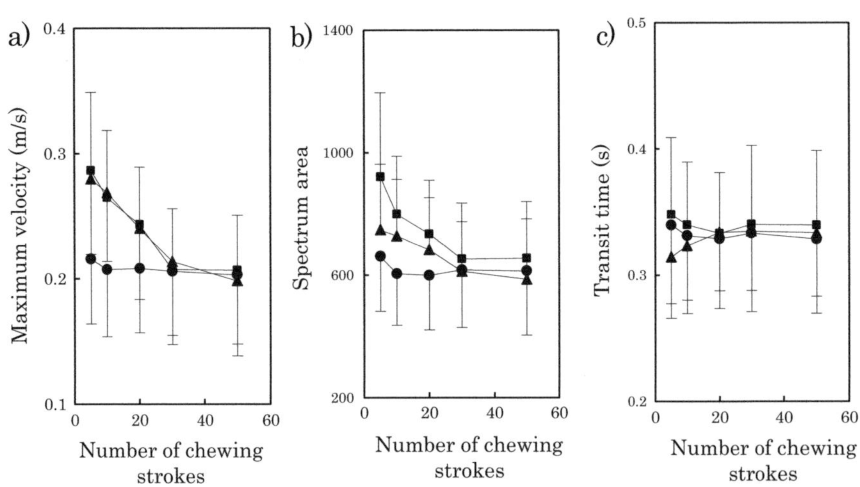

Fig. 4.21 Mobility features of food boluses at the pharynx [18]. (**a**) Maximum velocity, (**b**) area of a spectrum, (**c**) transit time. ■3.5-mm gel, ▲15-mm gel, ●pre-crushed gel

tension of boluses is low. It is clear that different initial sizes of gel samples have distinct influences on mastication and swallowing, and thus these influences need to be investigated further in the future.

References

1. Lo GS, Goldbeq AP, Lim A, Grundhauser J, Anderson C, and Schonfeld G, Soy fiber improves lipid and carbohydrate metabolism in primary hyperlipidemic subjects. *Atherosclerosis*, **62**, 239–248 (1986).
2. Tsai AC, Vinik AI, Lasichak A, and Lo GS, Effects of soy polysaccharide on postprandial plasma glucose, insulin, glucagon, pancreatic polypeptide, somatostatin, and triglyceride in obese diabetic patients. *Am. J. Clin. Nutr.*, **45**, 596–601 (1987).
3. Marlett JA, McBurney MI, and Slavin JL, Position of the American Dietetic Association: health implications of dietary fiber. *J. Am. Dietetic Assn.*, **102**, 993–1000 (2002).
4. Shahidi F, "Functional foods: their role in health promotion and disease prevention," *J. Food Sci.*, **69**, 146–149 (2004).
5. Cichero JAY, Jackson O, Halley PJ, and Murdoch BE, Which One of these is not like the others? An inter-hospital study of the viscosity of thickened fluids. *J. Speech Lang. Hear. Res.*, **43**, 537–547 (2000).
6. Cichero J, Jackson O, Halley PJ, and Murdoch BE, How thick is thick? Multicenter study of the rheological and material property characteristics of mealtime fluids and videofluoroscopy fluids. *Dysphagia,* **15**, 188–200 (2000).
7. Glassburn DL, and Deem JF, Thickener viscosity in dysphagia management: variability among speech-language pathologists. *Dysphagia*, 13, 218–222 (1998).
8. Nicosia MA, and Robbins JA, The fluid mechanics of bolus ejection from the oral cavity, *J. Biomech.*, **34**, 1537-1544 (2001).
9. Pelletier CA, Management of Adult Neurogenic. *Dysphagia*, **12**, 74–78 (1997).
10. Moritaka H, Sawamura S, Kobayashi M, Kitade M, and Nagata N, Relation between the Rheological Properties and the Swallowing Characteristics of Vegetable Juices Fortified with Carrot Puree. *Biosci. Biotechnol. Biochem.*,**76**(3), 429–435 (2012).
11. Tashiro A, Hasegawa A, Kohyama K, Kumagai H, and Kumagai H, Relationship between the Rheological Properties of Thickener Solutions and Their Velocity through the Pharynx as Measured by the Ultrasonic Pulse Doppler Method. *Biosci. Biotechnol, Biochem.*, **74**, 1598–1605 (2010).
12. Kumagai H, Tashiro A, Hasegawa A, Kohyama K, and Kumagai H, Relationship between flow properties of thickener solutions and their velocity through the pharynx measured by ultrasonic pulse method. *Food Sci. Technol. Res.*, **15**, 203–210 (2009).
13. Moritaka H, and Nakazawa F, The rheological and swallowing properties of rice starch. *Food Sci. Technol. Res.*, **15**, 133–140 434 (2009).
14. Green BG, Alvarez-Reeves M, George P, and Akirav C, Chemesthesis and taste: evidence of independent processing of sensation intensity. *Pysiol Behav.*, **86**(4), 526–537 (2005).
15. Moritaka H, Kitade M, Sawamura S, Takihara T, Awano I, Ono T, Tamine K, and Hori K, Effect of carbon dioxide in carbonated drinks on linguapalatal swallowing pressure, *Chem. Senses*, **39**, 133–142 (2014).
16. Ono T, Hori K, and Nokubi T. Pattern of tongue pressure on hard palate during swallowing. *Dysphagia*, **19**(4), 259–264 (2004).
17. Dessirier JM, Simons CT, Carstens MI, O'Mahony M, and Carstens E, Psychophysical and neurobiological evidence that the oral sensation elicited by carbonated water is of chemogenic origin, *Chem Senses,***25**, 277–284 (2000).

18. Kitade M, Kashiki H, Kobayashi N, and Moritaka H, Chewing and swallowing properties of different-sized agar gels. *Nippon Shokuhin Kagaku Kaishi*, **59**, 369–377 (2012). (*in Japanese*)
19. Kitade M, Kobayashi N, and Moritaka H, Relationship between number of chewing cycles and fragment size of agar gels, *Nippon Shokuhin Kagaku Kaishi*, **60**, 554–562 (2013). (*in Japanese*)
20. Kobayashi N, Kohyama K, Sasaki Y, and Matsushita M, Statistical laws for food fragmentation by human mastication. *J. Phys. Soc. Jpn.*, **75**, 083001 (2006).
21. Kobayashi N, Kohyama K, and Shiozawa K, Fragmentation of a viscoelastic food by human mastication, *J. Phys. Soc. Jpn.*, **79**, 044801 (2010).
22. Matsushita M, and Sumida K, How do thin glass rods break-Stochastic models for one-dimensional brittle fracture. *Bull. Fac. Sci. & Eng. Chuo Univ.*,**31**, 69–79 (1997).
23. Kitade M, Sagawa A, Fuwa M, and Moritaka H, Properties of the masseter and digastric muscles during swallowing of agar gels of different sizes. *Nippon Shokuhin Kagaku Kaishi*, **61**, 293–301 (2014). (*in Japanese*)
24. Ishihara S, Nakamura M, Funami T, Odake S, and Nishinari K, Viscoelastic and fragmentation characters of model bolus from polysaccharide gels after instrumental mastication. *Food Hydrocolloids*, **25**, 1210–1218 (2011).

Chapter 5
Moisture Distributions and Properties of Pasta Prepared or Cooked Under Different Conditions

Takenobu Ogawa and Shuji Adachi

Abstract A method using an image processing technique was developed to measure the moisture profile in pasta during its rehydration process. The method has the higher spatial resolution and can measure the lower moisture content than currently used methods. A very unique profile was recognized. Possible reasons for the profile will be discussed. The moisture distributions within pastas prepared at different temperatures were measured by the method, and the water sorption kinetics and texture of the pasta were also measured. The pasta prepared at higher temperature exhibited better textural properties. The effects of cooking temperature or salt concentration in cooking water on the kinetics and properties were also examined for the pastas prepared under different temperature-programmed conditions.

Keywords Pasta • Moisture distribution • Drying • Rehydration • Water migration

5.1 Introduction

An industrial food-making process is often designed and operated on the basis of significant experience. In many cases, the science behind the process is not well-understood. Drying is a commonly used process for improving the shelf life and transportation efficiency of food and is applied in the manufacturing of various foodstuffs. The primary objective of food drying is to decrease the moisture content of the food to a level that minimizes microbial spoilage to ensure quality preservation for a longer duration. Dried foods are typically rehydrated prior to use or consumption to improve the taste and digestibility, i.e., the water

T. Ogawa
Division of Agronomy and Horticultural Science, Graduate School of Agriculture Kyoto University, Kyoto, Japan
e-mail: ogawat@kais.kyoto-u.ac.jp

S. Adachi (✉)
Division of Food Science and Biotechnology, Graduate School of Agriculture, Kyoto University, Kyoto, Japan
e-mail: adachi@kais.kyoto-u.ac.jp

I. Kaneda (ed.), *Rheology of Biological Soft Matter*, Soft and Biological Matter,
DOI 10.1007/978-4-431-56080-7_5

molecules in food are removed and added during the drying and rehydration processes, respectively. Because water migration behavior during these processes largely affects the quality of dried and rehydrated foods, a better understanding of water migration kinetics would facilitate the efficient manufacture good-quality dry food and improve the texture, taste, and digestibility. However, the key mechanisms controlling water migration into the food remain unclear.

The Italian word "pasta" means "dough" and is generally used to describe products fitting the "Italian" style of extruded foods, such as spaghetti or lasagna. Pasta is healthy because of its relatively low fat content, high carbohydrate content, and a good protein composition. The main ingredients of pasta are principally durum wheat semolina and water. Durum wheat (*Triticum durum*) is the hardest wheat among wheat varieties, and durum milling produces a coarse particle known as semolina, which is ideal for making pasta because of its hardness, intense yellow color, and nutty taste [69]. A proportion of 18–25 % of water is mixed with dry raw durum semolina at 35–40 °C, and the mixture is kneaded for 10–20 min to produce fresh dough with an average moisture content of 30–32 % [69]. Next, pasta is produced by extruding the stiff durum semolina dough through a die using a vacuum extruder [27, 69]. A die made of bronze is traditionally used. However, a die made of Teflon has recently been used because of longer lifetime due to reduced wear, a smoother pasta surface, and improved general appearance of dried pasta [19, 20, 26]. Pasta prepared using the die made of Teflon has a smooth surface, while that prepared using a bronze die has a rough surface. Additionally, pasta prepared using a bronze die has higher porosity, lower density, lower rupture strength, and larger effective diffusion coefficient of water during drying compared to that prepared using a Teflon die [41, 44].

Because pasta is typically distributed in the dry state, the moisture content of fresh pasta is reduced to approximately 11 % by drying. Pasta is dried under various conditions, where both the temperature and the humidity are dexterously changed with time. The process takes several days at a drying temperature of 30 °C. Therefore, dried pasta is presently prepared on an industrial production scale at temperatures above 30 °C. Depending on the maximum temperature during processing, production processes can be classified into low temperature (LT), high temperature (HT), and very high temperature (VHT) levels. The maximum temperatures of the LT, HT, and VHT processes are approximately 50 °C, 70 °C, and 85 °C, respectively, and the drying times are 20, 13, and 6 h, respectively. Although pasta has traditionally been dried using the LT process, the VHT process is most commonly used by manufacturers because of the short production time. An ultrahigh-temperature process has recently been described at a drying temperature of 95 °C.

The drying conditions, which include the temperature, humidity, and duration, largely affect pasta quality traits, such as texture and appearance. In particular, the maximum temperature during drying has the largest effect on the properties of rehydrated pasta. Pasta dried under high-temperature conditions had better texture after rehydration compared to pasta dried under low-temperature conditions [55]. Drying conditions affect the inner structure of the pasta during drying [14, 73], including the states of starch and protein, the major components of pasta. Guler et al. [31] examined the characteristics of starch gelatinization in HT and VHT pasta using a rapid viscoanalyzer, a differential scanning calorimeter, an X-ray diffractometer, and a polarization microscope. Zweifel et al. [73] reported that drying under high-temperature conditions enhanced protein denaturation and suppressed starch granule swelling and collapse. The effect of temperature on the progress of the Maillard reaction, which affects the red-color development of pasta, was also studied [1, 6]. Baiano et al. [7] measured the leakage of amylose from LT, HT, and VHT pasta during rehydration and showed that LT pasta leaked the highest amounts of amylose.

Dried pasta is typically consumed after rehydration by cooking to recover its properties. Therefore, it is important to fully understand the processes occurring during the rehydration of dried pasta, which is a complicated mass transport process governed by several migration mechanisms of water into the pores [56].

Empirical and theoretical approaches can be used to formulate equations describing the rehydration kinetics [30]. Because the development of empirical equations requires considerably less effort, the equations can be useful for predicting and optimizing the rehydration kinetics [25]. Six types of empirical or semiempirical equations are often utilized to describe rehydration kinetics [43]: the exponential equation [45], Peleg's model [54], first-order kinetics [10], Becker's model [8], Weibull distribution function [15], and normalized Weibull distribution function [42]. Peleg's model and the Weibull distribution function enable accurate prediction of the moisture content of pasta for any rehydration time [18]. However, in empirical equations, the rehydration process is treated as a "black box," varying specific input setup parameters, measuring output quantities, and deriving adequate correlations. Therefore, it is necessary to determine the coefficients of the equation by varying the specific input setup parameters in detail, such as the physical-chemical characteristics of pasta manufactured under various conditions, rehydration temperature, and concentration of NaCl in an immersion solution. In contrast, theoretical equations are typically complicated and are not convenient for practical purposes [18, 43]. However, they provide insight into the mechanistic relevance of an observed phenomenon [56]. The numerous theoretical equations describing rehydration kinetics are based on Fick's first and second laws of diffusion, where the difference in the moisture content of pasta is considered a driving force for water migration [9, 29, 30, 58, 59]. The rehydration process of fresh penne pasta at 20–80 °C was characterized by two effective diffusion coefficients using the law [18]. In addition to water diffusion, starch crystalline domains melting, macromolecular matrix relaxation, and "residual deformation" release occur during pasta rehydration [22]. However, some researchers showed that the actual process of moisture migration is not diffusion-controlled and proposed several other mechanisms, such as water imbibition, capillarity, and flow in porous media [39, 56].

Although the moisture distribution is obtained by numerically solving the equations, a proposed model is typically validated by measuring the average moisture content of an entire sample [21] because of the absence of an adequate method for obtaining a precise moisture distribution of the sample and for verifying the numerical results. The absence of an adequate method makes it difficult to discern or interpret the mechanism controlling water migration during rehydration. Rehydration curves, which express changes in the average moisture content over time and are obtained by numerically solving the various models, are of the hyperbolic type and satisfactorily fit the experimental data [18, 21, 22, 30, 59]. However, agreement of an empirical curve to a theoretical curve is insufficient for judging the appropriateness of the models. The actual measurement of a precise moisture distribution is necessary for verification of the model.

Nuclear magnetic resonance (NMR) and magnetic resonance imaging (MRI) are powerful techniques for measuring the moisture distribution of foods. These techniques were applied to measure the moisture distribution of rice grains during cooking [34], noodles during drying or rehydration [32, 33, 35, 61], and cheese during brining [5]. However, the accuracy of the measurement is insufficient to verify the numerically calculated distribution because of the following four limitations. The minimum moisture content measureable by the techniques is high. For example, fast water proton relaxation cannot allow for measurement of a moisture content less than 0.67 kg-H_2O/kg-dry matter (d.m.) [35], although the dried pasta begins rehydrating from a moisture content of approximately 0.11 kg-H_2O/kg-d.m. Another limitation is the low spatial resolution. MRI measurement evaluates the moisture content every 65 μm at best [34]. This resolution indicates that only approximately 12 points of data can be obtained from pasta with a radius of 0.8 mm. The third limitation is that the measurement is time consuming. A few minutes are necessary for the MRI technique to measure the moisture distribution of a sample, although the moisture distribution of a sample, such as pasta, can change within a few minutes. Finally, very expensive equipment is required for NMR and MRI. Therefore, the development of a method without these limitations is necessary for determining the mechanism controlling water migration in dried foods.

Dried pasta is yellowish or yellowish brown and becomes lighter as the moisture content increases. We evaluated the color change of pasta during rehydration to develop a new method using a digital camera to precisely measure the moisture content (0.1 kg-H_2O/kg-d.m. or higher) of pasta. A sample is cut to measure the cross-sectional moisture distribution. A digital camera can acquire the color distribution of the sample, and digital cameras with high pixels are currently available and provide high-resolution images. The moisture distribution measured using this method will enable the rehydration process to be evaluated.

In this chapter, the kinetic analysis of the rehydration process of pasta and the factors affecting the kinetics are discussed. Next, a method for measuring the moisture distribution within pasta is introduced. Based on the measured moisture distributions measured, the mechanism of rehydration is discussed. Finally, the relationship between the moisture distribution within rehydrated pasta and its texture is described.

5.2 Rehydration Kinetics of Dried Spaghetti

5.2.1 *Effect of Rehydration Temperature on Rehydration Kinetics of Dried Spaghetti*

Cylindrical pasta (spaghetti) with carbohydrate content of 72 wt% was purchased from a local supermarket. The initial diameters of the spaghetti were 1.4 mm, 1.6 mm, and 1.8 mm. A 9-cm-long sample was immersed into a tube containing approximately 50 cm^3 of distilled water equilibrated at 20 °C, 40 °C, 50 °C, 55 °C, 60 °C, 70 °C, 80 °C, or 90 °C. Approximately 15 tubes were prepared under specific conditions, and the samples were removed at a given time from the tubes, immediately blotted to remove any superficial water, and weighed to find W_0. Samples were dried in a convection drying oven at 105 °C for 4 days and weighed to find W_1. The moisture content based on the net mass of spaghetti, X_t, was calculated using Eq. 5.1.

$$X_t = (W_0 - W_1)/W_1 \tag{5.1}$$

The spaghetti showed a reduction in mass because of leakage of the components into the immersion water during rehydration. The loss of spaghetti mass (initial matter (i.m.) basis), M_t, was greater with longer rehydration times and at higher temperatures of the immersion water. Loss of mass reached approximately 0.2 kg/kg-i.m., corresponding to approximately 20 % of the spaghetti mass. The relationship between the amount of loss of spaghetti mass and the quotient of time by the square of the diameter, t/d^2, where d is the initial diameter of the spaghetti, is expressed by Eq. 5.2 for spaghetti samples with three different diameters (1.4, 1.6, and 1.8 mm):

$$M_t = M_e\left[1 - \exp\left(-kt/d^2\right)\right] \tag{5.2}$$

where M_e is the equilibrium amount of the loss of spaghetti mass and k is the rate constant.

Figure 5.1 shows the rehydration curves for the spaghetti with an initial diameter of 1.6 mm at 20–90 °C [46]. Instead of the actual rehydration time, the quotient of time by the square of the diameter, t/d^2, is shown in Fig. 5.1. The rehydration curve could be expressed by a hyperbolic-type equation (Eq. 5.3) using the t/d^2 as an independent variable rather than using time t.

$$X_t = \frac{a \times \left(t/d^2\right)}{b + \left(t/d^2\right)} + X_0 \tag{5.3}$$

where a and b are constants. The constants, a and b, were estimated to best fit the calculated X_t values to the experimental ones. The solid curves in Fig. 5.1 were drawn using the estimated a and b values for each temperature.

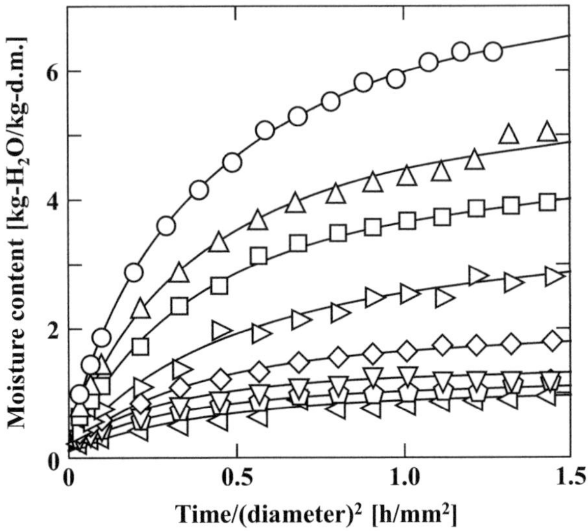

Fig. 5.1 Rehydration of spaghetti at 90 °C (○), 80 °C (△), 70 °C (□), 60 °C (▷), 55 °C (◇), 50 °C (▽), 40 °C (⬠), and 20 °C (◁). Pasta with an initial diameter of 1.6 mm was used. The water content was expressed as kg-water rehydrated per kg-dry matter (d.m.)

The equilibrium moisture content at $t = \infty$, X_e, is given by

$$X_e = \lim_{t \to \infty} X_t = a + X_0 \tag{5.4}$$

and can be approximated by a because of $X_e << a$ value. A derivative of X_t at $t = 0$, which corresponds to the initial rate of rehydration, v_0, gives the a/b value (Eq. 5.5).

$$v_0 = d^2\, dX_t/dt|_{t=0} = a/b \tag{5.5}$$

Figure 5.2 shows the temperature dependences of the a and a/b values [46]. The temperature dependence of the a value was analyzed based on van't Hoff's equation.

$$\frac{d \ln a}{d(1/T)} = -\frac{\Delta H}{R} \tag{5.6}$$

where ΔH is the change in the enthalpy of the rehydration, T is the absolute temperature, and R is the gas constant. The plots of a values versus $1/T$ were separated at high-temperature, transition, and low-temperature regions. a values were low in the low-temperature region but increased in the high-temperature region. The ΔH values in the low- and high-temperature regions were evaluated to be 1.44 kJ/mol and 25.1 kJ/mol, respectively. The a values at $T = \infty$ were calculated to be 3.16×10^{-4} and 2.06 kg-H_2O/kg-d.m., respectively. The transition occurred at approximately 50 °C, which was the middle of the onset temperature of 45.3 °C and the peak of 60.5 °C. The temperature dependence of the initial rate of rehydration

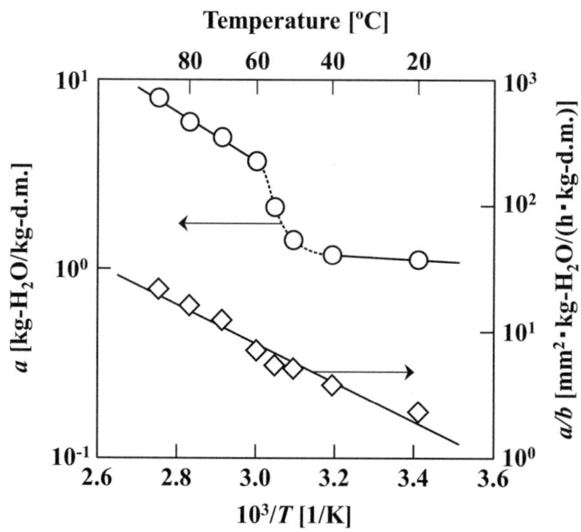

Fig. 5.2 Temperature dependencies of equilibrium moisture content *a* (○) and initial rate of rehydration *a*/*b* (◇)

obeyed the Arrhenius equation [43].

$$v_0 = a/b = A_0 \exp(-E_a/RT) \tag{5.7}$$

where A_0 is the frequency factor and E_a is the activation energy. The plots according to Eq. 5.7 lie on a line in the entire temperature range, and the E_a and A_0 values were estimated to be 30.5 kJ/mol and 1.36×10^{-4} $m^2 \cdot$kg-H_2O/(s·kg-d.m.), respectively. There was no temperature dependence of *a*/*b*, indicating that the initial rate of rehydration was not affected by starch gelatinization.

Figure 5.3 shows the relationship between the X_t and t/d^2 values for the spaghetti with diameters of 1.4, 1.6, and 1.8 mm at 50 °C and 80 °C [46]. The plots of X_t versus t/d^2 for all spaghetti were curved at all temperatures, suggesting that water diffusion plays an important role in the rehydration of spaghetti.

5.2.2 *Effect of Salts on Rehydration Kinetics of Dried Pasta*

Because dried pasta is typically cooked in water containing 0.1–1 % (w/v) NaCl, the effect of salts on the rehydration kinetics of dried spaghetti was examined [47]. The rehydration process was measured at 20–90 °C in 1.83 mol/L NaCl, equivalent to 10 % (w/v), as well as in 1.83 mol/L LiCl, KCl, NaBr, and NaI. Although salts other than NaCl and extremely high salt concentration are unrealistic for cooking, the conditions were evaluated to understand the role of NaCl and other salts in the

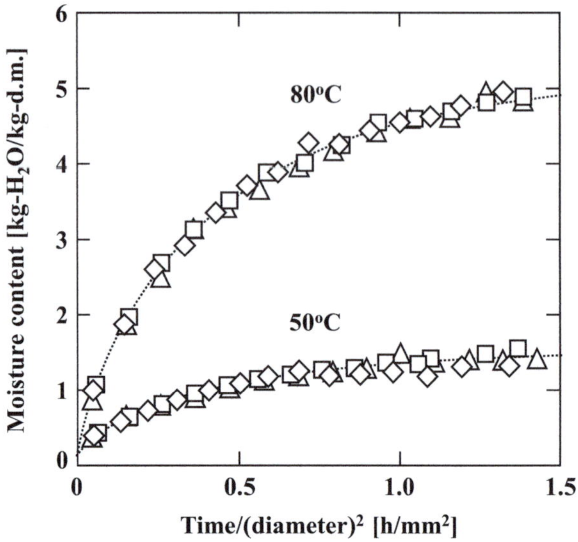

Fig. 5.3 Changes in the moisture content at 50 °C and 80 °C for spaghetti with initial diameters of 1.4 mm (△), 1.6 mm (□), and 1.8 mm (◇)

rehydration process of pasta. The moisture content, X_t, at any time t was estimated by taking into consideration the loss of spaghetti mass under the assumptions that the salt concentration in spaghetti was the same as that in the bulk solution and that only water was evaporated and salt remained in the spaghetti when the spaghetti was dried.

The rehydration processes in 1.83 mol/L NaCl at every temperature could be expressed by Eq. 5.3, and the *a* and *a*/*b* values were estimated at each temperature. The temperature dependences of *a* and *a*/*b* values were similar to those in water as in Sect. 5.3.1 and were analyzed using Eqs. 5.6 and 5.7, respectively. The ΔH values in the low- and high-temperature regions were estimated to be 1.08 and 33.1 kJ/mol. The ΔH value in the NaCl solution was nearly equal to that in water in the low-temperature region, but that in the NaCl solution was higher than that in water in the high-temperature region. Therefore, the difference between ΔH in the high-temperature range in the NaCl solution and that in water reflected the difference in gelatinization enthalpy, as Chiotelli et al. [11] reported that gelatinization enthalpy in approximately 2.0 mol/L NaCl was larger than that in water. The temperature dependence of the *a*/*b* value, which corresponds to the initial rehydration rate as shown by Eq. 5.5, in 1.83 mol/L NaCl was also expressed by Eq. 5.7 in the tested temperature range. The activation energy was found to be 25.6 kJ/mol, which was slightly lower than the value in water.

The equilibrium moisture content, *a*, was estimated for rehydration at 55 °C and 60 °C (Fig. 5.4) in various salt solutions. The *a* value for rehydration in salt solutions of the alkaline metal ions with chloride was in the order of LiCl < NaCl < KCl, and

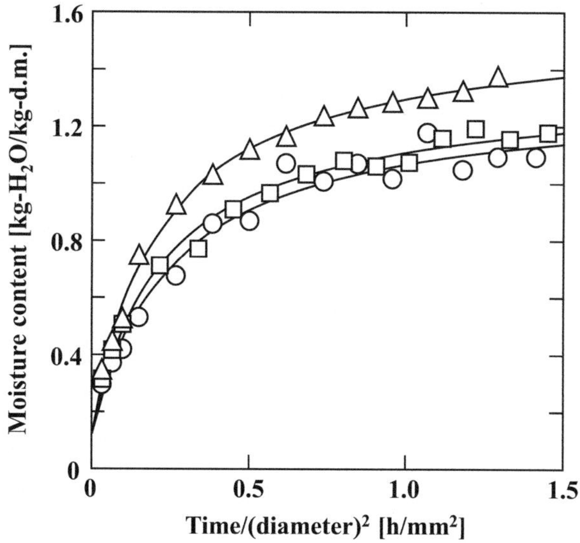

Fig. 5.4 Rehydration curves at 60 °C for 1.83 mol/L LiCl (○), NaCl (□), and KCl (△) solutions

that in sodium salt solutions containing different halogen ions was in the order of NaCl < NaBr < NaI. These orders are opposite of the Hofmeister (lyotropic) series [2, 36, 57, 72], where $Li^+ > Na^+ > K^+$ for alkaline metal ions with the same anion counterpart and $Cl^- > Br^- > I^-$ for halogen ions coupled with a constant cation.

Based on Fick's law of diffusion, an apparent diffusion coefficient D_a can be estimated for cylindrical pasta according to the following equation [12]:

$$\frac{X_t - X_e}{X_0 - X_e} = \sum_{n=1}^{\infty} \frac{4}{\beta_n{}^2} \exp\left[-\frac{D_a \beta_n{}^2}{(d/2)^2} t\right] \tag{5.8}$$

where β_n is the Bessel function root ($\beta_1 = 2.4048$) and X_0, X_e, and X_t are the initial moisture content, the equilibrium moisture content, and the moisture content at any time t, respectively. Because the equation can be approximated by only the first term in the series for long times or small diameters, the D_a value was estimated by plotting $(X_t–X_e)/(X_0 –X_e)$ versus time t for rehydration [18]. The D_a values obtained at 60 °C in the 1.83 mol/L LiCl, NaCl, and KCl solutions were plotted versus the crystalline radii of the alkaline metal ions in Fig. 5.5. Although the D_a value was higher for the salt of an alkaline metal ion with chloride having a larger crystal radius, this trend is not reasonably acceptable. Because the alkaline metal ions appear to migrate in water in the hydrated state, the D_a values were also plotted versus the Stokes radii of the hydrated ions [66] as shown in Fig. 5.5. The D_a value in the salt solution with the larger hydrate radius was lower. This indicates that hydration of the ions plays an important role in the water migration of pasta.

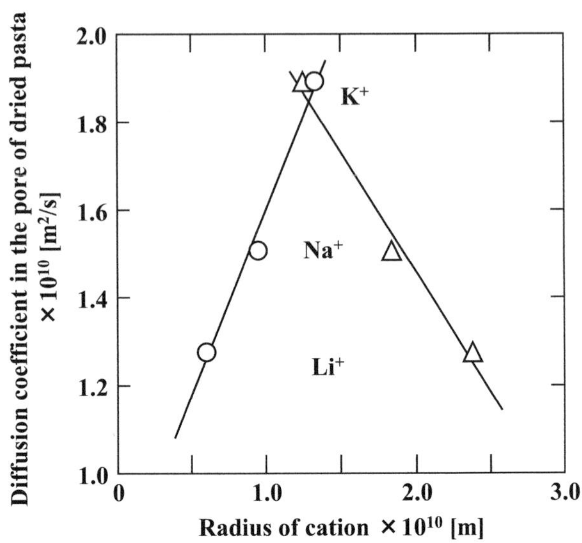

Fig. 5.5 Relationships between the apparent diffusion coefficient at 60 °C and crystalline radius (○) and Stokes radius (△) for 1.83 mol/L LiCl, NaCl, and KCl solutions

5.2.3 *Measurement of Moisture Distribution in Spaghetti Based on Image Processing*

Spaghetti increases in color brightness with increasing moisture content. We developed a method for measuring the moisture distribution in spaghetti during its rehydration process by using an image processing technique [49]. The method consists of the following five steps.

In the first step, two sets of rehydrated spaghetti samples were prepared. One was used to measure the moisture distribution, and the other was used to prepare a calibration curve correlating color brightness and moisture content. The samples for constructing the calibration curve were wrapped in polyethylene sealing film to prevent moisture loss and then stored at 70 °C for 3 days to equalize the moisture distribution in the samples.

In the second step, the rehydrated sample was cut crosswise using a sharp stainless steel blade and covered with a light shield with the same diameter or thickness as that of the sample. Cross-sectional images of the samples were acquired using a digital camera (Fig. 5.6). Each image contained 3888 × 2592 pixels (1.6 μm/pixel), indicating that the spatial resolution of the proposed method was approximately 3.2 μm, which was about 20 times higher than that obtained using MRI methods [34, 35, 61].

The third step involved digital image processing using two software packages: Mathematica 7 and Origin 8.1J. The original 24-bit RGB color image was pixelated into red, green, and blue images, and the blue image was used. The image was converted into an 8-bit gray-scale format using Origin 8.1J. In order to visually clarify the gray level of the image, the original gray level G_0 of each pixel was

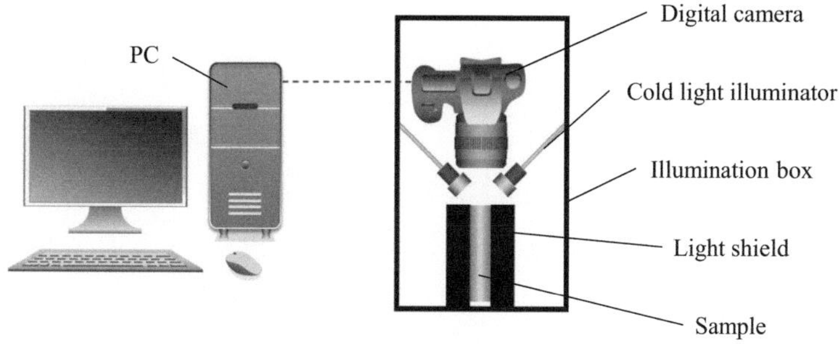

Fig. 5.6 Equipment used in the analysis

converted to the level G_γ by gamma correction using Eq. 5.9, and G_γ was further converted to the level G_c through contrast correction using Eq. 5.10 [62].

$$G_\gamma = 255(G_0/255)^{0.5} \tag{5.9}$$

$$G_c = 2 \times G_\gamma - 255/2 \tag{5.10}$$

In the fourth step, the calibration curve correlating the corrected gray level G_c (color brightness) with the moisture content was prepared using samples with equalized moisture distributions.

In the final step, the moisture distribution of the rehydrated spaghetti was obtained by converting the corrected gray level G_c of each pixel in the sample image to the moisture content using the calibration curve.

Figure 5.7 shows the average moisture distributions of spaghetti rehydrated for different periods. The rehydration time of 0 min represents the dried spaghetti. The precise moisture distribution in the al dente state at a rehydration time of 10.2 min was measured by our method because the moisture content is too low to be measured by commonly used methods.

Swelling occurred in an approximately 0.2-mm region near the surface during the first 1 min. During this period, the moisture content remained at the same level as that of the dried spaghetti in the inner region. The change in the distribution between 0 and 1 min indicated that water quickly penetrated only near the surface. Because many small holes and cracks are present on the surface of the pasta [17, 23], water likely entered the spaghetti through these small surface holes and cracks. A flat moisture distribution near the surface suggested that water imbibition during the early stage was not attributable to water diffusion but rather to water filling the holes and cracks.

Gelatinization was reported to be complete within 1 min of assessment of the extent of gelatinization by observing the birefringence in the starch/water system during heat treatment [67], indicating that starch gelatinization is a fast process. However, the moisture content on the surface gradually increased over time. As

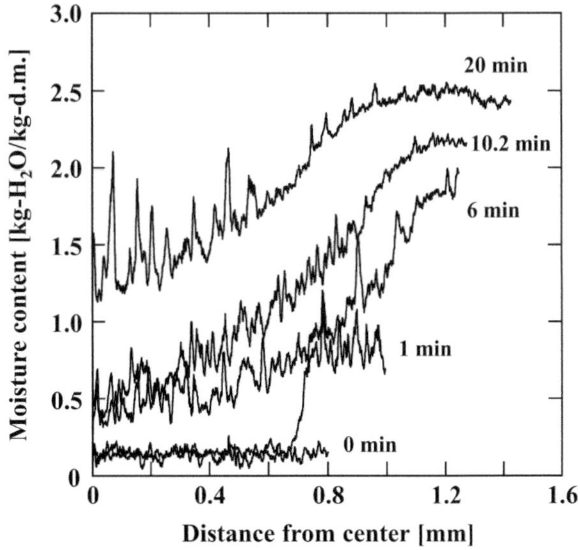

Fig. 5.7 Moisture distributions of spaghetti rehydrated for 0, 1, 6, 10.2, and 20 min

shown in Sect. 5.2.1, the average moisture content of spaghetti increased up to approximately 9 kg-H_2O/kg-d.m. Therefore, the gelatinization of starch granules in the sample did not reach equilibrium even after 20 min rehydration, and swelling of the granules was restricted. Although dried pasta showed a homogeneous internal structure in which the starch granules were deeply embedded in a protein matrix, the structure below the surface appeared to be a honeycomb-like structure after a 4-min rehydration [17, 23]. The protein matrix, which had shrunk during drying, relaxed in structure during rehydration, and gradual swelling of the starch granules occurred because of gelatinization. Structural relaxation of the protein matrix plays an important role in moisture content changes during water migration.

The moisture distributions observed for the samples rehydrated for 10.2 min or longer showed a flat moisture distribution near the surface and parabolic distribution of the moisture content in the inner region. The water migration in the inner region is thought to be diffusion-controlled, while the diffusion of water was not the rate-limiting step for rehydration near the surface. Because the moisture content near the surface of the sample rehydrated for 20 min was much lower than the equilibrium content, the starch granules in the sample may sorb more water. Therefore, most of the water supplied from the bulk phase would be sorbed by the starch granules to increase the moisture content near the surface. In the inner region, the gradient in the moisture content operates diffusion of water and the distribution gradually shifted to the higher level. However, the consumption of water by starch granules sorbed near the surface restricted the penetration of water into the inner region. Water migration near the surface may bring about the expansion of the region with a flat moisture distribution.

The high accuracy of using our method for moisture content determination in spaghetti was confirmed by comparing the experimentally observed average moisture content of the sample with that calculated from the moisture distribution within the sample.

5.2.4 Rehydration of Infinitely Thin Pasta

As shown above, the moisture content on the surface of pasta gradually increased rather than quickly reaching equilibrium.

The non-Fickian diffusion of a low-molecular-weight molecule into a matrix of high-molecular-weight molecules is attributed to the slow reconfiguration of high-molecular-weight molecule segments after accepting the penetrant [63], which has been observed in many high-molecular-weight polymer-penetrant systems [13]. Therefore, the diffusion of low-molecular-weight molecules (water) and the relaxation of high-molecular-weight molecules (components of wheat flour) play important roles in water migration within pasta. Nevertheless, previous studies examining the water migration of dried food have extensively focused on water diffusion, but little attention has been given to the relaxation of high-molecular-weight molecules. We kinetically examined the rehydration process of pasta with an infinitely small diameter, where the effects of water diffusion on rehydration were negligible, in order to elucidate the role of the relaxation of high-molecular-weight molecules in rehydration.

To estimate the moisture content of infinitely thin durum pasta, designated as 0-mm durum pasta, the average moisture content at a specific time was plotted against pasta diameters and extrapolated to 0 mm with a straight line as shown in the inset of Fig. 5.8 [53]. The moisture content of 0-mm durum pasta was estimated from the y-intercept of the line, and the normalized moisture content for 0-mm durum pasta, defined by the left-hand side of Eq. 5.8, was plotted versus the rehydration time in Fig. 5.8.

Gluten was isolated from durum wheat flour [28] to prepare gluten pasta. The gluten isolated from the durum wheat flour was freeze-dried. Water was then added to the freeze-dried gluten. The moistened gluten was placed in a rolling noodle machine and cut into pieces 80 mm in length, 3 mm in width, and 4 mm in depth to produce raw gluten pasta. The gluten pasta was placed in a drying chamber to prepare the dried pasta. The rehydration process of the gluten pasta was also measured under the same conditions as that of the durum pasta. The transient change in the moisture content of gluten pasta is shown in Fig. 5.8. The transient changes in the moisture contents of both gluten pasta and 0-mm durum pasta were exponential and were expressed using Eq. 5.11 [40].

$$\frac{X_t - X_0}{X_e - X_0} = 1 - \exp(-kt) \tag{5.11}$$

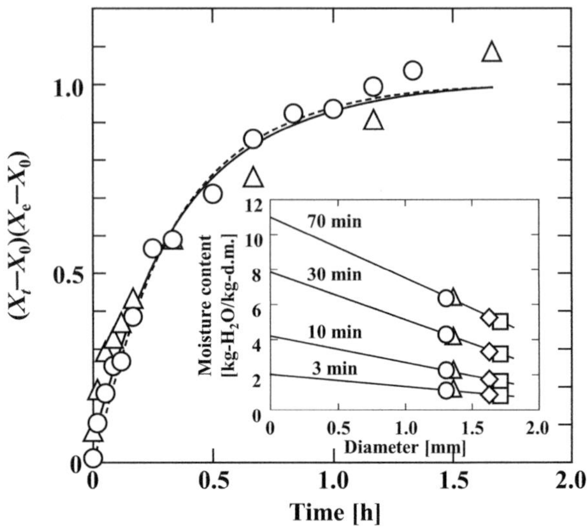

Fig. 5.8 Changes in the normalized moisture contents for 0-mm durum pasta (—○—) and gluten pasta (- -△- -). The curves represent the moisture contents calculated using Eq. 5.11. *Inset*: estimation of moisture content of infinitely thin durum pasta by extrapolation at rehydration times of 3 min, 10 min, 30 min, and 70 min. The initial diameters were 1.30 mm (○), 1.35 mm (△), 1.63 mm (◇), and 1.71 mm (□), respectively

where k is the rate constant. Although the X_e value of 0-mm durum pasta was much larger than that of the gluten pasta, the k value of 0-mm durum pasta (7.53 h $\times$ 10^{-4} 1/s) was nearly the same as that of gluten pasta (7.42×10^{-4} 1/s). Thus, the rehydration rates were the same for 0-mm durum pasta and gluten pasta even though the rehydration capacity of 0-mm durum pasta differed from that of gluten pasta.

Starch granules are embedded in the structural network of gluten in pasta. The similarity in the k value for 0-mm durum pasta and gluten pasta indicates that the rehydration kinetics at the pasta surface is governed by the gluten network. The starch granules embedded in the network swell via gelatinization immediately after water intake from their surroundings because of the high gelatinization rate of starch. However, the gluten network prevents the swelling of starch granules. Therefore, the moisture content near the surface gradually increased with gradual relaxation of the gluten network.

5.3 Properties of Spaghetti Prepared Under Various Conditions

5.3.1 Rehydration Characteristics of Spaghetti Prepared Using Various Dies

Pasta has been traditionally prepared using die made of bronze. However, dies made of Teflon have recently been increasingly used because of the longer lifetime of the die, smooth surface of pasta, and good appearance of dried pasta [19, 20, 26]. The surface roughness of pasta depends on the die material used to prepare the pasta, and the die material used appears to affect the properties of pasta. Therefore, the effects of the die material on the properties of raw pasta and rehydration behaviors of dried pasta were examined [70].

Spaghetti samples with varying surface roughness were prepared using aluminum, polypropylene, and polycarbonate dies as well as Teflon and bronze dies. The orifice diameter and length of each die were 1.8 mm and 5 mm, respectively. The raw spaghetti was dried in a temperature-humidity controllable chamber to produce the dried product with a moisture content of approximately 0.12 kg-H_2O/kg-d.m.

The surface of the spaghetti was observed using a digital microscope. The spaghetti was cut into 8-cm-long sticks, and the rehydration process of the spaghetti was measured in 0.5 % (w/v) NaCl at 99.7 ± 0.3 °C. The initial diameter and length of 50 spaghetti samples were measured at several locations on each stick by using a vernier caliper, and the surface area of the stick, S_s, was calculated from the mean diameter and length assuming a smooth and columnar shape. The amount of water rehydrated per unit surface area, x_t, was calculated using Eq. 5.12.

$$x_t = (W_2 - W_1)/S_s \tag{5.12}$$

where W_1 and W_2 are the sample weight before and after rehydration, respectively.

Figure 5.9 illustrates the surface images of the dried spaghetti prepared using the various dies at 200- and 1000-fold magnifications [70]. Based on the 200-fold magnification images, the surfaces of the spaghetti prepared using the Teflon, polypropylene, and polycarbonate dies were smooth, while those of the spaghetti prepared using the aluminum and bronze dies were rough. Observations at 1000-fold magnification revealed that the surface of the spaghetti prepared using the Teflon die was smooth. The spaghetti prepared using the Teflon die had the smoothest surface, while those prepared using the polypropylene, polycarbonate, aluminum, and bronze dies followed in this order.

Figure 5.10 shows the rehydration curves for spaghetti prepared using the various dies [70]. Because the early stage of rehydration could not be well-expressed by Eq. 5.3, only the a value was estimated using Eq. 5.3. Since the initial diameter (surface are) affects rehydration velocity, rehydration during the early stage within 60 s was characterized based on the x_t value defined by Eq. 5.12. Transient changes in the x_t value with the rehydration time t are shown in the inset of Fig. 5.10. The x_t value was correlated with the t by a quadratic equation, and the momentarily rehydrated amount of water when the sample was immersed in the boiling water, x_0, was estimated by extrapolating the equation to $t = 0$.

Geometry measurements of the spaghetti surface were carried out using a laser microscope [48]. The surface roughness and surface area were calculated from the height in the depth direction. Surface roughness was expressed by the average

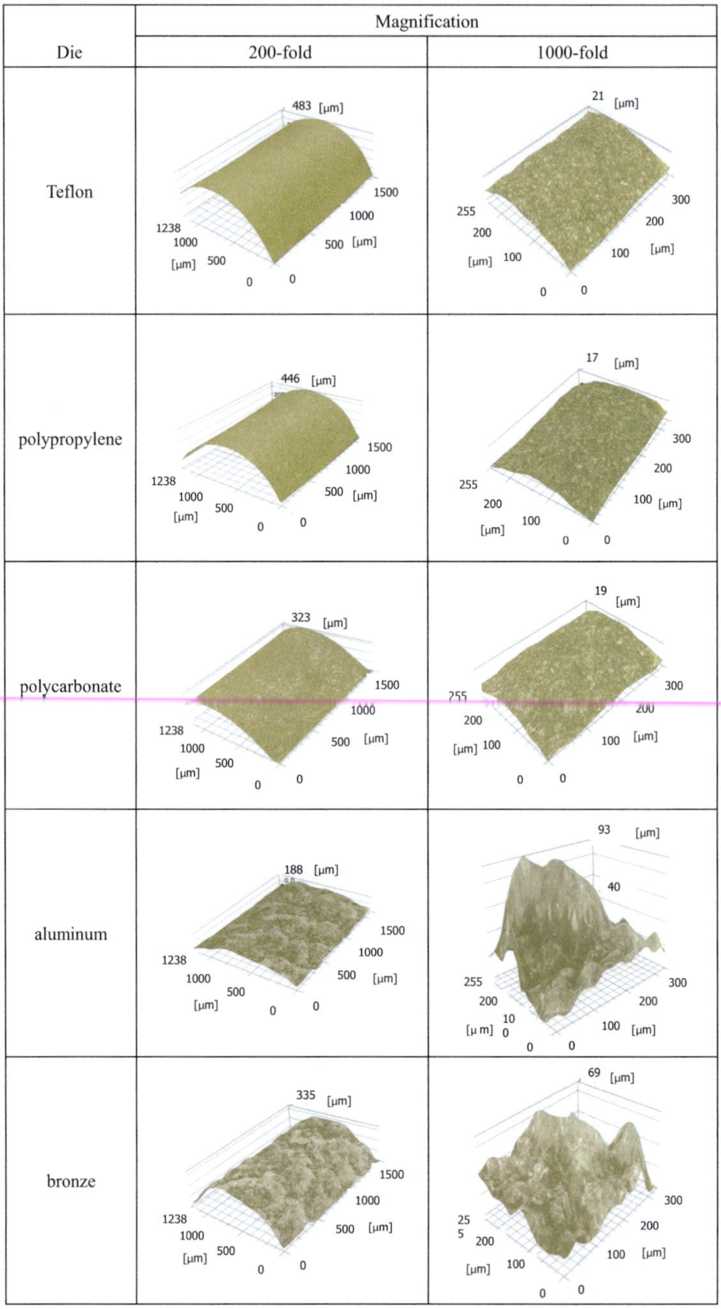

Fig. 5.9 Surface images at 200- and 1000-fold magnifications of spaghetti prepared using the dies made of Teflon, polypropylene, polycarbonate, aluminum, and bronze. Images were obtained using a digital microscope

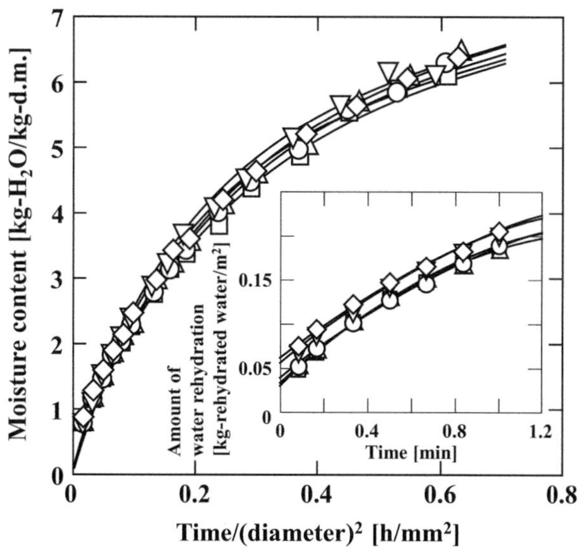

Fig. 5.10 Rehydration of spaghetti prepared using the dies made of Teflon (□), polypropylene (△), polycarbonate (○), aluminum (▽), and bronze (◇). *Inset*: rehydration during early stage of rehydration of spaghetti

roughness, R_a, determined using Eq. 5.13.

$$R_a = \frac{1}{l}\int_0^l f(x)dx \tag{5.13}$$

where l is the reference length and $f(x)$ is the roughness curve. Figure 5.11 shows the relationship between x_0 and R_a for spaghetti prepared using different dies [48]. The x_0 value for the spaghetti with a rougher surface was larger. This indicates that surface roughness play an important role in rehydration immediately after placing spaghetti into the rehydration solution.

5.3.2 Rehydration Process and Texture of Spaghetti Dried Under Different Conditions

Dried spaghetti is produced under various temperature-humidity controlled conditions. Spaghetti has traditionally been dried under low-temperature conditions, while drying under high- and very-high-temperature conditions has been adopted in industrial production of dried spaghetti. The effects of drying conditions on the rehydration kinetics of dried spaghetti, moisture distribution during rehydration, and texture of rehydrated spaghetti were examined.

Three types of spaghetti processed under different programmed drying conditions and supplied by Nisshin Foods Co., Ltd. (Tokyo, Japan) were used. The maximum temperature and duration for the production under low-, high-, and very-high-temperature conditions were 50 °C and 20 h, 70 °C and 11 h, and 85 °C

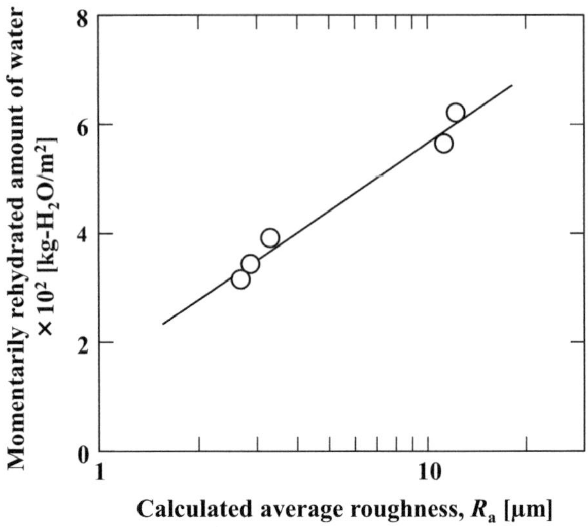

Fig. 5.11 Dependence of momentarily rehydrated amounts of water per unit area on calculated average roughness, R_a

and 6 h, respectively. The spaghetti samples were designated as LT, HT, and VHT spaghetti, respectively. The initial diameters of the spaghetti were approximately 1.6 mm.

The rehydration processes of the LT, HT, and VHT spaghetti were measured at temperatures from 20 °C to 90 °C at 10 °C intervals and at 97 °C [3]. The rehydration process of any spaghetti at various rehydration temperatures was well characterized by Eq. 5.3, and a and b values were evaluated. For any spaghetti, the plots of a values versus the reciprocal of the rehydration absolute temperature were separated into low-temperature, transition, and high-temperature regions. The temperature dependence of the equilibrium moisture content was analyzed separately in the low- and high-temperature regions according to the van't Hoff equation (Eq. 5.6) to evaluate the change in enthalpy of rehydration, ΔH. ΔH values in the low-temperature region were 4.12, 1.66, and 3.09 kJ/mol for LT, HT, and VHT spaghetti, respectively, and those in the high-temperature region were 20.5, 20.9, and 27.1 kJ/mol for LT, HT, and VHT spaghetti, respectively. The ΔH value of VHT spaghetti was larger than that of LT and HT spaghetti. The temperature dependence of the v_0 value was analyzed using the Arrhenius equation (Eq. 5.7) to estimate the activation energy, E_a. The E_a values for LT, HT, and VHT spaghetti were 29.8, 32.2, and 27.9 kJ/mol, respectively. There were no significant differences in E_a values among LT, HT, and VHT spaghetti, suggesting that the maximum temperature during drying has little effect on the initial water intake of spaghetti.

The moisture distribution of spaghetti prepared under three different conditions was measured to investigate the effects of moisture distribution in rehydrated spaghetti on the texture [50]. Figure 5.12 shows the moisture distributions of the

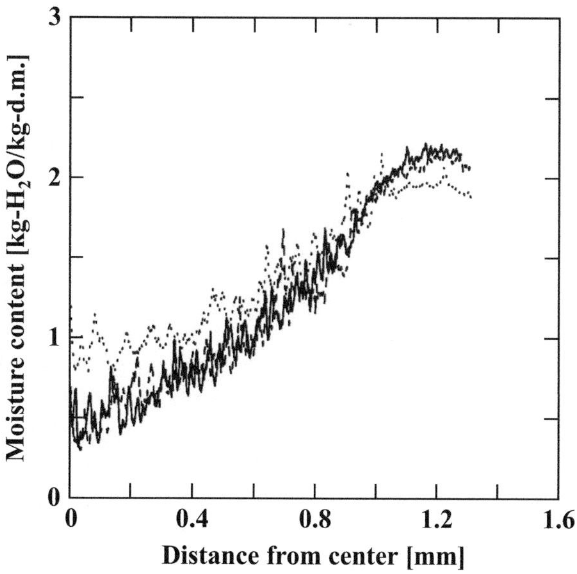

Fig. 5.12 Moisture distributions of LT spaghetti (•••), HT spaghetti (--), and VHT spaghetti (—)

three spaghetti samples cooked to the al dente state with an average moisture content of 1.71 kg-H_2O/kg-d.m. The moisture distribution in the VHT spaghetti was similar to that of HT spaghetti, but the distribution in LT spaghetti showed different characteristics from those of HT and VHT spaghetti. The moisture content of LT spaghetti was lower near the surface and higher at the center than those of HT and VHT spaghetti. We predict that the water migration mechanism was the same for spaghetti dried under different conditions, and that diffusion of water was the predominant mechanism for moisture migration near the center, but relaxation of the gluten network controlled moisture migration near the surface. Because amylase is slowly inactivated during drying at low temperatures, enzymatic damage of starch lowers the starch gelatinization temperature [71]. Starch with a low-gelatinization temperature easily sorbs more water, and the apparent diffusivity of water in the spaghetti is higher. Easy migration of water into the spaghetti likely produces a relatively flat moisture distribution. Because starch granules are embedded in a gluten network in pasta and starch gelatinization is a rapid process, the network controls the swelling of the granules by gelatinization. No formation of a sufficiently strong network because of weak thermal denaturation of gluten [73] suggests that starch granules embedded in the gluten network swell easily in LT spaghetti.

The breaking force and strain of dry spaghetti were measured using a creep meter equipped with a wedge-shaped plunger and a 20-N load cell at 0.5 mm/s. Spaghetti cut into 5-cm lengths was rehydrated in 0.5 % (w/v) NaCl at 100 °C to a moisture content of 1.70 kg-H_2O/kg-d.m. After removing the excess water on the sample surface by blotting with tissue wipers, the texture of the sample was measured using the creep meter.

The breaking forces and strains of the dry LT, HT, and VHT spaghetti were 4.07, 4.68, and 5.05 N and 24.7 %, 24.0 %, and 26.6 %, respectively [52]. The breaking force of the VHT spaghetti was the highest, although there was no significant difference in the breaking strain among the differentially dried spaghetti. The measurement of gluten leakage into an acetic acid solution showed that drying at high temperatures increased the extent of protein denaturation [4]. Wheat protein denaturation was promoted by the cross-linkage of glutenin and gliadin via a disulfide bond [60, 68]. This suggests that the formation of additional disulfide bonds at higher temperatures strengthens the protein network, resulting in increased structural stiffness. The difference in the breaking force of spaghetti dried at different temperatures can be explained by the changes in structure stiffness.

Figure 5.13 shows typical examples of texture-profile analysis curves for LT, HT, and VHT spaghetti. H_1 and H_2 indicate hardness and stickiness, respectively [52]. Cohesiveness was calculated as the ratio of A_2/A_1, and A_3 indicates adhesiveness. The hardness, cohesiveness, adhesiveness, and stickiness of LT, HT, and VHT spaghetti rehydrated to the al dente state are illustrated in Fig. 5.14. This figure also shows the breaking force and strain. The hardness of HT spaghetti was greater than that of LT and VHT spaghetti. LT spaghetti was the stickiest among the three samples, although there were no significant differences in cohesiveness and stickiness among the samples. Furthermore, the adhesiveness of LT spaghetti was significantly higher than that of HT and VHT spaghetti. Previously studies also reported that the drying of pasta at high temperatures decreased both stickiness and adhesiveness [4, 16, 24, 73]. The formation of large and insoluble protein aggregates caused an increase in firmness and a decrease in stickiness [37]. The strength of the gluten network at the center of the pasta governing the firmness and the stickiness was ascribed to the structural properties of the pasta surface [73]. Additionally, stickiness of the pasta surface was related to the extensive starch swelling and amylose leaching onto the surface. Although the breaking force of

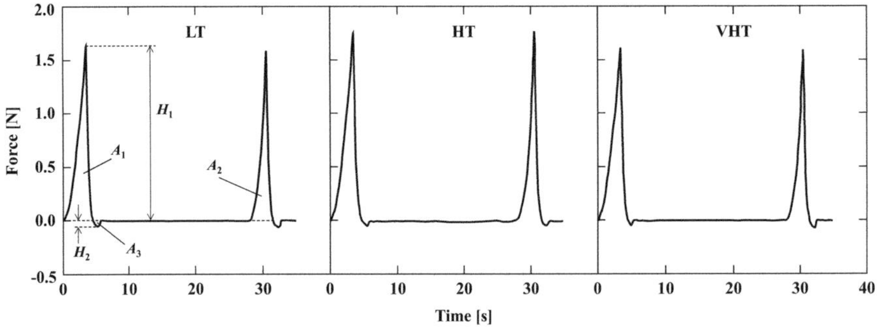

Fig. 5.13 Representative example of typical texture-profile analysis curve for LT, HT, and VHT spaghetti

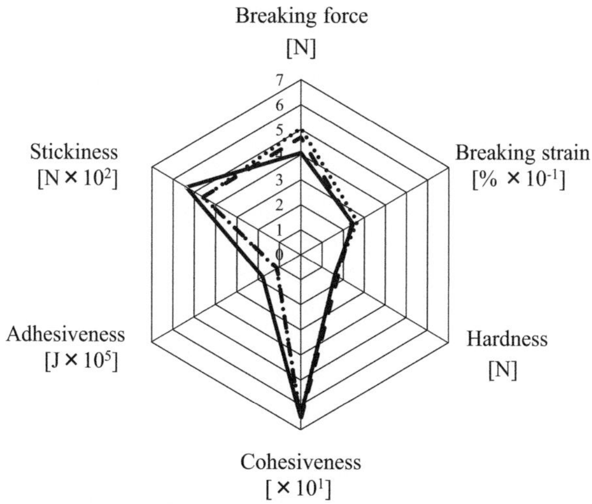

Fig. 5.14 Properties of differentially dried spaghetti: LT (—), HT (– –), and VHT (• • •)

the spaghetti dried at higher temperatures was greater, the drying temperature did not affect the hardness of rehydrated spaghetti. Therefore, rehydration may offset the effect of drying temperature on spaghetti firmness. The amount of amylose leaked during rehydration from pasta dried at higher temperature was lower than that from pasta dried at the low temperature [7]. Based on these findings, the stickiness and adhesiveness of the rehydrated LT, HT, and VHT spaghetti is governed by the leakage of amylose onto the surface, although our results did not show significant differences in stickiness among the differentially dried spaghetti samples. The programmed drying conditions, in which both temperature and humidity were controlled, showed limited effects on hardness and cohesiveness.

5.4 Effects of Rehydration Conditions on the Properties of Rehydrated Spaghetti

5.4.1 Effects of Rehydration Temperature

The palatability of food is affected by its flavor, taste, and texture. Texture plays a particularly important role in the palatability of noodles. In order to identify the major factors governing the texture of rehydrated spaghetti, the moisture distributions and stress-strain curves of spaghetti rehydrated to its optimal cooking state, known as al dente, at 60 °C, 80 °C, and 100 °C, in distilled water or 0.1, 1.0, and 2.0 mol/L NaCl solutions were examined [51]. The texture of al dente pasta is characterized by a soft external zone and a firm core, with the firm core providing a mouthfeel of resistance while biting. The relationship between the moisture distribution of pasta, as measured using MRI, and its mechanical properties, as measured by a texture analyzer, was examined. The factor affecting the texture of rehydrated pasta was previously reported to be moisture distribution [35]. Although average moisture content is lower at lower rehydration temperatures [46] and higher NaCl concentrations [47], the effects of rehydration temperature and NaCl concentration on the moisture distribution and texture of rehydrated pasta remain unclear. The moisture distribution and stress-strain curve of spaghetti (initial diameter of approximately 1.6 mm) rehydrated to the al dente state (average moisture content of 1.70 ± 0.05 kg-H_2O/kg-d.m.) at various temperatures and NaCl concentrations were examined to determine the major factors governing the texture of rehydrated pasta. The moisture distribution was measured using a digital camera with our method, and the stress-strain curve was measured using a texture analyzer.

Figure 5.15 shows the moisture distribution in spaghetti rehydrated at 60 °C, 80 °C, and 100 °C to the al dente state [51]. The spaghetti rehydrated at 80 °C and 100 °C had lower moisture contents near the center than in the peripheral region, with the moisture distribution exhibiting a concave shape. The moisture content of the spaghetti rehydrated at 60 °C exhibited a nearly flat distribution across the entire cross-section, although moisture content decreased slightly around the peripheral region, exhibiting a weakly convex shape. The rehydration temperature determines the gelatinization state of starch, and the spaghetti material leaks into the immersion solution during rehydration, resulting in a loss of mass. Loss of spaghetti mass increases exponentially with time [46]. Rehydration at 60 °C proceeded more slowly to reach the al dente state than rehydration at other temperatures. Therefore,

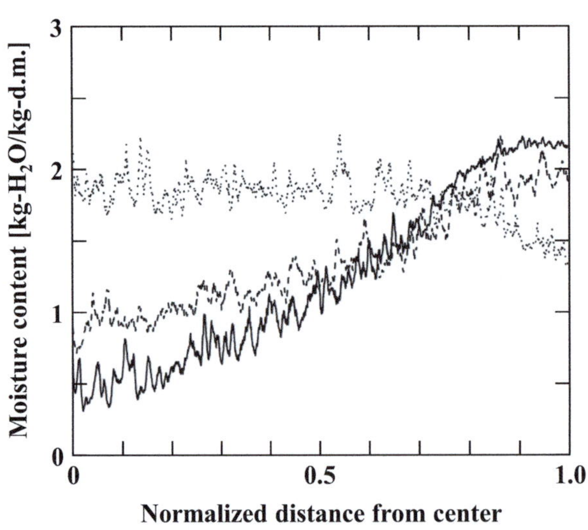

Fig. 5.15 Moisture distributions of spaghetti rehydrated at 60 °C (• • •), 80 °C (– –), and 100 °C (—) in distilled water

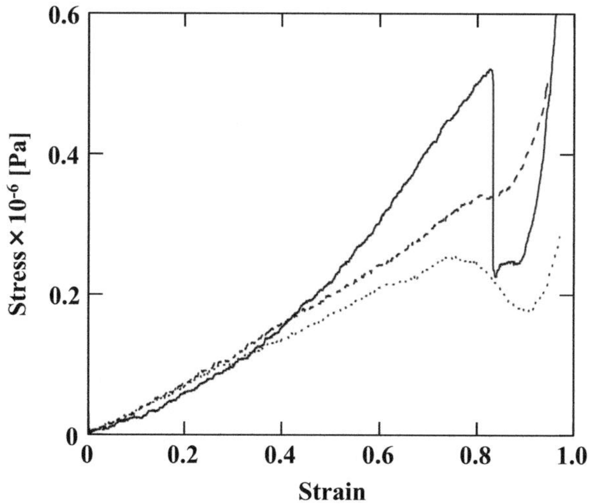

Fig. 5.16 Representative examples of stress-strain curves for spaghetti rehydrated at 60 °C (• • •), 80 °C (– –), and 100 °C (—) in distilled water

the gelatinization state of the starch and/or increased loss of spaghetti mass during rehydration at 60 °C resulted in a slightly convex or nearly flat moisture distribution. The difference in moisture content between the inner and peripheral regions was greater in spaghetti rehydrated at higher temperatures.

Figure 5.16 shows the stress-strain curves of spaghetti rehydrated at 60 °C, 80 °C, and 100 °C. For the spaghetti rehydrated at higher temperatures, stress was smaller and larger under lower and greater strain, respectively. The breaking stress was affected by rehydration temperature, and spaghetti rehydrated at higher temperatures exhibited higher breaking stress. However, breaking strain was not affected by rehydration temperature. The stress-strain curves generated for low and greater strain reflect the mechanical properties of the spaghetti surface and of the material near the spaghetti's core, respectively. The characteristics of the curve for the spaghetti rehydrated at higher temperatures was ascribed to higher and lower moisture contents near the surface and at the center, respectively. Because of the difference in moisture content, the spaghetti was compressed by lower and greater stresses under lower and greater strains, respectively. The similarity in the breaking strain of all spaghetti samples with different moisture distributions indicates that moisture distribution does not affect breaking strain but rather controls breaking stress.

5.4.2 Effects of NaCl Concentration

Figure 5.17 shows the moisture distributions of spaghetti rehydrated at 100 °C in 0, 0.1, 1.0, and 2.0 mol/L NaCl solutions to the al dente state [49, 51]. No significant effect of salt concentration on the moisture distribution of al dente spaghetti was observed, although the rehydration temperature had a much greater effect as described above. Water migration in the spaghetti is determined by starch gelatinization, water diffusion, and gluten matrix relaxation [49, 53], all of which are affected by rehydration temperature. Salt concentration affects the gelatinization temperature but does not directly affect the diffusion coefficient of the water. The difference in the moisture distribution of spaghetti rehydrated at different salt concentrations can be ascribed to the difference in the sensitivity of the salt concentration to starch gelatinization and diffusion of water.

Figure 5.18 shows the stress-strain curves of spaghetti rehydrated at 100 °C in the various NaCl solutions. The slope of the curve in the region of small strains was larger at higher salt concentrations, indicating that spaghetti rehydrated at higher salt concentrations is more elastic. The breaking stress was only minimally affected by salt concentration, while breaking strain was lower in spaghetti rehydrated at higher salt concentrations. The low strain may reflect a reduction in a property equivalent to the stickiness of the spaghetti. Because all spaghetti samples rehydrated in the salt solutions showed nearly the same moisture distributions, no significant difference in breaking stress was observed among the spaghetti samples. A gluten framework of spaghetti embeds starch granules [17, 23], and salt changes the

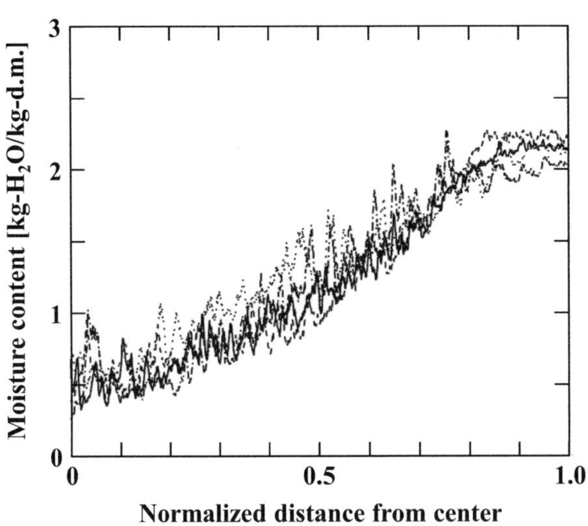

Fig. 5.17 Moisture distributions of spaghetti rehydrated at 100 °C in distilled water (—), 0.1 mol/L NaCl solution (– –), 1.0 mol/L NaCl solution (• • •), and 2.0 mol/L NaCl solution (– •)

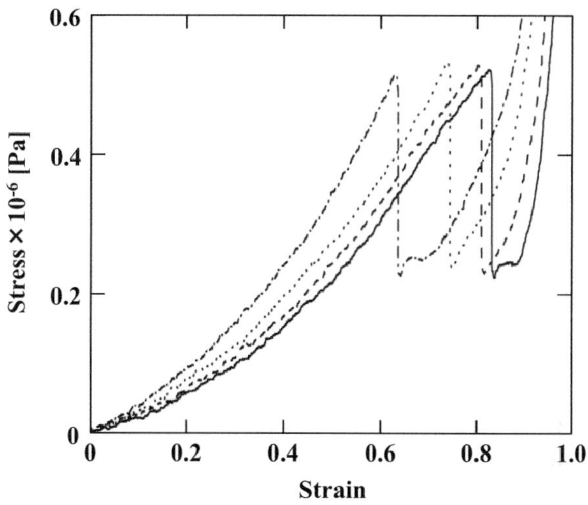

Fig. 5.18 Representative examples of stress-strain curves for spaghetti rehydrated at 100 °C in distilled water (—), 0.1 mol/L NaCl solution (– –), 1.0 mol/L NaCl solution (• • •), and 2.0 mol/L NaCl solution (– •)

molecular conformation of the gluten to a fibrous network during dough preparation [64]. Both the interactions and distances between the glutenin and gliadin in gluten were altered by the addition of NaCl during dough preparation, with the distances being shortened [65]. The reduction in gluten volume was caused by the addition of NaCl [38]. Although the effects of NaCl were observed during dough preparation, a similar effect may be observed during rehydration in the presence of NaCl. Rehydration at higher salt concentrations may contract the gluten network to lower the breaking strain.

In conclusion, the moisture distribution within spaghetti and its material properties govern its compressive stress and strain, respectively.

5.5 Conclusions

The rehydration curve of dried pasta can be empirically expressed using a hyperbolic equation for any type of pasta under any rehydration conditions, and two parameters reflecting the equilibrium moisture content and initial rate of rehydration can be estimated. The temperature dependence of the initial rate of rehydration obeyed the Arrhenius equation over the entire range of tested temperatures. This indicates that there was no significant effect of starch gelatinization on the rehydration rate and that the rate was affected by the surface structure of pasta because the rate for pasta with a rougher surface was greater. The temperature dependence of equilibrium moisture content was divided into the low- and high-temperature regions near the gelatinization temperature of starch. The dependence in the two regions could be individually expressed by the van't Hoff equation, and the change in the enthalpy of rehydration was much greater in the high-temperature region than in the low-temperature region. The difference in the enthalpy change indicated that the equilibrium moisture content was governed by starch gelatinization. The equilibrium moisture content was decreased by dissolving NaCl at different concentrations or with different salts in the immersion solution because of the lowering of the gelatinization temperature by the presence of salt.

A simple but precise method for measuring the moisture distribution was developed in order to understand the water migration mechanism in pasta. The method was based on the fact that the color brightness of pasta increases with increasing moisture content. The moisture distribution was measured using a digital camera, which is a more common and cheaper instrument than instruments used in current methods such as MRI at a higher spatial resolution than MRI. The moisture distribution observed for pasta using this method was quite different from that calculated by the diffusion model for a cylindrical material. The transient changes in the distribution during rehydration suggested that small holes and cracks near the pasta surface were quickly filled with water, the region near the surface gradually expanded, water migration occurred via diffusion in the inner region, and gelatinization of the starch granules restricted water diffusion.

Pasta with different surface roughness properties was prepared using dies made from various materials. The die material affected surface roughness and bulk density but did not affect gelatinization temperature. Pasta prepared using a bronze die has a rougher surface, and its momentarily sorbed amount of water is larger than that for pasta prepared using the Teflon die, which is widely used in the industrial production of pasta. However, no significant difference in the equilibrium moisture content was observed among the pasta prepared using five different dies. However, drying conditions affected the rehydrated pasta. Pasta dried at higher temperatures had a rougher surface and larger breaking stress. The difference between the moisture content near the surface and that near the center was larger for pasta rehydrated at higher temperatures to the al dente state. The pasta rehydrated at higher temperature also showed larger compressive stress in the low moisture range. The moisture distribution of pasta rehydrated in the solution containing NaCl showed a low dependence on salt concentration, but an increase in salt concentration significantly lowered the breaking strain. These observations suggested that the moisture distribution and material properties independently affected the breaking stress and strain, which governs the mouthfeel of resistance while biting.

Acknowledgments This study was carried out during the project study of The Cereal Science Consortium by the Graduate School of Agriculture, Kyoto University and the Nisshin Seifun Group, Inc. This study was also supported by a grant from the Japan Society for the Promotion of Science for a research fellow (T.O.).

References

1. Acquistucci, R. (2000) Influence of Maillard reaction on protein modification and colour development in pasta. Comparison of different drying conditions. *LWT-Food Sci. Technol.*, **33**, 48–52.
2. Ahmad, F.B.F.A. (1999) Effect of salts on the gelatinization and rheological properties of sago starch. *J. Agric. Food Chem.*, **47**, 3359–3366.
3. Aimoto, U., Ogawa, T., and Adachi, S. (2013) Water sorption kinetics of spaghetti prepared under different drying conditions. *Food Sci. Technol. Res.*, **19**, 17–22.
4. Aktan, B. and Khan K. (1992) Effect of high-temperature drying of pasta on quality parameters and on solubility, gel electrophoresis, and reversed-phase high-performance liquid chromatography of protein components. *Cereal Chem.*, **69**, 288–295.
5. Altan, A., Oztop, M.H., McCarthy, K.L., and McCarthy, M.J. (2011) Monitoring changes in feta cheese during brining by magnetic resonance imaging and NMR relaxometry. *J. Food Eng.*, **107**, 200–207.
6. Anese, M., Nicoli, M.C., Massini, R., and Lerici C.R. (1999) Effects of drying processing on the Maillard reaction in pasta. *Food Res. Int.*, **32**, 193–199.
7. Baiano, A., Conte, A., and Nobile, M.A.D. (2006) Influence of drying temperature on the spaghetti cooking quality. *J. Food Eng.*, **76**, 341–347.
8. Becker, H.A. (1960) On the absorption of liquid water by the wheat kernel. *Cereal Chem.*, **37**, 309–323.
9. Bilbao-Sáinz, C., Andrés, A., and Fito, P. (2005) Hydration kinetics of dried apple as affected by drying conditions. *J. Food Eng.*, **68**, 369–376.
10. Chhinnan, M.S. (1984) Evaluation of selected mathematical models for describing thin-layer drying of in-shell pecans. *Trans. Am. Soc. Agric. Biol. Eng.*, **27**, 610–615.
11. Chiotelli, E., Pilosio, G., and Meste, M.L. (2002) Effect of sodium chloride on the gelatinization of starch: a multimeasurement study. *Biopolymers*, **63**, 41–58.
12. Crank, J. (1975) The mathematics of diffusion. 1st ed., Clarendon Press, Oxford, UK.
13. Crank, J. and Park, G.S. (1951) Diffusion in high polymers: some anomalies and their significance. *Trans. Faraday Soc.*, **47**, 1072–1084.
14. Cubadda, R.E., Carcea, M., Marconi, E., and Trivisonno, M.C. (2007) Influence of gluten proteins and drying temperature on the cooking quality of durum wheat pasta. *Cereal Chem.*, **84**, 48–55.
15. Cunha, L.M., Oliveira, F.A.R., and Oliveira, J.C. (1998) Optimal experimental design for estimating the kinetic parameters of processes described by the Weibull probability distribution function. *J. Food Eng.*, **37**, 175–191.
16. Cunin, C. (1995) Investigations on starch and starch-emulsifier interactions in durum wheat pasta. PhD dissertation 11389. Swiss Federal Institute of Technology (ETH), Zurich, CH.
17. Cunina, C., Handschina, S., Waltherb, P., and Eschera, F. (1995) Structural changes of starch during cooking of durum wheat pasta. *LWT-Food Sci. Technol.*, **28**, 323–328.
18. Cunningham, S.E., McMinn, W.A.M., Magee, T.R.A., and Richardson, P.S. (2007) Modelling water absorption of pasta during soaking. *J. Food Eng.*, **82**, 600–607.
19. Dalbon, G., Grivon, D., and Pagani, M.A. (1996) Continuous manufacturing process. In "Pasta and noodles technology" ed. by Kruger, J.E., Matsu, R.B., and Dick, J.W. Am. Assoc. Cereal Chem., MN, USA.
20. Dawa, P.R. (2001) Pasta shape design. In "Pasta and semolina technology" ed. by Kill, R.C. and Turnbull, K. Blackwell Science Ltd., Oxford, UK.
21. De Temmerman, J., Verboven, P., Nicolaï, B., and Ramon, H. (2007) Modelling of transient moisture concentration of semolina pasta during air drying. *J. Food Eng.*, **80**, 892–903.
22. Del Nobile, M.A., Buonocore, G.G., Panizza, A., and Gambacorta, G. (2003) Modeling the spaghetti hydration kinetics during cooking and overcooking. *J. Food Sci.*, **68**, 1316–1323.
23. Dexter, J.E., Dronzek, B.L., and Matsuo, R.R. (1978) Scanning electron microscopy of cooked spaghetti. *Cereal Chem.*, **55**, 23–30.

24. Dexter, J.E., Matsuo, R.R., and Morgan, B.C. (1981) High temperature drying: effect on spaghetti properties. *J. Food Sci.*, **46**, 1741–1746.
25. Djomdi, E.R. and Ndjouenkeu, R. (2007) Soaking behaviour and milky extraction performance of tiger nut (*Cyperus esculentus*) tubers. *J. Food Eng.*, **78**, 546–550.
26. Donnelly, B.J. (1982) Teflon and non-Teflon lined dies: effect on spaghetti quality. *J. Food Sci.*, **47**, 1055–1058.
27. Feillet, P. and Dexter, J.E. (1996) Quality requirements of durum wheat for semolina milling and pasta production. In "Pasta and noodle technology" ed. by Kruger, J.E., Matsuo, R.R., and Dick. J.W. AACC Int., MN, USA.
28. Fuwa, H., Komaki, T., Hidukuri, S., and Kainuma, K. (2003) Handbook of starch science (in Japanese; Denpun Kagaku no Jiten). 1st ed., Asakura Shoten, Tokyo, Japan.
29. García-Pascual, P., Sanjuán, N., Bon, J., Carreres, J.E., and Mulet, A. (2005) Rehydration process of *Boletus edulis* mushroom: characteristics and modelling. *J. Sci. Food Agric.*, **85**, 1397–1404.
30. García-Pascual, P., Sanjuán, N., Melis, R., and Mulet, A. (2006) *Morchella esculenta* (morel) rehydration process modelling. *J. Food Eng.*, **72**, 346–353.
31. Güler, S., Köksel, H., and Ng, P.K.W. (2002) Effects of industrial pasta drying temperatures on starch properties and pasta quality. *Food Res. Int.*, **35**, 421-427.
32. Hills, B.P., Babonneau, F., Quantin, V.M., Gaudet, F., and Belton, P.S. (1996) Radial NMR microimaging studies of the rehydration of extruded pasta. *J. Food Eng.*, **27**, 71–86.
33. Hills, B.P., Godward, J., and Wright, K.M. (1997) Fast radial NMR microimaging studies of pasta drying. *J. Food Eng.*, **33**, 321–335.
34. Horigane, A.K., Takahashi, H., Maruyama, S., Ohtsubo, K., and Yoshida, M. (2006) Water penetration into rice grains during soaking observed by gradient echo magnetic resonance imaging. *J. Cereal Sci.*, **44**, 307–316.
35. Irie, K., Horigane, A.K., Naito, S., Motoi, H., and Yoshida, M. (2004) Moisture distribution and texture of various types of cooked spaghetti. *Cereal Chem.*, **81**, 350–355.
36. Jay-Lin, J. and Ames, I.A. (1993) Mechanism of starch gelatinization in neutral salt solutions. *Starch/Stärke*, **45**, 161–166.
37. Lamacchia, C., Di Luccia, A., Baiano, A., Gambacorta, G., la Gatta, B., Pati, S., and La Notte, E. (2007) Changes in pasta proteins induced by drying cycles and their relationship to cooking behaviour. *J. Cereal Sci.*, **46**, 58–63.
38. Larsson, H. (2002) Effect of pH and sodium chloride on wheat flour dough properties: Ultracentrifugation and rheological measurements. *Cereal Chem.*, **79**, 544–545.
39. Lee, K.T., Farid, M., and Nguang, S.K. (2006) The mathematical modelling of the rehydration characteristics of fruits. *J. Food Eng.*, **72**, 16–23.
40. Long, R.A. and Richman, D. (1960) Concentration gradients for diffusion of vapors in glassy polymers and their relation to time dependent diffusion phenomena. *J. Am. Chem. Soc.*, **82**, 513–519.
41. Lucisano, M., Pagani, M.A., Mariotti, M., and Locatelli, D.P. (2008) Influence of die material on pasta characteristics. *Food Res. Int.*, **41**, 646–652.
42. Marabi, A., Livings, S., Jacobson, M., and Saguy, I.S. (2003) Normalized Weibull distribution for modeling rehydration of food particulates. *Eur. Food Res. Technol.*, **217**, 311–318.
43. Maskan, M. (2002) Effect of processing on hydration kinetics of three wheat products of the same variety. *J. Food Eng.*, **52**, 337–341.
44. Mercier, S., Des Marchais, L.P., Villeneuve, S., and Foisy, M. (2011) Effect of die material on engineering properties of dried pasta. *Proc. Food Sci.*, **1**, 557–562.
45. Misra, M.K. and Brooker, D.B. (1980) Thin-layer drying and rewetting equations for shelled yellow corn. *Trans. Am. Soc. Agric. Biol. Eng.*, **23**, 1254–1260.
46. Ogawa, T., Kobayashi, T., and Adachi, S. (2011) Water sorption kinetics of spaghetti at different temperatures. *Food Bioprod. Process.*, **89**, 135–141.
47. Ogawa, T. and Adachi, S. (2013) Effect of salts on the water sorption kinetics of dried pasta. *Biosci. Biotechnol. Biochem.*, **77**, 249–252.

48. Ogawa, T. and Adachi, S. (2014a) Effect of surface roughness on rehydration kinetics of spaghetti. *Jpn. J. Food Eng.*, **15**, 101–104.
49. Ogawa, T. and Adachi, S. (2014b) Measurement of moisture profiles in pasta during rehydration based on image processing. *Food Bioprocess Technol.*, **7**, 1465–1471.
50. Ogawa, T. and Adachi, S. (2014c) Effects of drying conditions on moisture distribution in rehydrated spaghetti. *Biosci. Biotechnol. Biochem.*, **78**, 1412–1414.
51. Ogawa, T. and Adachi, S. (2016) Moisture distribution and texture of spaghetti rehydrated under different conditions. *Biosci. Biotechnol. Biochem.*, **80**, 769–773.
52. Ogawa, T., Chuma, A., Aimoto, U., and Adachi, S. (2015) Characterization of spaghetti prepared under different drying conditions. *J. Food Sci.*, **80**, E1959–E1964.
53. Ogawa, T., Hasegawa, A., and Adachi, S. (2014) Effects of relaxation of gluten network on rehydration kinetics of pasta. *Biosci. Biotechnol. Biochem.*, 78, 1930–1934.
54. Peleg, M. (1988) An empirical model for the description of moisture sorption curves. *J. Food Sci.*, **53**, 1216–1219.
55. Petitot, M., Brossard, C., Barron, C., Larre, C., Morel, M.H., and Micard, V. (2009) Modification of pasta structure induced by high drying temperatures. Effect on the in vitro digestibility of protein and starch fractions and the potential allergenicity of protein hydrolysates. *Food Chem.*, **116**, 401–412.
56. Saguy, I.S., Marabi, A., and Wallach, R. (2005) New approach to model rehydration of dry food particulates utilizing principles of liquid transport in porous media. *Trends Food Sci. Technol.*, **16**, 495–506.
57. Sandstedt, R.M., Kempf, W., and Abbott, R.C. (1960) The effect of salts on the gelatinization of wheat starch. *Starch/Stärke*, **12**, 333–337.
58. Sanjuán, N., Bon, J., Clemente, G., and Mulet, A. (2004) Changes in the quality of dehydrated broccoli florets during storage. *J. Food Eng.*, **62**, 15–21.
59. Sanjuán, N., Simal, S., Bon, J., and Mulet, A. (1999) Modelling of broccoli stems rehydration process. *J. Food Eng.*, **42**, 27–31.
60. Schofield, J.D., Bottomley, R.C., Timms, M.F., and Booth, M.R. (1983) The effect of heat on wheat gluten and the involvement of sulphydryl-disulphide interchange reactions, *J. Cereal Sci.*, **1**, 241–53.
61. Sekiyama, Y., Horigane, A.K., Ono, H., Irie, K., Maeda, T., and Yoshida, M. (2012) T2 distribution of boiled dry spaghetti measured by MRI and its internal structure observed by fluorescence microscopy. *Food Res. Int.*, **48**, 374–379.
62. Takagi, M. and Shimoda, H. (ed) (2004) Handbook of image analysis (revised edition). University of Tokyo Press, Tokyo, Japan.
63. Toi, K., Odani, H., and Nakagawa, T. (1995) High-molecular-weight molecule and water (in Japanese; Koubunsi to Mizu), 1 ed. Kyoritsu Pab., Tokyo, Japan.
64. Tuhumury, H.C.D., Small, D.M., and Day, L. (2014) The effect of sodium chloride on gluten network formation and rheology. *J. Cereal Sci.*, **60**, 229–237.
65. Ukai, T., Matsumura, Y., and Urade, R. (2008) Disaggregation and reaggregation of gluten proteins by sodium chloride. *J. Agric. Food Chem.*, **56**, 1122–1130.
66. Uedaira, H. (1977) What is water? (in Japanese). 1st edn., Kodansha, Tokyo, Japan.
67. Watanabe, H. (2004) The factor which governs water migration in starchy foods. *Jpn. J. Food Eng.*, **5**, 143–151.
68. Weegels, P.L. and Hamer, R.J. (1998) Temperature-induced changes of wheat products. In "Interactions: The keys to cereal quality" ed. by Hamer, R.J. and Hoseney, R.C. Am. Assoc. Cereal Chem., St. Paul, MN, USA, p95–130.
69. Wrigley, C., Corke, H., and Walker, C.E. (2004) Encyclopedia of grain science. 1st ed., Oxford: Elsevier.
70. Yoshino, M., Ogawa, T., and Adachi, S. (2013) Properties and water sorption characteristics of spaghetti prepared using various dies. *J. Food Sci.*, **78**, E520–525.

71. Yue, P., Rayas-Duarte, P., and Elias, E. (1999) Effect of drying temperature on physicochemical properties of starch isolated from pasta. *Cereal Chem.*, **76**, 541–547.
72. Zhang, Y. and Cremer, P.S. (2006) Interactions between macromolecules and ions: the Hofmeister series. *Curr. Opin. Chem. Biol.*, **10**, 658–663.
73. Zweifel, C., Handschin, S., Escher, F., and Conde-Petit, B. (2003) Influence of high-temperature drying on structural and textural properties of durum wheat pasta. *Cereal Chem.*, **80**, 159–167.

Chapter 6
Rheological Studies on Gelation Kinetics of Powdered Soybean in the Presence of Glucono-δ-Lactone

Miki Yoshimura

Abstract Rheological properties on gelation kinetics of powdered soybean in the presence of glucono-δ-lactone (GDL) were studied by dynamic viscoelastic measurements, compression tests, differential scanning calorimetry (DSC), and observation of the network structure by confocal laser scanning microscopy (CLSM). The gelation time became shorter, and the rate constant of gelation increased with increasing concentration of powdered soybean and GDL. The rupture strain, rupture stress, rupture energy, storage modulus, and loss modulus increased with increasing the concentration of powdered soybean. In protein concentration of 4.4–6.9 % range, the storage modulus was proportional to 2.3 power of protein concentrations. It is indicated from CLSM that the network structure of the powdered soybean curd increased with the increasing concentration of powdered soybean. The network structure of the powdered soybean curd could be correlated well with the results obtained from rheological measurements. A non-heated dispersion of powdered soybean showed two endothermic peaks in the temperature range studied by heating DSC (20–140 °C). It is suggested that a heated dispersion of powdered soybean formed a stronger gel with increasing concentration of powdered soybean. The powdered soybean curd was coarser and more heterogeneous than soybean curd (tofu) by CLSM. The powdered soybean (SP) curd showed a smaller rupture strain and rupture energy than soybean curd (tofu) at the almost same protein concentration. It was suggested that ingredients such as insoluble dietary fiber other than soybean protein don't contribute to the network structure.

Keywords Powdered soybean • Dispersion • Curd • Gelation • Viscoelasticity • CLSM

M. Yoshimura (✉)
University of Hyogo, Kobe, Japan
e-mail: miki@shse.u-hyogo.ac.jp

I. Kaneda (ed.), *Rheology of Biological Soft Matter*, Soft and Biological Matter,
DOI 10.1007/978-4-431-56080-7_6

6.1 Introduction

Soybean products have been consumed by Asian people such as in China, Japan, and Korea for more than 2,000 years. Soybean curd (tofu) is a food made by coagulating soy milk. Soy milk is produced by soaking dried soybeans and grinding them in water and then squeezing. The soybean grounds squeezed from soy milk are called *okara*. Tofu contains relatively large amounts of protein. Additionally, it has low energy, saturated fats, and cholesterol. It is high in iron and calcium. The soybean grounds contain a lot of dietary fiber. Many researchers reported that components of soybean, heating temperature, and time of soy milk and coagulants have effects on the texture and flavor of soybean curds [1–4]. In most studies on soybean curd formation, soybean protein isolates or purified proteins were used. Generally, the gelation kinetics of soybean protein consists of two steps: (1) Soybean protein is heated and denatured. This leads to the exposure of hydrophobic bonds located inside the native protein molecules. (2) Different coagulants are added to stimulate gelation under incubation [5]. Most studies on soybean curd gelation have used glucono-δ-lactone (GDL) or $CaSO_4$ as the coagulant because the rates of curd gelation are relatively slow and more homogeneous gels are formed [6, 7]. Recently, there are many different varieties of soybean curd [8, 9]. Soybean curd made from powdered soybean is higher in insoluble dietary fiber than that made from soy milk [10]. In the present work, rheological properties of powdered soybean dispersion and curd in the presence of glucono-δ-lactone (GDL) are studied by dynamic viscoelastic measurements, compression measurements, differential scanning calorimetry (DSC), and observation of the network structure by confocal laser scanning microscopy (CLSM).

6.2 Materials and Methods

6.2.1 Materials

Soybean was hulled and grinded using a pin mill (Hosokawa Alpine). Powdered soybean and glucono-δ-lactone (GDL) were obtained from local companies in Japan. Particle size distribution of powdered soybean (SP) was measured by using a Microtrac FRA laser (Leeds and Northrup), and average particle diameter was 31.24 μm. The contents of moisture, protein, oil, and dietary fiber were 7.3 %, 40.3 %, 17.5 %, and 10.6 %, respectively.

6.2.1.1 Preparation of Powdered Soybean Dispersion

Powdered soybean (SP) was dispersed in distilled water at room temperature. The dispersion was heated in a water bath to 90 °C for 5 min and then cooled to 20 °C for 1 h. The solution of glucono-δ-lactone (GDL) was also prepared. SP concentrations used were 11–17 w/w%. GDL concentrations used were 0.2–0.5 w/w%. GDL solution was added to the SP dispersion and mixed, and then immediately the storage modulus G' and loss modulus G'' were measured as a function of time.

6.2.1.2 Preparation of Powdered Soybean Curd

Powdered soybean (SP) was dispersed in distilled water by a motorized stirrer with a mixing blade at 250 rpm for 30 min at room temperature. Then the dispersion was heated in a water bath to 90 °C and then cooled to 20 °C for 1 h. Distilled water was added to SP dispersion, and the concentration of SP was adjusted to 11, 13, 15, and 17 w/w%. GDL solution was added to the SP dispersion and mixed. The mixture was poured into cylindrical molds and kept at 90 °C for 30 min and then cooled to 20 °C for 1 h.

6.2.1.3 Preparation of Soybean Curd (Normal Tofu)

Dried soybeans were soaked at 10 °C for 16 h and grinded in distilled water and then squeezed. The soy milk was heated in a water bath to 90 °C and then cooled to 20 °C for 1 h. GDL (0.3 %) was added to the soy milk and mixed. The hot mixture was poured into cylindrical molds and kept at 90 °C for 30 min and then cooled to 20 °C for 1 h. The protein concentration of soybean curd (tofu) was 12 w/w%.

6.2.2 *Methods*

6.2.2.1 Dynamic Viscoelastic Measurement

The samples of powdered soybean (SP) dispersions were prepared as described above. Dynamic viscoelastic measurements for SP dispersion were carried out using a rheolograph sol (Toyoseiki Seisakusho Co., Japan) at 1 Hz with an amplitude of 50 μm corresponding to a shear strain of 0.05. The SP dispersion (1.5 ml) containing 0.3 % GDL was poured directly in the cell of the instrument, which had been kept at constant temperatures (40, 50, 60, 70, 80, and 90 °C). The surface of sample was covered with silicon oil to prevent the evaporation of water. Immediately the storage shear modulus G' and the loss shear modulus G'' were measured as a function of time for 60 min.

The sample of powdered soybean (SP) curd and soybean curd (tofu) were prepared as described above. Dynamic viscoelastic measurements for SP curd and tofu were carried out using a rheolograph gel (Toyoseiki Seisakusho Co., Japan) at 3 Hz with an amplitude of 100 μm corresponding to a shear strain of 0.1. G' and G'' are proportional to the energy stored and lost in a gel per cycle, respectively.

6.2.2.2 Compression Measurement

The samples of powdered soybean (SP) curds and soybean curd (tofu) were prepared as described above. Compression tests of SP curd and tofu were carried out using a Rheoner RE-3305 (Yamaden Co. Ltd., Japan) with a 2 kg load cell and at a crosshead speed of 10 mm/min. The diameter of the plunger was 40 mm. The rupture stress was calculated from the load value at a breaking point divided by the initial cross-sectional area of the curd. Rupture strain was determined as the ratio of the deformation at breaking point to initial height. The rupture energy was calculated from the area under the stress-strain curve, and it was normalized per unit volume. The measurement was repeated more than eight times, and the mean value and standard deviation were determined.

6.2.2.3 Differential Scanning Calorimetry

Differential scanning calorimetry (DSC) was made with a DSC apparatus DSC6100 (Seiko Instruments Inc., Japan). Each 15 mg of powdered soybean and 35 mg of distilled water were directly weighed into a 70 μl silver pan. A silver pan containing an equal amount of distilled water was used as reference. The temperature was raised from 20 to 140 °C at 2 °C/min.

6.2.2.4 Syneresis Measurement

The samples of powdered soybean (SP) curds were prepared as described above. The SP curds were put on the filter paper (diameter, 24 cm; no. 4, Whatman). The extent of syneresis was estimated by the expanded area of syneresis of filter paper with time.

6.2.2.5 Confocal Laser Scanning Microscopy

Confocal laser scanning microscopy (CLSM) was used to observe the network structure. Dispersion of powdered soybean was stained with rhodamine B, which is a fluorescent dye. SP dispersions were put on special glass slides with a shallow hole; the sample was covered with a cover glass, sealed with nail polish to prevent evaporation, kept at room temperature for 5 min, and put in a heat-resistant package.

The sample was heated in a water bath at 80 °C for 30 min, cooled at 10 °C for 30 min, and then kept at 20 °C for 1 h. The samples of soybean curds were observed using confocal laser scanning microscopy (Radiance 2000, Japan Bio-Rad Laboratory Corporation, Japan). Laser excitation of the fluorescent samples was at 543 nm. Digital images were acquired in 1280 × 1024 pixel resolution. The soybean curd (normal tofu) was also observed.

6.2.2.6 Sensory Evaluations

The samples of powdered soybean (SP) curds were prepared as described above. The panel consisted of 30 university students aged 20–22 years. In this investigation, the use of subjects was conducted based on the Helsinki Treaty. The following were estimated on a scale of nine points: "color" (+1, white; +9, yellow), "hardness evaluated with a teaspoon" (1, soft; +9, hard), "springness evaluated with a teaspoon" (1, none; +9, springy), "hardness" (+1, soft; +9, hard), "smoothness" (+1, none; +9, smoothly), "sloppy" (+1, less; +9, more), "body" (+1, little; +9, much), "aftertaste" (+1, little; +9, much), "color" (+1, dislike; +9, like), "hardness" (+1, dislike; +9, like), "smoothness" (+1, dislike; +9, like), "body" (+1, dislike; +9, like), and "overall" (+1, dislike; +9, like). Data analysis was conducted using SPSS 12J.

6.3 Results and Discussion

6.3.1 The Gelation of Powdered Soybean (SP) Dispersion

6.3.1.1 Effects of Heating Temperature

Figure 6.1 shows the time dependence of the storage shear modulus G', the loss shear modulus G'', and tan δ for 15 w/w% powdered soybean dispersions with 0.3 % GDL at different heating temperatures. The storage modulus G' and the loss modulus G'' increased with time and attained the plateau value. The mechanical loss tangent tan $\delta = G''/G'$ decreased with time. The plateau values of tan δ for SP dispersions of heating temperature at 40 and 50 °C were larger than 0.1, and these dispersions formed weak gels. While the plateau values of tan δ for SP dispersions of heating temperature at 80 and 90 °C were smaller than 0.1, these SP dispersions formed elastic gels. Some gelation processes at a constant temperature have been treated by an equation of first-order kinetics or other modified equations [11]. Dynamic viscoelastic measurement is a useful method to study the gelation processes of soybean protein [12], as well as dietary fibers such as konjac glucomannan [13] and other biopolymers [14], because it can be carried out at small strain without breaking the structure being formed during gelation. These curves could be well approximated by the following equation:

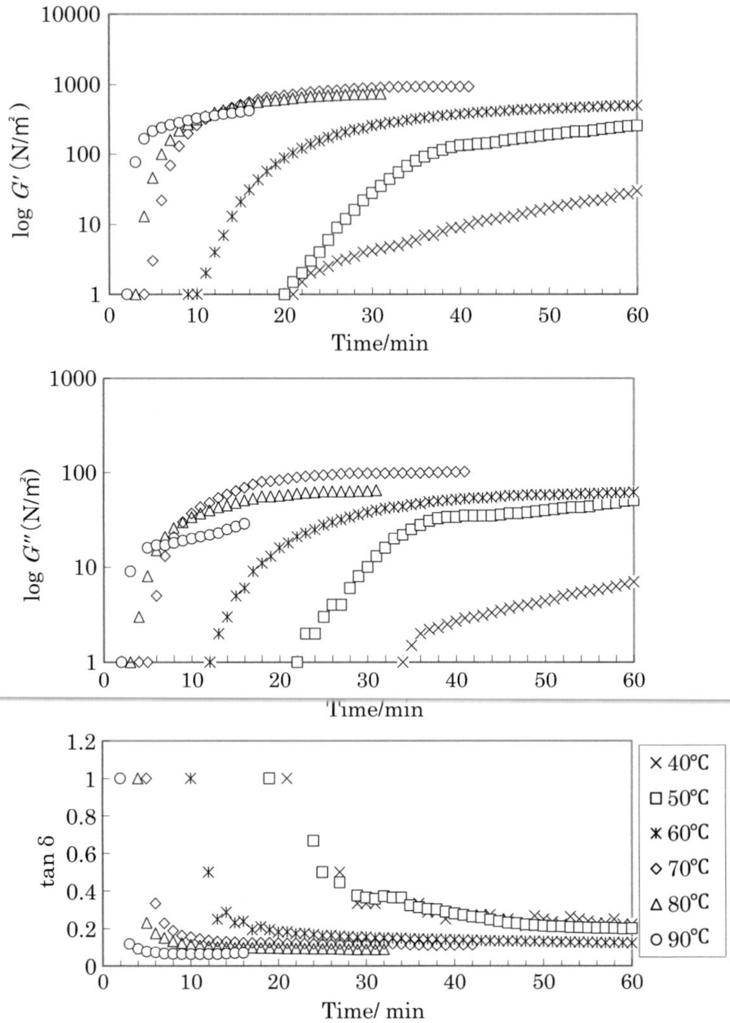

Fig. 6.1 Time dependence of G', G'', and tan δ for 15 % powdered soybean dispersions with 0.3 % GDL solution at different heating temperatures

$G'(t) = G'\text{sat}\,[1 - \exp\{-k\,(t - t_0)\}]$, where G'sat is the plateau value of G' after a long time, k is the rate constant of gelation of powdered soybean, and t_0 is the gelation time. The gelation time t_0 is taken as the time at which G' begins to deviate from the baseline. Table 6.1 shows parameters of first-order kinetic model for the gelation of 15 w/w% powdered soybean dispersion with 0.3 % GDL solution at different heating temperatures. The gelation time t_0 became shorter, the rate constant k became larger, and the plateau value G'sat increased with increasing heating temperature up to 70 °C and then decreased at 80 °C and 90 °C. Generally,

Table 6.1 Parameters for gelation of powdered soybean dispersion at different heating temperatures

Sample	t_0/min	k/min^{-1}	G'(N/m^2)	G''(N/m^2)	tan δ
15 % dispersion of SP, 40 °C	21		30	7	0.23
15 % dispersion of SP, 50 °C	20		254	51	0.20
15 % dispersion of SP, 60 °C	11	0.0150	504	62	0.12
15 % dispersion of SP, 70 °C	5	0.0560	932	102	0.11
15 % dispersion of SP, 80 °C	4	0.0570	741	65	0.09
15 % dispersion of SP, 90 °C	3	0.0690	420	29	0.07

the gelation proceeds faster at higher temperature for heat-setting systems such as methyl cellulose [15], xyloglucan [16], glycinin, and β-conglycinin [6, 17, 18]. The reason why the plateau value G'sat decreased with increasing the temperature beyond a certain temperature (70 °C) may be attributed to one of the following possibilities: (1) dispersion of powdered soybean formed gel so rapidly at higher temperature that disordered networks were formed, and (2) the slip between the plate and sample solution may occur during measurement because of the exuded water from powdered soybean curd.

Figure 6.2 shows the time dependence of G', G'', and tan δ for 15 w/w% non-heated SP dispersions with 0.3 % GDL at different heating temperatures (40, 50, 60, 70, and 80 °C). Table 6.2 shows parameters of first-order kinetic model for the gelation of 15 w/w% non-heated SP dispersions. Powdered soybean (SP) was dispersed in distilled water at room temperature and not heated. GDL solution was added to the non-heated SP dispersion and mixed, and then immediately the storage modulus G' and loss modulus G'' were measured as a function of time. The non-heated SP dispersions did not show any gelation curves at 40 and 50 °C, and the curves were shown above 60 °C. The gelation time t_0 became shorter, the rate constant k became larger, and the plateau value G'sat increased with increasing heating temperature from 60 to 90 °C. However, the plateau values of tan δ for non-heated SP dispersions at 60, 70, 80, and 90 °C were larger than 0.1, and these dispersions formed weak gels. Gelation of globular proteins is induced by the denaturation unfolding and the subsequent aggregation of the proteins [19]. It was suggested that the non-heated SP dispersion is not denatured and formed weak gels.

6.3.1.2 Effects of Powdered Soybean (SP) Concentration

Figure 6.3 shows the time dependence of G', G'', and tan δ for 11–17 w/w% SP dispersions with 0.3 % GDL at 80 °C. Table 6.3 shows parameters of first-order kinetic model and pH for the gelation of 11–17 w/w% powdered soybean dispersions. The gelation time t_0 became shorter, the rate constant k became larger, and the plateau value G'sat increased with increasing concentration of powdered

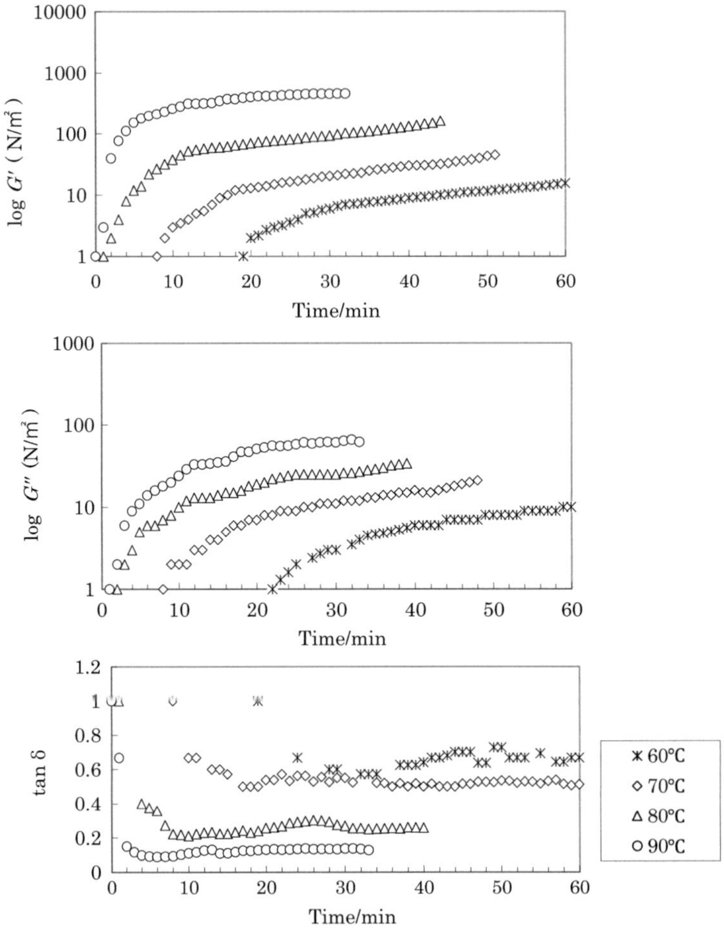

Fig. 6.2 Time dependence of G', G'', and tan δ for 15 % non-heated dispersions with 0.3 % GDL solution at different heating temperatures

Table 6.2 Parameters for non-heated powdered soybean dispersion at different heating temperatures

Sample	t_0/min	k/min^{-1}	G'(N/m^2)	G''(N/m^2)	tan δ
15 % non-heated SP dispersion, 60 °C	20	0.0003	16	10	0.67
15 % non-heated SP dispersion, 70 °C	9	0.0010	45	21	0.53
15 % non-heated SP dispersion, 80 °C	2	0.0036	162	34	0.21
15 % non-heated SP dispersion, 90 °C	2	0.0299	462	62	0.13

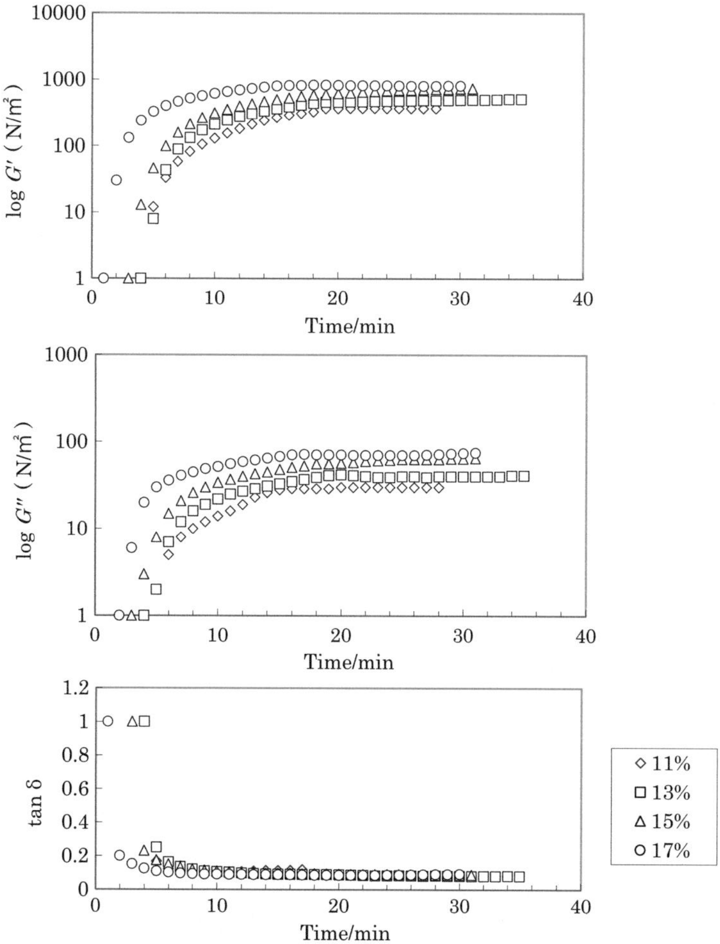

Fig. 6.3 Time dependence of G', G'', and tan δ for powdered soybean dispersions in various concentration with 0.3 % GDL

Table 6.3 Parameters for gelation of powdered soybean dispersion in various concentrations

Sample	t_0/min	k/min^{-1}	G'(N/m^2)	G''(N/m^2)	tan δ	pH
11 % dispersion of SP, 0.3%GDL	4	0.0215	369	30	0.11	5.4
13 % dispersion of SP, 0.3%GDL	4	0.0457	514	41	0.08	5.5
15 % dispersion of SP, 0.3%GDL	4	0.0570	741	65	0.09	5.6
17 % dispersion of SP, 0.3%GDL	2	0.1086	809	75	0.09	5.6

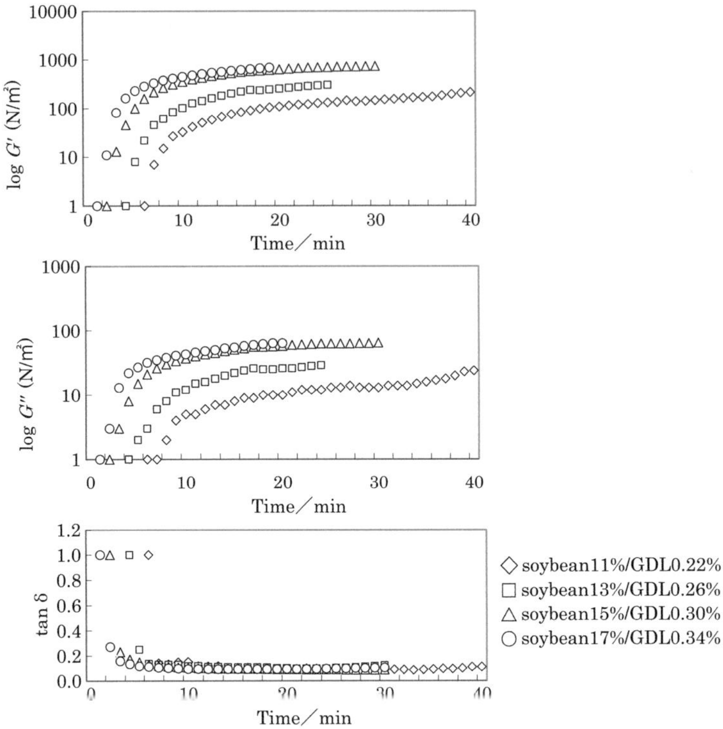

Fig. 6.4 Time dependence of G', G'', and tan δ for powdered soybean dispersions in various concentrations with 0.22 % – 0.34 % GDL

soybean. Kohyama [6] reported that the parameters of first-order kinetic model for the gelation of 7S and 11S proteins of soybean with 0.4 % GDL and the rate constant k became smaller with increasing concentration over the range of protein concentration 1–6 %. The pH of 7S and 11S proteins of soybean increased with increasing concentration, resulting in that the rate constant k became smaller, whereas the pH of 11–17 w/w% powdered soybean dispersions with 0.3 % GDL was in the 5.4–5.6 range, resulting also in that the rate constant k became smaller. The reason why the pH of 11–17 w/w% powdered soybean dispersions was almost the same value may be attributed to that the 11–17 w/w% powdered soybean dispersion contains a lot of insoluble dietary fiber, the protein content of it was in the 4.4–6.9 w/w% range, and its protein range was small.

The time dependence of G', G'', and tan δ for 11–17 w/w% powdered soybean dispersions with 0.22–0.34 w/w% GDL is shown in Fig. 6.4. Table 6.4 shows parameters of the gelation of powdered soybean dispersion. The pH of 11–17 w/w% powdered soybean dispersions were almost same. The gelation time t_0 became shorter, the rate constant k became larger, and the plateau value G'sat increased

Table 6.4 Parameters for gelation of powdered soybean dispersion in various GDL concentrations

Sample	t_0/min	k/min^{-1}	G'(N/m^2)	G''(N/m^2)	tan δ	pH
11 % dispersion of SP, 0.20 % GDL	8	0.0080	215	24	0.11	5.6
13 % dispersion of SP, 0.26 % GDL	6	0.0200	402	50	0.12	5.6
15 % dispersion of SP, 0.30 % GDL	4	0.0570	741	65	0.09	5.6
17 % dispersion of SP, 0.34 % GDL	3	0.0660	750	78	0.10	5.4

with increasing concentration of powdered soybean. It was suggested that the concentration of SP dispersion affected the gelation process of SP curd.

6.3.1.3 Effects of GDL Concentration

Figure 6.5 shows the time dependence of G', G'', and tan δ for 15 w/w% powdered soybean dispersions with various GDL concentrations (0.2, 0.3, 0.4, and 0.5 %) at 80 °C. Table 6.5 shows parameters of first-order kinetic model and pH for the gelation of 15 w/w% powdered soybean dispersions. The gelation time t_0 became shorter, the rate constant k became larger, and the plateau value G'sat increased with increasing the concentration of GDL. The plateau value of tan δ for SP dispersion of 0.2 % GDL was larger than 0.1, and this dispersion formed a weak gel. While the plateau values of tan δ for SP dispersion of 0.3, 0.4, and 0.5 % GDL were smaller than 0.1, these SP dispersions formed elastic gels. GDL is partially cleaved into gluconic acid in water, and then some gluconic acid molecules dissociate. Since lactone is chemically inactive, it should not react with soybean protein. The lowered pH shows the existence of dissociated gluconic acid. Nishinari et al. reported that the gelation mechanism of soybean protein in the presence of GDL is as follows: after the exposure of hydrophobic groups by heat denaturation, soluble aggregates are formed, and then the protons from GDL shield the electrostatic repulsion of negative charges on the surface of soluble aggregates which leads to the network formation [6].

6.3.2 *Properties of Powdered Soybean Curd and Soybean Curd (Tofu)*

6.3.2.1 Effects of Powdered Soybean (SP) Concentration

Figure 6.6 shows the changes in behaviors of compression measurements for 11–17 w/w% SP curd and 12 w/w% soybean curd (tofu). The rupture strain, rupture stress, and rupture energy increased with increasing the concentration of powdered soybean. This tendency corresponded well with the observation within

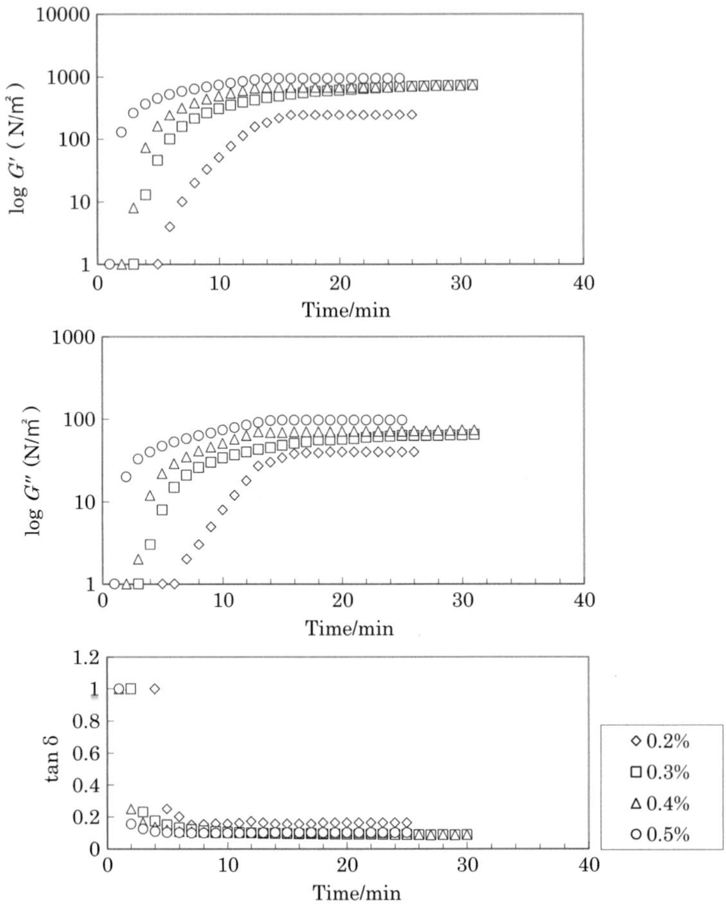

Fig. 6.5 Time dependence of G', G'', and tan δ for powdered soybean dispersions with various GDL concentrations

Table 6.5 Parameters for gelation of powdered soybean dispersion in various GDL concentrations

Sample	t_0/min	k/min^{-1}	G'(N/m^2)	G''(N/m^2)	tan δ
15 % dispersion of SP, 0.2 % GDL	5	0.022	349	40	0.16
15 % dispersion of SP, 0.3 % GDL	4	0.057	741	65	0.09
15 % dispersion of SP, 0.4 % GDL	2	0.086	751	74	0.10
15 % dispersion of SP, 0.5 % GDL	2	0.153	946	97	0.10

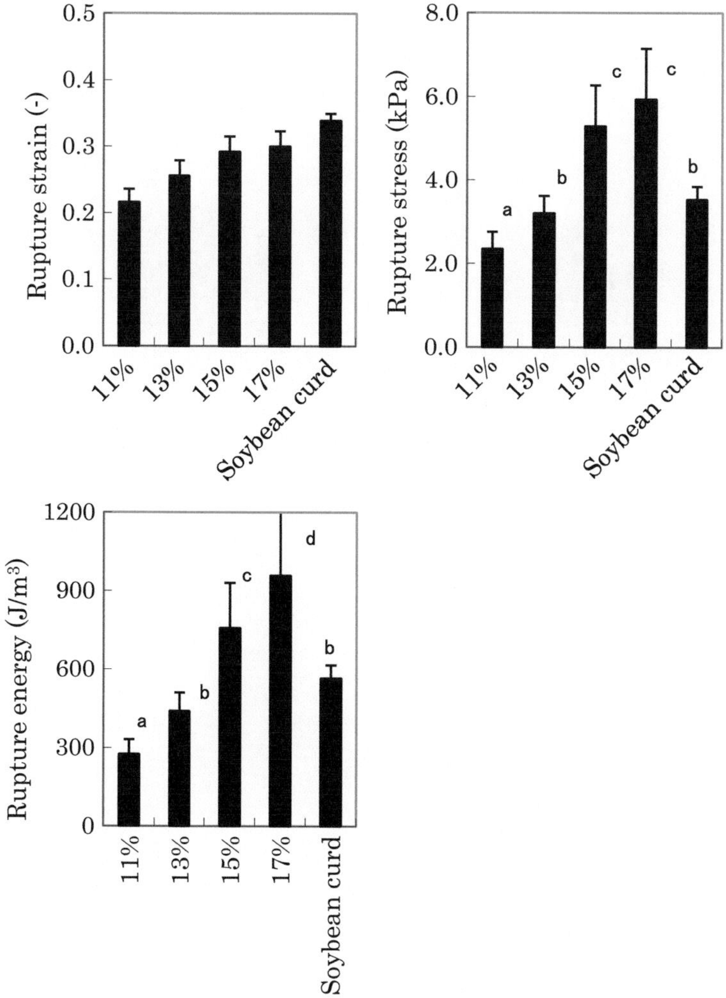

Fig. 6.6 Rupture properties of 11–17 % powdered soybean curd and tofu. Different letters in the same column show significant difference ($P < 0.05$), $n > 10$

the linear viscoelastic regime that G' and G'' of SP dispersions increased with SP concentration as shown in Fig. 6.4.

Figure 6.7 shows the dynamic viscoelasticity of 11–17 w/w% powdered SP curd and 12 w/w% soybean curd (tofu). The storage modulus and loss modulus increased with increasing the concentration of powdered soybean.

The values of storage modulus G' of SP curd were plotted against protein concentration on a logarithmic scale as shown in Fig. 6.8. The G' of SP curd exhibited a power-law relationship with the protein concentration that can be fitted to the form G'–C^n, where C is the protein concentration and n is the power-law

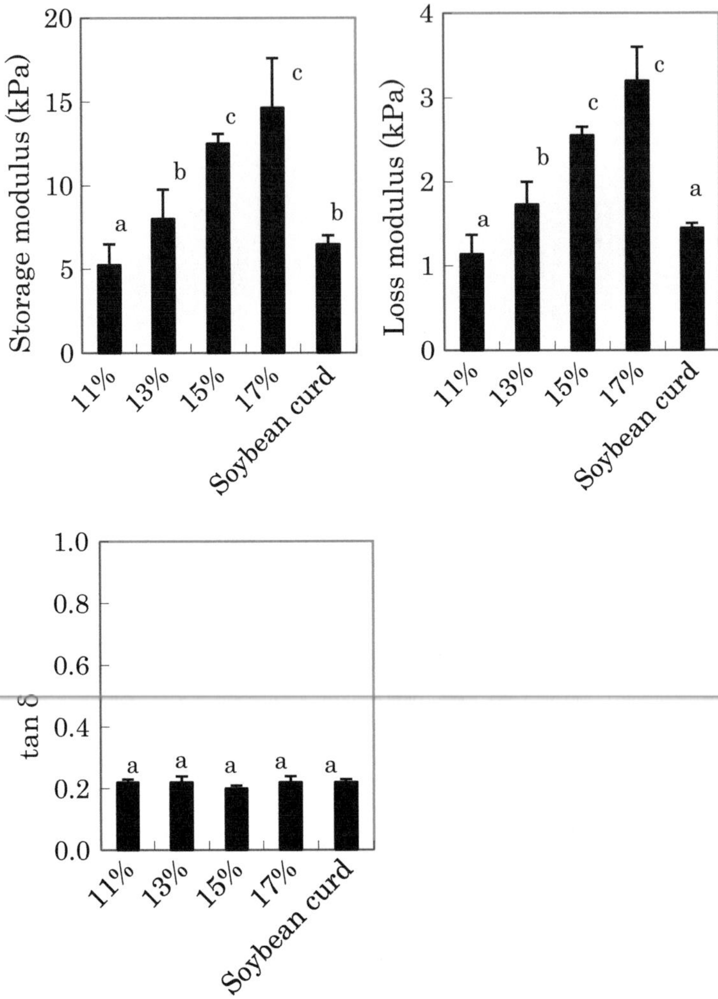

Fig. 6.7 Dynamic viscoelasticity of 11–17 % powdered soybean curd and tofu. Different letters in the same column show significant difference ($P<0.05$), $n>10$

exponent. In protein concentration of 4.4–6.9 w/w% range, the storage modulus G' was proportional to 2.3 power of protein concentrations. Kohyama reported that in protein concentration of 2–6 w/w% range, storage modulus on soybean 11S protein gel with GDL was 3.4 power of protein concentration [6]. Nagano reported that in protein concentration of 3–5 w/w% range, the value of n was 2.9 for soybean 11S protein gel with magnesium chloride [20]. Many researchers reported that G' of other biopolymer gels (e.g., gelatin, agar, casein micelle) showed a square power dependence on concentration. The exponent n close to two for various polymers is reported, but the reported value of n for soybean protein is much larger than two.

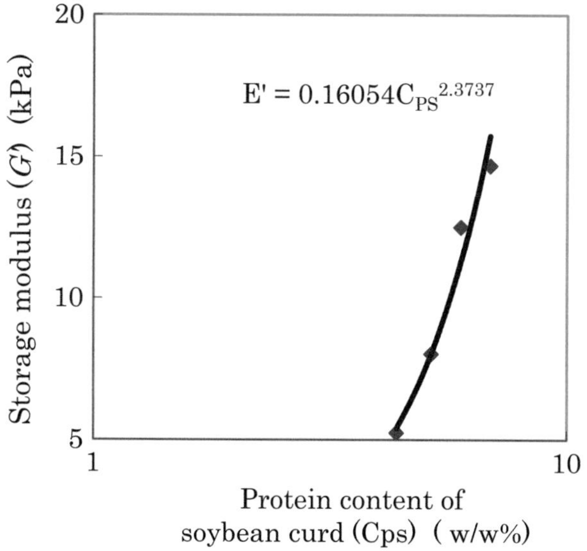

Fig. 6.8 Double logarithmic plot of storage modulus vs protein content for powdered soybean curd

The exponent of soybean protein was five, which was reported by Bikbov et al. for heat-induced soybean globulin gels of concentration of 7.5–58.4 w/w% range [21]. If the relationship between storage modulus G' and protein concentration at higher concentrations by GDL could be examined, the concentration dependence of G' should tend to be less pronounced. In the gel formation of globular proteins, some portion which is not completely denatured may be incorporated in gel network, and elasticity may be lower than the total protein concentration [5].

Figure 6.9 shows the change in syneresis of 11–17 w/w% powdered soybean (SP) curd. For powdered soybean curd, syneresis proceeded as a function of time. The degree of syneresis decreased with increasing the powdered soybean concentration. And also, it may be attributed that the insoluble dietary fiber of powdered soybean absorbs water and prevents syneresis of SP curd.

The DSC curves of SP plus distilled water (15 %) are shown in Fig. 6.10. SP plus distilled water showed a small endothermic peak at 72.1 °C and a large endothermic peak at 94.4 °C. Heated SP dispersion did not show any exothermic and endothermic peaks at the temperature range studied.

Soybean 7S globulin (β-conglycinin) and 11S globulin (glycinin) are two major storage protein components. They account for about 70 % of total protein in soybean seed [22, 23]. Soybean proteins contain two major globulins, 7S and 11S, which show different thermal transition temperatures [24, 25]. The thermal denaturation temperature of soybean 11S globulin in water was reported as 84.5 °C, and the denaturation was completed at around 93 °C by differential adiabatic scanning

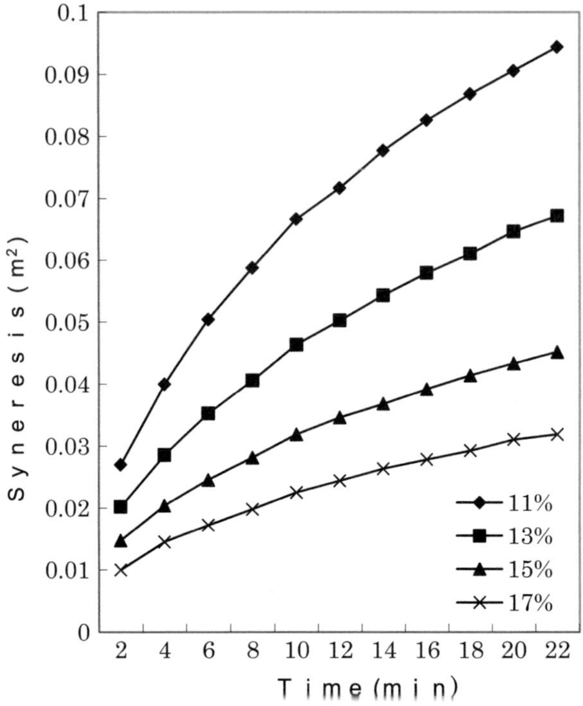

Fig. 6.9 Syneresis of 11–17 % soybean curd prepared from soybean powders

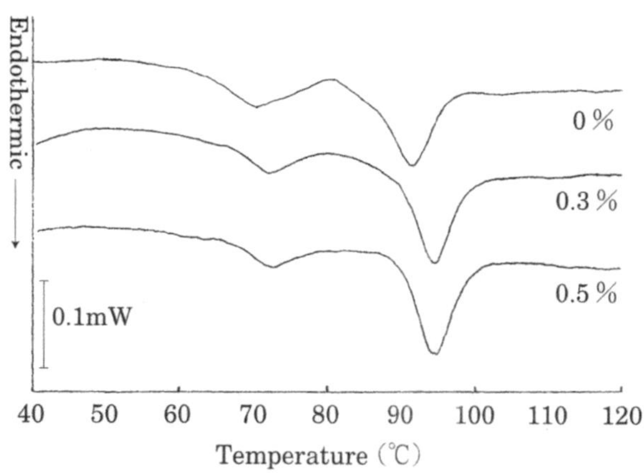

Fig. 6.10 Heating DSC curves for powdered soybean dispersions

calorimetry [26]. The two endothermic peaks of the non-heated SP dispersion attributed to denaturation unfolding of powdered soybean protein.

The peak temperature (Tp) shifted to higher temperatures with increasing GDL concentration (0.0–0.5 %). It was suggested that GDL concentration affected the denaturation unfolding of powdered soybean protein.

6.3.2.2 Comparison Between Powdered Soybean (SP) Curd and Soybean Curd (Normal Tofu)

Figure 6.11 shows CLSM of 11–17 w/w% powdered soybean curd. A white strand network represents powdered soybean. The white strand network increased with increasing powdered soybean concentration (A–E). It is indicated from CLSM that the network structure of the soybean curd increased with increasing the concentration of powdered soybean. The microstructures of powdered soybean curds were highly affected by SP concentration. The network structure of the powdered soybean curd could be correlated well with the results obtained from rheological measurements. At the almost same protein concentration of 12 w/w%, the powdered soybean (SP) curd (B) was coarser and more heterogeneous than normal tofu (F). The influence of dietary fibers on the enzymatic reaction for soy protein isolate (SPI) was studied by dynamic viscoelastic measurement [27]. The rate constant of SPI degradation decreased with increasing dietary fiber concentration; it may be attributed to that dietary fibers inhibited degradation of SPI [27]. Tran et al. reported that soybean soluble polysaccharides were applied into soybean protein isolate stabilized solution to improve the emulsion stability, because soluble polysaccharide can coat the protein to prevent protein-protein interactions between two fat droplets [28]. Gelation characteristics of soy milk and rice powder mixture were studied; rice powder inhibited soy milk gelation with acid coagulant [29]. The powdered soybean contains a lot of insoluble dietary fiber [10]. The powdered soybean (SP) curd showed a smaller rupture strain and rupture energy than soybean curd (tofu) at the almost same protein concentration (Fig. 6.6). It was suggested that ingredients such as insoluble dietary fiber other than soybean protein don't contribute to the network structure.

6.3.2.3 Sensory Evaluation of Powdered Soybean Curds

Figure 6.12 shows sensory evaluation of SP curds prepared from 11 to 17 w/w% powdered soybean. The main effect of the powdered soybean concentration for "color" (+1, white; +9, yellow), "hardness evaluated with a teaspoon" (1, soft; +9, hard), "hardness" (+1, soft; +9, hard), "sloppy" (+1, less; +9, more), "color" (+1, dislike; +9, like), "hardness" (+1, dislike; +9, like), "body" (+1, dislike; +9, like), and "overall" (+1, dislike; +9, like) became significant at the level of 0.01 %. As the SP concentration increased, the SP curds became yellow, became hard, and became less moist. As the sensory evaluation indicated that the soybean curd with a higher concentration of powdered soybean (15 or 17 w/w%) was observed to be harder, less moist, and fuller bodied, it was judged preferable to the powdered soybean

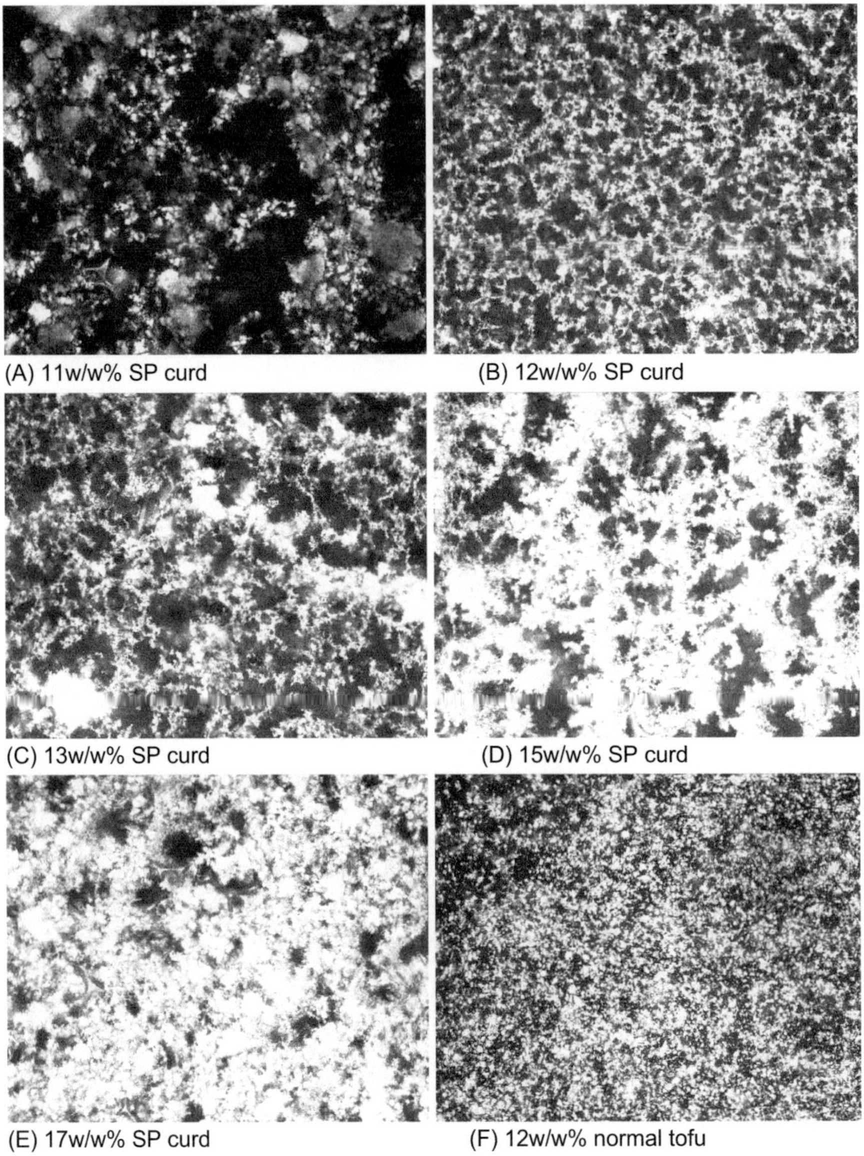

Fig. 6.11 Confocal laser scanning microscopy of 11–17 % SP curd prepared from soybean powders (**a–e**) and 12 % normal tofu (**f**). Each image width is 87.5 μm. **a** 11 w/w% SP curd, **b** 12 w/w% SP curd, **c** 13 w/w% SP curd, **d** 15 w/w% SP curd, **e** 17 w/w% SP curd, **f** 12 w/w% normal tofu

curd with lower concentration (11 w/w%). The sensory evaluation of the powdered soybean curd could be correlated well with the results obtained from rheological measurements and CLSM observation.

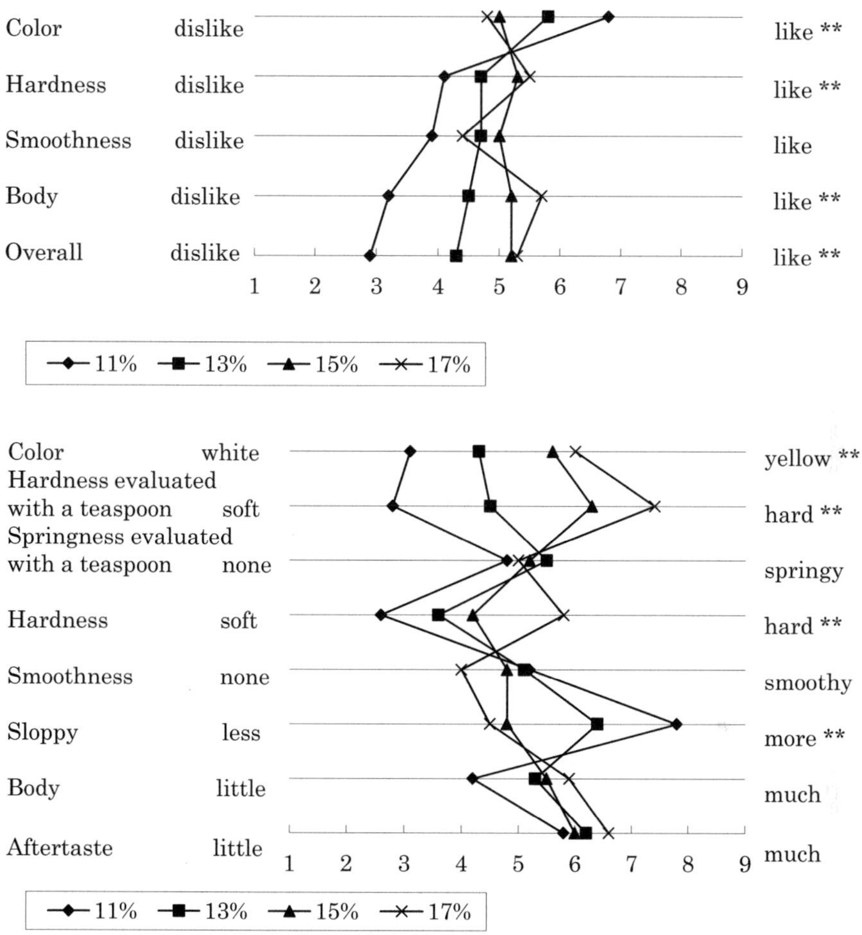

Fig. 6.12 Sensory evaluation of SP curds prepared from 11 to 17 w/w% powdered soybean

6.4 Conclusion

In this study, viscoelastic properties, compression properties, and microstructures of powdered soybean (SP) dispersion and SP curd with GDL were quantified. The gelation time became shorter and the rate constant of gelation increased with increasing the concentration of SP and GDL. The rupture strain, rupture stress, rupture energy, storage modulus, and loss modulus increased with increasing SP concentration. The powdered soybean (SP) dispersion contains a lot of insoluble dietary fiber, the protein content of it was in the 4.4–6.9 w/w% range, and its protein range was small. In protein concentration of 4.4–6.9 w/w% range, the storage modulus was proportional to 2.3 power of protein concentrations. The network structure of the soybean curd increased with increasing the concentration of powdered soybean by CLSM. The SP curd was coarser and more heterogeneous than soybean curd (tofu). The SP curd showed a smaller rupture strain and rupture energy than soybean curd (tofu) at the almost same protein concentration. It was suggested that ingredients other than soybean protein don't contribute to the network structure.

References

1. K. Saio, Tofu-relationships between texture and fine structure, *Cereal Foods World*, **24**, 342–354(1979).
2. J.M.deMan, L.deMan, S.Gupta, Texture and microstructure of soybean curd (Tofu) as affected by different coagulants, *Food Microstructure*, **5**, 83–89(1986)
3. Y. Mine, K. Murakami, K. Azuma, S. Yoshihara, K. Fukunaga, T. Saeki, E. Sawano; A comparison of various coagulants in tofu-forming properties, *Nippon Shokuhin Kagaku Kogaku Kaishi*, **52**, 114–119 (2005).
4. E.J.Noh, S.Y. Park, J.I. Pak, S.T. Hong, S.E. Yun, Coagulation of soymilk and quality of tofu as affected by freeze treatment of soybeans, *Food Chem.*, **91**, 715–721 (2005).
5. K. Nishinari, Y. Fang, S. Guo, G.O. Phillips, Soy proteins: A review on composition, aggregation and emulsification. *Food hydrocolloids*, **39**, 301–318 (2014).
6. K. Kohyama, K. Nishinari 1993, Rheological studies on the gelation process of soybean 7S and 11S properties in the presence of glucono-δ-lactone., *J.Agric. Food Chem.*, **41**, 8–14 (1993).
7. Z.S. Liu, S.K.C. Chang, L.T. Li, E. Tasumi, Effect of selective thermal denaturation of soybean proteins on soymilk viscosity and tofu's physical properties, *Food Research International*, **37**, 815–822 (2004).
8. T. Soeda, T. Ishii, K. Yamazaki, K. Murase, The effect of transglutaminase on texture of tofu (the functionalities of microbial transglutaminase for food application part 1). *Nippon Shokuhin Kagaku Kogaku Kaishi*, **42**, 254–261 (1995).
9. F. Kao, N. Su, M. Lee, Effect of calcium sulfate concentration in soymilk on the microstructure of firm tofu and the protein constitutions in tofu whey. *J. Agric. Food. Chem.*, **51**, 6211–6216 (2003).
10. M. Yoshimura, F. Shibata, M. Eto, K. Nishinari, Rheological studies on gelation process of dispersion of powdered soybean in the presence of glucono -δ-lactone, *Nippon Shokuhin Kagaku Kogaku Kaishi*,**51**, 115–122 (2004).
11. K. Nishinari, Rheological and DSC study of sol-gel transition in aqueous dispersions of industrially important polymers and colloids, *Colloids Polym.Sci.*, **275**, 1093–1107 (1997).
12. K. Nishinari, K. Kohyama, Y. Zhang, K. Kitamura, T. Sugimoto, K. Saio, Y. Kawamura, Rheological study on the effect of the A5 subunit on the gelation characteristics of soybean proteins, *Agric. Boil. Chem.*, **55,** 351–355 (1991).
13. M. Yoshimura, K. Nishinari, Dynamic viscoelastic study on the gelation of konjac-glucomannan with different molecular weights, *Food Hydrocolloids*, **12**, 227–233 (1999)
14. K. Nishinari, Rheological and DSC study of sol-gel transition in aqueous dispersions of industrial important polymers and colloids., *Colloid Polym.Sci.*, **275**, 1093–1107 (1997).
15. K. Nishinari, K.E. Hofmann, H. Moritaka, K. Kohyama, K. Nishinari,N. Gel-sol transition of methylcellulose., *Macromol.Chem.Phys.*,**198**, 1217–1226 (1997).
16. J.S.G. Reid, M. Edwards, I.C.M. Dea, Enzymatic modification of natural seed gums. In G.O. Phillips, D.J. Wedlock, P.A.Williams (Eds.) Gums and Stabilizers for the Food Industry 4, pp391–398, IRL Press.Oxford (1988).
17. T. Nagano, H. Mori, K. Nishinari, Effect of heating and cooling on the gelation kinetics of 'S globulin from soybeans, *Journal of Agricultural and Food Chemistry*, **42**, 1415–1419 (1994).

18. T. Nagano, H. Mori, K. Nishinari, Rheological properties and conformational states of beta-conglycinin gels at acidic pH., *Biopolymers*, **34**, 293–298 (1994)
19. E. Doi, Gels and gelling of globular protein, Trends in Food Science and Technology, **4**, 1–5 (1993).
20. T. Nagano, M. Masayuki, Viscoelastic properties and microstructures of 11S globulin and soybean protein isolates gels: Magnesium chloride-induced gels, *Food hydrocolloids*, **25**, 1647–1654 (2011).
21. T.M. Bikbov, V.Ya, Grinberg, Yu.A. Antonov, V.B. Tolstoguzov,, H. Schmandke, On the concentration dependence of the elasticity modulus of soybean globulin gels, *Polymer Bull.*, **1**, 865–869 (1979).
22. I. Koshiyama, A newer method for isolation of the 7S globulin in soybean seeds, *J.Agric. Food Chem.*, **43**, 1802–1812 (1972).
23. V.H.Thanh, K.Shibasaki, Major proteins of soybean seeds. Subunit structure of B-conglycinin., *J. Agric.Food Chem.*, **26**, 692–695 (1976).
24. B.German, S,Damodaran, J.E Kinsella, Thermal dissociation behavior of soy proteins., *J.Agric.Food Chem.*, **30**, 807–811 (1982).
25. S.Damodaran , Refolding of thermally unfolded soy proteins during the cooking regime of the gelation process, Effect on gelation., *J.Agric.Food Chem.*, 36, 262–269 (1988)
26. T.M.Bikbov, V.Ya.Grinberg, A.N.Danilenko, T.S.Chaika, J.A.Vaintraub, V.B.Tolstoguzov, Studies on gelation of soybean globulin solutions. Part 3. Investigation into thermal denaturation of soybean globulin fraction by the method of differential adiabatic scanning calorimetry: Interpretation of thermograms, The effect of protein concentration and sodium chloride, *Colloid Polym. Sci.*, **261**, 346–358 (1983).
27. M.Yoshimura, K.Nishinari, Rheological studies of influence of dietary fibers on the enzymatic reaction for soy protein isolate, *Foods Food Ingredients J.Jpn.*,**210**,954–962 (2005).
28. T.Tran, D.Rousseau, Stabilization of acidic soy protein-based dispersion and emulsion by soy soluble polysaccharides, *Food Hydrocoll.*, **30**, 382–392 (2013)
29. M.Yoshimura, S.Eguchi, A.Toura, K.Nakagawa, Gelation characteristics of oil in water emulsion stabilized by soy milk and rice powder mixture. Part 1 A study on gelation characteristics on addition of an acetic acid coagulant, *Nihon Shokuhin Kougakkaishi,***15**, 243–249 (2014)

Chapter 7
Linkage Between Food Rheology and Human Physiology During Oral Processing: Human Eating Behavior Deduced by Instrumental Compression of Food on a Soft Substrate

Takahiro Funami

Abstract In the food industry, development of in vitro instrumental evaluation system has been demanded for objective assessment of food textural acceptability for specific consumer groups. The instrumental evaluation system which the author's research team developed is consisted of an artificial tongue fabricated from a soft deformable elastomer and a hard non-deformable metal platen on a conventional uniaxial compression apparatus to mimic human tongue–palate compression. When apparent Young's modulus of artificial tongue is ca. 55 kPa and a crosshead speed is 10 mm/s, fracture profile of agar gels (as a test food) relates to human oral strategy for size reduction; gel samples which fracture on the evaluation system are processed by tongue–palate compression, whereas gel samples which do not fracture on the evaluation system are processed by mastication. Validation of the evaluation system using gellan gum gels shows the necessity of modification of the instrumental operation condition in some cases, and this can relate to the change in the physiology of tongue–palate compression by food texture.

Keywords Artificial tongue • Gel • Human eating behavior • Tongue–palate compression • Instrumental uniaxial compression

7.1 Introduction

Texture is a sensory property and thus should be assessed by human senses [54]. However, instrumental methods are preferably used for texture assessment due to reasons of cost saving and the complication of ethic issues involved in human sensory tests. Instrumental assessments of food texture have been studied for a long period using various foods [4, 11, 30, 34, 35, 60]. However, instrumental assessments cannot express exactly food texture which human perceives.

T. Funami (✉)
San-Ei Gen F.F.I., Inc., 1-1-11 Sanwa-cho, Toyonaka, Osaka 561-8588, Japan
e-mail: tfunami@saneigenffi.co.jp

I. Kaneda (ed.), *Rheology of Biological Soft Matter*, Soft and Biological Matter,
DOI 10.1007/978-4-431-56080-7_7

Instrumental methods may oversimplify the reality of oral experience, and results may not necessarily reflect the precise food texture which human perceives [3, 13]. This can arise from a difference in the motion between simple instrumental measurements and complex human oral processing, including deformation mode (compression or shear), deformation speed, clearance or the minimum gap, dynamic change in the size of food fragments, heat transfer, and salivation [45]. Additional difference between man-made machines and human is the deformability of the geometry and organ. In instrumental measurements, geometries used are generally made of metal and are much stiffer than food, completely dissimilar from human tissues like the tongue, which is highly deformable and flexible [43, 44]. This is believed to be one of the main causes of the deviation between instrumental measurements and sensory evaluation [42].

Human tongue plays a crucial role throughout food oral processing, including recognition, transportation of food to the molar or cheek teeth, compressing or squeezing food against the hard palate (i.e., tongue–palate compression), mixing food particles with saliva (i.e., bolus formation), and transporting bolus from the oral cavity to the pharynx [6]. In addition to mechanical device, the tongue works also as a sensory organ for taste and temperature [6]. In food oral processing, a food (solid or semisolid) is compressed slightly by the tongue against the hard palate (anterior part around the incisive papilla) for texture recognition [1], and this determines the subsequent oral strategy for size reduction. When a food is judged as soft enough to be broken down by tongue–palate compression, tongue compression continues. When a food is too strong to deal with by tongue–palate compression, oral strategy changes to the mastication, and a food is transported to the posterior part of the oral cavity via the so-called stage I transport [1, 16]. Thus, the tongue's compression of food against the hard palate plays a dominant role to initiate food oral processing [16].

Among a series of oral strategy, size reduction by tongue–palate compression should be investigated more extensively as there is an increasing number of the elderlies and patients with masticatory difficulty in aged society. Under this social circumstance, texture design of food products for specific consumer groups is increasingly important in the industry. Gels like jellies and puddings are preferably consumed by the consumer groups [47] because these foods are soft enough to be consumed by tongue–palate compression without the need of mastication. This can lead to perception of easiness and comfortableness of consumption. When oral strategy for size reduction is decided during tongue–palate compression must be of another research interest. Results from the cineradiographic visualization during oral processing suggest that 12 % strain is the critical point for the decision. This boundary strain is consistent when the size of food samples is the same and is independent of subjects [1]. Although this finding is invaluable to know the detail of the physiological mechanism of texture recognition, observation on tongue deformation is lacking.

Both the tongue and food deform during oral processing. To realize this physiological situation in instrumental assessment, usage of artificial tongues fabricated from a deformable soft elastomer should be one of the ideas instead of non-deformable hard probe on a conventional uniaxial compression apparatus. Recent trials by the author's research team for development of in vitro texture evaluation system and for the deduction of human oral strategy for size reduction will be reviewed in this chapter based on previous papers [22, 23]. In developing the system, practical usage for industrial purpose was prioritized rather than reproducing precise human physiology. Modification of the operation condition is necessary in some cases to improve correlation with human test, and this may give some hints for better understanding of the physiology of tongue–palate compression during oral processing. Also, mechanical criterion of food which determines the oral strategy for size reduction should be suggested on this evaluation system.

7.2 Artificial Tongue

As an artificial tongue, silicone rubbers of different apparent elastic moduli were prepared using a room temperature vulcanization (RTV) rubber (KE-12, Shin-Etsu Chemical Co., Ltd., Tokyo, Japan). The KE-12 was mixed with silicone oil (RTV thinner, Shin-Etsu Chemical Co., Ltd.) at ratios of 40:60, 50:50, and 60:40, onto which a curing catalyst (CAT-RM, Shin-Etsu Chemical Co., Ltd.) was added at 0.5 % (w/w) at 20 °C (Table 7.1). After vacuum deaeration, the mixture was placed into cylindrical glass molds of 20 mm in diameter and 10 mm in height and cured at 20 °C for 2 days. The silicone rubbers obtained were termed as S40, S50, and S60 in the increasing order of silicone rubber concentration or consistency. As a result, apparent Young's modulus of the artificial tongue ranged between a relaxed state (or at rest) and a tension state (or excitation) of the human tongue.

Table 7.1 Formulation and mechanical property of artificial tongues made from silicone rubber

Artificial	Formulation (% w/w)			Mechanical property		
Tongue	Silicone rubber	Silicone oil	Curing agent	Apparent Young's modulus (kPa)		
S40	40	60	0.5	18.3	±	3.5
S50	50	50	0.5	54.9	±	10.3
S60	60	40	0.5	113.0	±	9.6

Each artificial tongue was shaped into a cylinder of 20 mm in diameter and 10 mm in height or 58 mm in diameter and 10 mm in height (only for S50). Apparent Young's modulus of the smaller size artificial tongue was determined at 20 °C as the stress/strain ratio at 20 % nominal strain using a micrometer. Data were presented as means ± standard deviation in triplicate

7.3 Mechanical Characteristics of Human and Artificial Tongues

Apparent Young's modulus of the human tongue was measured directly using healthy subjects. Upon informed consent, six males and two females ranging from 25 to 39 in age (32.4 years old on average) and having normal dental status participated. Apparent Young's modulus of each subject's tongue was determined by the stress/strain ratio at $20 \pm 5.1\,\%$ strain using a digimatic micrometer (type MDC-25MJ, Mitutoyo Corporation, Kanagawa, Japan) through compression. The strain of 20 % was used due to some experimental limitations: insufficient sensitivity of the apparatus at lower strains and physical pain for subjects at higher strains. Here, 0 % strain was defined as that giving rise to 1 mN, which is equivalent to the sensitivity of the pressure sensor. As the configuration, stainless probe of 16 mm in diameter and 5 mm in height was on each side of the micrometer, and a sheet-type pressure sensor of 3 mm in diameter and 0.1 mm in thickness (Swallow Scan, Nitta Corporation, Osaka, Japan) [19] was adhered to the upper probe (Fig. 7.1). Thickness of the tongue was measured either in a relaxed state (i.e., at rest where no intentional force was applied to the tongue) or in a tension state (at excitation where the maximum intentional force was applied to the tongue) by tucking protruded tongue between probes at a distance of 10–20 mm from the tip. Stress against that strain was measured at the same time. The stress captured for 10 s was averaged for

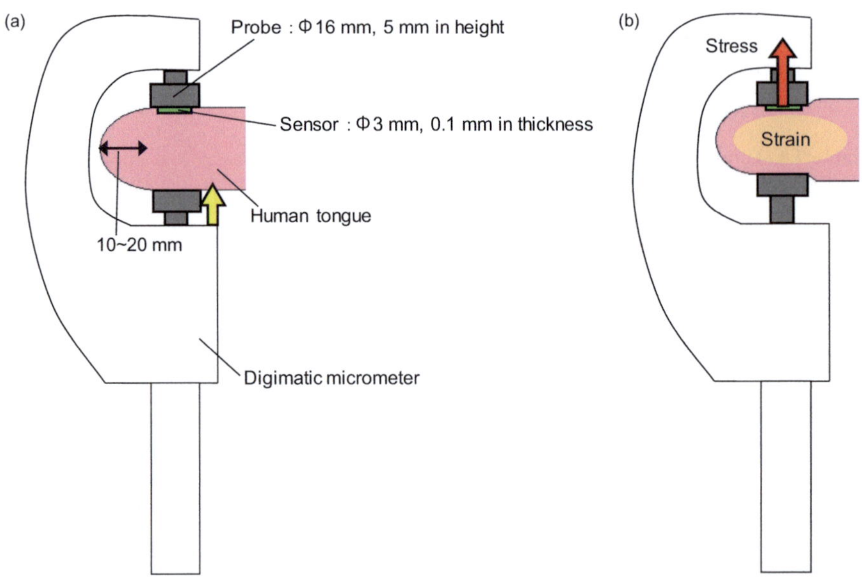

Fig. 7.1 Schematic drawing for the elasticity measurements of human tongue (illustrated as a lateral view). (**a**) At the initial (0 % strain); (**b**) at 20 % strain. See the text for the definition of 0 % strain

determination of apparent Young's modulus in a relaxed state, whereas the stress captured for 2–4 s around the maximum was averaged for determination of apparent Young's modulus in a tension state. That is, these measurements were for the tongue in a static condition. Measurements were carried out on three separate days for each subject. Similar measurements were also applied to artificial tongues. Central part of an artificial tongue was compressed between probes up to 20 % strain to determine apparent Young's modulus.

The thickness of the human tongue was 7.53 ± 0.99 mm in a relaxed state and 11.09 ± 2.18 mm in a tension state. Apparent Young's modulus (at ca. 20 % strain) of the human tongue was 12.2 ± 4.2 kPa in a relaxed state and 122.5 ± 58.5 kPa in a tension state. Apparent Young's modulus (at ca. 20 % strain) of artificial tongue ranged from 18.3 ± 3.5 to 113.0 ± 9.6 kPa (Table 7.1). Thus, artificial tongue S40 corresponded to the human tongue in a relaxed state, and artificial tongue S60 corresponded to the human tongue in a tension state. It is reasonable to hypothesize that apparent Young's modulus of human tongue should be between a relaxed and a tension states during food oral processing and that such physiological conditions can be mimicked by the selection of the artificial tongue from S40 to S60.

Elastic modulus of the human tongue has been measured directly (in situ) or simulated on some computational models. As in situ data, ca. 10 kPa and 60 kPa have been reported for human tongue in a relaxed state and in a tension state, respectively [49]. On the other hand, 60 kPa and 120 kPa for human tongue in paralyzed and in sleeping status were predicted based on a finite element model using the closing pressure in the male upper airway [33]. Similar results have been also simulated by a separate study for passive tongue using the closing pressure in the male airway [20]. These reported values verify our measurements of the elastic modulus of the human tongue, supporting approximately tenfold increase in the elastic modulus of the human tongue upon excitation.

From anatomical point of view, the human tongue is a large bundle of muscle fibers, consisting of extrinsic and intrinsic muscles which work cooperatively [6, 25]. Actions of these muscles, particularly intrinsic muscles with highly complex and three-dimensional arrangement, contribute to the versatility of its postures and movements. Contraction of the muscle fibers hardens the tongue as a result of the increase in internal pressure, and contraction of longitudinal muscle fibers with controlled release of transverse and vertical muscle fibers shortens and thickens the tongue [6]. Increase in tongue thickness upon stress (from relax to tension) indicates greater contribution of longitudinal fibers relative to transverse fibers. Artificial tongue used is a simple soft material which resembles the mechanical characteristics of the human tongue. In reality, human tongue is composed of multiple muscles, including styloglossus, genioglossus, palatoglossus, hyoglossus, and geniohyoid [14], and its architecture is much more complicated as has been described by several three-dimensional models [10]. The human tongue has the anisotropy and functional segments, which enable it to transform into complex shape upon deformation. Units of the tongue may deform in the same direction or in different direction depending on task [51], and the motion of the human tongue could not be uniform upon deformation [39, 40]. Also, the axis of tongue compression is not only in vertical during food oral processing but more in three dimensional. It is for sure that all these characters cannot be necessarily recreated by artificial tongue, but nevertheless the use of mechanical homogeneous artificial tongue can ensure physical meaning and objectivity of data and should be rather preferable for industrial purpose due to its simplicity and reproducibility of experiments.

7.4 Hydrocolloid Gels as a Model Food

Two types of polysaccharide gel were prepared as a model food. One is a series of agar gels having equivalent fracture strain but different fracture forces, and the other is a series of gellan gum gels in a wide mechanical range by controlling both fracture strain and fracture force. For gellan gum gels, a mixture of low-acylated and high-acylated gellan gums was used. Low-acylated gellan gum forms less deformable and more brittle gels than high-acylated gellan gum, and diverse textures can be obtained through blending of the two [53]. To mask subtle flavor derived from these polysaccharides, sucrose was added at 10 % (w/w) to all gel samples. Mixture of sucrose and each gelling agent was added to deionized water at 90 °C and stirred for 10 min at the same temperature. In the case of gellan gum, calcium lactate (0.1 % as pentahydrate) and a food color (0.2 %, SAN GREEN GC-EM, San-Ei Gen F.F.I., Inc.) were added. The reason for the use of food color was to clarify the boundary between the gel sample and artificial tongue on the image. Solutions obtained were poured into cylindrical glass molds of 20 mm in diameter and 10 mm in height, heated at 85 °C for 30 min, and refrigerated at 8 °C for 2 h. Cylindrical gel samples obtained were subjected to instrumental compression test and human test (i.e., sensory evaluation) at 20 °C after curing at 5 °C overnight. A series of agar gels was termed as from A1 to A7 in the increasing order of agar concentration, whereas a series of gellan gum gels was denominated as a combination of fracture strain and fracture force. For more experimental details, Ishihara et al. [22, 23] should be referred to.

Again, gels are often used as a base material for the elderly and dysphagia foods [21, 23, 24]. Also, texture control of gels is relatively easy by changing the formulation and/or the concentration of gelling agents, which is another reason for preferable use of gel samples in texture study. However, soft foods including gels which require tongue–palate compression for size reduction [15, 21, 23, 37, 38, 55–57] have been investigated less frequently than hard foods which require mastication for size reduction [2, 5, 7–9, 12, 15, 26–29, 31, 37, 38, 55, 56, 58, 59]. Human eating behavior of viscoelastic gels can be different from that of solid foods with low fluidity and deformability, including vegetables and nuts, or of cohesive foods like meat [24].

7.5 Mechanical Characteristics of Gel Samples (Determination by Instrumental Compression Between Non-deformable Hard Platens)

Mechanical characteristics of hydrocolloid gels were measured using a TA XT-plus texture analyzer (Stable Micro Systems, Surrey, UK). Mechanical characteristics, including fracture strain, fracture force, fracture stress, and apparent Young's modulus, were determined by compression of the gels on a hard metal stage using an aluminum plate of 50 mm in diameter basically at a crosshead speed of 10 mm/s at 20 °C. This crosshead speed was based on the finding in human [17] that average speed of tongue surface movement is 10.34 mm/s with a range from 2.10 to 32.43 mm/s in liquid swallowing. Size of the gels was 20 mm in diameter and 10 mm in height. No marked changes are assumed in mechanical properties of the gels in the temperature range from 20 °C to body temperature (ca. 37 °C). Also, temperature of the gels cannot change markedly within a short period during oral processing. To prevent slippage between the gel and the geometry, waterproof sandpaper (type DCC, grit number 2,000, Sankyo-Rikagaku Co., Ltd., Saitama, Japan) was adhered onto the surface of the aluminum probe. Data were presented as means ± standard deviation in triplicate. Mechanical characteristics of the gels determined in this way are called the initial mechanical parameters. Strain and stress represent nominal ones. Apparent Young's modulus was determined from the strain–stress curve at 20 % strain to ensure the consistency with artificial tongue. Also, the linearity between strain and stress was not necessarily secured for some gel samples below 10–15 % strain due to the deviation from the horizontality of the cross-section surface, some syneresis at the surface, deformation by the self-weight, etc.

All agar gels from A1 to A7 showed a fracture point through instrumental compression. It was confirmed not only by visual inspection but also by calculation on the image under an assumption of isovolumetric change that the gel samples retained their cylindrical shape during instrumental compression till fracture. Fracture strain ranged between 57.20 % (A2) and 62.16 % (A1), whereas fracture force ranged from 4.25 N (A1) to 50.99 N (A7) (Table 7.2). Apparent Young's modulus (at ca. 20 % strain) ranged from 5.4 kPa (A1) to 144.6 kPa (A7) (Table 7.2). All gellan gels also showed a fracture point through instrumental compression. It was confirmed that the gel samples retained their cylindrical shape during instrumental compression till fracture as in the case of agar gels. Gellan gel samples were categorized into four groups by a combination of fracture strain (in four steps from series A to D in the increasing order) and fracture force (in five steps from ca. 10 to 30 N in 5 N increments) (Table 7.3). When compared at equivalent fracture strain, fracture force and fracture stress both increased with gum concentration, and significant difference ($p < 0.05$) was detected between any set of two gel samples. Also, when compared at equivalent fracture force, fracture strain increased with the ratio of high-acylated gellan gum, and significant difference ($p < 0.05$) was detected between any set of two gel samples. Apparent Young's modulus generally increased with

Table 7.2 Mechanical properties of agar gels through instrumental compression between non-deformable hard platens

Agar gel	Fracture strain (%)			Fracture force (N)			Apparent Young's modulus (kPa)		
A1	62.16	±	1.74	4.25	±	0.16	5.4	±	0.3
A2	57.20	±	0.94	5.41	±	0.28	12.5	±	2.1
A3	57.30	±	0.41	6.92	±	0.46	14.7	±	2.3
A4	58.92	±	0.93	8.03	±	0.84	17.5	±	0.4
A5	59.35	±	0.68	22.54	±	0.33	53.5	±	3.0
A6	59.32	±	1.02	36.21	±	2.21	97.9	±	1.7
A7	60.76	±	0.79	50.99	±	2.42	144.6	±	7.1

Each agar gel was shaped into a cylinder of 20 mm in diameter and 10 mm in height and was compressed uniaxially at a crosshead speed of 10 mm/s with a cylindrical probe of 50 mm in diameter at 20 °C on a hard stage to determine fracture strain, fracture force, and apparent Young's modulus. Apparent Young's modulus was determined by the stress/strain ratio at 20 % nominal strain through compression. Data represent means ± standard deviation in triplicate

increasing fracture force when compared at equivalent fracture strain but decreased with increasing fracture strain when compared at equivalent fracture force.

7.6 Instrumental Compression Test of Hydrocolloid Gels on Deformable Artificial Tongue

Deformation and fracture profiles of each type of hydrocolloid gel were observed during instrumental compression on artificial tongue (Fig. 7.2). Size and shape of the gel sample were the same as those of artificial tongue (i.e., cylinder of 20 mm in diameter and 10 mm in height). When human initiates compression of food with a relatively small size between the tongue and hard palate, the whole dorsal surface area of the tongue is not contacted to food. To approximate this oral physiology, size of the gel sample was the same to that of artificial tongue, which helps with theoretical analyses of the mechanical behavior for each entity during instrumental compression. The combination of the gel sample (upper) and artificial tongue (bottom) was compressed uniaxially by the aluminum probe of 50 mm in diameter at a fixed crosshead speed up to 50 % strain of the combination (i.e., 10 mm) at 20 °C. Standard operation condition of the instrumental test consists of crosshead speed at 10 mm/s, size of artificial tongue at 20 mm in diameter and 10 mm in height, and the maximum strain at 50 % of the combination. These may change on some occasions to improve correlation with human test. Deformation profile till gel fracture or the maximum strain was pictured continuously by a digital camera as a movie, which was synchronized with the record of the stress–strain curve to ensure the fracture point. Snapshot images were captured from the movie at a certain interval for the determination of the vertical strain for either the gel sample or artificial tongue using a WinROOF image software (ver. 5.9, Nihon Denkei Co.,

Table 7.3 Mechanical characteristics of gellan gels through instrumental compression between non-deformable hard platens

Gellan gel	Fracture strain (%)			Fracture force (N)			Fracture stress (kPa)			Apparent Young's modulus (kPa)		
A10	45.5	±	0.6	10.7	±	0.7	34.1	±	2.2	63.8	±	4.9
A15	47.6	±	1.3	15.9	±	0.4	50.7	±	1.4	88.5	±	3.3
A20	46.7	±	1.4	21.1	±	0.8	67.1	±	2.7	112.9	±	5.8
A25	48.0	±	0.9	26.6	±	1.1	84.6	±	3.4	134.1	±	3.2
A30	47.6	±	1.6	30.9	±	1.6	98.7	±	5.1	161.9	±	2.7
B10	55.7	±	1.7	11.0	±	0.9	35.0	±	2.8	34.9	±	3.6
B15	57.9	±	0.7	15.5	±	0.5	49.2	±	1.5	44.4	±	6.6
B20	57.6	±	0.3	20.3	±	0.4	64.6	±	1.2	51.3	±	3.1
B25	59.9	±	0.6	25.7	±	1.3	81.9	±	4.1	61.7	±	5.3
B30	59.8	±	0.3	30.4	±	1.4	98.5	±	4.3	66.6	±	5.6
C10	62.1	±	0.9	11.5	±	2.0	36.7	±	6.4	22.0	±	0.4
C15	64.9	±	0.4	16.0	±	0.9	50.8	±	2.7	24.8	±	2.2
C20	64.4	±	0.5	20.5	±	0.9	65.3	±	2.8	31.8	±	0.1
C25	65.4	±	1.0	25.8	±	1.5	82.2	±	4.7	36.9	±	2.3
C30	65.9	±	1.4	32.0	±	0.6	102.1	±	2.0	39.5	±	4.9
D10	71.4	±	0.9	11.5	±	2.2	36.5	±	7.0	6.5	±	1.0
D15	72.2	±	1.3	16.0	±	1.0	51.0	±	3.3	9.5	±	0.4
D20	72.9	±	0.6	21.0	±	1.3	67.0	±	4.3	13.4	±	0.7
D25	73.8	±	1.0	26.0	±	1.0	82.7	±	3.3	13.5	±	0.6
D30	74.6	±	1.0	31.3	±	1.1	99.8	±	3.5	15.1	±	3.0

Each gellan gel sample was shaped into a cylinder of 20 mm in diameter and 10 mm in height and was compressed uniaxially at a crosshead speed of 10 mm/s with a cylindrical aluminum probe of 50 mm in diameter at 20 °C on a hard stage to determine fracture strain, fracture force, fracture stress, and apparent Young's modulus. Apparent Young's modulus was determined by the stress/strain ratio at 20 % nominal strain. Data were presented as means ± standard deviation in triplicate

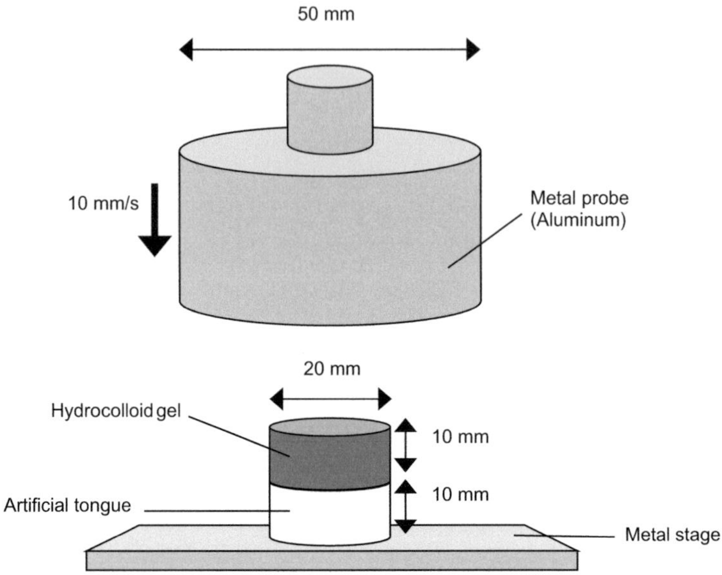

Fig. 7.2 Setup and standard operation condition of instrumental compression test

Ltd., Tokyo, Japan). When the vertical strain could not be determined directly on the image, it was calculated using the horizontal strain by assuming incompressibility [41]. The measurements were carried out in triplicate for agar gels. For gellan gum gels, the fracture probability was determined by ten repeated measurements, whereas the image analysis of the deformation profile was carried out in triplicate.

Force–strain curves of agar gel were recorded up to 50 % strain of the combination of agar gel and artificial tongue at a crosshead speed of 10 mm/s, from which the fracture point was identified from a peak or rapid decrease in force (Fig. 7.3). When S40 was used as artificial tongue, A3 and A4 fractured, while A5 and A6 did not. When S50 was used as artificial tongue, A3, A4, and A5 fractured, while A6 did not. When S60 was used as artificial tongue, all gels from A3 to A6 fractured. The same results were obtained at a crosshead speed of 1 mm/s. Deformation profile for either agar gel or artificial tongue was analyzed on the image when crosshead speed was 10 mm/s (Fig. 7.4). When S40 was used as artificial tongue, strain of A3 or A4 was larger than that of artificial tongue throughout the strain range tested (Fig. 7.5). This resulted in gel fracture. On the other hand, strain of A5 or A6 was equivalent to that of artificial tongue at the initial stage below ca. 10 % strain and was smaller than that of artificial tongue at larger strains. This did not result in gel fracture. When S50 was used as artificial tongue, strain of A3 or A4 was larger than that of artificial tongue throughout the strain range tested, whereas A5 showed a crossover at ca. 30 % strain, below which strain of agar gel was larger than that of artificial tongue. These resulted in gel fracture. On the other hand, strain of A6 was smaller than that of artificial tongue throughout the strain range tested except the initial stage below

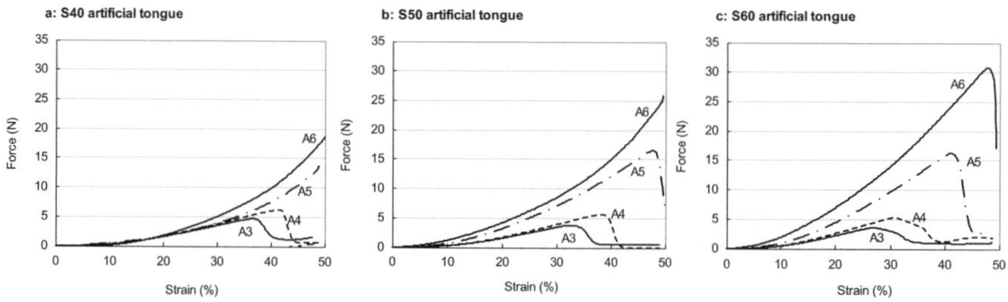

Fig. 7.3 Force–strain curves of agar gels during instrumental compression on artificial tongue. Agar gels of 20 mm in diameter and 10 mm in height were compressed uniaxially on each artificial tongue, (**a**) S40, (**b**) S50, or (**c**) S60, with a cylindrical aluminum probe of 50 mm in diameter at a crosshead speed of 10 mm/s at 20 °C. Size of artificial tongue was the same as agar gels

ca. 5 % strain. This did not result in gel fracture. When S60 was used as artificial tongue, strain of A3, A4, or A5 was larger than that of artificial tongue throughout the strain range tested, whereas A6 showed a crossover at ca. 25 % strain, below which strain of agar gel was larger than that of artificial tongue. These resulted in gel fracture. Results were similar when crosshead speed was decreased to 1 mm/s with some changes found. These were findings that the initial strain range where agar gel and artificial tongue showed equivalency was almost half for the combination of S40 and A5 or A6 and that the crossover was not detected for the combination of A5 and S50 and A6 and S60. In conclusion, it was generally recognized that agar gel fractured when its strain was larger than that of artificial tongue, while agar gel did not fracture when its strain was equivalent to or smaller than that of artificial tongue.

Force–strain curves of gellan gel were recorded up to 50 % of the combination of gellan gel and artificial tongue at a crosshead speed of 10 mm/s, from which the fracture point was identified from a peak or rapid decrease in force. The curves when S50 was used as artificial tongue were shown in Fig. 7.6 as a representative, and data from the compression test were summarized in Fig. 7.7. All gel samples except A10 did not fracture at all when S40 was used as artificial tongue as presented by closed circles in Fig. 7.7, whereas most gel samples except C30 and D25–D30 fractured at 100 % when S60 was used as artificial tongue as presented by closed squares in Fig. 7.7. When S50 was used as artificial tongue, fracture probability of the gel samples was in-between; A10–A20, B10–B15, C10–C15, and D10 fractured at more than 70 % probability, whereas A30, B25–B30, C25–C30, and D15–D30 hardly fractured as presented by closed triangles in Fig. 7.7. When crosshead speed was decreased from 10 to 5 mm/s for the combination of series D and S50, D10–D20 fractured at equal to or more than 70 % probability, whereas D25–D30 did not fracture at all as presented by closed diamonds in Fig. 7.8. Similar to this, when the size of S50 artificial tongue was increased from 20 to 58 mm in diameter, D10–D20 fractured at 100 % even at crosshead speed of 10 mm/s, whereas D25 and D30 fractured at less than 40 % probability as presented by crosses in Fig. 7.8. Deformation profile for either gellan gel or artificial tongue was analyzed on the image when crosshead speed was 10 mm/s. Representative data were shown in

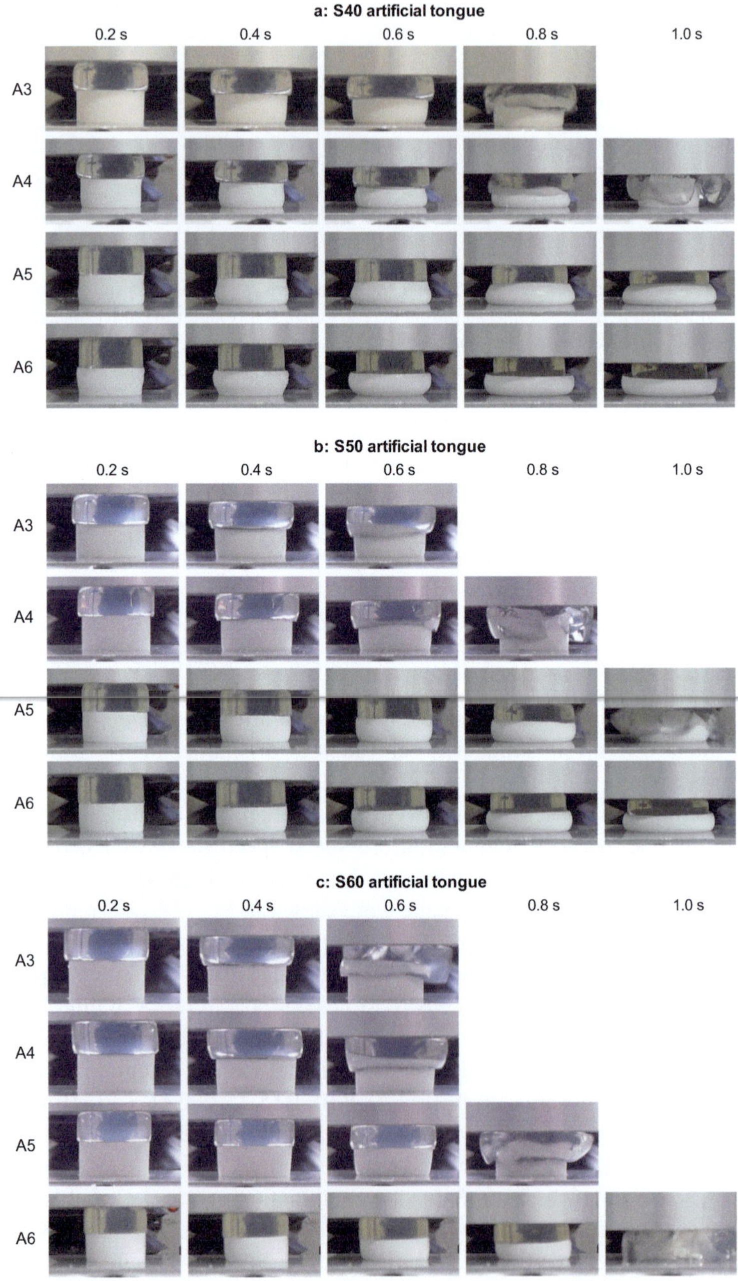

Fig. 7.4 Snapshot images representing the deformation behavior of agar gels during instrumental compression on artificial tongue. Images were captured at a time interval of 1/15 s during instrumental compression of agar gels with a cylindrical aluminum probe of 50 mm in diameter at a crosshead speed of 10 mm/s. Deformation profile of agar gels A3–A6 was shown during instrumental compression on each artificial tongue, (**a**) S40, (**b**) S50, or (**c**) S60, at 20 °C. The initial size was 20 mm in diameter and 10 mm in height for either the gel or artificial tongue

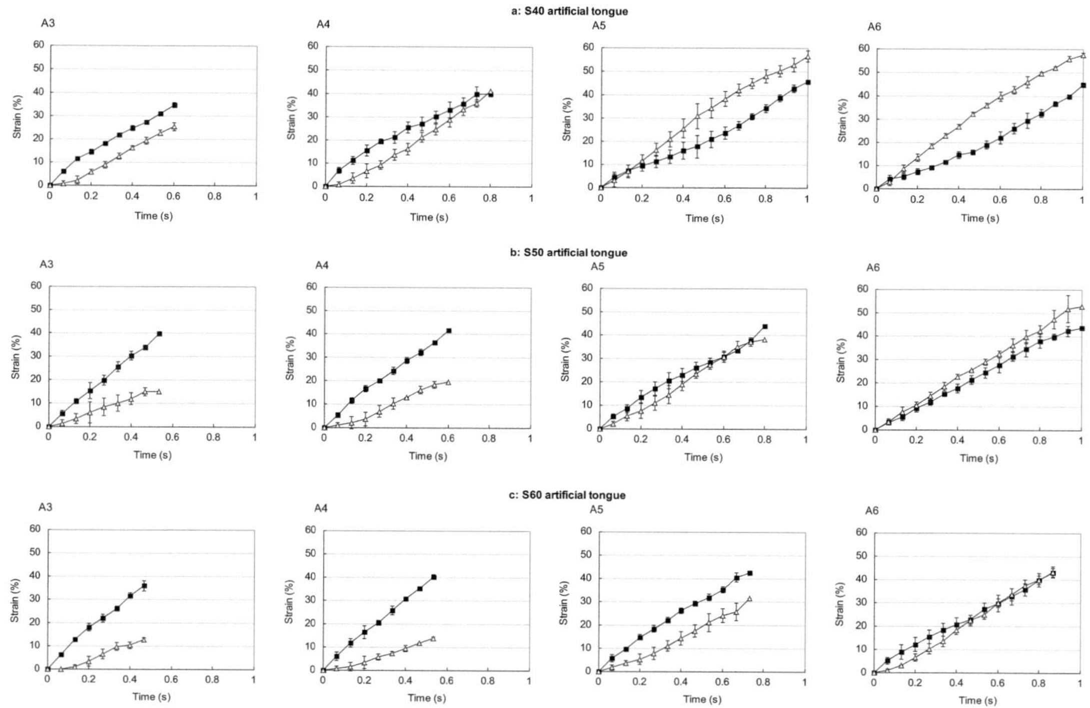

Fig. 7.5 Time course of the strain development for agar gel and artificial tongue during instrumental compression. Nominal strain in the vertical direction was determined on the image shown in Fig. 7.4, and the time course of the strain development was presented for either agar gel or artificial tongue. The initial size was 20 mm in diameter and 10 mm in height for each entity, and its combination was compressed with a cylindrical aluminum probe of 50 mm in diameter at a crosshead speed of 10 mm/s at 20 °C. Data were presented as means ± standard deviation in triplicate. Results before the fracture point were shown. (**a**) Compression on S40 artificial tongue. (**b**) On S50 artificial tongue. (**c**) On S60 artificial tongue. *Closed squares*, agar gel; *open triangles*, artificial tongue

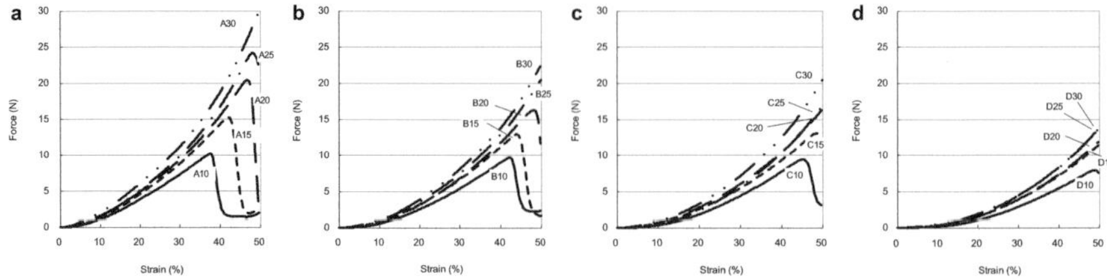

Fig. 7.6 Force–strain curves of gellan gels during instrumental compression on S50 artificial tongue. Gellan gels of 20 mm in diameter and 10 mm in height were compressed uniaxially on S50 artificial tongue with a cylindrical aluminum probe of 50 mm in diameter at a crosshead speed of 10 mm/s at 20 °C. Size of artificial tongue was the same as gellan gels. (**a**) Gellan gels from series A. (**b**) From series B. (**c**) From series C. (**d**) From series D

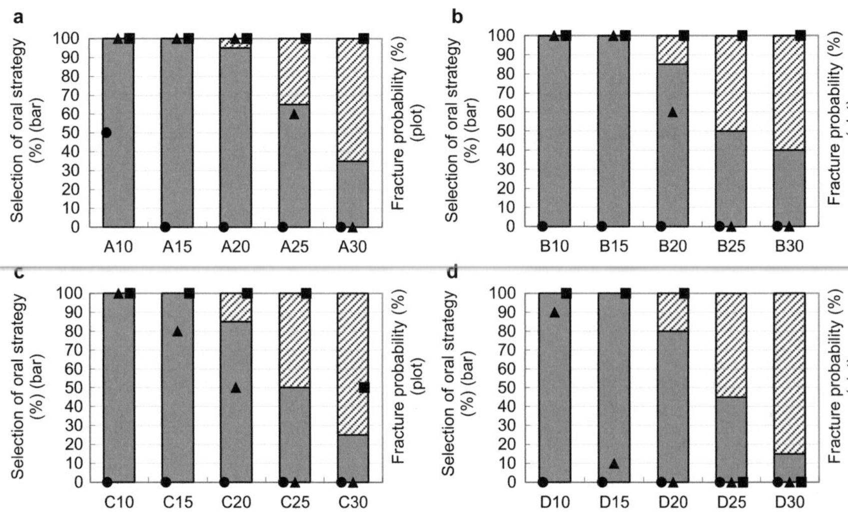

Fig. 7.7 Relationship between the fracture probability of gellan gels in instrumental compression test using artificial tongue and selection of oral strategy for the first size reduction in sensory evaluation. Gellan gels were compressed uniaxially on each artificial tongue with a cylindrical aluminum probe of 50 mm in diameter at a crosshead speed of 10 mm/s at 20 °C, and fracture probability of the gel was determined by ten repeated measurements (presented by *plots*). Sensory evaluation was carried out using ten subjects in duplicate, and the ratio of subjects who used tongue–palate compression or mastication for the first size reduction was determined using twenty data (presented by *bars*). (**a**) Gellan gels from series A. (**b**) From series B. (**c**) From series C. (**d**) From series D. *Closed circles*, gel fracture probability when S40 was used as artificial tongue; *closed triangles*, gel fracture probability when S50 was used as artificial tongue; *closed squares*, gel fracture probability when S60 was used as artificial tongue. *Plain dark bars*, ratio of subjects who used tongue–palate compression for the first size reduction; *slashed bars*, ratio of subjects who used mastication for the first size reduction

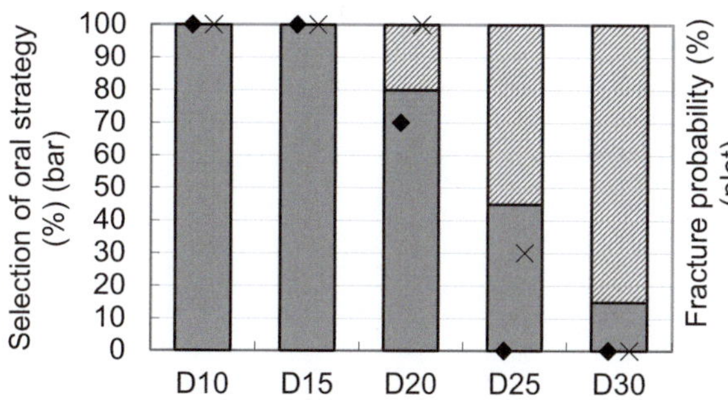

Fig. 7.8 Relationship between the fracture probability of gellan gels in instrumental compression test using S50 artificial tongue and selection of oral strategy for the first size reduction in sensory evaluation for the gels from series D. Tests were carried out principally in the same manner as Fig. 7.7, except crosshead speed or the size of artificial tongue. *Closed diamonds*, gel fracture probability at a crosshead speed of 5 mm/s; *crosses*, gel fracture probability using larger size (58 mm in diameter) S50 artificial tongue at a crosshead speed of 10 mm/s. *Plain dark bars*, ratio of subjects who used tongue–palate compression for the first size reduction; *slashed bars*, ratio of subjects who used mastication for the first size reduction

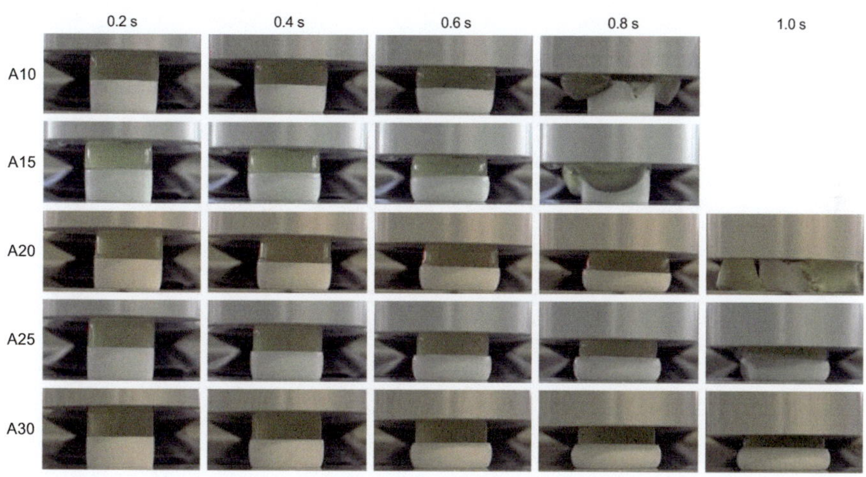

Fig. 7.9 Snapshot images representing the deformation profile for either gellan gel or artificial tongue during instrumental compression. Images were captured at a time interval of 1/5 s during instrumental compression of gellan gels with a cylindrical aluminum probe of 50 mm in diameter at a crosshead speed of 10 mm/s. Deformation profile of gellan gels from series A was shown as a representative during instrumental compression on S50 artificial tongue at 20 °C. The initial size was 20 mm in diameter and 10 mm in height for either the gel or artificial tongue

Figs. 7.9 and 7.10, where S50 was used as artificial tongue. For the gel samples from series A to C and showing more than 70 % fracture probability (i.e., A10–A20, B10–B15, and C10–C15), strain of the gel samples was always larger than that of S50 artificial tongue throughout the test (Fig. 7.10). For the gel samples from

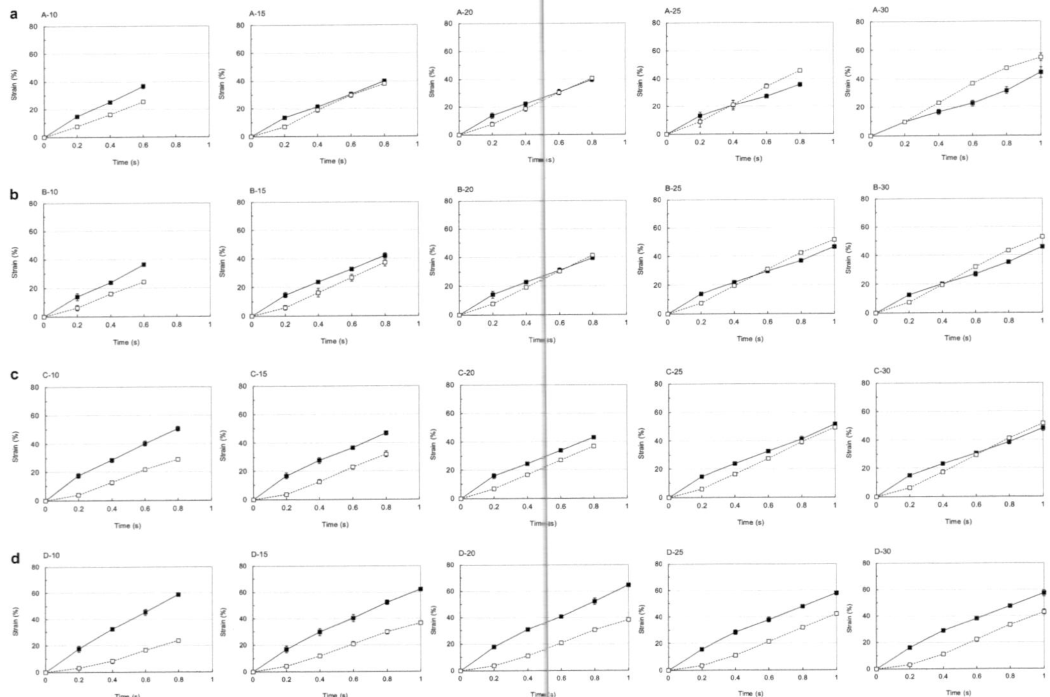

Fig. 7.10 Time course of the strain development for gellan gel and artificial tongue during instrumental compression. Nominal strain in the vertical direction was determined on the image shown in Fig. 7.9, and the time course of the strain development was presented for either gellan gel or artificial tongue. The initial size was 20 mm in diameter and 10 mm in height for each entity, and its combination was compressed with a cylindrical aluminum probe of 50 mm in diameter at a crosshead speed of 10 mm/s at 20 °C. Data were presented as means ± standard deviation in triplicate. Results before the fracture point were shown. (**a**) Gellan gels from series A. (**b**) From series B. (**c**) From series C. (**d**) From series D. *Closed squares*, gellan gel; *open squares*, S50 artificial tongue

the same series but showing less than 40 % fracture probability (i.e., A30, B25–B30, and C30), strain of the gel samples was larger than that of S50 artificial tongue below ca. 10–20 % strain for series A and B and ca. 30 % strain for series C but vice versa at larger strains. That is, a crossover was detected at ca. 10–20 % strain for series A and B and at ca. 30 % strain for series C. For C25, which also showed less than 40 % fracture probability, such a crossover was not detected, and the deformation profile was rather similar to that of series D. For the gel samples from series D and showing more than 70 % fracture probability (i.e., D10), strain of the gel sample was always larger than that of S50 artificial tongue throughout the test. This is similar to the behaviors of series A to C and also to that of agar gel. However, even for the gel samples showing less than 40 % fracture probability (i.e., D15–D30), strain of the gel samples was always larger than that of S50 artificial tongue throughout the test. This is quite different from the behaviors of series A to C and also from agar gel.

7.7 Human Test (Sensory Evaluation Asking Subjects Oral Strategy for Size Reduction)

In the agar study, the same subjects in the tongue elasticity measurements participated (see Sect. 3). They had been trained for gel texture evaluation. The gel samples (ca. 3 g) of 20 mm in diameter and 10 mm in height were served at 20 °C in duplicate in increasing order of consistency (i.e., from A1 to A7) to each subject. Subjects were allowed to process the gel samples orally without any restriction. After oral processing, subjects were asked which oral strategy was used for size reduction: tongue–palate compression or mastication.

In the gellan study, five males and five females ranging from 25 to 41 in age (31.7 years old on average) and having normal dental status participated. They had been trained for gel texture evaluation. The gel samples were served to and evaluated by subjects in the same manner as agar study. After oral processing, subjects were asked which oral strategy was used for the first size reduction (i.e., the first oral strategy for size reduction). Sensory evaluation was performed in four separate sessions. Two sample series were selected randomly from series A to D, and in each sample series, gel samples were provided to subjects in increasing order of consistency. Tests were repeated twice for each sample series. Ratio of subjects who used tongue–palate compression for the first size reduction was determined using twenty data for each gel sample. Subjects were also asked to choose the best reason for changing oral strategy for the first size reduction from tongue–palate compression to mastication from the following options: reluctance of deformation in compressing the gel sample to relatively smaller degree, reluctance of deformation in compressing the gel sample to relatively larger degree, force renitency in compressing the gel sample to relatively smaller degree, and force renitency in compressing the gel sample to relatively larger degree.

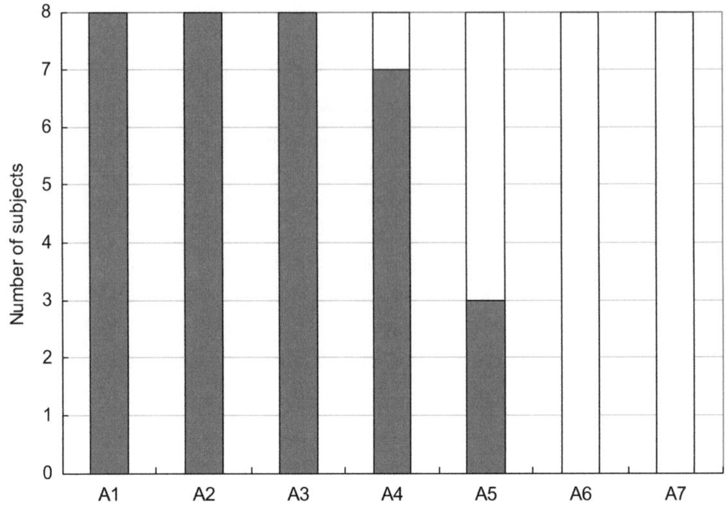

Fig. 7.11 Sensory evaluation asking subjects of oral strategy for size reduction using agar gels. *Closed*, the number of subjects who used the tongue–palate compression for size reduction; *open*, the number of subjects who used mastication for size reduction

In the agar study, oral strategy for size reduction depended on gel consistency (Fig. 7.11). Tongue–palate compression was used by all subjects for A1–A3, whereas mastication was used by all subjects for A6 and A7. For the gel samples of medium consistency (A4 and A5), both strategies were observed. One out of eight subjects used mastication for A4, and five out of eight subjects used mastication for A5. EMG studies revealed very different muscle activities by oral strategy. For subjects using tongue–palate compression for size reduction, there was hardly any detectable EMG signal from the masseter muscles, but EMG signals were detected easily from the suprahyoid muscle (Fig. 7.12a). For subjects using mastication for size reduction, signals from the masseters became much more apparent, indicating jaw closing as a result of chewing actions (Fig. 7.12b). The maximum EMG amplitudes from the masseters' activity during oral processing were measured for two representative subjects. For subject who changed the oral strategy from tongue–palate compression to mastication from the sample A5, the maximum amplitude showed a rapid increase at a boundary between A4 and A5 (Fig. 7.13a). Similarly, for subject who changed the oral strategy from the sample A6, there was also a sudden increase in the maximum amplitude at a boundary between A5 and A6 (Fig. 7.13b). Results show very good agreements between sensory evaluation and the EMG recording.

In the gellan study, all subjects used tongue–palate compression for the gel samples of relatively smaller fracture forces (10 and 15 N) regardless of fracture strain (Fig. 7.7). The number of subjects who used mastication increased with increasing fracture force. This is more apparent with increasing fracture strain,

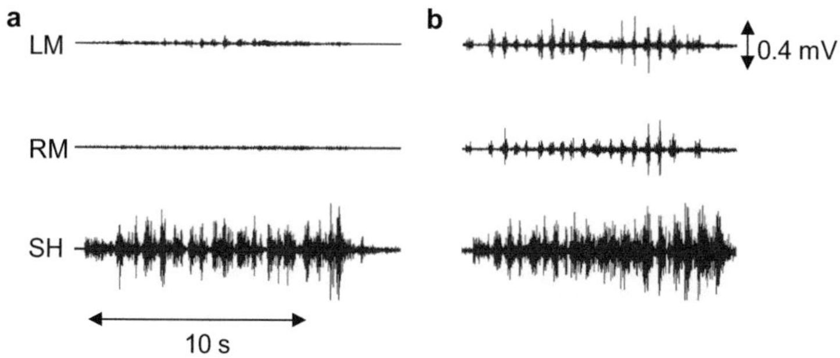

Fig. 7.12 Representative electromyograms during oral processing. The EMG signals from the left (*LM*) and right (*RM*) masseter muscles and the suprahyoid musculature (*SH*) were presented for the same subject when the subject used the tongue–palate compression (**a**) and mastication (**b**) as an oral strategy for size reduction. Results from the subject during consumption of different gels are shown

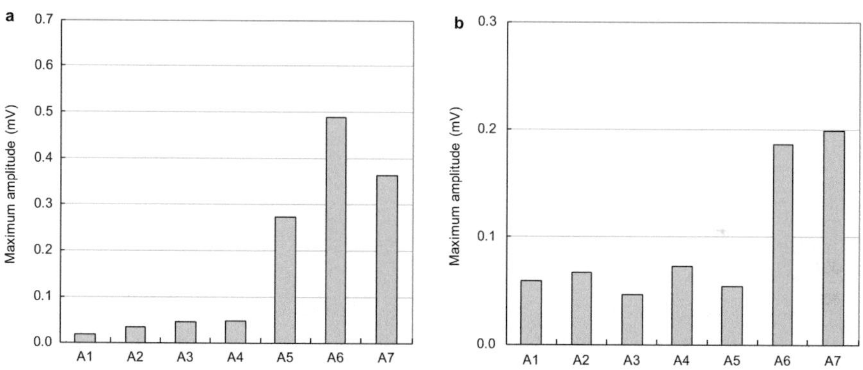

Fig. 7.13 Maximum electromyogram amplitude from the masseters' activity during oral processing. Results from two representative subjects are shown; one changed the oral strategy from the tongue–palate compression to mastication from the sample A5 (**a**), and the other did so from the sample A6 (**b**)

and series D was more apparent than series A. Equal to or more than 50 % of subjects used mastication for A30, B25–B30, C25–C30, and D25–D30. Regarding the cue for the change of oral strategy, most subjects (equal to or more than 80 %) decided to use mastication instead of tongue–palate compression by reluctance of deformation in compressing the gel samples to relatively smaller degree for series A and B, whereas most subjects (equal to or more than 90 %) did so by force renitency in compressing of the gel samples to relatively larger degree for series D. Series C was in-between; 70 % subjects determined the oral strategy by force renitency in compressing the gel sample to relatively larger degree for C25, whereas 60 % subjects did so by reluctance of deformation in compressing the gel sample to relatively smaller degree for C30. It is suggested that dominant parameter for both decision of oral strategy and its timing should depend on the initial mechanical property of the gel sample. That can be "early" strain for samples with brittle texture like agar and with low initial fracture stress whereas "later" stress for samples with deformable texture and with high initial fracture stress.

7.8 Correlation Between the Oral Strategy for Size Reduction and the Deformation and Fracture Profiles of Gels Through Instrumental Compression on Artificial Tongue

In the agar study, oral strategy changes from tongue–palate compression to mastication with increasing fracture force or apparent Young's modulus of the gels. Agar gels from A3 to A6 fracture at an equivalent strain (ca. 60 % as in Table 7.2) determined by instrumental compression between non-deformable hard platens. However, when these gel samples are compressed on S60 artificial tongue, fracture strain increases with the initial fracture force or apparent Young's modulus of the gels (Fig. 7.3c). Similarly, in compression on S50 artificial tongue, fracture strain of A3–A5 increases with the initial fracture force or apparent Young's modulus (Fig. 7.3b). In compression on S40 artificial tongue, fracture strain of A4 is larger than that of A3 (Fig. 7.3a). Metal material is much stiffer than agar gels, and its deformation relative to the gels is completely negligible. In contrast, deformation of artificial tongue during instrumental compression depends on the mechanical properties of agar gels placed on it. It has been pointed out that the force–displacement curve of the combination of food and sensory organs like the fingers and the tongue should be different from that of food alone on man-made testing machines built from very stiff materials [42]. This difference is attributed to viscoelastic characters of sensory organs, causing relaxation upon time. Also, mechanical sensitivity of sensory organs is a function of the stiffness ratio between specimen and the tissues involved. Mechanical sensitivity decreases with increasing relative stiffness of the specimen to the tissues, and when mechanical resistance is equivalent between the specimen and the tissues, the effect will be the most. With decreased apparent Young's modulus for artificial tongue from S60 to S50 or S40, its mechanical resistance comes closer to that of agar gels, thus affecting greatly the deformation behavior of the gels. In relation to sensory evaluation, it would be logical to suggest that a majority of subjects use tongue–palate compression for size reduction of agar gels when the gels fracture on an artificial tongue but use mastication when the gels do not fracture on an artificial tongue. This could be a very useful criterion in selecting artificial tongue for instrumental evaluation system for food texture. Similarity between human oral strategy (Fig. 7.11) and the fracture profile of agar gels during instrumental compression on artificial tongue (Fig. 7.3) suggests that apparent Young's modulus of human tongue should be intermediate between S40 and S50 during size reduction. As some subjects use tongue–palate compression for size reduction of A5, S50 with apparent Young's modulus of ca. 55 kPa may be the best choice for the system. It has been reported that the oral strategy for size reduction is decided at ca. 12 % strain [1]. This suggests that food texture before fracture should be a criterion for decision-making. From the strain development (Fig. 7.5), strain of agar gels dominates over that of artificial tongue at very early stage of compression in most cases. For some agar gels which do not fracture, strain of the gels is equivalent to that of artificial tongue, or strain of artificial tongue dominates over that of the gels below ca. 10 % strain (Fig. 7.5). This boundary is almost consistent with the previous report [1], validating results with each other. From a physiological point of view, the oral strategy for size reduction may be determined by sensing the difference in the deformation between food and the tongue up to ca. 10 % strain of food. In relation to the initial fracture force (determined by compression between non-deformable hard platens), agar gels with fracture force smaller than 8 N (i.e., A1–A3) are fragmented by tongue–palate compression for all subjects, whereas agar gels with fracture force larger than 36 N (i.e., A6 and A7) are fragmented by mastication for all subjects (Fig. 7.11). The boundary to change the oral strategy can be in the initial fracture force ranging between 10 and 35 N, which is also consistent with previous reports [1, 56]. Human oral strategy is determined at early stage of oral processing before fracture, and there is no way to know the fracture force at that time. From this perspective, it is reasonable to hypothesize that apparent Young's modulus should be more important as a mechanical input than fracture force for determination of the oral strategy for size reduction. The reason why fracture force relates well to the oral strategy is explained by good correlation between fracture force and apparent Young's modulus in the case of agar gels.

In the gellan study, results from sensory evaluation relate the best to the fracture profile of the gel samples through instrumental compression on S50 artificial tongue for series A to C (Fig. 7.7). For the gel samples which fracture, the ratio of subjects who use tongue–palate compression as an oral strategy for the first size reduction corresponds to the fracture probability of the gel samples (Fig. 7.7). For the gel samples which do not fracture, equal to or more than 50 % of subjects use mastication instead of tongue–palate compression (Fig. 7.7). These validate the use of S50 as artificial tongue on the instrumental evaluation system to deduce oral strategy for size reduction. For series D, on the other hand, results from sensory evaluation relate the best to the fracture profile of the gel samples through instrumental compression on S60 artificial tongue (Fig. 7.7). When S50 is used as alternative artificial tongue, correlation with sensory evaluation can be improved by decreasing crosshead speed from 10 to 5 mm/s and also by increasing the size of the artificial tongue from 20 to 58 mm in diameter (Fig. 7.8). It is suggested that oral physiology during size reduction should be changed by food texture, which requires modification of some operation conditions of the instrumental evaluation system.

7.9 Validation of Conclusion and Hypothesis

Under standard operation condition for instrumental test, results from the gellan study are comparable with those from the agar study when S50 is used as artificial tongue, and the gel fracture can relate to the oral strategy for the first size reduction when the initial fracture strain of gellan gels is smaller than 60 %. This is the case of less deformable gels from series A and B. Results are different from the agar study, on the other hand, when the initial fracture strain of gellan gels is larger than 70 %. This is the case of highly deformable gels from series D. These validate conclusions from the agar study but show a limitation at the same time for evaluation of highly deformable gels such as from series D.

Regarding the boundary strain and stress to change the oral strategy for size reduction from tongue–palate compression to mastication, hypothesis drawn by the agar study is in accordance with previous videofluorography inspection [1], both elucidating the critical strain at around 10–12 %. The hypothesis can be true when the initial fracture strain of gellan gum gels is smaller than 60 % with a shift of the critical strain to a larger one, ca. 20–30 %, while cannot be true when the initial fracture strain is larger than 70 %. From sensory evaluation, all subjects use tongue–palate compression for the first size reduction when the initial fracture force of gellan gum gels is equal to or lower than 15 N, whereas some subjects use mastication when the initial fracture force is equal to or higher than 20 N. This is also consistent with the hypothesis drawn by the agar study that oral strategy for size reduction should change from tongue–palate compression to mastication at a boundary fracture force of 10–35 N. A new finding from the gellan study is that a threshold force to change the oral strategy should decrease with increasing the initial fracture strain.

In the case of agar, gels always fracture through instrumental compression on artificial tongue when apparent Young's modulus of the gel is smaller than that of artificial tongue, while they do not when apparent Young's modulus of the gel is larger than that of artificial tongue. That is, the difference in apparent Young's modulus between the gel and artificial tongue determines the gel fracture on the instrumental evaluation system. However, this is not necessarily true for gellan gum gels, and the fracture profile of the gels cannot be always explained by the difference in apparent Young's modulus between both entities. Results should be discussed in relation to the rheological characters representing a large deformation behavior rather than a small deformation behavior such as Young's modulus. From this perspective, it may be worth comparing the strain-hardening index between gellan gum gels and artificial tongue, and the comparison was thus tried. However, the strain-hardening index was difficult to determine experimentally due to poor data reproducibility and did not work well for explaining the fracture profile of the gels (data not shown). This should be considered in the future work.

As many macroscopic mechanical parameters are not necessarily useful, microscopic difference in the gel structure can be associated with the inconsistency between agar gels and gellan gum gels. Gellan gum gels, which were prepared using two types of gellan gum with different acyl contents at various mixing ratios, form the interpenetrated network structure, where loosely packed inter-helical associations from high-acylated gellan gum and tightly packed inter-helical associations from low-acylated gellan gum exist together [50]. Structural homogeneity of the gel network structure can be different between agar and gellan gum, and this may explain the inconsistency.

7.10 Modification of Operation Condition for Improved Correlation with Human Test

Some behavioral modulation should be considered during consumption of highly deformable gels, including the tongue–palate compression speed, tongue excitation, shear loading by the tongue, and increase in the maximum deformation (i.e., decrease in the clearance). The first factor is recreated by decreasing the crosshead speed from 10 to 5 mm/s, improving in fact the correlation between instrumental and human tests in the case of highly deformable gels such as gellan gum gels from series D. This is associated with the stress relaxation, a phenomenon of energy dissipation enhanced at lower deformation speed [32]. Stress relaxation occurs preferably in soft plastic samples rather than in hard elastic samples due to low capacity of energy storage by viscous component [15]. Physiologically, decrease in the deformation speed should be for increased effectiveness of eating effort as found in cheese consumption, where decrease in the biting speed contributes to lower the biting force and the work required for consumption [8]. The second factor is recreated by using S60 artificial tongue instead of S50, improving in fact the correlation between instrumental and human tests in the case of highly deformable gels such as gellan gum gels from series D. It has been elucidated from in vivo tongue pressure measurements that the slope of the curve (time versus tongue pressure) during the first size reduction is almost independent of physical properties (from elastic to plastic) or hardness (from soft to hard) of hydrocolloid gels [18]. Based on this finding, it is reasonable to think that decrease in the tongue–palate compression speed should give rise to increased tongue pressure. This may validate the usage of lower crosshead speed or artificial tongue of higher Young's modulus for evaluation of highly deformable samples. Similar physiological behavior has been elucidated during mastication, where one bite of plastic products takes much longer time than that of elastic products [36]. Also, plastic foods like caramel confectionaries are bitten at a slower frequency than elastic foods like gelatin jellies [12], and mastication adjusts to increased adhesiveness in caramel confectionaries by decreasing chewing frequency and also by prolonging the duration [5].

The initial fracture stress of all gellan gum gels tested is larger than the critical tongue pressure to change the oral strategy from tongue–palate compression to mastication reported previously (20–30 kPa) [55] although the initial fracture stress of the gels whose initial fracture force is below 20 N is smaller than the maximum tongue pressure during size reduction reported previously (50–70 kPa) [37, 38]. These suggest the necessity to incorporate the shear force (force in a lateral axis) to a degree into the instrumental evaluation system as found in mastication [46]. It has been recognized that the jaw moves to a much bigger distance in both vertical and lateral directions in chewing plastic food than in chewing elastic food [12]. This should be considered in the future work. Reason for choosing 50 % as the maximum deformation is based on a physiological condition reported previously for assessment of perceived food hardness, sensing the biting force by keeping its deformation at 50 % strain, particularly for harder food [46]. Increasing the maximum deformation can be instrumentally feasible but makes it difficult to determine the strain of the gel samples on the image under such an extremely large deformation condition. Risk of overloading, which may damage seriously the machine, should be also avoided. Usage of artificial tongue of larger size is another option to improve the correlation with human test. In reality, it is assumed that surface contact area between gel and the tongue should be larger in consuming highly deformable gels.

7.11 Key Mechanical Parameters for Determination of Size Reduction

Texture is perceived dynamically via stress and/or strain which are generated on the oral mechanoreceptors during food oral processing. Because of the complexity of oral processing, it is not still known whether or not the mechanoreceptors are stimulated by force, deformation (or strain), energy density, and the combinations of these [52]. Physiological mechanism in assessing hardness for solid foods [46] and thickness for liquid foods [48] has been indicated, and dominant mechanical factors for texture perception are identified. However, such kind of information has been lacking so far regarding the oral strategy for size reduction. Studies by the author's research team suggest that the mode to decide oral strategy for size reduction should depend on food texture, characterized by the initial fracture strain at the boundary of ca. 60–70 %. Also, human should decide the oral strategy for size reduction by sensing the difference in strain of food relative to the tongue perceived dynamically during oral processing. Fracture strain has been emphasized as a dominant mechanical parameter for decision of the oral strategy for size reduction [1], and this is the same as the determination of the biting speed [36]. These all validate our results. It has been also shown [1] that mastication is not necessarily used for size reduction even if the initial fracture stress of gels is beyond the threshold tongue pressure (20–30 kPa), suggesting that the initial fracture stress should not be an exclusive mechanical factor for the decision. From sensory evaluation in the gellan study, oral strategy for the first size reduction for C25 can be decided at a larger deformation than that for C30. This explains the similarity of C25 to D15–D30 in the cue of oral strategy change and in the deformation profile (as shown in Fig. 7.10) although the reason for such a reversion between C25 and C30 is not clear. Gellan gum gels from series D, except D10, do not fracture through instrumental compression on artificial tongue even if strain of the gels is larger than that of artificial tongue although the same geometric change is the cause for gel fracture for series A to C, where penetration of artificial tongue into the gels triggers gel fracture. It was confirmed experimentally that artificial tongue hardened to a greater degree than gellan gum gels, particularly when the gel samples did not fracture through instrumental compression on artificial tongue (data not shown). This mimics physiological response by human, who tries to process food by exciting the tongue. Thus, strain applied in the instrumental compression test does not reach to fracture strain of each gel sample, and this is only a possible reason why no fracture of the gels occurs for series D under the standard operation condition.

It is worth investigating whether or not simulated saliva placed at the interface between gel sample and artificial tongue could affect results due to the change of lubrication at the interface. Preliminary study showed that simulated saliva placement at the interface did not affect results greatly for some gellan gum gels, but this should be investigated further in the future work. Also, the surface of artificial tongue used is hydrophobic, and usage of more hydrophilic substrates should be of another research interest because compatibility with simulated saliva can be increased.

7.12 Conclusion

There is an increasing demand for texture-controlled foods which can be consumed by tongue–palate compression without the need of mastication as the number of the elderly is expanding. An in vitro evaluation system of food texture developed provides food manufactures with a simple method for assessing food textural acceptability without using human subjects. The point is to use artificial tongue fabricated from a soft deformable elastomer instead of hard non-deformable metal platen as a substrate. The evaluation system can contribute to not only the industry but also the fundamental food texture study, leading to better understanding of the oral physiology of eating, particularly size reduction, during consumption of solid or semisolid foods.

References

1. ARAI, E. and YAMADA, Y. 1993. Effect of the texture of food on the masticatory process. Jpn. J. Oral Biol. 35, 312–322.
2. BARRANGOU, L.M., DRAKE, M.A., DAUBERT, C.R. and FOEGEDING, E.A. 2006. Sensory texture related to large-strain rheological properties of agar/glycerol gels as a model food. J. Texture Studies 37, 241–262.

3. BOURNE, M.C. 2000.Why so many tests to measure texture? In Hydrocolloids, Part 2: Fundamentals and Applications in Food, Biology, and Medicine (K. Nishinari, ed.) pp. 425–430, Elsevier Science B.V., Amsterdam, The Netherlands.
4. BREUIL, P. and MEULLENET, J.-F. 2001. A comparison of three instrumental tests for predicting sensory texture profiles of cheese. J. Texture Studies 32, 41–55.
5. CAKIR, E., KOC, H., VINYARD, C.J., ESSICK, G., DAUBERT, C.R., DRAKE, M. and FOEGEDING, E.A. 2012. Evaluation of texture changes due to compositional differences using oral processing. J. Texture Studies 43, 257–267.
6. CHEN, J. 2009. Food oral processing – a review. Food Hydrocoll. 23, 1–25.
7. DAN, H. and KOHYAMA, K. 2007. Interactive relationship between the mechanical properties of food and the human response during the first bite. Arch. Oral Biol. 52, 455–464.
8. DAN, H., HAYAKAWA, F. and KOHYAMA, K. 2008. Modulation of biting procedures induced by the sensory evaluation of cheese hardness with different definitions. Appetite 50, 158–166.
9. ENGELEN, L., FONTIJN-TEKAMP, F.A. and VAN DER BILT, A. 2005. The influence of product and oral characteristics on swallowing. Arch. Oral Biol. 50, 739–746.
10. ENGWALL, O. 2003. Combining MRI, EMA and EPG measurements in a three-dimensional tongue model. Speech Commun. 41, 303–329.
11. EVERARD, C.D., O'CALLAGHAN, D.J., HOWARD, T.V., O'DONNELL, C.P., SHEEHAN, E.M. and DELAHUNTY, C.M. 2006. Relationships between sensory and rheological measurements of texture in maturing commercial cheddar cheese over a range of moisture and pH at the point of manufacture. J. Texture Studies 37, 361–382.
12. FOSTER, K.D., WODA, A. and PEYRON, M.A. 2006. Effect of texture of plastic and elastic model foods on the parameters of mastication. J. Neurophysiol. 95, 3469–3479.
13. FUNAMI, T., ISHIHARA, S., NAKAUMA, M., KOHYAMA, K. and NISHINARI, K. 2012. Texture design for products using food hydrocolloids. Food Hydrocolloids 26, 412–420.
14. GAIGE, T.A., BENNER, T., WANG, R., WEDEEN, V.J. and GILBERT, R.J. 2007. Three dimensional myoarchitecture of the human tongue determined in vivo by diffusion tensor imaging with tractography. J. Magn. Reson. Imaging 26, 654–661.
15. GUO, Q., YE, A., LAD, M., DALGLEISH, D. and SINGH, H. 2013. The breakdown properties of heat-set whey protein emulsion gels in the human mouth. Food Hydrocolloids 33, 215–224.
16. HEATH, M.R. and PRINZ, J.F. 1999. Oral processing of foods and the sensory evaluation of texture. In Food Texture. Measurement and Perception (A.J. Rosenthal, ed.) pp. 18–29, Aspen Publishers, Gaithersburg, MD.
17. HIIEMAE, K.M. and PALMER, J.B. 2003. Tongue movements in feeding and speech. Crit. Rev. Oral Biol. Med. 14, 413–429.
18. HORI, K., HAYASHI, H., YOKOYAMA, S., ONO, T., ISHIHARA, S., MAGARA, J., TANIGUCHI, H., FUNAMI, T., MAEDA, Y. and INOUE, M. 2015. Comparison of mechanical analysis and tongue pressure analyses during squeezing and swallowing gels. Food Hydrocolloids 44, 145–155.
19. HORI, K., ONO, T., TAMINE, K., KONDO, J., HAMANAKA, S., MAEDA, Y., DONG, J. and HATSUDA, M. 2009. Newly developed sensor sheet for measuring tongue pressure during swallowing. J. Prosthodont. Res. 53, 28–32.
20. HUANG, Y., WHITE, D.P. and MALHOTRA, A. 2007. Use of computational modeling to predict responses to upper airway surgery in obstructive sleep apnea. Laryngoscope 117, 648–653.
21. ISHIHARA, S., NAKAUMA, M., FUNAMI, T., TANAKA, T., NISHINARI, K. and KOHYAMA, K. 2011. Electromyography during oral processing in relation to mechanical and sensory properties of soft gels. J. Texture Studies 42, 254–267.
22. ISHIHARA, S., ISONO, M., NAKAO, S., NAKAUMA, M., FUNAMI, T., HORI, K., ONO, T., KOHYAMA, K., and NISHINARI, K. 2014. Instrumental uniaxial compression test of gellan gels of various mechanical properties using artificial tongue and its comparison with human oral strategy for the first size reduction. J. Texture Studies 45, 354–366.

23. ISHIHARA, S., NAKAO, S., NAKAUMA, M., FUNAMI, T., HORI, K., ONO, T., KOHYAMA, K., and NISHINARI, K. 2013. Compression test of food gels on artificial tongue and its comparison with human test. J. Texture Studies 44, 104–114.
24. ISHIHARA, S., NAKAUMA, M., FUNAMI, T., ODAKE, S., NISHINARI, K. 2011. Viscoelastic and fragmentation characters of model bolus from polysaccharide gels after instrumental mastication. Food Hydrocolloids 25, 1210–1218.
25. KIESER, J., BOLTER, C., RANIGA, N., WADDELL, J.N., SWAIN, M. and FARLAND, G. 2011. Tongue-palate interactions during swallowing. J. Texture Studies 42, 95–102.
26. KOHYAMA, K., SAKAI, T. and AZUMA, T. 2001. Patterns observed in the first chew of foods with various textures. Food Sci. Technol. Res. 7, 290–296.
27. KOHYAMA, K., SAKAI, T., AZUMA, T., MIZOGUCHI, T. and KIMURA, I. 2001. Pressure distribution measurement in biting surimi gels with molars using a multiple-point sheet sensor. Biosci. Biotechnol. Biochem. 65, 2597–2603.
28. KOHYAMA, K., SASAKI, T. and DAN, H. 2003. Active stress during compression testing of various foods measured using a multiple-point sheet sensor. Biosci. Biotechnol. Biochem. 67, 1492–1498.
29. KOHYAMA, K., SASAKI, T. and HAYAKAWA, F. 2008. Characterization of food physical properties by the mastication parameters measured by electromyography of the jaw-closing muscles and mandibular kinematics in young adults. Biosci. Biotechnol. Biochem. 72, 1690–1695.
30. LEE, S.-Y., LUNA-GUZMAN, I., CHANG, S., BARRETT, D.M. and GUINARD, J.-X. 1999. Relating descriptive analysis and instrumental texture data of processed diced tomatoes. Food Qual. Prefer. 10, 447–455.
31. LUCAS, P.W., PRINZ, J.F., AGRAWAL, K.R. and BRUCE, I.C. 2002. Food physics and physiology. Food Qual. Prefer. 13, 203–213.
32. LUYTEN, H. and VAN VLIET, T. 1995. Fracture properties of starch gels and their rate dependency. J. Texture Studies 26, 281–298.
33. MALHOTRA, A., HUANG, Y., FOGEL, R.B., PILLAR, G., EDWARDS, J.K., KIKINIS, R., LORING, S.H. and WHITE, D.P. 2002. The male predisposition to pharyngeal collapse: Importance of airway length. Am. J. Respir. Crit. Care Med. 166, 1388–1395.
34. MEULLENET, J.-F. and GROSS, J. 1999. Instrumental single and double compression tests to predict sensory texture characteristics of foods. J. Texture Studies 30, 167–180.
35. MEULLENET, J.-F., LYON, B.G., CARPENTER, J.A. and LYON, C.E. 1998. Relationship between sensory and instrumental texture profile attributes. J. Sensory Studies 13, 77–93.
36. MIOCHE, L. and PEYRON, M.A. 1995. Bite force displayed during assessment of hardness in various texture contexts. Arch. Oral Biol. 40, 415–423.
37. MORITA, A. and NAKAZAWA, F. 2002. Palatal pressure and electromyography while eating gelatin, agar and carrageenan jelly (in Japanese with English summary and figure captions). J. Home Econ. Jpn. 53, 7–14.
38. MORITA, A. and NAKAZAWA, F. 2005. Representation of mastication and swallowing of gellan jelly by palatal pressure measurements (in Japanese with English summary and figure captions). J. Home Econ. Jpn. 56, 425–434.
39. NAPADOW, V.J., CHEN, Q., WEDEEN, V.J. and GILBERT, R.J. 1999. Biomechanical basis for lingual muscular deformation during swallowing. Am. J. Physiol. Gastrointest. Liver Physiol. 277, G695–G701.
40. NAPADOW, V.J., CHEN, Q., WEDEEN, V.J. and GILBERT, R.J. 1999. Intramural mechanics of the human tongue in association with physiological deformations. J. Biomech. 32, 1–12.
41. PELEG, M. 1984. A note on the various strain measures at large compressive deformations. J. Texture Studies 15, 317–326.
42. PELEG, M. 2006. On fundamental issues in texture evaluation and texturization – a view. Food Hydrocolloids 20, 405–414.
43. PELEG, M. and CAMPANELLA, O.H. 1988. On mathematical form of psychophysical relationships with special focus on the perception of mechanical properties of solid objects. Percept. Psychophys. 44, 451–455.

44. PELEG, M. and CAMPANELLA, O.H. 1989. The mechanical sensitivity of soft compressible testing machines. J. Rheol. 33, 455–467.
45. PELEG, M. and CORRADINI, M.G. 2012. Soft machine mechanics and oral texture perception. In Food Oral Processing. Fundamentals of Eating and Sensory Perception (J. Chen and L. Engelen, eds.) pp. 319–336, Blackwell Publishing Ltd., Chichester, UK.
46. PEYRON, M.A., MIOCHE, L. and CULIOLI, J. 1994. Bite force and sample deformation during hardness assessment of viscoelastic models of foods. J. Texture Studies 25, 59–76.
47. QUINCHIA, L.A., VALENCIA, C., PARTAL, P., FRANCO, J.M., BRITO-DE FUENTE, E. and GALLEGOS, C. 2011. Linear and non-linear viscoelasticity of puddings for nutritional management of dysphagia. Food Hydrocolloids 25, 586–593.
48. SHAMA, F. and SHERMAN, P. 1973. Identification of stimuli controlling the sensory evaluation of viscosity. II: Oral methods. J. Texture Studies 4, 111–118.
49. SHIBATA, A., HIGASHIMORI, M., RAMIREZ-ALPIZAR, I.G. and KANEKO, M. 2012. Tongue elasticity sensing with muscle contraction monitoring. Proceedings of the 2012 ICME International Conference on Complex Medical Engineering (CME2012, July 2, 2012, Kobe, Japan).
50. SHINSHO, A. and MATSUKAWA, S. 2012. Evaluation of gels and elucidation of the gelation mechanism for mixed gellan solutions (in Japanese with English summary). Foods & Food Ingredients J. Jpn. 217, 170–176.
51. STONE, M., EPSTEIN, M. and ISKAROUS, K. 2004. Functional segments in tongue movement. Clin. Linguist. Phon. 18, 507–521.
52. STRASSBURG, J., BURBIDGE, A. and HARTMANN, C. 2009. Identification of tactile mechanisms for the evaluation of object sizes during texture perception. Food Qual. Prefer. 20, 329–334.
53. SWORN, G. 2000. Gellan gum. In Handbook of Hydrocolloids (G.O. Phillips and P.A. Williams, eds.) pp. 117–135, Woodhead Publishing, Cambridge.
54. SZCZESNIAK, A.S. 2002. Texture is a sensory property. Food Qual. Prefer. 13, 215–225.
55. TAKAHASHI, J. and NAKAZAWA, F. 1991. Palatal pressure patterns of gelatin gels in the mouth. J. Texture Studies 22, 1–11.
56. TAKAHASHI, J. and NAKAZAWA, F. 1992. Effects of dimensions of agar and gelatine gels on palatal pressure patterns. J. Texture Studies 23, 139–152.
57. VAN DEN BERG, L., VAN VLIET, T., VAN DER LINDEN, E., VAN BOEKEL, M.A.J.S. and VAN DE VELDE, F. 2007. Breakdown properties and sensory perception of whey proteins/polysaccharide mixed gels as a function of microstructure. Food Hydrocolloids 21, 961–976.
58. VINYARD, C.J., WALL, C.E., WILLIAMS, S.H. and HYLANDER,W.L. 2008. Patterns of variation across primates in jaw-muscle electromyography during mastication. Integr. Comp. Biol. 48, 294–311.
59. WODA, A., FOSTER, K., MISHELLANY, A. and PEYRON, M.A. 2006. Adaptation of healthy mastication to factors pertaining to the individual or to the food. Physiol. Behav. 89, 28–35.
60. YUAN, S. and CHANG, S.K.C. 2007. Texture profile of tofu as affected by Instron parameters and sample preparation, and correlations of Instron hardness and springiness with sensory scores. J. Food Sci. 72, S136–S145.

Chapter 8
Gelation Characteristics of Heat-Induced Gels Mixed Meat with Fish Proteins

Yasuhiro Funatsu and Tomohito Iwasaki

Abstract The aim of this study was to make a comprehensive assessment of the quality of fish surimi (FS) from scale-eye plaice (SP) and surimi-like products from culled hen and FS from sand eel (SE) and Arabesque greenling (AG), which are regarded as low-valued food resources. The thermal gelation properties of a variety of the gels mixed meat with fish proteins were investigated by the use of the rheological and protein chemistry approaches, and the obtained results are as follows:

(1) SP surimi has advantages such as high whiteness and *suwari* phenomenon at low temperature.
(2) A variety of heat-induced gel products were able to produce with different mixing ratios of ground culled hen (GCH) and FS regardless of adding egg albumen powder (EAP).
(3) Disulfide bond (S-S bond) was related to an improvement in gel properties of the directly heated gel mixed FS with GCH by addition of EAP. In particular, cleavage of myosin heavy chains in the gels, which have a high mixing ratio of SE surimi, was inhibited by addition of EAP.
(4) AG can be used as ingredients of heat-induced gel products mixed fish with hen, while SP is inappropriate for the product because of syneresis phenomenon by use of microbial transglutaminase as a reagent for gelation. These findings revealed that adequate understanding of gel-forming ability of FS could be essential to improve gel properties of the heat-induced gels mixed meat with FS.

Keywords Fish surimi • Culled hen • Egg albumen powder • Heat-induced gel • Gel property

---

Y. Funatsu (✉) • T. Iwasaki
Rakuno Gakuen University, Ebetsu, Japan
e-mail: funatsu@rakuno.ac.jp

I. Kaneda (ed.), *Rheology of Biological Soft Matter*, Soft and Biological Matter,
DOI 10.1007/978-4-431-56080-7_8

8.1 Introduction

There are variety of processed marine products such as dry products, surimi-based products, fermented products, and canned products in Japan now because the Japanese people have been eating fish and shellfish since ancient times. *Kamaboko* is one of the surimi-based products and it has a unique elastic food texture. *Kamaboko* is produced by heating after grinding with salt and fish meat and its main material has been used as frozen surimi. Frozen surimi was developed by Mr. K. Nishiya's study group from the Hokkaido fisheries experimental station (Abashiri Fisheries Research Institute, Hokkaido Research Organization, called now) in Japan in 1960 [1]. The common practice for making frozen surimi is as follows [2]. The fish is filleted by removing the head, bones, and entrails, and then the fillet is chopped into minced meat. The dewatered meat is produced by dewatering with a screw press after leaching the minced meat. Before the final dewatering under a screw press, impurities such as the skin, fine bones, scales, and connective tissues are removed by the refiner. The dewatered meat after refining is mixed with carbohydrates such as sucrose, sorbitol, and sodium polyphosphates using mixer and then is quickly frozen at −25 °C with a contact freezer. It has become possible to control fish protein denaturation and keep neutral pH during frozen storage by adding carbohydrates and sodium polyphosphates, respectively [3]. Fishes, which were caught in large amounts all at once, were provided with the main materials of surimi-based products throughout the year because of long-term frozen storage. Due to reasons like these, surimi has been produced from various kinds of fishes not only in Japan but also around the world in recent years. Total world production came up to about 500,000 tons.

Domestic production of frozen surimi in Japan has been considerably decreasing since a 200-mile exclusive fishery zone was enacted in 1977, although the production reached to 450,000 tons in 1976. Imported surimi from America, Thailand, and China, etc., have come to account for 70 % of the domestic supply today. So, it is necessary to develop new domestic surimi for the purpose of efficient utilization of unutilized and underutilized fishes that live in coastal areas to reduce the degree of the imported surimi in the fish paste manufacturing industry inside and outside Hokkaido, Japan [4].

Hens (20 months old) declined egg-laying rate have been changing to culled hen at tens of thousands of units since a whole group updating system was adopted at a unit of hen house in Japan. It is necessary to develop a technology capable of enhancing the added value of the culled hen due to the inferior quality of meat, low performance of meat production, and limits to usage applications in comparison with a broiler [5].

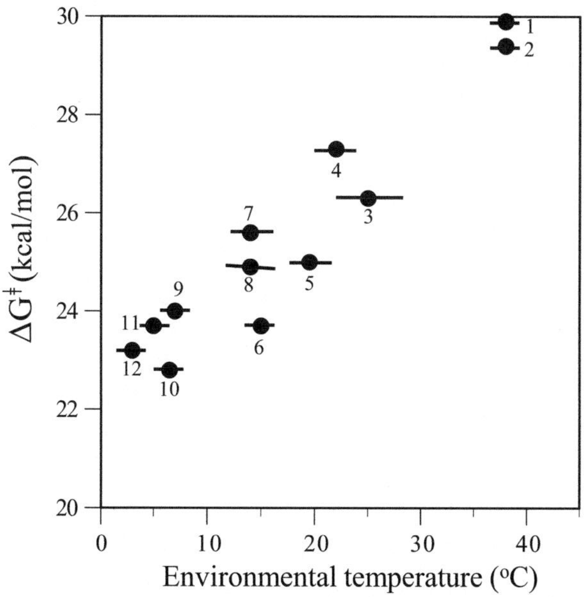

Fig. 8.1 The relationship between thermodynamic parameters for inactivation of fish myofibrillar Ca-ATPase and the environmental temperatures at which the fish lives. *1* Rabbit, *2* little piked whale, *3* yellowfin tuna, *4* skipjack tuna, *5* bigeye tuna, *6* pacific mackerel, *7* yellowtail, *8* chum salmon, *9* Atka mackerel, *10* Atka mackerel, *11* Pacific halibut, *12* Alaska pollack (Adapted from reference [6])

According to UN reports, the world population is expected to reach about 9.6 billion by 2050 because it is on an upward trend. There has been a continued scarcity of food as protein sauce along with a rise in the world population. It is well-known that thermal stability of the protein is quite different in land animals from marine ones, namely, species specificity [6], although both are important as protein sauce (Fig. 8.1). If it possible to produce a variety of heat-induced gel products with different mixing ratios of meat and fish proteins, it could be very useful in the securing of a stable food supply as a protein source in the future. It is necessary to investigate the role of the proteins concerning gelation characteristics of the heat-induced gel products in order to make it realized.

Due to those circumstances, in this study, the development of various heat-induced gel products mixed meat with fish proteins was conducted, and the analysis results from the rheological and protein chemistry approaches concerning their gelation characteristics will be discussed in Chap. 8.

8.2 Analysis of Gelation Characteristics of Heat-Induced Gels Prepared with Fish Proteins

8.2.1 Analysis Method for Gel-Forming Ability of Surimi

The quality of frozen surimi was previously evaluated with an indirect procedure concerning gel-forming ability. The thermal gelation property of frozen surimi was reflected very minimally due to the heating condition which was restricted when measurement of the physical properties was conducted. The temperature vs. time curve was determined by calculations from the relationship between temperature and gel property of products prepared after heating salt-ground meats under constant temperature in the range from 30 to 90 °C for 20 and 120 min [7]. Some information concerning *suwari* (this phenomenon is said to be caused by reduction of viscosity and gelation of salt-ground meat in the case of it being left at room temperature) and *modori* (this phenomenon is said to be caused by the degradation of the fully formed gel structure at the time of passing the temperature zone in the range from 50 to 70 °C by preheating with salt-ground meat) phenomena are able to obtain by use of the temperature vs. time curve. However, there is a little information concerning gel-forming ability of the frozen surimi in comparison with data measured to reach maximum value of the gel property. Abe et al. [8] and Kitakami et al. [9] investigated the change in gel property using a penetration test for heating gel from frozen surimi during the preheating time. So, the breaking strength (BS: g) and breaking strain (bs: cm) of the heated gel were measured with a rheometer by using a spherical plunger of 0.5 cm in diameter. It was defined that BS divided by bs was gel stiffness or stiffness (Gs: g/cm). The gelation characteristics of surimi using the relationship between Gs and BS were analyzed. Gs involves force per unit area at break. The slope of a regression line and localization of plot graphics differ from quality of surimi and the method for its preparation. It is shown that characteristics of gels are prepared from various kinds of surimi by comparing rheological properties as mentioned above. In this study, an analysis of the gelation characteristics was conducted by the method of Kitakami et al. [9].

8.2.2 Quality Assessment of Surimi from Scale-Eye Plaice

This section describes the case study with scale-eye plaice (SP) (*Acanthopsetta nadeshnyi*). SP has been bycaught in the central area in Hokkaido, Japan. However, it is returned to the sea straight after catching it so the food-processing applications are poor. It is estimated that the annual haul will see about 300 tons.

8.2.2.1 Experimental Protocol Concerning Preparation of Fish Surimi (FS)

Frozen surimi from SP (SP surimi) was prepared on a small industrial scale and stored at −30 °C before use. Additives of SP surimi were used as 6 % adjusted sugar (the composition ratio was sorbitol and sucrose, 83 % and 17 %, respectively) and 0.2 % polyphosphate (the composition ratio was sodium polyphosphate and sodium pyrophosphate, 50 % and 50 %, respectively). The mixture of SP surimi and water with various ratios was ground with 3 % salt using a silent cutter (SCP-2, Hanaki-mfg Co., ltd., Tokyo) under 10 °C for 15 min after partially thawing it up to −5 °C. The salt-ground meat was stuffed into a polyvinylidene chloride (PVDC) casing of 4.8 cm in folding diameter, and it was preheated at 10–50 °C for 96 h (preheated gel) and subsequently heated at 90 °C for 30 min (two-step heated gel). The gels were sliced 3.0 cm in diameter and 2.5 cm in length after cooling with running water and storing them at 25 °C in one night. The Bs and bs were measured using a rheometer (NRM2002J, Fudo industrial Co., Ltd., Tokyo) with a spherical plunger of 0.5 cm in diameter. The velocity of approach of the plunger was set at 60 mm/min. The number of determinations was 6–12 times per one sample, and the data obtained in the experiment were expressed as mean.

8.2.2.2 Proximate Composition, Color, and Impurity of FS

Proximate composition, color, and impurity of SP surimi were measured by the quality inspection standard [10], and the results are shown in Table 8.1. The protein level was higher in SA, A, and 2nd grades of walleye pollack (WP) surimi than in SP surimi, while the moisture level was in the reverse order [11]. The pH value in SP surimi was the same as that in 2nd grade of surimi [11]. Hunter whiteness (HW) was higher in SP surimi (37.0) than in FA (30.0) and 2nd (19.0) grades of WP surimi. The impurity was 39 number/10 g of surimi (Score: 1), suggesting that a small-sized epidermis could be contaminated during the production process.

8.2.2.3 Protein Concentration Dependence of Gel-Forming Ability of Heat-Induced Gel Mixed FS with Water

In research for sample concentration dependency on the texture of heat-induced gel, elastic constant of the gel was considerably dependent on its concentration, and it was indicated by the following formula, i.e., $\mathrm{G} = K\mathrm{C}^{\mathrm{a}}$ (K and a are constant, which are determined by the substance) from the fact that modulus of elasticity (G) was often proportional to the square of the concentration [12]. Kitakami et al. [13] reported that the relationship between BS and protein concentration (C) of two-step heated gel from several kinds of frozen surimi was expressed as the following formula, i.e., $\mathrm{BS} = KC^{a}$ (in particular, the higher in grade of surimi, the higher in a value). Protein concentration dependence of maximum BS of two-step heated gels

Table 8.1 Proximate composition, color, and impurity of SE surimi

Product	
Proximate composition	
Moisture (g/100 g)	81.3
Protein (g/100 g)	11.3
Lipid (g/100 g)	1.0
Ash (g/100 g)	0.4
Carbohydrate (g/100 g)	6.0
pH	7.6
Color	
L*	71.0
a*	−0.9
b*	6.5
W*	70.3
HW	37.0
Impurity	
Number/10 g	39
Score	1

Hunter whiteness, Z/1.18. W*, whiteness degree
Score: 1, many; 0, few. (Source: From unpublished data)

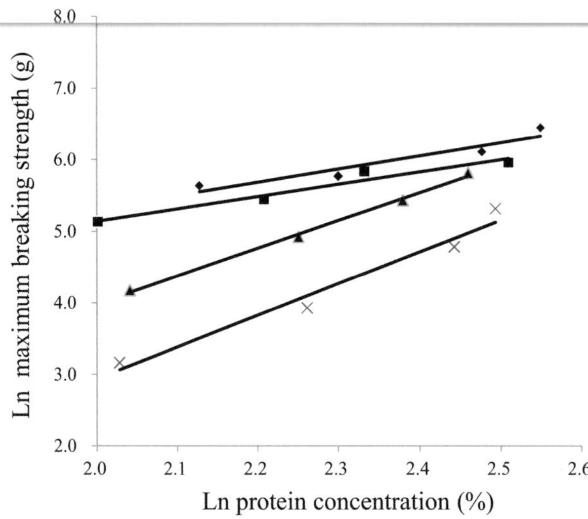

Fig. 8.2 Protein concentration dependence of maximum breaking strength of two-step heated gel mixed fish surimi with water. Preheating temperature: 10 °C (◆), 25 °C (■), 40 °C (▲), and 50 °C (×) (Source: From unpublished data)

mixed FS with water is shown in Fig. 8.2. Protein concentration used the data of the directly heated gel. Compared to a slope of each line and maximum value of BS, the former is lower in the preheated gel prepared at low temperature (10 and 25 °C) than

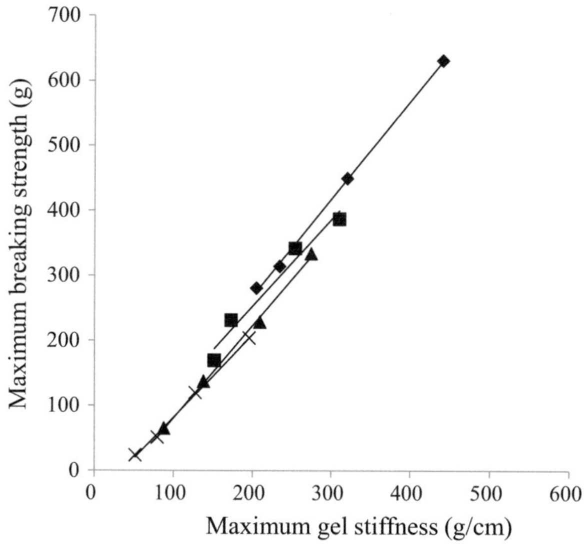

Fig. 8.3 Relationship between maximum breaking strength and maximum gel stiffness of two-step heated gels formed. Refer to the legend of Fig. 8.2 for symbols (Source: From unpublished data)

in the preheated gel prepared at high temperature (40 and 50 °C), while the latter is in the reverse order at low protein concentrations. Therefore, the main materials could be economized on in the manufacturing process of the product owing to the *suwari* gel that is formed at low temperature.

8.2.2.4 Relationship Between Maximum Breaking Strength (BS) and Maximum Gel Stiffness (Gs) of the Heat-Induced Gels

Figure 8.3 shows that the relationship between maximum BS and maximum Gs of two-step heated gels was formed. A positive correlation was recognized in spite of preheating temperature, and a linear equation, i.e., Max.BS $= a \cdot$ Max.Gs $+ b$ (a and b are constant of each gel), was recognized. Table 8.2 shows the relational expression of various two-step heated gels at different heating temperature. According to the relational expression, not only slope of a line but also maximum BS was high at low temperature, suggesting that *Kamaboko* having a specific texture with chewability could be produced at low temperatures. Table 8.3 shows the comparison relationship between maximum breaking strength and maximum gel stiffness of two-step heated gels from different grades of frozen surimi. In comparing the gel properties of SP surimi with that of WP surimi, the gel-forming ability of the former was almost the same as that of two grades of the latter judging from a slope of the line and maximum BS. From the above, SP surimi might utilize the material of fish

Table 8.2 Relational expression of various two-step heated gels at different heating temperature

Heating temperature (°C)	y=ax+b (y: Max.BS, x: Max.Gs) a	b	r^2
10	1.504	−32.9	0.999
25	1.422	−61.6	0.998
40	1.337	−14.8	0.966
50	1.277	−46.3	0.998

A regression line was fitted using the least squares method on the basis of the data shown in Fig. 8.3 (Source: From unpublished data)

Table 8.3 Comparison relationship between maximum breaking strength and maximum gel stiffness of two-step heated gels from different grades of frozen surimi

Grade	y=ax+b (y: Max.BS, x: Max.Gs) a	b	r^2	Max.BS
SA	1.700	21.8	0.998	1708
FA	1.679	−64.7	0.999	1313
A	1.527	−45.7	0.998	1375
2nd	1.384	−84.4	1.000	479
SP (10 °C)	1.504	−32.9	0.999	625

Data for SA, FA, A, and 2nd grades of frozen surimi from walleye pollack and SP surimi were quoted from reference [2] and Fig. 8.3, respectively (Source: From unpublished data)

meat paste by means of improvement of surimi production technology in the near future because it has advantages such as high whiteness and *suwari* phenomenon at low temperature, although it has disadvantages such as low protein level and high impurity concentration.

8.2.3 Gelation Characteristics of Variety of Heat-Induced Gels Mixed Ground Chicken with FS

Sand eel (*Ammodytes personatus*) is one of the fishes, which is caught in the northern area in Hokkaido, Japan. The catch was 8,000 tons in 2013, and a large amount of the catch was seen in the summer season from May to August [14]. Most of the catch is used as fish culture feed on account of its low usage application as foods [14]. So, the effective use of sand eel as food materials has been required. This section describes that case study concerning quality characteristics of surimi-like products from sand eel and culled hen, which are regarded as low-valued food resources.

8.2.3.1 Experimental Protocol Concerning Preparation of Surimi-Like Products from Ground Culled Hen (GCH) and FS

Frozen surimi from sand eel (SE) surimi was purchased from a commercial product (Kusach Shoten Co., ltd., Wakkanai). A ground culled hen (GCH) was prepared as follows. A breast muscle of the culled hen was collected after slaughter. The subcutaneous fat in the breast muscle (*M. superficial* and *M. deep pectoral*) was removed, and the muscles were minced with a mincer. The minced meat was strained with a strainer (Type 41-AS-NF; Akashi Steel Mill Co., Ltd., Tokyo). Strained meat that was put through a strainer of 1.5 mm in diameter was mixed with additives, i.e., 4 % sorbitol and 0.3 % sodium polyphosphate. The mixed meat was divided into 1 kg per piece and packaged with a freeze pack. After packaging, the ground culled hen was frozen at −30 °C before use.

Table 8.4 shows the proximate composition of GCH and SE surimi. The moisture, lipid, and pH levels were 6.1 %, 0.6 %, and 0.5 lower in GCH than in SE surimi, respectively, while the protein level was higher in GCH than in SE surimi. And then, egg albumen powder (K type, kewpie Co., Ltd., Tokyo) was used due to the reinforced gel-forming ability of a variety of meat products in this study. The moisture, protein, and pH levels in EAP were 11.6 %, 76.0 %, and 7.1 %, respectively [15].

Table 8.5 shows the mixing ratio of water into GCH mixed with SE surimi. The mixture of GCH, SE surimi, and water with various ratios was ground with 3 % NaCl with and without 2 % EAP using a silent cutter (SCP-2, Hanaki-mfg Co., ltd., Tokyo) under 5 °C for 15 min after partially thawing it up to −5 °C. The mixture of GCH and water, or SE surimi and water with various ratios, was ground in a same manner as mentioned above. In general, heat-induced gel from FS was prepared by use of two-step heating after kneading. However, in case of GCH, the preheating temperature was selected 50 °C in this study due to heat-induced gelation which was insufficient in the range from 25 to 40 °C. The salt-ground meat was stuffed into PVDC casing of 4.8 cm in folding diameter, and it was preheated at 50 °C for 6 h (preheated gel) and subsequently heated at 90 °C for 30 min (two-step heated gel). The directly heated gel was prepared at 90 °C for 30 min without preheating. Method for measurement of the gel property was carried out in a same manner as in Chap. 8.1.

Table 8.4 Proximate composition of GCH and SE surimi

		Content (%)					
Product	Materials	Moisture	Protein	Lipid	Sugar	Na-PPi	pH
Ground culled hen	Breast muscle	69.2	23.6	1.9		0.30	6.3
(GCH)					Sorbitol, 4.0		
Fish surimi	Sand ell	75.3	15.5	2.5	Sucrose, 3.0	0.25	6.8
(SE)					Sorbitol, 3.0		

Na-PPi: Sodium polyphosphate (Source: From reference [4] with permission)

Table 8.5 Mixing ratio of water into GCH mixed with SE surimi

Mixing ratio (w/w)			Water	Albumen
Ground culled hen	SE surimi	Water	(%)	(%)
50	50	0	0	0
42.5	42.5	15	18	0
35	35	30	43	0
27.5	27.5	45	82	0
20	20	60	150	0
50	50	0	0	2
42.5	42.5	15	18	2
35	35	30	43	2
27.5	27.5	45	82	2
20	20	60	150	2

The frozen GCH and the frozen SE surimi were thawed, and various ratios of water were added to the mixtures of GCH and SE surimi (1:1). The mixtures were ground with 3.0 % NaCl with and without 2 % albumen powder
(Source: From reference [4] with permission)

8.2.3.2 Effect of Egg Albumen Powder on Gel-Forming Ability of Heat-Induced Gels Mixed GCH with FS

Figure 8.4 shows that the preheating time is dependent on changes in the breaking strength (BS) and breaking strain (bs) of the two-step heated gels with or without albumen as a function of mixing with GCH, SE surimi, and water. In the absence of EAP (A), BS (a) of gel from GCH alone was almost constant during preheating although it deteriorated along with an increase in the mixing ratio of water. The change in bs (b) of the two-step heated gel from GCH alone during preheating was similar to that of BS (a) except for the degree of deterioration along with an increase in the mixing ratio of water which was smaller in bs than in BS. BS (c) and bs (d) were both decreased along with an increase in the mixing ratio of water and decreased with preheating time. The higher in mixing ratio of water, the larger the gel property was deteriorated. The BS (e) and bs (f) of the heat-induced gels from SE surimi alone were able to measure those of the directly heated gel. However, gel properties of the heat-induced gels were not able to be measured due to considerably progress of the deterioration (e, f).

In the presence of EAP (B), changes in BS (a) and bs (b) during preheating were almost the same as the changes in those without EAP (a) although physical property of the two-step heated gel from GCH alone was almost evenly elevated (a, b). BS (c) and bs (d) of the two-step heated gel mixed GCH with SE surimi were decreased along with an increase in the mixing ratio of water. However, deterioration of the gel with preheating time was restrained by addition of EAP in spite of the mixing ratio of water. Compared with the bs of the gel without EAP (a), the bs of the gel with EAP (b) did not see a decrease in preheating time although the bs of the gel decreased with an increase in mixing ratio of water (d). On the other

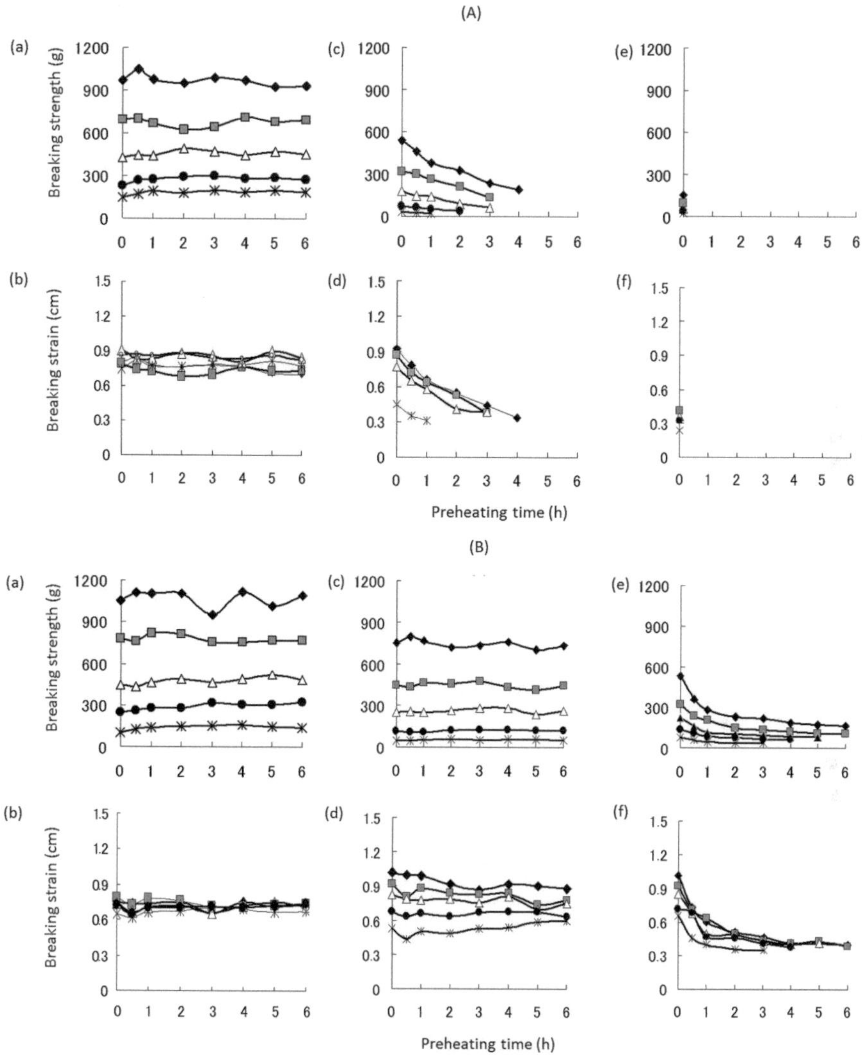

Fig. 8.4 Preheating time-dependent changes in breaking strength and breaking strain of two-step heated gels with or without EAP as a function of mixing with ground culled hen, fish surimi, and water. The weight ratio of water: 0 (♦), 18 (■), 43 (△), 82(●), and 150 (*) % of heated gel (wet weight). (**A**), without 2 % EAP; (**B**), with 2 % EAP. Mixing ratios (w/w) of ground culled hen and fish surimi from sand eel are ((a), (d)) 100:0, ((b), (e)) 50:50, ((c), (f)) 0:100 (Source: From reference [4] with permission)

hand, BS (e) and bs (f) of the two-step heated gel with EAP were both decreased with preheating time and showed low values although deterioration of the gel was restrained. From the above, tendencies of gel properties of SE surimi alone (e, f) were different from those of GCH alone (a, b) and mixture GCH with SE surimi (c, d). Therefore, enhancing the effects by addition of EAP could be most obvious on the heat-induced gels from mixtures of GCH and SE surimi, or SE surimi alone in comparison with the mixture of GCH alone.

8.2.3.3 Relationship Between Protein Concentration Dependence of Gel-Forming Ability of Directly Heated Gel Mixed GCH and FS

It is obvious that *suwari* gel is not formed in all the samples with and without EAP (Fig. 8.4). So, in this study, protein concentration (PC) dependence on BS of directly heated gel mixed with GCH, SE surimi, and water was investigated (Fig. 8.5), and the parameter was calculated with a regression line. According to proximate composition of the directly heated gels, additive water was retained in the gels, and moisture in the gels was changed dependent on addition. The pH values in the gels were increased to some extent with increment of addition of SE surimi, and the values was in the range of 6.4–6.9 because the pH value in SE surimi was high (6.8) (Table 8.4).

The relationship between Ln BS and Ln PC was expressed as a liner regression, and a regression line was fitted using the least squares method. In absence of EAP (A), BS of the directly heated gel was considerably lower in SE surimi alone and mixture (GCH: SE surimi = 25:75, w/w) than in the other mixtures although a slope of a line was different from each other. In SE surimi alone and mixture (GCH: SE surimi = 25:75, w/w), relationship between Ln BS and Ln PC was approached.

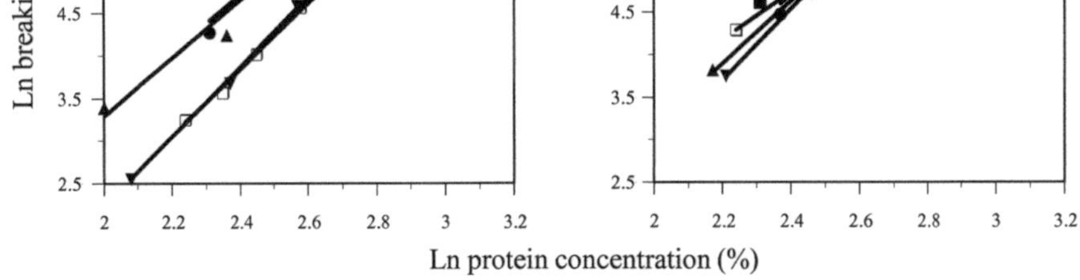

Fig. 8.5 Protein concentration dependence of breaking strength of directly heated gel mixed with ground culled hen, fish surimi, and water. The value of breaking strength was obtained in a same manner as shown in Fig. 8.4 by using five different mixing ratios of ground culled hen and fish surimi. The protein mixing ratios (w/w) of ground culled hen and fish surimi are 100:0 (■), 75:25 (●), 60:40 (▲), 25:75 (▼), and 0:100 (□). (**A**), without 2 % EAP; (**B**), with 2 % EAP (Source: From reference [4] with permission)

On the other hand, in the presence of EAP (B), BS values of all the samples were almost the same although a slope of a line was slightly different in all the samples. Regression lines of the gels from the mixtures (GCH: SE surimi = 75:25 and 60:40, w/w) were approached, and BS values of the gels from the mixture (GCH: SE surimi = 25:75, w/w) were almost the same regardless of the presence or absence of EAP.

The gel properties of the mixtures, which have a high mixing ratio of SE surimi, were almost the same as those of GCH alone by means of usage of the mixture mixed GCH with about 40 % SE surimi or 2 % EAP, suggesting that a low-valued food resources could be efficiently utilized. The relationship between Ln Gs and Ln PC was also expressed as a liner regression in this study (data not shown).

8.2.3.4 Relationship Between BS and Gel Gs of the Directly Heated Gel Mixed GCH with FS

Kitakami [15] reported that the change in the relationship between BS and Gs was investigated in case of adding to various ratios of water into surimi because the effect of the preheating time on BS and Gs was not recognized in directly heated gels. This evaluation method was applied in this study, and the results obtained are shown in Fig. 8.6 and Table 8.6. A positive correlation was recognized regardless of the presence or absence of EAP, and a liner equation, i.e., BS $= a \cdot \mathrm{Gs} + b$ (a and b are constant of each gel), was recognized. In the absence of EAP (A), slopes ($a = 0.485$ and 0.657) of regression lines of the gels from SE alone and mixture (GCH: SE surimi = 25:75, w/w) were smaller than those ($a = 0.786$ and 0.982) of the other samples. Maximum BS values of the former two were considerably

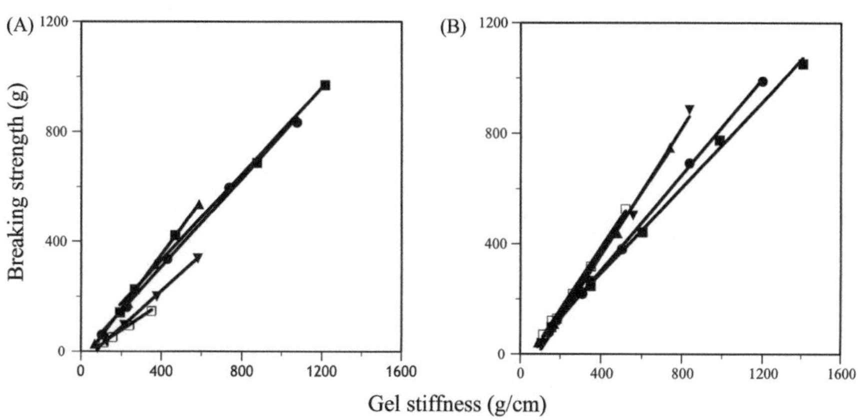

Fig. 8.6 Relationship between breaking strength and gel stiffness of directly heated gels formed. From the data shown in Fig. 8.4, the gel stiffness was calculated as (breaking strength/breaking strain) and plotted against the breaking strength. Refer to the legend of Fig. 8.5 for symbols. (**A**), without 2 % EAP; (**B**), with 2 % EAP (Source: From reference [4] with permission)

Table 8.6 Relationship between breaking strength (BS) and gel stiffness (Gs) of directly heated gels with or without 2 % EAP mixed with GCH) and SE surimi

Additive	Mixing ratio of protein concentration of GCH and SE surimi	y=ax+b (y: BS, x: Gs) a	b	r^2
None	100:0	0.786	−18.5	0.995
	75:25	0.804	−12.4	0.998
	60:40	0.989	−43.4	0.999
	25:75	0.657	−42.2	0.999
	0:100	0.485	−19.0	0.999
2.0 % albumen	100:0	0.773	−14.3	0.998
	75:25	0.863	−43.3	0.999
	60:40	1.092	−67.0	0.997
	25:75	1.153	−101.3	0.994
	0:100	1.112	−60.6	0.996

A regression line was fitted using the least squares method on the basis of the data shown in Fig. 8.6 (Source: From reference [4] with permission)

low. Maximum BS value of the gel from the mixture (GCH: SE surimi = 60:40, w/w) remained at a low level, although a slope ($a = 0.989$) of the regression line of the gel was large. Maximum BS values of the gels from the mixture (GCH: SE surimi = 75:25, w/w) and GCH alone were about twice as high as that of the gel from the mixture (GCH: SE surimi = 60:40, w/w) although the slopes of regression lines were smaller in the former two than in the latter. On the other hand, in the presence of EAP (B), the slopes of regression lines ($a = 0.863$ and 0.773) of the gels from the mixture (GCH: SE surimi = 75:25, w/w) and GCH alone were rather small and were almost the same as those ($a = 0.804$ and 0.786) of the gels without EAP, but maximum BS values were slightly higher in the former two than in the latter two. It was estimated that an improvement effect of the gel property by addition of EAP was most effective because the slopes of the regression lines of the gels from the mixture (GCH: SE surimi = 25:75, w/w) and SE surimi alone were larger than those of the gels without EAP and the maximum BS value was considerably higher in the gel from the mixture (GCH: SE surimi = 25:75, w/w) than in the gel without EAP. In other words, the gels having a specific texture with chewability were formed by the addition of EAP in comparison with the gels having the same BS. However, improvement in the effects of gel properties by addition of EAP could be slightly different from the gels prepared with the mixtures (GCH: SE surimi = 60:40, 75:25, 100:0, w/w) from the other gels. It would take advantage of a useful database concerning usage of EAP as reinforcement for the purpose of improvement of the gel property.

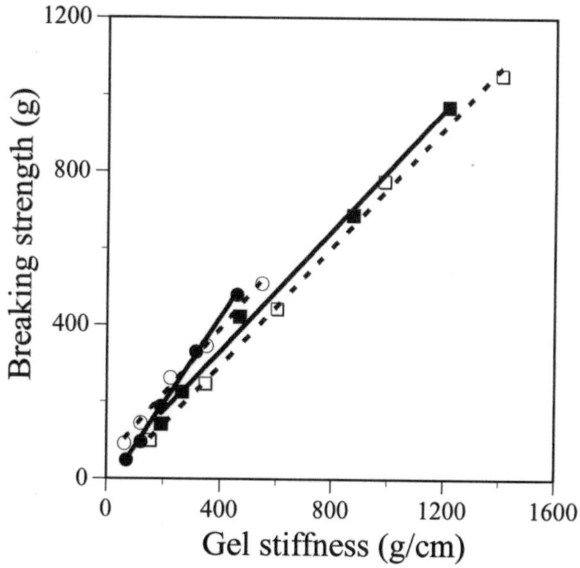

Fig. 8.7 Comparison of relationship between breaking strength and gel stiffness of directly heated gel from ground culled hen (■, □) and chicken surimi (●, ○). The directly heated gel of chicken surimi was prepared in a same manner as shown in Fig. 8.4. Without 2 % EAP (*closed symbol, solid line*), with 2 % EAP (*open symbol, dotted line*) (Source: From reference [4] with permission)

The gel-forming ability could be different GCH from chicken surimi under the same protein concentration because SE surimi was a salt-soluble protein concentrate, which has removed the water-soluble component by watering and dewatering. So, chicken surimi was prepared on a laboratory scale from the breast muscle from culled hens by the method of Kitakami et al. [9]. The additives, moisture, protein concentration, and pH of chicken surimi were 4 % sorbitol and 0.3 % sodium polyphosphate, 79.6 %, 14.8 %, and 7.0 %, respectively. The methods for preparation of heat-induced gel and measurement of gel properties were the same as mentioned above. Figure 8.7 shows a comparison of the relationship between BS and Gs of directly heated gel from GCH and chicken surimi with and without EAP. A positive correlation was recognized between both values regardless of the addition of EAP. The structure of gel formed was more flimsy in GCH than in chicken surimi regardless of the addition of EAP in comparison with same protein concentration judging from slopes of regression lines. However, the supplemental effect of EAP on the gel property was slightly different in the former from the latter.

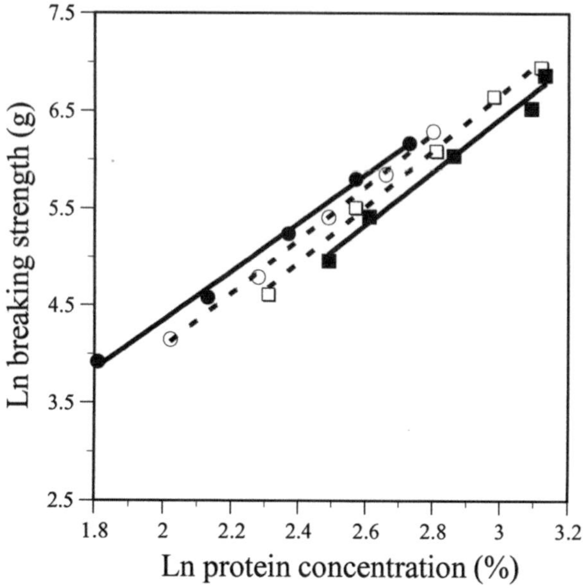

Fig. 8.8 Protein concentration dependence of breaking strength and breaking strain of directly heated gel mixed with ground culled hen (■, □) or chicken surimi (●, ○). The values of breaking strength were obtained in a same manner as shown in Fig. 8.7 by using five different mixing ratios of ground culled hen or chicken surimi. Without 2 % EAP (*closed symbol, solid line*), with 2 % EAP (*open symbol, dotted line*) (Source: From reference [4] with permission)

Figure 8.8 shows protein concentration dependence of BS of directly heated gel from GCH or chicken surimi with and without EAP. In the absence of EAP, a slope of the regression line was almost the same as GCH and chicken surimi. However, the BS value was higher in the latter than in the former in comparison with same protein concentration. BS value of the heat-induced gel from the latter was below 500 g (Ln 6.2) because water was retained in the gel and the moisture content was increased, but then, the protein content was decreased on account that there was a limit on dewatering [15]. The protein concentration dependence of BS of the heat-induced gel with and without EAP was similar except for the BS value which was slightly approximated between the two. Therefore, the chicken surimi is able to prepare the heat-induced gel having a specific texture with chewability in case of protein concentration below 15 % (Ln 2.7). However, it is apparent that GCH is more favorable than chicken surimi to prepare the gel having high gel property because it is impossible for the latter to prepare the gel, which has 18 % (Ln 2.9) of protein concentration.

8.3 Relationship Between Physicochemical and Morphological Properties of Heat-Induced Gel Mixed Meat with FS

In Chap. 8.2, it was quoted in the case study that mixed culled hen with sand eel, which are regarded as low-valued food resources, can develop unique characteristics of surimi-like products. As a result, it has become clear that a variety of heat-induced gel products can be produced with different ratios of GCH and FS regardless of adding EAP. This section describes physicochemical properties such as gel property, protein solubility, SDS-PAGE analysis, and microscopic observations of directly heating gels to understand the differences in structural properties of heat-induced gels mixed GCH with two kinds of FS.

8.3.1 Quality Assessment of Heat-Induced Gels Mixed GCH with FS

Ground culled hen (GCH) and commercial fish surimi from sand eel (SE) and Arabesque greenling (AG) (*Pleurogrammus azonus*) were used as main raw materials. AG has also been bycaught in eastern Hokkaido area, Japan. However, it is returned to the sea straight after catching it so the food-processing applications are poor. It is estimated that an annual haul of catch will see about 500–1,000 tons. Table 8.7 shows the proximate compositions and colors of GCH, SE, and AG surimi. The protein level was higher in GCH than in SE and AG surimi, while moisture and pH levels were in the reverse order. L* and W* values were lower in the former than in the latter two, while a* and b* values were in the reverse order.

Table 8.7 Proximate compositions and colors of GCH, SE, and AG surimi

Items	Products		
	GCH	SE surimi	AS surimi
Proximate composition			
Moisture (g/100 g)	72.8	75.5	78.3
Protein (g/100 g)	21.7	15.5	15.1
Lipid (g/100 g)	2.7	2.5	0.8
pH	6.3	6.8	7.0
Color			
L*	48.1	52.2	59.2
a*	6.5	−1.0	−2.1
b*	14.1	12.4	10.8
W*	47.9	50.6	59.1

Source: From unpublished data

GCH ground culled hen, *SE* sand eel, *AS* Arabesque greenling

Egg albumen powder (K type, kewpie Co., Ltd., Tokyo) was used in this study, and the moisture, protein, and pH levels were the same as those in Chap. 2.

GCH and fish surimi (FS) were thawed, and various ratios of water were added to the mixtures of GCH and FS (100:0, 75:25, 50:50, 25:75, 0:100, protein concentration ratio). Total protein concentrations of the samples were adjusted to 15 % (w/w). The mixtures were ground with 3.0 % NaCl with and without 2 % EAP using food processor (HFP-B3, Hitachi, Ibaraki) at 4 °C for 5 min. The salt-ground meats were stuffed into a stainless ring (diameter, 3 cm; length, 3 cm) and covered with cling film and tied at both ends and then heated at 90 °C for 30 min in a water bath (directly heated gel). After heating, the gels were subsequently cooled in an ice water and stored at 4 °C.

The breaking strength (BS, gf) and breaking strain (bs, cm) of the samples were measured with a rheometer (RE-33005, Yamaden, Tokyo) by using a spherical plunger of 0.5 cm in diameter after leaving them out for one night. Assay conditions are as follows: load cell, 2 kgf; distortion factor, 100 %; and plunger speed, 1 mm/s. The measurement was triplicates, and the obtained data was analyzed using a turkey HSD test (JMP ver. 8.0, Tokyo).

Figure 8.9 shows the effect of various mixing ratios of FS and GCH on BS and bs of the directly heating gels. In the absence of EAP, the BS value was significantly ($p < 0.05$) increased along with increment in mixing ratio of GCH in spite of the fish species (A, C). However, the bs value slightly increased along with an increment in the mixing ratio of GCH in spite of the fish species (B, D). In the presence of EAP, BS values of the samples with SE surimi were almost the same except for the mixing ratio of FS and GCH (50:50), while BS of the samples with AG surimi significantly ($p < 0.05$) increased along with an increment in the mixing ratio of GCH (A, C). The bs values were almost the same in all the samples in spite of the fish species (C, D). In the case of SE surimi, BS and bs values of the samples with EAP, which have mixing ratios of FS and GCH (100:0, 75:25, 50:50), were significantly ($p < 0.05$) higher than those of the samples without EA (a, c). In case of AG surimi, BS values of the samples, which have mixing ratios of FS and GCH (100:0, 75:25, 50:50, 25:75), were significantly ($p < 0.05$) higher than those of the samples without EA, while bs value of the sample with EA, which has a mixing ratio of FS and GCH (100:0), was significantly ($p < 0.05$) higher than that of the sample without EAP (B, D). In particular, the effect of EAP on the gel properties of the directly heated gels was different for fish species, and its effect was higher in SE surimi than in AG surimi.

8.3.2 *Protein Solubility of Heat-Induced Gels Mixed GCH with FS by Use of Protein Denaturation Reagents*

Four kinds of solvents, i.e., 0.6 M NaCl (S1), 0.6 M NaCl+1.5 M urea (S2), 0.6 M NaCl+8.0 M urea (S3), and 0.6 M NaCl+8.0 M urea+2-mercaptethanol (S4), were prepared by the method of Perez-Mateos et al. [16]. The directly heated gel

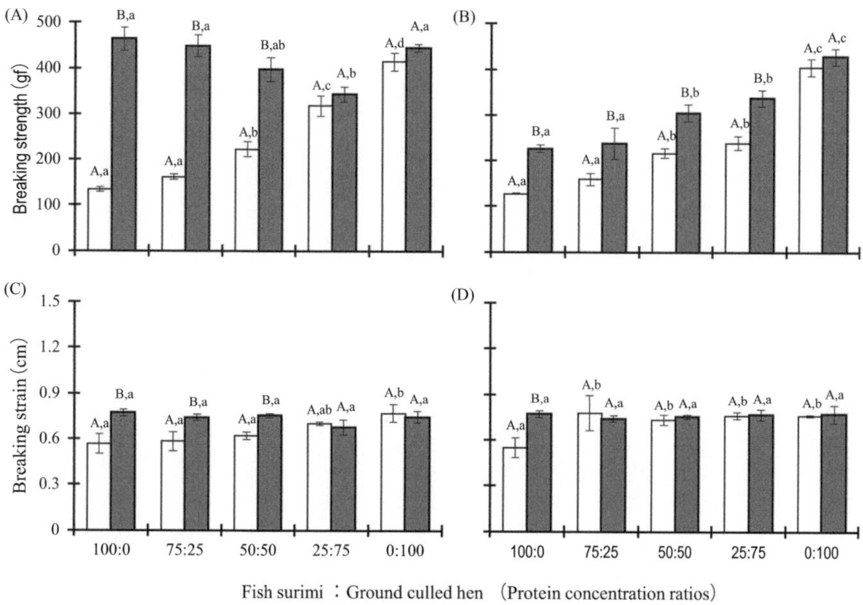

Fig. 8.9 Effect of various mixing ratios of fish surimi and ground culled hen on breaking strength (**A**, **B**) and breaking strain (**C**, **D**) of directly heating gels. (**A**), (**C**), sand ell; (**B**), (**D**), Arabesque greenling. (□) with 2 % egg albumen powder, (■) without 2 % egg albumen powder. The data are expressed as mean ± standard deviation. Means with different capital letters in the presence and absence of egg albumen are significantly different ($p<0.05$). Means with different lowercase letters in the contribution ratio of FS and GCH are significantly different ($p<0.05$) (Source: From unpublished data)

was minced and solubilized in four solvents by continuously shaking it at room temperature for about 1 day. The soluble protein in the solvent was recovered by centrifugation (3,000 rpm, 30 min). The recovered protein was precipitated on addition of 7.5 % TCA. The precipitation was recovered by centrifugation again. After air-drying, the precipitation was dissolved in 1 N NaOH. The protein determined quantitatively by biuret method [17]. Protein solubility was expressed as a relative value (%) of the soluble protein in solvent per total protein in salt-ground meat. In case of biuret reaction inhibition by lipid, its removal was conducted by the addition of petroleum ether. The measurement was conducted triplicate, and the obtained data was analyzed using the turkey HSD test (JMP ver. 8.0, Tokyo).

An interaction and binding species among proteins contributing heat-induced gel formation were estimated by the soluble protein levels in various solvents. The soluble protein level in S1, differences in the soluble protein levels between S2 and S1, S3 and S2, and S4 and S3, could be related to the degree of intensity of ionic bond, hydrogen bond, hydrophobic interaction, and disulfide bond, respectively, because each protein-protein interaction in the heat-induced gel could be cleaved by each solvent. The content of insoluble protein in the gel was calculated by

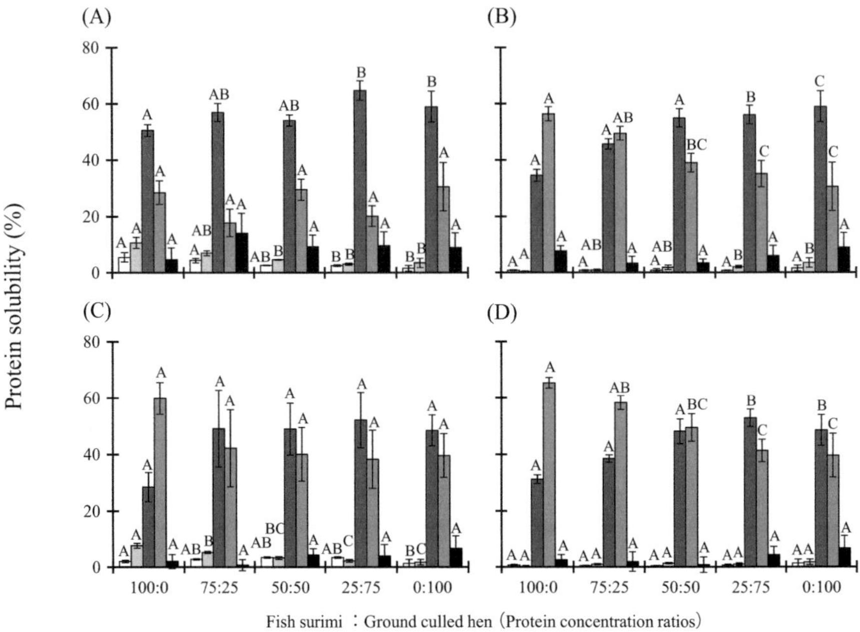

Fig. 8.10 Effect of various mixing ratios of fish surimi and ground culled hen on protein solubility of the directly heating gels. (**A**), (**B**), without 2 % EAP; (**C**), (**D**), with 2 % EAP. (**A**), (**C**), sand eel, (**B**), (**D**), Arabesque greenling. (□) S1, (■) S2–S1, (■) S3–S2, (■) S4–S3, (■) insoluble protein. Each solvent was prepared by the method of Perez-Mateos et al. [16]. The data are expressed as mean ± standard deviation. Means with different capital letters in the contribution ratio of FS and GCH are significantly different ($p < 0.05$) (Source: From unpublished data)

subtracting S4 from 100. Kunimoto et al. [18] reported that binding species contributing the insoluble protein was expressed as "more intensive bond" because it was almost soluble in 2%SDS-8 M urea-2-mercaptethanol (pH 7.5). This word was also used in this study because a similar trend was recognized.

Figure 8.10 shows the effect of various mixing ratios of FS and GCH on protein solubility of the directly heating gels. It is obvious that a hydrophobic interaction and a disulfide bond were the main binding species concerning gel formation in all the samples in spite of the fish species judging from protein solubility. In the absence of EAP, the rate of S3-S2 and S4-S3 fractions in the gels with SE surimi was slightly increased and almost the same, respectively, along with increment in mixing ratios of GCH (A). On the other hand, the rate of S3-S2 and S4-S3 fractions in the gels with AG surimi was increased and decreased, respectively, along with an increment in the mixing ratios of GCH (B). In comparison with binding species in the gels with SE and AG surimi alone, there was a difference of binding species contributing gel formation between the samples, although gel properties of both sides were similar (Figs. 8.9 and 8.10A, B). The difference could be caused by the quality of FS, i.e., protein denaturation during storage [19], because hydrophobic interaction was strongly related to the gel properties of the directly heated gel from AG surimi, which was prepared immediately after catch at Kushiro, Hokkaido, Japan (data not shown). The rate of S4-S3 fraction increased more in the sample with EAP than in the sample without EAP (C, D). However, the accumulation level of insoluble protein was in the reverse order (C, D). In other words, more intensive bond in the directly heated gel was decreased by addition of EAP. In the presence of EAP, the rate of S3-S2 and S4-S3 fractions in the gels with SE surimi was increased and decreased, respectively, up to mixing ratios of FS and GCH (75:25), and they continued approximately constant with an increment in the mixing ratios of GCH (C). On the other hand, the rate of S3-S2 and S4-S3 fractions in the gels with AG surimi was increased and decreased, respectively, along with an increment in the mixing ratios of GCH (D). Therefore, it is clear that an increment in disulfide bond (S-S bond) could be related to an improvement of gel properties of the directly heated gel mixed FA with GCH by addition of EAP. These results were slightly different from those of Kunimoto et al. reports [19] regarding the rate of hydrophobic interaction and disulfide bond participation. The reason might be considered to be the differences in quality of FS and fish species etc.

8.3.3 SDS-PAGE Analysis of Heat-Induced Gels Mixed GCH with FS

Solubilizing solution (8 M urea, 2 % SDS, 2 % 2-mercaptoethanol, and 20 mM Tris-Bicine, pH 8.2) was added to the directly heated gel, heated at 100 °C for 2 min, and then solubilized by continuous shaking at room temperature for about 1 day. After shaking, the supernatant was collected by centrifugation, and the amount of protein was determined in a same manner as in Chap. 8.3.2. Solubilized protein was analyzed by SDS-PAGE. SDS-PAGE analysis was carried out using Laemmli method [20] with a slight modification and used as 7.5 % polyacrylamide gel. The staining of the gel was conducted with the solution containing 0.2 % CBB R-250, 50 % methanol, and 10 % acetic acid. The destaining of the gel was conducted with the solution containing 10 % methanol and 10 % acetic acid until the gel surface was clear.

Protein solubility in this solvent was considerably high (about 95 % over) in all the samples. Figure 8.11 shows the SDS-PAGE pattern of soluble proteins in various kinds of directly heated gels mixed FS with GCH. In the absence of EAP, myosin heavy chain (MHC) was decreased along with an increment in the mixing ratios of SE surimi (A). In particular, this phenomenon was notably recognized in the directly heated gels mixed FS with GCH (100:0, 75:25). On the other hand, the band of MHC had slightly disappeared in the directly heated gel with AG surimi alone (B). In the presence of EAP, the decrease in MHC along with an increment in the mixing ratios of SE surimi and the disappearance in band of MHC in the directly heated gel with AG surimi alone were not recognized (C, D). These findings

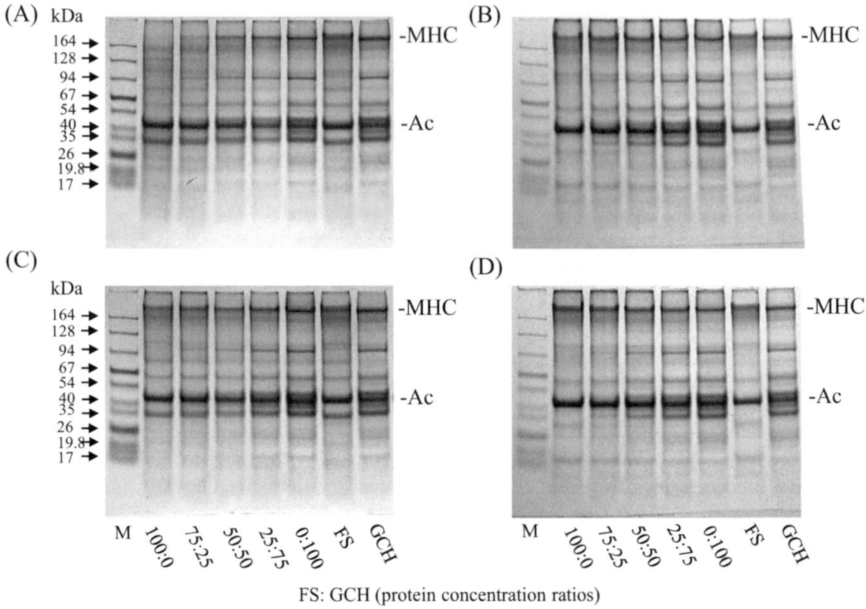

Fig. 8.11 SDS-PAGE pattern of soluble proteins in various kinds of directly heated gels mixed fish surimi and ground culled hen. (**A**), (**B**), without 2 % EAP; (**C**), (**D**), with 2 % EA. **A**, **C**, sand eel; **B**, **D**, Arabesque greenling. *M* molecular weight maker, *MHC* myosin heavy chain, *Ac* actin. FS and GCH are fish surimi and ground culled hen before heating, respectively. Each sample (20 μg) was applied to the gel slot (Source: From unpublished data)

suggest that cleavage of the MHCs in the directly heated gels, which have a high mixing ratio of FS, could be inhibited by addition of EAP. It is well-known that an ovomucoid and cystatin are serine [21] and cysteine [22] proteinase inhibitors in EAP, respectively, and then ovostatin, which has a unique action of enfolding proteinase, inhibits a variety of proteinase activities [23]. Liu et al. [24, 25] reported that 150 kDa component followed by degradation of the MHC was generated by the action of serine and cysteine proteinases in the preheated gel from WP surimi during setting process. The proteinase inhibitors, which act in near neutral pH range, could inhibit the degradation of MHC during the directly heated gel production in this study. Konno and Imamura [26] reported that 150 kDa and 70 kDa fragments, which were generated during incubation of fish meat paste from WP surimi at 25 °C, identified heavy meromyosin (HMM)- and light meromyosin (LMM)-like fragments, respectively. The band near the former was detected in SE and AG surimi alone samples, while the band near the latter was detected in SE surimi alone sample in this study (Fig. 8.11A, B). However, the details concerning the source are still unclear.

8.3.4 Microstructure Characteristics of Heat-Induced Gels Mixed GCH with FS

From the results of Figs. 8.9, 8.10, and 8.11, it was found that the supplemental effect of EAP on gel properties and reducing protein degradation of the directly heated gel was higher in SE surimi than in AG surimi. So, the microstructure in the directly heated gels mixed GCH and SE surimi was observed by scanning electron microscope (SEM) in order to investigate the role of EAP on the gel structure formed. In the absence of EAP, an aggregate in the actomyosin-like network structure increased along with a rise in the mixing ratio of SE surimi (Fig. 8.12A–C). On the other hand, in the presence of EAP, it was observed that the aggregates, which were formed in the network structure, were reduced (Fig. 8.12D–F). This phenomenon was only notable in SE surimi. Therefore, the improvement of gel properties of the heat-induced gels mixed GCH with SE surimi by addition of

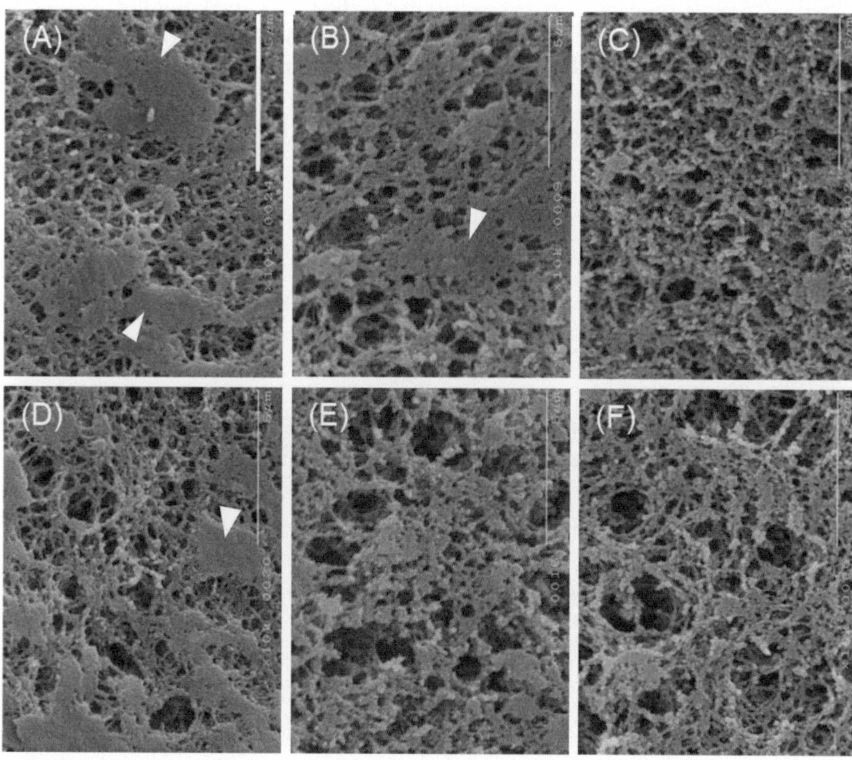

Fig. 8.12 Scanning electron micrographs of directly heated gels mixed GCH with SE surimi. Protein concentration of the gels was 8 %. (**A**–**C**), without; (**D**–**F**), with 2 % EAP. Protein concentration ratios of FS/GCH were 100:0 (**A**, **D**); 50:50 (**B**, **E**); and 0:100 (**C**, **F**), respectively. *Arrowhead* indicates the aggregate. Scale *bar* indicates 5 μm (Source: From unpublished data)

EAP could be related to multiple factors such as increment in disulfide binding between proteins, decreasing protein degradation and the reduction of aggregates in the network structure.

Based on these set of facts, it is obvious that an improvement of the gel properties of the direct heated gel from FS could be effective to mix livestock products such as GCH and EAP. In particular, this method would be more effective to utilize fishes, which are not recognized in a *suwari* phenomenon.

8.4 Improvement of Thermal Gelation Property of Mixed Actomyosin by Use of Cross-Linking Reaction Between Proteins

In order to facilitate effective utilization of culled hen meat and also unvalued fishes, Iwasaki et al. [27] attempt to develop novel surimi-like products. In their preliminary study [28], no satisfactory gel was formed by simple mixing of hen and fish actomyosins because of the difference of optimal temperature of gelation among those actomyosins (Fig. 8.13). In other words, each peak temperature of relative storage modulus of the mixed actomyosins from hen and fishes was separately detected because an interaction between hen and fish actomyosins did not occur.

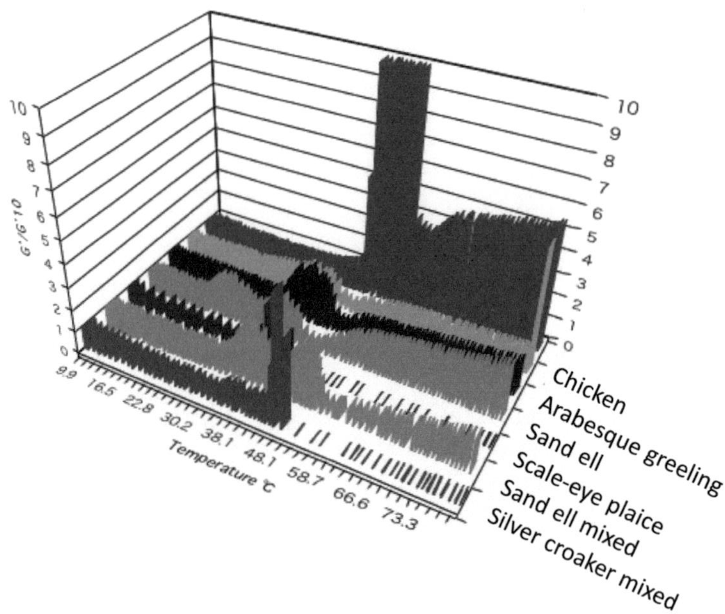

Fig. 8.13 Relative storage modulus (G′/G′10) of several actomyosin solution. Actomyosin in 0.3 M NaCl and 10 mM K phosphate (pH 7.0) was heated with a heating rate at 2 °C/min. G′; storage modulus, G′10; storage modulus at 10 °C (Source: From reference [28] with permission)

Transglutaminase (TG) is an enzyme that catalyzes the formation of a covalent bond between a free amine group of lysine and an amide of glutamine, and it is used in a variety of food production, including processed meat and fish products [29]. This section describes that effect of microbial transglutaminase (MTG) on the properties of heat-induced gel of mixed actomyosin from culled hen and fish.

8.4.1 Experimental Protocol Concerning Preparation of Actomyosins from Culled Hen and Fish

Actomyosin was extracted from the pectoral muscle Weber-Edsall solution and purified further. Three kinds of fresh fishes, i.e., plain sculpin (PS) (*Myoxocephalus jaok*), scale-eye plaice (SP) (*Acanthopsetta nadeshnyi*), and Arabesque greenling (AG) (*Pleurogrammus azonus*), were used. Minced fish meats were washed with low ionic strength buffer; then extracted with 0.45 M NaCl, 3.38 mM Na_2HPO_4 (pH 7.5) for 24 h; and purified further.

Hen and fish actomyosins were mixed (1:1, w/w) in 0.3 M NaCl and 10 mM Bis-Tris (pH 7.0) to give a protein concentration of 5 or 10 mg/ml. Microbial transglutaminase (Activa TG-K, Ajinomoto Co., Ltd., Japan) was added to actomyosin solution at 1 U/g of actomyosin and incubated for 3–4 h at 37 °C.

Solubilizing solution (8 M urea, 2 % SDS, 2 % 2-mercaptoethanol, and 20 mM Tris-HCl, pH 8.0) was added to MTG-treated actomyosin, heated at 100 °C for 2 min, and then incubated at room temperature for 20 h. After centrifugation, the supernatant was collected and the amount of protein was determined by the biuret method [17]. Solubilized protein was analyzed by SDS-PAGE in the same manner as in Chap. 8.3.4.

8.4.2 Changes in Solubility of Microbial Transglutaminase-Treated Actomyosin by Heating

Solubilizing solution, which contains urea-SDS-mercaptoethanol, was added to MTG-treated actomyosin, and solubilized protein was determined. The change in solubility of actomyosin is shown in Fig. 8.14. Individual actomyosin of hen, AG, and SP without MTG showed over 90 % of solubility during incubation at 37 °C for 3 h, while the solubility of PS actomyosin decreased to 85 %. When MTG was added to actomyosins, the solubilities of hen, PS, and AG decreased to about 80 % and about 50 % for SP (Fig. 8.14A). This indicates that MTG forms more cross-linking in SP actomyosin than the others.

In the case of mixed actomyosin of hen and fish without MTG, the solubilities of mixed actomyosin of hen + AG and hen + PS remained at almost 100 %, while that of hen + SP decreased to 80 %. Addition of MTG induced a decrease of solubility

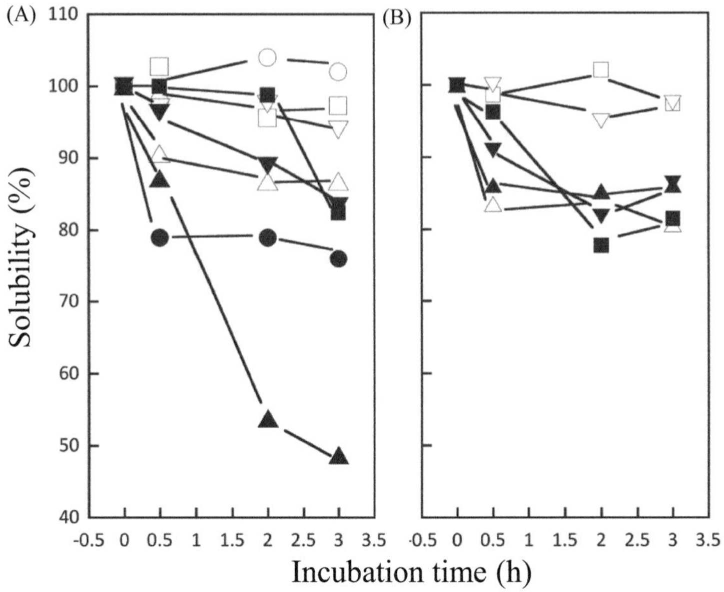

Fig. 8.14 Solubility of actomyosin incubated at 37 °C with or without MTG for various time. (**A**) individual actomyosin, ○, □, △, ▽: Hen, AG, SP, and PS. Open symbol represents without MTG and closed is with MTG. (**B**) mixed actomyosin of hen and fish. Symbols are the same as in (**A**), excluding hen actomyosin (Source: From reference [27] with permission)

to about 80 % in hen + AG and hen + PS actomyosins (Fig. 8.14B). It was found that there was no effect of MTG on the solubility of hen +SP actomyosin. It is well-known that decreasing of solubility by solubilizing solution and cross-linking of myosin heavy chains takes place in walleye pollack surimi gel [30]. So, it was estimated that cross-linking of the formation of the proteins could have occurred in a variety of actomyosin solutions with MTG in this study.

8.4.3 SDS-PAGE Analysis of Solubilized Protein

In order to investigate cross-linking formation of the proteins included in actomyosin during incubation at 37 °C with or without MTG, the solubilized fraction was analyzed by SDS-PAGE (Figs. 8.15 and 8.16). There was no additional band above myosin heavy chain (MHC) in individual actomyosin without MTG. MHCs of AG (B) and PS (D) decreased with time of incubation, and the new bands appeared below MHC band (Fig. 8.15), suggesting degradation of MHC by endogenous proteases during incubation [31]. No new bands appeared above MHC band in the mixed actomyosin. Furthermore, there was no decrease of MHC band (Fig. 8.16A–C).

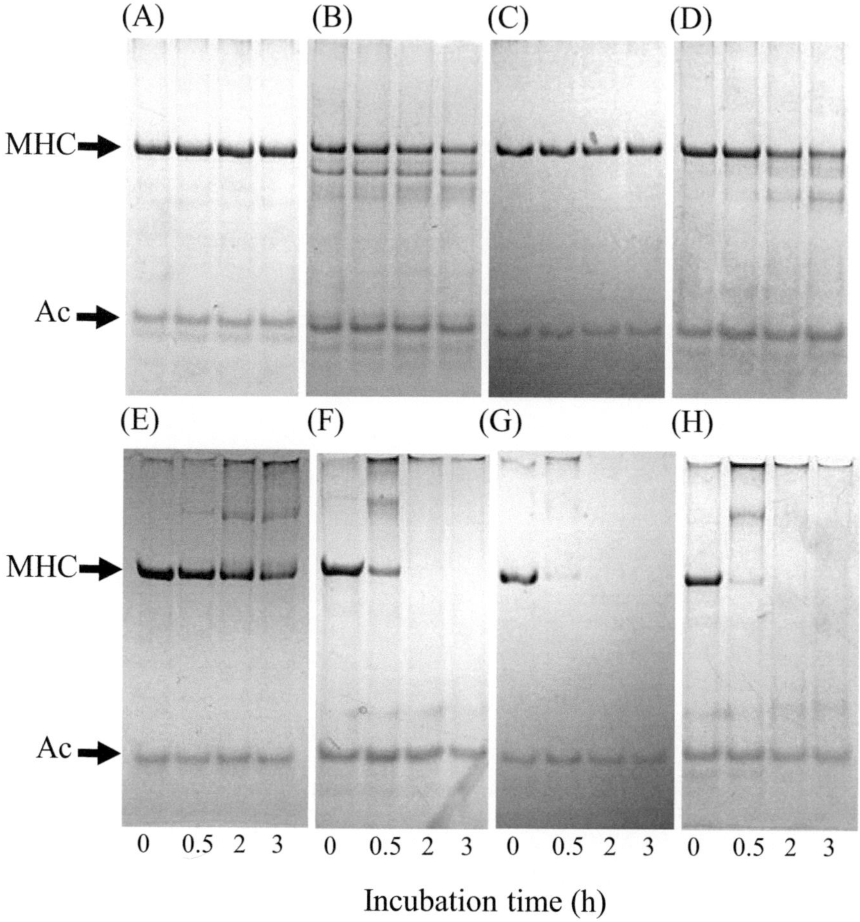

Fig. 8.15 Changes in SDS-PAGE pattern of solubilized fraction from the MTG-induced gel at 35 °C. (**A**, **E**), Hen; (**B**, **F**), AG; (**C**, **G**), SP; (**D**, **H**), PS. (**A**–**D**) without MTG. (**E**–**H**) with MTG. MHC and Ac show myosin heavy chain and actin, respectively (Source: From reference [27] with permission)

The MHC bands in all of MTG-treated actomyosins decreased, and the high molecular weight bands appeared, especially in fish actomyosins (Fig. 8.15E–H). Cross-linking of MHC in SP actomyosin was faster than the others (Fig. 8.15G). The MHC in mixed actomyosin also decreased, and high molecular weight components appeared with MTG although no complete disappearance of the MHC band within 3 h was observed (Fig. 8.16D–F). MHC decrease in SP actomyosin was the fastest among the fishes, while the decrease of MHC in hen-SP mixed actomyosin was slower compared to hen-PS and hen-AG.

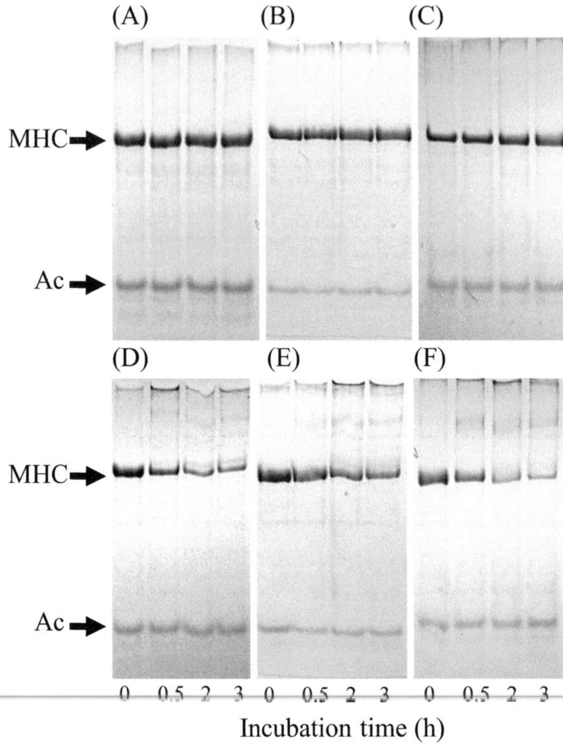

Fig. 8.16 Changes in SDS-PAGE pattern of solubilized fraction from the MTG-induced mixed actomyosin gel at 35 °C. (**A**, **E**), Hen+AG; (**B**, **F**), Hen+SP; (**C**, **G**), Hen+PS. (**A**–**C**), without MTG. (**D**–**F**), with MTG. MHC and Ac show myosin heavy chain and actin, respectively (Source: From reference [27] with permission)

8.4.4 Force vs. Indentation Curve and Gel Structure of a Variety of Heated Actomyosins

Two kinds of thermal treatments were employed to form a gel. One is one-step heating, that is, heating at 37 °C for 4 h or 90 °C for 30 min. The other is two-step heating, preheating at 37 °C for 4 h and then heating at 90 °C for 30 min. A penetration test using a spherical plunger of 0.5 cm in diameter was done for actomyosin gel.

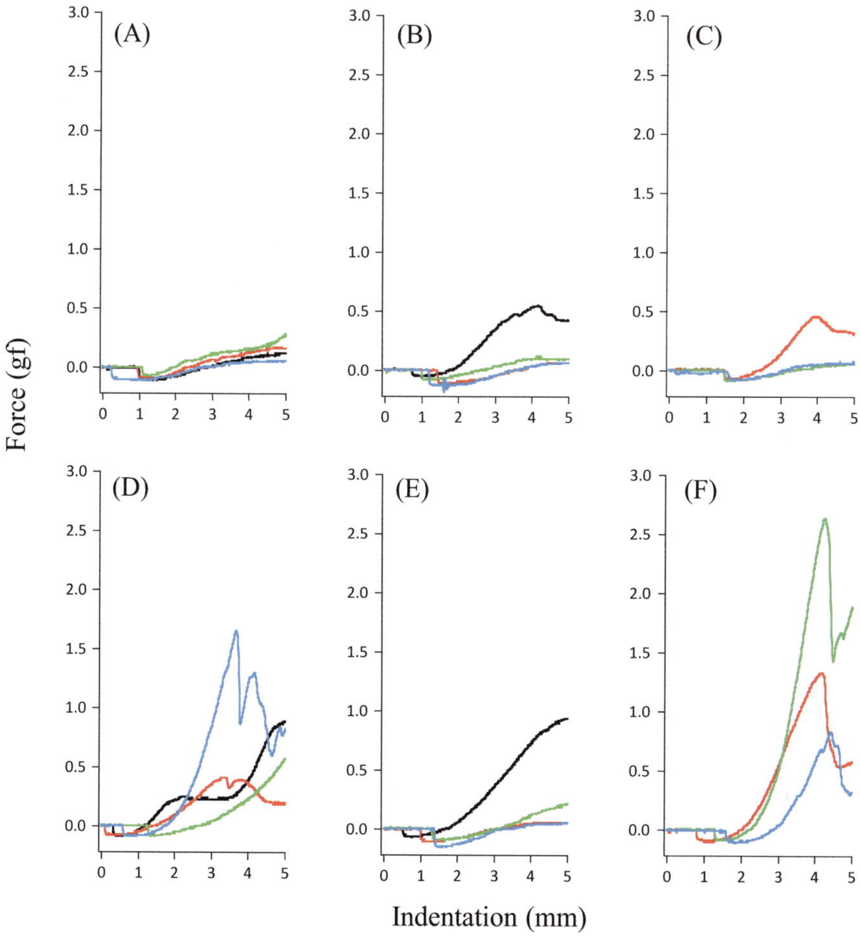

Fig. 8.17 A typical example of force vs. indentation curve of heated actomyosin suspension (10 mg/ml in 0.3 NaCl containing 10 mM Bis-Tris, pH 7.0). (**A–C**), without; (**D–E**), with MTG. The incubation was done at 37 °C for 4 h; (**A, D**), 90 °C for 30 min; (**B, E**), 37 °C for 4 h; and then 90 °C for 30 min (**C, F**). *Lines* show Hen (*black*), AG (*red*), SP (*green*), and PS (*blue*) (Source: From reference [27] with permission)

Rheological measurement was done for thermally treated actomyosin with or without MTG (Figs. 8.17 and 8.18). When individual actomyosin or mixed actomyosin was incubated at 37 °C for 4 h without MTG, no steady gel was formed (Figs. 8.17 and 8.18A), while they formed gel in the presence of MTG (Figs. 8.17 and 8.18D). Regardless of the presence or absence of MTG, fish actomyosins did not form gel after heating at 90 °C for 30 min, whereas hen actomyosin formed steady gel and the gel strength increased by addition of MTG (Figs. 8.17 and 8.18B, E). Mixed actomyosins of hen and fish also did not form a steady gel by heating two-step heating in the presence of MTG (Figs. 8.17 and 8.18C, F). Mixed actomyosin

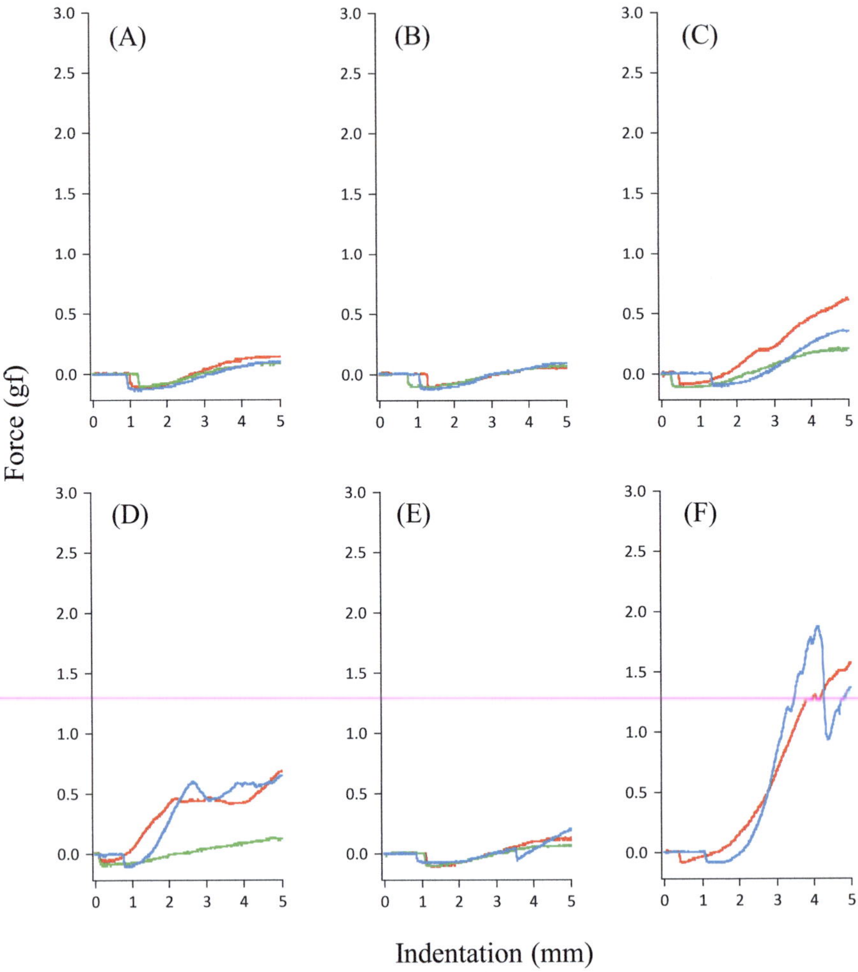

Fig. 8.18 A typical example of force vs. indentation curve of heated mixed actomyosin suspension (10 mg/ml in 0.3 NaCl containing 10 mM Bis-Tris, pH 7.0). (**A–C**), without MTG; (**D–E**), with MTG. The incubation was done at 37 °C for 4 h (**A**, **D**); 90 °C for 30 min (**B**, **E**); 37 °C for 4 h; and then 90 °C for 30 min (**C**, **F**). *Lines* show AG (*red*), SP (*green*), and PS mixed with Hen (*blue*) (Source: From reference [27] with permission)

of hen and AG or PS showed the same tendency, though the mixed actomyosin of hen and SP released water by heating and no steady gel was formed (Fig. 8.18F). Cross-linking of MHC is known to relate to gel strength [32], and the present result is in accordance with it.

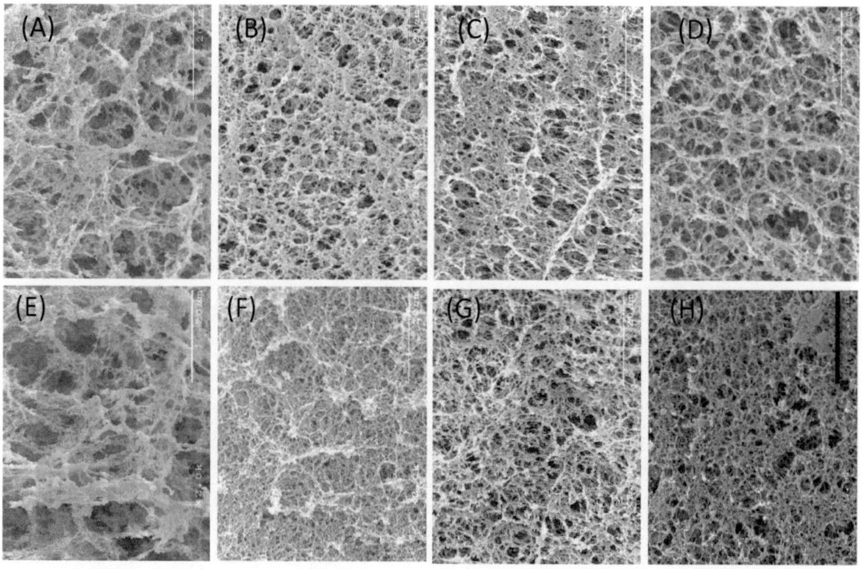

Fig. 8.19 Scanning electron micrographs of heat-induced actomyosin gels. Actomyosin solution was heated at 37 °C for 4 h and then 90 °C for 30 min. (**A–D**), without; (**E–H**), with MTG. (**A**, **E**), Hen; (**B**, **F**), AG; (**C**, **G**), SP; (**D**, **H**), PS. Scale bar indicates 20 μm (Source: From reference [27] with permission)

Scanning electron microscopy was performed to observe the microstructure of actomyosin gel. Actomyosin gel formed by two-step heating with or without MTG was observed by SEM (Figs. 8.19 and 8.20). In hen actomyosin, the thickness of fibers in the network structure, which formed gels, was larger with MTG (E) than without MTG (A) (Fig. 8.19). However, the gels that formed with MTG showed a finer network structure compared to those without MTG. These fine network structures are possibly responsible for high gel strength in the MTG-treated actomyosin gel (Figs. 8.19 and 8.20). The species specificity in thermal stability of the protein [6] might contribute to the difference in the rate of cross-linking reaction and thickness of the fiber in the network structure of the gel formed with MTG between hen and fish actomyosins, but the details remain unclear.

From the above, plain sculpin and Arabesque greenling can be used as ingredients of heat-induced gel products mixed fish with hen, while scale-eye plaice is inappropriate for the product in cases where MTG is utilized for the purpose of improving the gel property of the products. These findings revealed that difference in the gel-forming ability of the fish protein itself such as easy and hard to form *suwari* gel could be related to the quality of the products because the improvement effect of gel properties by the use of MTG is different in fish species.

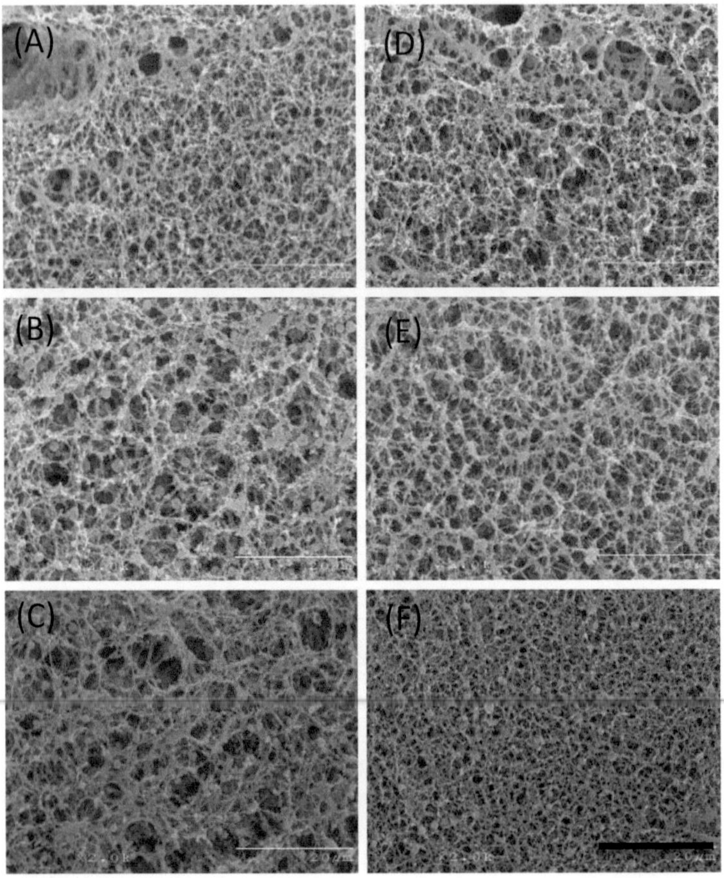

Fig. 8.20 Scanning electron micrographs of heat-induced mixed actomyosin gels. Actomyosin solution was heated at 37 °C for 4 h and then 90 °C for 30 min. (**A**–**C**), without; (**D**–**F**), with MTG. (**A**, **D**), AG; (**B**, **E**), SP; (**C**, **F**), PS mixed with Hen. Scale bar indicates 20 μm (Source: From reference [27] with permission)

8.5 Conclusion

In this study, thermal gelation properties of a variety of gels mixed meat with fish proteins were investigated by use of the rheological and protein chemistry approaches. It is necessary to pay attention to these properties that might be considerably changed by the quality of frozen surimi such as fish species and standard, muscle region of land animals and food additives such as egg albumen, polyphosphates, and transglutaminase. In particular, adequate understanding of the gel properties of fish surimi could be essential to produce high-quality gel products mixed with livestock products. Moreover, research and study concerning the quality of surimi would be a novel database to effectively use the underutilized marine resources as foods because they are influenced by fluctuations of catch in marine environments.

Acknowledgment This research was partly supported by a fund from Hokkaido Research Organization, for the Scientific Research on Priority Areas during fiscal years 2010–2012. The authors would like to express their gratitude to Drs. K. Yamamoto and M. Ishioroshi, Rakuno Gakuen University, for their encouragement and kind support. The authors would like to express their hearty thanks to Drs. K. Arai and S. Kitakami and Ms. Y. Murakami of National Surimi Manufactures Association; Mr. T. Imamura, Mr. M. Kitagawa, and Dr. K. Ebitani of the Hokkaido Research Organization; Mr. K. Hasegawa of Nunokawa Limited; Mr. H. Kon of Maruha Hashimoto Shokai Inc; and Mr. K. Takayanagi of Rakuno Gakuen University for their helpful discussion. The authors also thank Mr. D.A. Miller, a part-time teacher of Rakuno Gakuen University, for his helpful proofreading.

References

1. Nishiya, K., Takeda, F., Tamamoto, K., Tanaka, O., Kubo, T. (1960). Studies on freezing of surimi (fish paste) and its application. III. Influence of salts on quality of fish meat. Monthly Report of Hokkaido Fisheries Research Laboratory, Fisheries Agency, Japan, 17, 373–383.
2. Park, J.W., Lin, T.M.J. (2005). Surimi; Manufacturing and evaluation. In Surimi and Surimi seafoods, CRC press, Boca Raton, FL.
3. Yamaguchi, A., Abe, Y., Ishioroshi, M., Samejima, K., Arai, K. (2000). Effect of polyphosphate salt on gel forming ability of walleye pollack frozen surimi in relation to frozen surimi. 66: 481–488.
4. Funatsu, Y., Yamamoto, K., Iwasaki, T., Kaneda, I., Ishioroshi, M., Yamamoto, K., Oohori, T., Kitakami, S., Arai, K. (2010). Effect of albumen powder on physical property of heat-induced gels formed from ground chicken mixed with fish surimi. *Nihon Chikusan Gakkaiho*, 81: 169–180.
5. Miyaguchi, Y., Sakamoto, T., Hayashi, Y., Nagayama, K. (2005). Effective use of culled hen: The use of chicken salcoplasmic proteins for rheological improvement of model sausage. *Nippon Shokuhin Kagaku Kogaku Kaishi,* 52: 572–577.
6. Hashimoto, A., Kobayashi, A., Arai, K. (1981). Thermostability of fish myofibrillar Ca-ATPase and adaptation to environmental temperature. *Bulletin of the Japanese Society of Scientific Fisheries,* 48: 671–684.
7. Shimizu, Y., Machida, R., Takenami, T. (1981). Species variations in the gel-forming characteristics of fish meat paste. *Bulletin of the Japanese Society of Scientific Fisheries,* 47: 95–104.
8. Abe, Y., Yasunaga, K., Kitakami, S., Murakami, Y., Ohta, T., Mihori, T., Arai, K. (1996). Characteristics of two-Step heating gels from frozen surimi with a food additive containing TGase or a bovine plasma powder. *Nippon Suisan Gakkaishi,* 62: 446–452.
9. Kitakami, S., Yasunaga, K., Murakami, Y., Abe, Y., Arai, K. (2003). pH-dependency of gel forming ability of walleye pollack surimi and effect of polyphosphate salt. *Nippon Suisan Gakkaishi,* 69: 405–413.
10. Nishioka, F. (1994). Quality inspection standard of frozen surimi. *Nippon Suisan Gakkaishi,* 60: 281–285.
11. Kitakami, S., Murakami, Y., Yasunaga, K., Abe, Y., Katoh, N., Arai, K. (2009). Heated gel forming ability of walleye pollack frozen surimis of various grades, as measured by physical properties of heated gel and its dependence on protein concentration. *Nippon Suisan Gakkaishi,* 75: 250–257.
12. Nishinari, K. (1978). Rheology of the gel, In Food texture (Y. Matsumoto ed.), pp.41–74, Shokuhin Shizai Kenkyukai, Tokyo (in Japanese).

13. Kitakami, S., Murakami, Y., Yasunaga, K., Katoh, N., Arai, K. (2009). Gel forming ability of walleye Pollack surimi-protein and its concentration dependency. *Nippon Suisan Gakkaishi,* 71: 957–964.
14. Miyake, H. 2005. Shinkitanosakanatachi (T. Mizushima and M. Torisawa eds.), Second edition, pp.220–223, Hokkaido News Paper, Hokkaido (in Japanese).
15. Kitakami, S., Murakami, Y., Yasunaga, K., Katoh, N., Arai, K. (2005). Gel forming ability of walleye Pollack frozen surimi-protein and its concentration dependency. *Nippon Suisan Gakkaishi,* 71, 957–964.
16. Perez-Mateos, M., Lorenco, H., Montero, P., Borderias, A.J. (1997). Rheological and biochemical characteristics of high-pressure and heat induced gels from blue whiting (*Micromesistius poutassou*) muscle. *J. Agric. Food Chem.*, 45: 44–49.
17. Gornall, A.G., Bardawill, C.J., David, M.M. (1949). Determination of serum protein by means of the biuret reaction. *J. Biol., Chem.*, 177: 763–765.
18. Kunimoto, M., Okumura, T., Kato, N., Arai, K. (2014). Characteristics Properties of heated gels formed from frozen surimi in terms of protein solubility in various solvents and effect of addition of albumen powder on gel-formation. *Nippon Shokuhin Kagaku Kaishi,* 61: 19–26.
19. Kato, N., Kinimoto, M., Koseki, S., Kitakami, S., Arai, K. (2009). Freshness and quality of fish and shellfish (Supplementary edition). *Journal of The School of Marine Science and Technology, Tokai University*, 7: 87–99.
20. Laemili, U.K. (1970). Cleavage of structural proteins during the assembly of the head of bacteriophage T4. *Nature*, 227: 680–685.
21. Lineweaver, H., Murray, CW. (1947). Identification of the trypsin inhibitor of egg white with ovomucoid. *J. Biol. Chem.*, 171: 565–581.
22. Niclin M.J.H., Barrett, A.J. (1984). Inhibition of cysteine proteinases and dipeptidase I by egg-white cystatin. *Biochem. J.*, 223: 245–253.
23. Matsuda, M. (1998). Egg. In protein science (A. Suzuki, S. Watabe, H. Nakagawa eds.), Asakura Shoten, Tokyo, pp.66–84 (in Japanese).
24. Liu, D., Nowsad, A.A., Kanoh, S, Niwa, E. (1996). Effect of serine proteinase inhibitor on the setting of Alaska Pollack surimi paste. *Nippon Suisan Gakkaishi*, 62: 791–795.
25. Liu, D., Kanoh, S, Niwa, E. (1996). Effect of cysteine proteinase inhibitor on the setting of Alaska Pollack surimi paste. *Nippon Suisan Gakkaishi,* 62: 275–279.
26. Konno, K., Imamura, K. (2000). Identification of the 150 and 70 kDa fragments generated from during the incubation of salted surimi paste of walleye pollack. *Nippon Suisan Gakkaishi*, 66: 869–875.
27. Iwasaki, T., Watanabe, T., Moriya, N., Funatsu, Y., Yamamoto, K. (2012). Transglutaminase affects gel properties of actomyosin from chicken and fishes. *J. Rakuno.Gakuen Univ.,* 36: 263–270.
28. Iwasaki, T., Kadoma, I., Suzuki, K., Okumura, M., Sakurai, K., Funatsu, Y., Kaneda, I., Ishioroshi, M., Yamamoto, K., Kato, N., Arai, K. (2010). Development of novel food stuff from muscle proteins of different species and its application: Effect of egg white protein on thermal gelation of mixture of chicken and fish actomyosins. *J. Rakuno Gakuen Univ.,* 34: 197–209.
29. Abe, Y. (1994). The quality of transglutaminase-treated Kamaboko. *Nippon Suisan Gakkaishi* 36: 381–387.
30. Numakura, T., Seki, N., Kimura, I., Toyoda, K., Fujita, T., Takama, K., Arai, K. (1985). Gel formation of fish meat paste and myosin cross-linking reaction during suwari. *Bulletin of the Japanese Society of Scientific Fisheries,* 36: 169–172.
31. Toyohara, H., Kinoshita, M., Shimizu,Y. (1990). Proteolytic degradation of threadfin bream meat gel. *J. Food Sci.*, 55: 259–260.
32. Hemung, B.J., Yongsawatdigul, C.Li, Eunice, C.Y. (2008). Thermal stability of fish natural actomyosin affects reactivity to cross-linking by microbial and fish transglutaminases. *Food Chemistry,* 111: 439–436.

Chapter 9
Rheology Modifiers for the Management of Dysphagia

Graham Sworn

Abstract Dysphagia is the medical term used to describe swallowing problems resulting from a disorder in the mechanics of swallowing which can lead to impairment in the safety, efficiency or quality of eating and drinking. Dysphagia can be caused by many disorders including neurological disorders, stroke, traumatic brain injury, Huntington's disease, multiple sclerosis, Parkinson's disease and cerebral palsy. Many of these conditions are associated with the elderly and, with the continuing demographic shift, dysphagia is a growing problem.

The rheological properties of the bolus significantly influence the effectiveness and safety of the swallowing process, and the use of hydrocolloids to control the rheology can greatly help in the management of dysphagia. To be an effective thickener for this application, the following properties are important:

- Easy dispersion at low mixing speeds
- Fast hydration (cold and hot)
- Fast hydration in different media (water, tea, coffee, fruit juices)
- Stable viscosity as a function of:
  - Time
  - Temperature

Creating instant viscosity under suboptimal mixing conditions and choosing the right rheological properties pose a number of challenges for the use of hydrocolloids in this application, and the factors influencing their choice and performance will be reviewed.

Keywords Dysphagia • Swallowing • Hydrocolloids • Xanthan • Viscosity • Rheology

---

G. Sworn (✉)
DuPont, 20 rue Brunel, 75017 Paris, France
e-mail: graham.sworn@dupont.com

I. Kaneda (ed.), *Rheology of Biological Soft Matter*, Soft and Biological Matter,
DOI 10.1007/978-4-431-56080-7_9

9.1 Introduction

Dysphagia, the medical term used to describe swallowing problems resulting from a disorder in the mechanics of swallowing, can lead to impairment in the safety, efficiency and quality of eating and drinking and ultimately serious health issues. Swallowing is one of the most complex neuromuscular interactions in the human body and involves four distinct phases [23]:

1. Oral preparatory phase
   In this phase the ingested food is held in the mouth and is manipulated by the teeth and tongue to produce a mixture of saliva and food called the bolus.
2. Oral transport phase
   The bolus is moved by the tongue towards the back of the mouth into the pharynx. The soft palate is retracted upwards to seal the nasal cavity and avoid nasal regurgitation, and peristaltic waves begin to transport the bolus downwards.
3. Pharyngeal phase
   Several protective reflexes occur to prevent the bolus from moving towards the airway. The hyoid bone is moved up and forwards, the epiglottis lowers and the vocal cords contract.
4. Final phase
   The bolus passes through the cricopharyngeal sphincter and descends via the oesophagus to the stomach with the help of peristaltic waves and gravity.

Dysphagia can be caused by many disorders including neurological, stroke, traumatic brain injury, Huntington's disease, multiple sclerosis, Parkinson's disease and cerebral palsy. Conservative estimates suggest that about 8 % of the world population experience difficulties in eating and drinking as a result of dysphagia. This equates to around 99 million individuals from the developed world and 560 million individuals worldwide [5]. Some conditions result in a permanent impairment in swallowing function, whilst others such as a stroke can be temporary. Many of these conditions are associated with the elderly and, with the continuing demographic shift, dysphagia is a growing problem. Dysphagia can also be a problem in sick infants, particularly preterm infants. The improvements in neonatal care and the associated increase in survival rates have resulted in an increase in this group [14, 36].

For those suffering from dysphagia, the speed of the bolus flow from the mouth into the pharynx can be so rapid that it does not allow enough time for the person to close the airway before the bolus arrives at the entrance to the larynx and airway. This can allow entry of food or liquids into the airway. This is known as aspiration and often manifests itself as coughing or choking during swallowing which, in turn, can lead to a reluctance to eat or drink with the subsequent risk of malnutrition and dehydration. In the longer term, if not controlled, aspiration can lead to complications with the respiratory system such as chronic obstructive pulmonary disease or aspiration pneumonia.

The use of foods and liquids with modified texture/rheology is an established clinical practice to address dysphagia. The practice has been based on the assumption that modifying the rheology of normal foods and liquids makes them easier and safer to swallow. It is generally accepted that thin liquids, such as water, tea, coffee, squashes and juices, flow quickly in the mouth during the swallowing process and can cause aspiration and the general practice is to increase the viscosity of the liquid to decrease the risk. Studies in Japan have shown that water, which is often aspirated by dysphagic patients, has a maximum velocity in the pharynx of about 0.5 ms^{-1}, whereas yoghurt, which is rarely aspirated, has a maximum of about 0.2 ms^{-1}. The reduction in velocity has been attributed to the higher viscosity of yoghurt [12, 13]. A recent systematic review which set out to answer the question of '. . . whether evidence supports or refutes the practices of thickening liquids and modifying food textures in the context of the clinical management of dysphagia' concluded that there was evidence to support the practice [51]. The evidence showed that thickened liquids provided a benefit in terms of reducing aspiration, but it is also associated with a risk of post-swallow residue in the pharynx with the thicker consistencies.

Hydrocolloids are widely used in the food industry to control the texture and rheology of foods. They are found, for example, in salad dressings, ice cream, confectionery, soups, sauces, meat, dairy and bakery products, beverages and desserts. Their ability to modify the rheology of both solid and liquid foods makes them a natural choice as rheology modifiers for the management of dysphagia.

There is a very wide variety of pre-prepared foods and liquids with modified rheology specifically designed for the management of dysphagia; however the focus of this chapter will be on the specific use of powders for instant thickening of liquids which are used in hospitals, in nursing homes and at home. Creating instant viscosity under suboptimal mixing conditions and choosing the right rheological properties pose a number of challenges for the use of hydrocolloids in this application, and the factors influencing their choice and performance will be reviewed.

9.2 Hydrocolloid Rheology Modifiers

Hydrocolloids are high molecular weight polymers composed of polysaccharides. Production processes vary from simple collection of tree exudates and milling in the case of gum arabic to more complex production by fermentation as in the case of xanthan gum. A number of these hydrocolloids are also derivatised in order to modify their properties. Table 9.1 provides a simple classification of these products by source.

A brief description of each class of hydrocolloids is given below, but for more detailed information on each, the reader is referred to a number of publications [9, 17, 42]. Modified starches, xanthan gum, guar gum, cellulose gum and tara gum are the main hydrocolloids used to produce powdered products for instant thickening of liquids for the management of dysphagia. A detailed description of these specific hydrocolloids is given in Sect. 9.3.

Table 9.1 Classification of hydrocolloids (Reproduced with modification from Sworn [52] with permission from John Wiley and Sons (Copyright © 2007 by Blackwell Publishing Ltd))

Marine	Botanical	Microbial	Chemically modified
Carrageenans	Guar gum	Xanthan gum	Cellulose derivatives
Agar-agar	Locust bean gum	Gellan gum	Modified starches
Alginates	Gum tragacanth	Pullulan	Propylene glycol alginate
	Konjac glucomannan	Curdlan	
	Tara gum	Dextran	
	Cassia gum	Succinoglycan	
	Gum arabic		
	Pectin		
	Starches		

Marine hydrocolloids include the carrageenans, a group of sulphated galactans, which are extracted from red seaweed (*Rhodophyceae*) species such as *Eucheuma cottonii*, *Eucheuma spinosum*, *Chondrus crispus* and *Gigartina*. The carrageenans are split into three main types according to their ester sulphate content. These are lambda, iota and kappa in order of decreasing ester sulphate content. The carrageenan type varies according to the seaweed source. Lambda carrageenan is a non-gelling thickener, whereas iota and kappa types are gelling. Iota forms soft, thixotropic gels in the presence of calcium, whereas kappa forms firm, brittle gels in the presence of potassium or to a lesser extent calcium. These are commonly used to produce specific textured foods such as water jellies for use by dysphagic patients.

Alginates are extracted from brown seaweed (Phaeophyceae) species such as *Macrocystis pyrifera*, *Laminaria hyperborea* and *Ascophyllum nodosum*. Alginates are block copolymers composed of mannuronic acid (M) and guluronic acid (G) residues. The ratio of these substituents, the M/G ratio, is dependent on the seaweed source and the part of the seaweed used. M/G ratio also governs the properties of the alginate. Sodium salts of alginate are soluble in water and are used as thickeners and gelling agents. Gelation occurs through addition of calcium. Alginates rich in mannuronic acid residues (high M) form softer more flexible gels with little or no syneresis compared to their guluronic acid-rich (high G) counterparts.

Agar is a collective term for a complex mixture of polysaccharides which are extracted from *Gelidium* and *Gracilaria* species of red seaweed. Agarose, a neutral polymer, and agaropectin, a charged sulphated polymer, are the two major fractions. Agar typically forms firm, brittle gels on cooling and shows thermal hysteresis. It is used extensively in microbiological media and confectionery products. Agar is also commonly used in Japan for the preparation of foods for dysphagia with a predefined texture.

Botanical hydrocolloids are perhaps the most diverse group of hydrocolloids. Many of these materials have been known to man for centuries. Guar gum, locust bean gum (LBG), tara gum and cassia gum are composed of a linear (1→4)-linked β-D-mannan chain with single galactose substituents and are therefore referred to as galactomannans. The galactose units are attached to the mannose backbone by an (1→6) α-glycosidic linkages. They differ in the degree of galactose substitution, guar typically containing one galactose per every two mannose residues, whereas locust bean gum typically has only one galactose every four to five residues. All the galactomannans are thickeners and their properties, such as solubility and interaction with xanthan or carrageenan, are governed by the galactose content. For example, guar is soluble in cold water, whereas LBG must be heated to ~90 °C to hydrate. LBG, under certain conditions, will form soft flexible gels with xanthan, whereas guar only shows a synergistic increase in viscosity. Tara and cassia gums have properties intermediate to those of guar and LBG, respectively.

Pectins are commercially extracted from a variety of sources including apples and citrus fruits. They are composed of galacturonic acid residues with occasional rhamnose interruptions. They are usually classified in terms of their degree of methyl esterification. Low-ester (<50 %) pectins gel in a similar way to alginates through reaction with calcium. High-ester (>50 %) pectins require low pH and highly soluble solids (>~55 %) to gel. Under these conditions intermolecular electrostatic repulsions are reduced. The type of solids has an effect on the gels. For example, sucrose is more effective at promoting gelation than corn syrup. This is related to the relative affinity of the sugars for the water and their influence on water activity. Different sugars at equivalent soluble solids give differing water activity, and this can result in differences in gel formation and quality. This class of hydrocolloids also includes the starches and gum arabic.

There are a number of microbial hydrocolloids produced by fermentation. In nature many microorganisms produce polysaccharides, often referred to as exopolysaccharides. The purpose of the exopolysaccharide in nature is not fully understood, but it is thought to be related to protection of the microorganism from changes in the environment, e.g. pH, humidity and temperature. It is also thought to play a role in adhesion to the host. Industrial production involves fermentation of the pure strains of the microorganism under very controlled conditions to maximise production of the exopolysaccharide. The hydrocolloid is then extracted by precipitation with alcohol. This group includes hydrocolloids such as xanthan gum from *Xanthomonas campestris*, gellan gum from *Sphingomonas paucimobilis*, succinoglycan gum from *Agrobacterium tumefaciens*, dextran from *Leuconostoc mesenteroides*, pullulan from *Aureobasidium pullulans* and curdlan from *Alcaligenes faecalis*.

The chemically modified hydrocolloids include cellulose products such as carboxymethyl cellulose (CMC), hydroxypropylmethyl cellulose (HPMC) and hydroxyethyl cellulose (HEC). The purpose of the modifications is primarily to render the basic cellulose backbone soluble. In this way a range of cellulose-based products are produced with a variety of functions from thickening in the case of CMC to thermogelation in HPMC. Similarly, there are a wide variety of chemically modified starches available including hydroxyethyl and hydroxypropyl. These modifications to the native starch improve stability to heat and acid, improve processing and reduce the tendency to retrogradation. Alginates can also be modified by esterification with propylene glycol to produce propane 1,2-diol alginate

(PGA). This modification makes the alginates less sensitive to precipitation by acid and calcium which enables the PGA to remain in solution below pH 4.0.

9.3 Rheology of Hydrocolloids

The science of rheology encompasses the flow and deformation of both solid and liquid materials. This extends from a perfectly elastic solid, defined by Robert Hooke in 1678 [15], to a perfectly viscous liquid, defined by Newton in 1687 [37]. In practice, few if any materials show this ideal behaviour and most are more appropriately described as viscoelastic. That is to say, they exhibit both viscous and elastic behaviour.

In general, polymer gels and networks can be divided into three categories according to the nature of the interactions between the polymer chains:

1. Covalent cross-linked networks
2. Physical networks
3. Entanglement networks

The majority of hydrocolloids fall into categories 2 and 3. In category 2 the polysaccharides are physically cross-linked through such mechanisms as hydrogen bonding or cation-mediated junction zones to form networks, and these polysaccharides are often referred to as gelling agents. Thickeners generally fall into category 3. In the case of entanglement networks, in which no enthalpic polymer-polymer interactions are observed, properties are very much related to the size and number of molecules in solution, i.e. molecular weight and concentration, and the concept of space occupancy (hydrodynamic volume) becomes important [32, 33]. In dilute solution the individual molecules (random coils) are free to move independently. As the concentration is increased, the molecules begin to come into contact with one another. The motion of the molecules becomes restricted and the system can be visualised as a sea of entangled spaghetti. This transition from free moving molecules to an entangled network is accompanied by a change in the concentration dependence of the viscosity as shown schematically in Fig. 9.1. The concentration of polysaccharide at which the change occurs is referred to as the critical coil overlap concentration and is denoted C*.

Intrinsic viscosity $[\eta]$ can be used to compare the dilute solution viscosities of hydrocolloids [29]. The intrinsic viscosity will depend on the size and shape of the polymer molecule. The onset of entanglement for a wide range of neutral and charged polysaccharides is found to occur when $C[\eta] \cong 4$ (i.e. $C^* \cong 4/[\eta]$). In other words, the higher the intrinsic viscosity of the polysaccharide, the lower the concentration at which C* is exceeded. Hydrodynamic volume is also important. The greater the hydrodynamic volume, the lower the concentration required to exceed C*. A number of factors will influence the hydrodynamic volume [29]. For example:

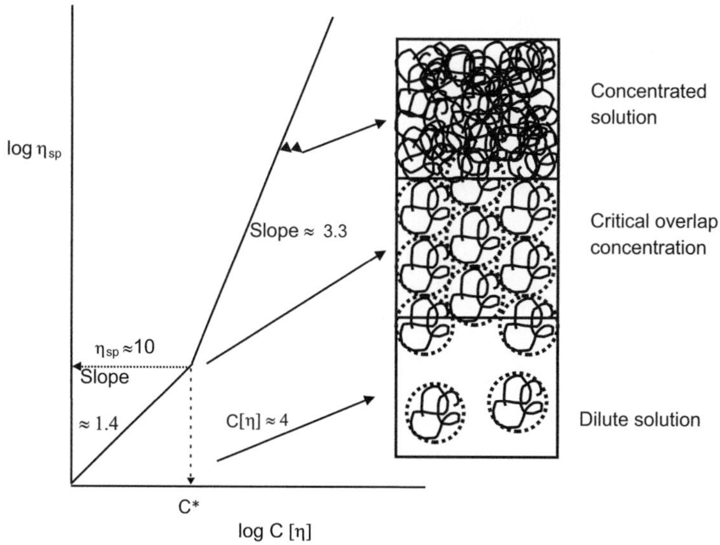

Fig. 9.1 Generalised concentration dependence of viscosity for random coil polysaccharides (Reproduced from Sworn [52] with permission from John Wiley and Sons (Copyright © 2007 by Blackwell Publishing Ltd))

- Polymer chain stiffness
  - The stiffer the chains, the larger the volume occupied.
- Polymer chain branching
  - Branched polymers will be more compact than linear polymers of the same molecular weight and will therefore have a lower hydrodynamic volume.
- Solvent
  - The higher the quality of the solvent for the polymer, the greater the hydrodynamic volume. In a good solvent, interactions between the polymer and the solvent will be favoured at the expense of polymer-polymer interactions.
- Polymer charge
  - Generally, ionic polymers will be more expanded than nonionic polymers due to electrostatic repulsion between like charges. Increasing the ionic strength of the solution will decrease this effect.

It is generally recognised that a further distinction can be made for hydrocolloid networks and gels, namely:

1. Entanglement network solutions
2. Weak gels
3. Strong gels

The above classification, which is based on the measurement of both steady and oscillatory rheological behaviours [32, 43], has been widely adopted by many in the field of hydrocolloid rheology.

9.3.1 Viscosity of Solutions

The viscosity of a fluid is a measure of the frictional resistance it offers to an applied shearing force which is expressed in the following equation:

$$\sigma = \eta\dot{\gamma} \tag{9.1}$$

where the shear stress (σ) is equivalent to the force per unit area and the shear rate ($\dot{\gamma}$) is the velocity gradient. In other words, the viscosity is equal to the shear stress divided by the shear rate.

There are generally two approaches to the measurement of viscosity, namely, controlled stress and controlled strain. In controlled stress, a known stress is applied to the fluid and the resultant shear rate is measured. Conversely, in controlled strain, a known strain or more accurately rate of strain (shear rate) is applied and the stress is measured. For example, a falling ball viscometer is a controlled stress viscometer in which the applied stress comes from the size of the ball and gravity, whereas the Brookfield viscometer is an example of a controlled strain instrument where the rate of strain is related to the rotational speed applied.

Fluids are primarily classified into two types depending on their response to an applied shearing force:

- Newtonian
  Viscosity is constant as a function of shear rate. Examples include water, honey, sugar solutions and oils.
- Non-Newtonian
  Viscosity changes as a function of shear rate. Examples include hydrocolloid thickeners and emulsions. This class of materials can be further classified depending on how the viscosity changes:
  - Pseudoplastic (shear thinning)
    Viscosity decreases with increasing shear rate.
  - Rheopectic (shear thickening)
    Viscosity increases with increasing shear rate.

Many hydrocolloid thickeners form entanglement networks and a distinction between the behaviour above and below C* must be made. Hydrocolloid solutions below C* will typically exhibit near Newtonian steady shear flow, and the increase in the viscosity of the solvent is roughly proportional to the number of molecules present. In general a doubling of concentration will increase viscosity by a factor of ~2.5. Above C* entanglement network solutions will exhibit shear-thinning (pseudoplastic) flow as a function of shear rate. The concentration dependence of viscosity increases, and generally a doubling of polymer concentration above C* will result in an approximately tenfold increase in viscosity [32, 33]. It is to be noted that the C* for hydrocolloids occurs typically at concentrations of <0.1 % and therefore the concentrations of thickeners used for the management of dysphagia are well in excess of this.

Figure 9.2 illustrates a generalised plot of the viscosity versus shear rate for a polysaccharide entanglement network (C > C*). At low shear rates, the shear rate dependence of viscosity is Newtonian since the disentanglement due to shear is slower than or equal to the formation of new entanglements. This is known as the zero shear or lower Newtonian viscosity, denoted η_0. As shear rate is increased, the rate of disruption also increases and exceeds that of formation. At this point viscosity begins to decrease sharply as a function of shear rate. This is known as the shear-thinning or pseudoplastic region. Eventually, at very high shear rates, no further disruption can take place, and the system once again exhibits Newtonian flow. This is called the infinite shear or upper Newtonian viscosity, denoted η_∞.

The flow behaviour in Fig. 9.2 can be accurately modelled using the Cross equation [6]:

$$\eta = \eta_\infty + (\eta_0 - \eta_\infty) / \left(1 + k\dot{\gamma}^n\right) \tag{9.2}$$

where η_0 and η_∞ are the zero and infinite shear viscosity, respectively, k is a constant parameter with the dimension of time and n is a dimensionless constant.

The origin of shear thinning in macromolecular solutions is believed to be caused by the following factors, one or more of which may apply to a particular solution:

- Increased orientation of asymmetric molecules with shear rate
- Change in the shape of flexible molecules with shear
- Effect of flow on intermolecular interactions

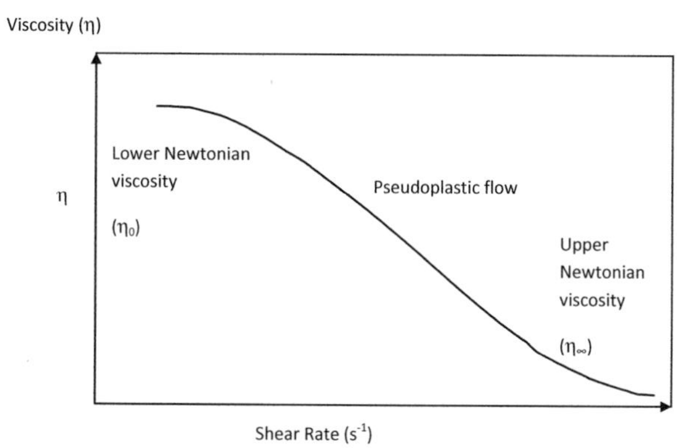

Fig. 9.2 Idealised flow curve of an entanglement solution of a polysaccharide (Reproduced from Sworn [52] with permission from John Wiley and Sons (Copyright © 2007 by Blackwell Publishing Ltd))

Entanglement solutions of the various hydrocolloids differ greatly in the magnitude of their zero shear viscosity and in the shear rate at which shear thinning commences. Generally, the degree of pseudoplasticity for a given hydrocolloid will increase with an increase in the concentration and the molecular weight. However, it has been demonstrated that the form of shear-thinning behaviour is a general phenomenon. Morris et al. [34] showed that, by expressing measured viscosity as a fraction of the zero shear viscosity (η/η_0) and applied shear rate relative to the shear rate required to reduce viscosity to a fixed fraction of η_0 $\left(\dot{\gamma}/\dot{\gamma}_f\right)$, a wide range of entanglement solutions of polysaccharides of differing primary structure, molecular weight and concentration collapse onto a single master curve.

The shear-thinning region of flow can be conveniently modelled with a power law equation of the form:

$$\tau = k\dot{\gamma}^n \tag{9.3}$$

where τ is the applied stress, $\dot{\gamma}$ is the shear rate and k and n are constants often referred to as the consistency index and flow index, respectively. The flow index (n), also known as the Power Law Index, can be particularly instructive since it is a measure of the degree of pseudoplasticity. A value of 1 would indicate Newtonian flow whilst values <1 indicate pseudoplastic flow. The constants k and n are very much dependent on the concentration and shear rate range within which the measurements are made. Whilst the power law model does have some utility in the practical interpretation of rheology clearly, it is not sufficient to fully characterise the flow behaviour of polysaccharides, and as indicated previously, the Cross model is a much more accurate representation of the entire flow behaviour of hydrocolloid solutions.

9.3.2 *Viscoelasticity of Solutions*

Few if any materials show the ideal behaviour of a perfectly elastic solid or a perfectly viscous liquid and are therefore more appropriately described as viscoelastic. That is to say, they exhibit both viscous and elastic behaviour.

Mechanical spectroscopy, which involves the application of a sinusoidal oscillation to the material, is an ideal technique for the investigation of the viscoelastic properties of fluids. The strain (γ), which is the amplitude of the rotational displacement divided by the height of the sample, and the oscillation frequency (ω) in units of radians per second, can be varied under controlled conditions to provide rheological information. A perfectly elastic material will have a stress wave exactly in phase with the applied strain wave. A purely viscous material will have a stress wave exactly 90° out of phase with the applied strain. Typically the stress wave will have a phase difference (δ) between 0 and 90°. The in-phase and out-of-phase components of the stress wave can be separated to give the elastic or storage modulus (G′) and the viscous or loss modulus (G″) as well as other characteristic measurements such as complex shear modulus (G*) and complex viscosity (η*).

Initially, for characterisation by mechanical spectroscopy, the strain dependence of, for example, the complex shear modulus (G*) is established. This experiment establishes the linear viscoelastic region of the system, within which the viscoelastic functions are independent of strain. In other words the applied strain does not perturb the sample. For entanglement networks the linear viscoelastic region extends to approximately 25 % strain.

Just as viscosity is dependent on shear rate, the viscoelasticity is dependent on frequency. At low frequency the response of the material will appear more viscous (liquid like), whereas at high frequency, the response will appear more elastic (solid like). Frequency can also be considered in terms of the timescale of the applied stress. If the timescale of the applied stress is relatively slow (low frequency), the material will appear liquid. Conversely, if the timescale is relatively fast (high frequency), the material will appear more solid. A good illustration of this is the property of water when entering a swimming pool. Entering the pool slowly (low frequency), the water is, of course, very liquid, and conversely jumping of the high diving board and entering pool at high speed (high frequency) give the water a more solid feel.

Hydrocolloid solutions can be characterised by measurement of the mechanical spectrum at a strain within the linear viscoelastic region defined by the strain sweep. Here the storage (G′) and loss (G″) modulus and complex viscosity (η*) are measured as a function of frequency (ω) and plotted on double logarithmic plots. Idealised mechanical spectra of entanglement solutions are shown in Fig. 9.3. Figure 9.3a and b compares the mechanical behaviour below and above C*, respectively. Below C* both moduli are strongly dependent on frequency (ω) with typically $G' \propto \omega^2$ and $G'' \propto \omega$. Complex viscosity (η*) shows little dependence on frequency. Above C* where entanglement networks are formed, the mechanical spectrum at the low frequency end is similar to that below C* since the timescale of the measurement is long by comparison to the timescale of the molecular motions. At higher frequencies where the experimental timescale is shorter than the molecular rearrangements, the mechanical spectrum resembles that of a gel, $G' > G''$, with little dependence on frequency and η* decreasing with increasing ω. These systems also obey the Cox-Merz superposition principle [3]. That is, complex viscosity as a function of frequency $\eta^*(\omega)$ rad·s^{-1} will superimpose on a plot of steady shear viscosity versus shear rate $\eta(\gamma)$ s^{-1} over the same range of ω and γ. In other words, the viscosity measured in a non-destructive way (oscillation) is the same as the viscosity measured in a destructive way (flow).

9.3.3 Viscoelasticity of Weak and Strong Gels

The strain dependence of the viscoelastic functions of strong gels is very similar to entanglement solutions with the linear viscoelastic region extending to

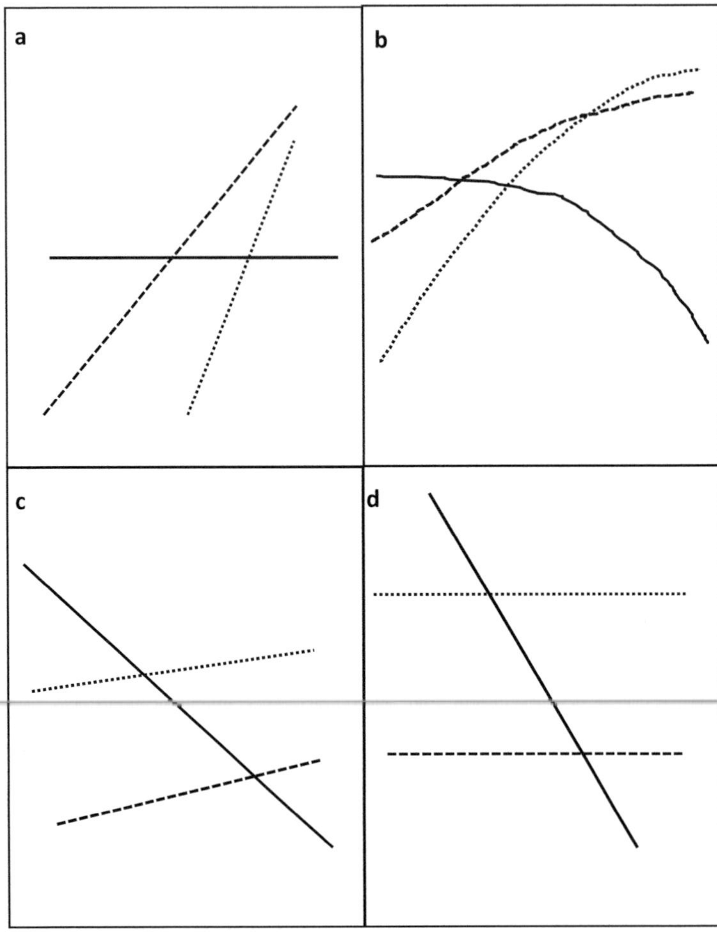

Fig. 9.3 Frequency dependence of storage modulus G′ (—-), loss modulus G″ (– – –) and complex viscosity η* (—) for aqueous hydrocolloid solutions. Idealised mechanical spectra of (**a**) dilute solution (below C*), (**b**) entanglement network (above C*), (**c**) weak gel and (**d**) strong gel (Reproduced from Sworn [52] with permission from John Wiley and Sons (Copyright © 2007 by Blackwell Publishing Ltd))

approximately 25 % strain. For weak gels this region is typically less than 5 % strain. Idealised mechanical spectra for weak and strong gels, as shown in Fig. 9.3c and d respectively, exhibit a very similar mechanical spectrum: G′ is significantly higher than G″ throughout the frequency range and both moduli show little or no dependence on frequency. The complex viscosity decreases with increasing frequency with a slope approaching −1. However, they can be distinguished on the basis of their strain dependence. Furthermore, in steady shear a strong gel will rupture completely, whereas a weak gel will flow, albeit without obeying the Cox-Merz superposition principle. In this case the complex viscosity (η^*) is higher than the flow viscosity (η). This indicates the presence of weak interactions between the molecules that contribute to the viscosity measured by the non-destructive oscillatory technique but not to the viscosity measured by the destructive flow technique.

9.4 What Makes a Good Rheology Modifier for Dysphagia?

Fundamentally, the ability to thicken a liquid is a prerequisite for a hydrocolloid to be effective for the management of dysphagia. In addition it must deliver the desired thickening effect at the right time and right place under the conditions of use. Important properties are:

- Easy dispersion at low mixing speeds
- Fast hydration (cold and hot)
- Fast hydration in different media (water, tea, coffee, fruit juices)
- Stable viscosity as a function of:
  - Time
  - Temperature

The first rheology modifiers to be used for products for dysphagia were starches and these are still popular today. Starch is one of the most widely used rheology modifiers in food products. It consists of two types of glucose polymer: amylose, which is linear, and amylopectin, which is branched. The relative composition of the two polymers is dependent on the origin of the starch. Starch occurs in the form of granules, is found in plant tissue and is particularly abundant in tubers such as potato and the endosperm of seeds such as corn and rice. Native starch granules are insoluble in cold water and need to be heated to create viscosity. As the temperature is increased above, what is termed the gelatinisation temperature, water is taken into the granule causing swelling, which ultimately bursts leading to the development of viscosity. The gelatinisation temperature is a characteristic of the origin of the starch and is typically between 55 and 70 °C. After gelatinisation, when starch solutions or pastes are cooled and allowed to stand for a few hours, they begin to show changes in their rheology. Known as retrogradation, this is normally seen as an increase in viscosity or a gelation effect caused by the reassociation of the amylose and amylopectin fractions.

Lack of cold solubility and retrogradation make native starches unsuitable as rheology modifiers for dysphagia. Modified starches are preferred. Starches suitable for use for dysphagia are usually modified in two ways. Firstly, to minimise rheological changes with time (retrogradation), the starches are cross-linked, and secondly, the starch is made cold soluble by a process known as pre-gelatinisation. A typical choice of starch would be a pre-gelatinised modified corn starch (E1442).

They are called hydroxypropyl distarch phosphate and are produced from the reaction of an aqueous starch suspension with a combination of propylene oxide to stabilise the starch by etherification and either sodium trimetaphosphate or phosphorus oxychloride to cross-link the starch. They are produced with a range of different levels of cross-linking and stabilisation. They produce thick, rich creamy pastes with excellent mouthfeel and cling and have good resistance to retrogradation. Pre-gelatinisation is a physical rather than chemical modification. Essentially, the starch is pre-cooked by simultaneously cooking and drying using drum-drying, extrusion or spray-drying technology. This process renders the starch cold swelling. These types of starches are readily dispersible in water and other liquids and as a result can be used directly. Many products for dysphagia are available as a pure starch or as a simple blend with other ingredients such as salts or maltodextrin, for example.

More recently other rheology modifiers such as guar gum and xanthan gum have become more common. These are often referred to as gum-based products. Also other hydrocolloids such as cellulose gum and tara gum can find occasional use. These gums are soluble in water and form viscous solutions. They are used in the food industry in convenience foods, dairy products, ice cream, beverages, baked product dressings and sauces, for example. Xanthan and guar gums are often used together in foods as they exhibit a synergistic interaction at a molecular level that results in modified rheological properties.

Guar gum is found naturally in the seeds of the guar plant *Cyamopsis tetragonoloba L.* More specifically it is found in the endosperm. This plant is native to India and Pakistan and has been cultivated commercially in this region for centuries. It can also be cultivated in the southern hemisphere in the semiarid zones found in Brazil, Australia and South Africa. Guar gum is produced by simple separation and grinding of the endosperm to produce a fine off-white powder. The galactose content of guar gum varies between 33 % and 40 % which equates to, on average, 1 galactose for every 1.6 mannose residues.

Like guar gum, tara gum is also found in the endosperm of the seeds of a plant (*Caesalpinia spinosa*) and is produced in a very similar way. The tara tree or shrub is native to the Cordillera region of Bolivia, Peru and northern Chile. Tara gum is also a galactomannan but differs from guar gum with respect to the galactose-to-mannose ratio. Tara gum has approximately 25 % galactose which equates to a galactose-to-mannose ratio of 1:3.

Cellulose gum, also known as carboxymethyl cellulose (CMC), is an anionic cellulose derivative. Cellulose from either wood pulp or cotton linters, which is insoluble in water, is treated with alkali to swell the polymer. The swollen cellulose is then reacted with monochloroacetic acid under controlled conditions to substitute the anhydroglucose monomers. The number of monomers substituted is known as the degree of substitution (DS). The permitted maximum DS for food grade in the EU is 1.5. Commercial products are typically in the range of between 0.6 and 0.95. Commercial products also vary according to their viscosity and particle size.

The polysaccharide xanthan gum is produced by the microorganism *Xanthomonas campestris*. Commercially it is manufactured by aerobic fermentation of the microorganism with a suitable media containing carbon, oxygen, nitrogen and trace elements. At the end of the fermentation stage, the viscous broth containing the xanthan gum is heat treated to kill the microorganism, and the polysaccharide is recovered by precipitation with alcohol, followed by drying and milling to produce a white to off-white powder. The primary structure of xanthan gum consists of a cellulosic backbone of ß 1,4-linked D-glucose units substituted on alternate glucose residues with a trisaccharide side chain. The trisaccharide side chain is composed of two mannose units separated by a glucuronic acid [18, 27]. Approximately half the terminal mannose units are linked to a pyruvate group and the nonterminal residue usually carries an acetyl group. More recently it has been shown that up to 20 % of the terminal mannose can be acetylated [21]. The carboxyl groups on the side chains render the gum molecules anionic.

9.4.1 Dispersion and Hydration

Dispersion and hydration of the hydrocolloids is a critical step in achieving the full functionality and can be defined as follows:

Dispersion: The ease of separation of the individual powder particles as the hydrocolloid is put into the liquid

Hydration: The ease with which the individual particles hydrate and dissolve to create viscosity

Good dispersion is a critical first step in order to achieve good hydration. The main factors that affect the dispersion and hydration are mixing speed, particle size and the composition of the solvent. The following general rules can be applied:

- The higher the mixing speed, the better the dispersion and the more rapid the hydration.
- Small particle size powders are difficult to disperse, but once dispersed are quick to hydrate.
- Large particle size powders are easier to disperse but hydrate more slowly.
- High temperature can speed up hydration but can cause lumping due to pre-wetting of the powder by the steam.
- High salt levels and low pH can slow down hydration.

Pre-thickened products can be produced at industrial scale with equipment that ensures the correct dispersion and full hydration of the hydrocolloid. This is not the case for products sold in dry powder form for preparation in hospitals and nursing homes or at home. The dispersion conditions for hydrocolloids in these types of products for dysphagia are particularly challenging. Products are usually added to the liquid in one dose often with a spoon and then stirred by hand perhaps for less than 1 min. Hydrocolloid concentrations can be in excess of 1 % and any lump formation is unacceptable.

In general, gums are more difficult to disperse than starches as they have a stronger tendency to lump. For a gum to hydrate efficiently, the individual powder particles must be well dispersed in the liquid. Poor dispersion leads to lumping of particles during mixing, which results in the formation of swollen lumps (sometimes referred to as 'fish eyes'). Severe lumping prevents complete hydration and this reduces viscosity, and the presence of lumps presents a risk to dysphagia sufferers and makes the beverage unpleasant to drink.

Dispersion and hydration of gums during the industrial production of beverages and other food products does not normally present a problem as high-shear mixing is often used. Dispersion and hydration can also be improved by simple blending of the powders with other ingredients in the formulation such as sugar, starch, salts or oil. These act as dispersants by physically separating the gum particles, preventing them from forming lumps during dispersion and thereby ensuring good hydration. Gums are also available in a range of particle sizes, and agglomerated products with large particle size are able to disperse more easily. Whilst these strategies for dispersion and hydration work well for the industrial food production processes, they are not effective for the very challenging dispersion and hydration conditions encountered by powdered products for dysphagia. Many early products were based on starches because, compared to gums, they are relatively easy to disperse.

More recently, technologies such as fluid bed and spray drying have been used to create readily dispersible gum products that meet the requirements for use as an instant thickener for the management of dysphagia. The gums are usually co-processed with a dispersant such as maltodextrin. Fluid bed technology is particularly suitable for this and a diagram of the process is shown in Fig. 9.4.

The basic process can be divided into two phases: addition of a binder (spraying phase) to agglomerate the gum/dispersant blend, followed by drying of the agglomerated material (drying phase). Additionally, there is a third, optional, cooling phase where the temperature of the dried, agglomerated product is reduced before the product is discharged from the fluid bed. The binder solution can be simply water or a solution of the dispersant. Products produced in this way are more easily dispersed than a simple dry blend of the same composition.

9.4.2 Viscosity Development

The fluid to be thickened, the time between preparation and consumption and the temperature of consumption can vary greatly, and it is important that the thickener is able to perform well under all conditions. Typical fluids to be thickened include water, milk, fruit juices and hot beverages such as tea and coffee.

The dispersion properties and the viscosity development as a function of time of a starch and a gum (GRINDSTED® Xanthan NTJ-1)-based product have been compared in different beverages [53]. The xanthan and the modified starch were tested by adding 3 g of product to 97 g of beverage in a single dose with no mixing. The dispersion was then stirred for 5 s at 200 rpm using a four bladed helix

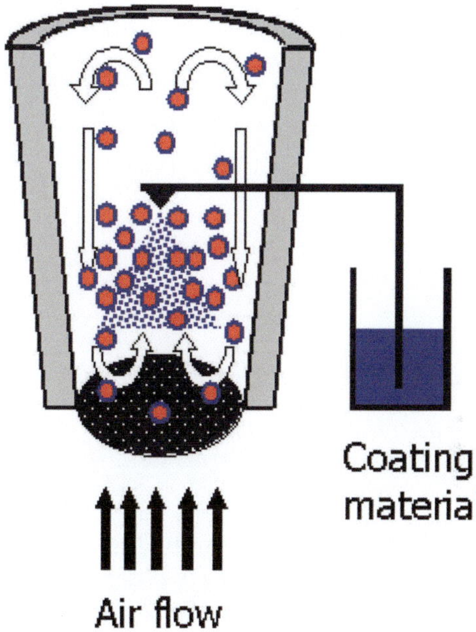

Fig. 9.4 Representation of a typical fluid bed process

Table 9.2 Comparison of the dispersion properties of starch and xanthan gum-based thickeners for dysphagia (Reproduced from Sworn et al. [53] with permission from the Royal Society of Chemistry)

Product	Dispersibility in tap water	
	20 °C	70 °C
GRINDSTED® Xanthan NTJ-1	Excellent	Excellent
	No lumps	No lumps
Modified starch	Acceptable	Poor
	Some small lumps which disappeared with mixing	Large lumps which did not disappear with mixing

paddle stirrer. The dispersion and hydration conditions were designed to reflect the actual conditions of use whilst providing a reproducible method to allow meaningful comparison of samples (e.g. a single, rapid addition of the powder to the liquid with no mixing, followed by a short mixing time at low speed). The dispersibility was assessed visually and is summarised in Table 9.2.

The hydration rate of the samples was measured by following the development of viscosity with time. Samples were dispersed as described above and then placed on a controlled stress rheometer (TA Instruments AR-1000) fitted with a 4 cm diameter flat plate with a gap of 1,500 μm (to avoid problems with lumps or micro-gels). Viscosity measurements were started 2 min after the initial addition of the powder to the water and were made at 20 °C at a shear rate of 50 s^{-1}. The results are shown in Fig. 9.5 and illustrate some important differences between the types of thickener. For example, in the three beverages tested, the xanthan gum has a more

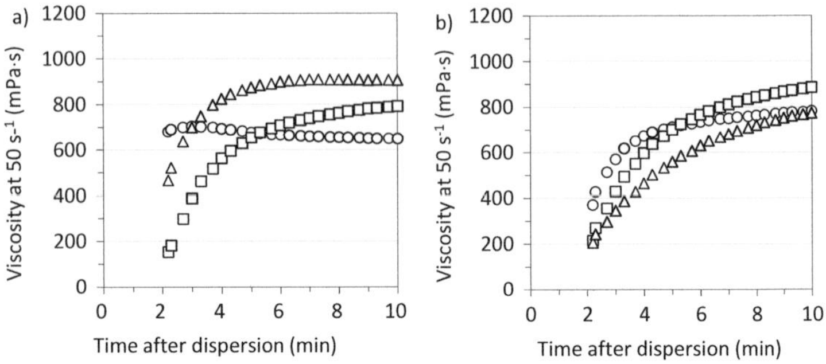

Fig. 9.5 Development of viscosity with time of (**a**) GRINDSTED® Xanthan NTJ-1 product at 3 % and (**b**) modified starch product at 2 % dispersed in tap water (○) orange juice (Δ) and milk (□) at room temperature and measured at 20 °C (Reproduced from Sworn et al. [53] with permission from the Royal Society of Chemistry)

rapid development of viscosity. This is particularly evident in tap water. The xanthan gum tends to reach a steady plateau in viscosity, whereas the starch continues to increase in viscosity even after 10 min. The starch has similar viscosity development profiles in the three beverages, whereas the xanthan gum varies. Also the viscosity of beverage thickened with starch is quite similar at the end of the test but is quite different for the beverages thickened with xanthan gum.

The differences in properties between starch- and gum-based thickeners are important to understand. For example, for xanthan gum, a correct dosage for the correct viscosity in water may not be the same dosage in a fruit juice. Similar effects have been reported by Garin et al. [7]. They showed that both starch- and gum-based products using a fixed dosage in different beverages resulted in different viscosities. Therefore dosage recommendations often need to vary according to the product to be thickened. The continued increase in viscosity seen in the starch will mean that the time between preparation and consumption could influence the viscosity and effectiveness of the product and will therefore need to be taken into account.

In a recent study, conducted in healthcare homes in the UK, a gum-based product was introduced alongside the existing starch-based product, and staff and patients were surveyed with regard to the performance of the two products [41]. In general, the staff found the gum-based product easier to mix with fewer problems and no lumping and found a lower incidence of drinks changing over time (e.g. continuing to thicken). Staff said that the residents preferred the texture, taste and appearance of drinks thickened with the gum-based product.

In another study four commercial thickener products for dysphagia were used to thicken a variety of liquids (milk, apple juice, orange juice, nutritional beverage and water), and the sensory properties were evaluated by a highly trained descriptive sensory analysis panel [26]. Two products were starch based and two products were gum based. The gum-based products were xanthan in one case and cellulose gum in the other. Results showed that all thickeners suppressed the main flavours of the base beverages and imparted slight off-flavours (bitter, sour, metallic or astringent) in some beverages. Starch-based thickeners imparted a starchy flavour and grainy texture, whereas gum-based thickeners gave added slickness to the beverages. Although thickeners mixed well with some beverages, they were difficult to disperse in others, making some thickened beverages lumpy. The authors concluded that additional development of thickening agents seems necessary for improved sensory properties.

9.4.3 Rheology

Thickened beverages are usually characterised by their viscosity, and in order to compare, measurements are often made at shear rates of 50–100 s^{-1}. This shear rate range is widely used as studies have suggested that this is broadly representative of the shear rate in the mouth [56]. This is, however, an oversimplification as it has been shown subsequently that the effective shear rate at which samples of similar perceived viscosity have the same measured viscosity decreased as viscosity increased [47]. Furthermore, the perceived thickness of xanthan gum solutions and food products with weak gel properties was underestimated [32]. In this case it was concluded that the complex viscosity measured using small deformation oscillatory measurements was better correlated.

The measurement of viscosity at a single shear rate for non-Newtonian beverages may be useful for comparative information, but it is not able to provide a full picture of the rheological profile. This approach does not take into account the pseudoplastic nature of the products. To characterise the flow properties of non-Newtonian liquids more completely, a flow curve of viscosity as a function of shear rate must be measured. Few if any commercial instruments are capable of measuring a broad enough shear rate range to see the whole flow curve of a polysaccharide solution illustrated in Fig. 9.2, and in practice it is more typical to see perhaps the lower Newtonian region and the onset of shear-thinning behaviour. An example of the flow curves of water thickened with different rheology modifiers is shown in Fig. 9.6. In this example, although the concentration of the rheology modifier has been adjusted in order to give equivalent viscosity at a shear rate of 100 s^{-1}, the resulting flow curves differ markedly. This illustrates the limitation of using a single point viscosity measurement to define the thickness of fluids with non-Newtonian rheological behaviour.

The viscoelastic properties of xanthan and modified starch at concentrations that give approximately equivalent viscosity at a shear rate of 50 s^{-1} are compared in Fig. 9.7. As with the viscosity/shear rate profiles, the viscoelastic profiles are very different even when the viscosity is matched at a given reference shear rate. This further illustrates the limitation of using a single point viscosity measurement to define the rheology of non-Newtonian fluids. The xanthan gum solution shows a mechanical spectrum typical of a weak gel, whereas the starch solution is more typical of an entanglement network. In fact, starch is more accurately described as a particulate thickener rather than an entanglement network since the viscosity is created by the overlap of swollen starch granules rather than entangled polymer chains.

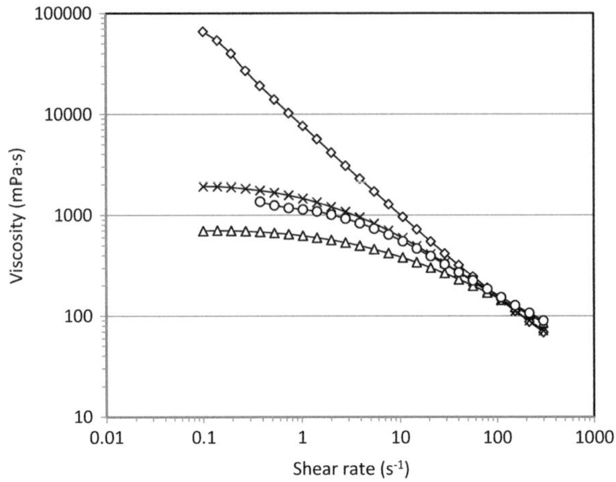

Fig. 9.6 Viscosity as a function of shear rate of solutions of 1 % xanthan (◇), 0.5 % cellulose gum (Δ), 0.6 % guar gum (x) and 4.5 % modified starch (○) in water with concentration adjusted to give equal viscosity at a shear rate of 100 s^{-1}

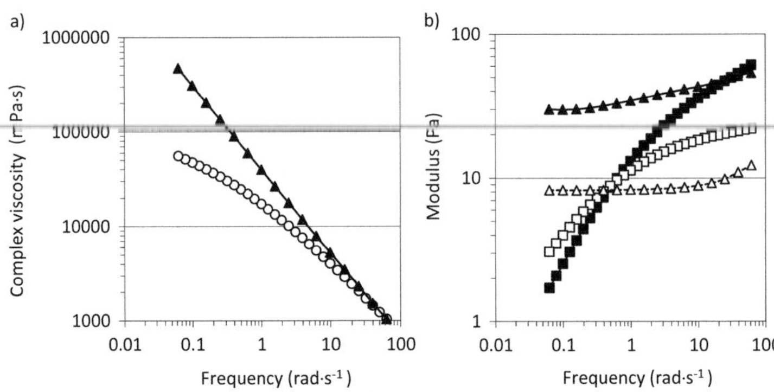

Fig. 9.7 Complex viscosity as a function of frequency for 1.2 % xanthan (▲) and 2 % modified starch (○) in tap water and (**b**) mechanical spectrum of 1.2 % xanthan (▲G′, Δ G″) and 2 % modified starch (■G′, □ G″) in tap water (Reproduced from Sworn et al. [53] with permission from the Royal Society of Chemistry)

9.5 How Thick Is Thick?

There seems to be little doubt among health professionals that thicker foods are easier and safer to swallow than thinner foods; however the problem remains: How do you define thick? This is not a new question. Cichero et al. [4] asked this question and highlighted the need to establish objective measurements of consistency. They also highlighted the challenges of characterising non-Newtonian fluids. The normal procedure in the management of dysphagia is for a speech-language pathologist (SLP) to complete a videofluorographic evaluation of swallowing during which the patient is asked to swallow barium sulphate solutions and the movement is observed using fluoroscopy. This radiographic technique uses x-rays to visualise the bolus. Based on the observations, the SLP will recommend foods of a certain consistency and a registered dietician (RD) will create a diet. This diet will be implemented by the nursing staff who will often select the liquids to be included or excluded. The use of the hydrocolloid-based powders by the nursing staff or carers is also common practice to thicken certain thin liquids for inclusion in the diet. As pointed out by Brown et al. [1], this system can only work if all involved share the same understanding of the consistency terms.

Many studies have tried to define how thick is thick and to implement simple consistency and/or viscosity measurement standards to help in the preparation of thickened foods for patients [1, 8, 10, 25, 28, 39, 49]. An example is the classification published in 2002 by the National Dysphagia Diet project conducted in the USA. This was based on a study in which ten orange drinks, prepared with viscosities (measured at 9 s^{-1}), ranging from 10 to 3,000 mPa·s by addition of modified starch, were presented to a number of speech-language pathologists (SLPs) and registered dieticians (RDs). They were asked to write down what they considered to be the best label for the liquid. They could stir the drinks but could not place the samples in their mouths. The study concluded that there was very poor agreement in the labels given. For example, one sample was given a total of 51 different labels [1]. The most common labels from this study were adopted by the National Dysphagia Diet, and corresponding viscosity ranges, measured at a shear rate of 50 s^{-1}, were also given to each category (Table 9.3). These consistency definitions were adopted by many US manufacturers of commercial thickeners for dysphagia.

Similar approaches have been taken by many countries which have led to the development of national standards (Table 9.4). Descriptors and measurement methods, however, still vary from country to country.

As described previously in Sect. 9.4.3, very often in order to compare thickeners, viscosity measurements are made at shear rates of 50–100 s^{-1}. The choice of this shear rate range is based on studies concerned with perceived thickness in the mouth in relation to the sensory properties of foods rather than with the swallowing mechanism. Recent studies have started to look at this in relation to swallowing. For example, Yamagata et al. [57] focused on swallowing in elderly people with normal swallow and found that measurement of viscosity at 100 s^{-1} correlated better with ease of swallowing than measurements made at 2–3 s^{-1}. Tashiro et al. [54] measured

Table 9.3 Consistency labels recommended by the National Dysphagia Diet project

Subjective consistency descriptor	Viscosity at 50 s^{-1} (mPa·s)
Thin	1–50
Nectar-like	51–350
Honey-like	351–1750
Spoon-thick	>1751

Table 9.4 List of national initiatives for the standardisation of descriptors for fluid consistency (data from IDDSI website)

Year	Initiative
2002	The USA publishes National Dysphagia Diet (adults); UK British National Descriptors for Texture Modification in Adults published
2007	Australian National Standardised Terminology and Definitions for Texture
	Modified Foods and Fluids (adult + paediatric) published; Australian Standards endorsed by New Zealand
2009	Irish Consistency Descriptors for Modified Foods and Fluids [adult + paediatric] (modified from Australian standards); Japanese Standards for Texture Modified Foods (adults) published
2010	Swedish National Descriptors published
209	National Patient Safety Agency – Dysphagia Diet Food Texture Descriptors. [multidisciplinary and industry involvement, suitable for adults and paediatrics]
2012	Canadian Dysphagia Diet Standardisation Task Force data collection for national terminology and definitions; European Task Force commencement for European Standardised Terminology and Definitions

the viscosity and the viscoelastic properties of carboxymethyl cellulose, xanthan gum and guar gum solutions and studied the effect of these rheological properties on the maximum velocity through the pharynx using the ultrasonic pulse Doppler method. Based on the dimensions of the pharynx and average velocities, they estimated the average shear rate during swallowing to be 27 s^{-1} and showed that the maximum velocity of the thickener solution correlated well with the viscosity especially when measured at similar shear rates of 20–30 s^{-1}.

Factors such as size, shape, volume, consistency, pH and temperature have been shown to be important in a person's evaluation of the readiness of the bolus for swallowing [2]. Once swallowed, other factors may be important, and it is likely that the different stages of swallowing will require different rheological measurements to obtain relevant correlations. For example, recent work on the miscibility of solutions thickened with different hydrocolloids has revealed some interesting differences that could be of relevance to this application. It has been shown that solutions prepared with particulate thickeners such as starch mix more easily with water than those thickened by polymer solutions such as guar gum and xanthan gum [30]. In the case of starch, easy mixing with the saliva could lead to dilution and loss of viscosity in the mouth prior to swallowing, whereas with xanthan the poorer mixing may mean that the gel-like rheology would be more likely to remain intact during swallowing. This could potentially offer better protection against aspiration. Starch systems are also susceptible to amylase degradation upon contact with saliva. This can result in a loss of viscosity and present a risk to the patient. Amylase-resistant systems have been developed by adding amylase-resistant gums such as xanthan to the starch [11, 48]. Although the mechanism is not clear, it is likely that the reduction in the miscibility due to the gums offers some protection to the starch from the amylase attack by limiting the contact time with the amylase prior to swallowing.

The differences in the rheological properties of the rheology modifiers typically used for management of dysphagia and discussion of the possible implications for their use are well documented [24, 31, 40, 46, 53, 54]. However, a fundamental question that is now being asked is: Do these differences really matter? A recently published comment on a paper studying rheological differences in thickeners captured the essence of this question very succinctly [45]:

> *...the question of real interest is: Do these differences impact swallowing success or influence any biomechanical variables during swallowing? It is irrelevant to the patient and the treating deglutologist if there is no improvement in swallowing success despite small but statistically significant differences in the steady flow and viscoelastic properties of various cold-thickened beverages. Now that we know what these differences are, it is time to bring them out of the laboratory and into the realworld setting to see what benefits patients best.*

Very few studies to date have directly addressed this question partly due to the complexity and ethics of running human trials with dysphagic patients. Nishinari et al. [38] compared xanthan gum to guar gum in a study involving healthy persons and patients suspected of having swallowing difficulties. Videofluorographic observations were performed and compared with rheological characteristics of the gum solutions. They concluded that the risk of aspiration might depend on the absolute value of the viscosity and the degree of shear thinning. However, all the results could not be explained only by the difference in the degree of shear thinning. Leonard et al. [22] made a dynamic fluoroscopic swallow study of patients given contrast agent thickened to nectar-thick consistency (51–350 mPa·s at 50 s^{-1}) with a starch (290–330 mPa·s) or a xanthan gum (150–170 mPa·s) thickener, and results were compared to the unthickened agent (4–5 mPa·s at 50 s^{-1}). Results showed a statistically significant reduction in aspiration for the xanthan-thickened contrast agent compared to the unthickened but no difference for the starch. The viscosity at 50 s^{-1} was not matched exactly between the thickeners. However, the higher viscosity for the starch fluid might be expected to favour the starch product, but this was not the case.

Nakauma et al. [35] compared liquids thickened with xanthan gum and locust bean gum using sensory evaluation and acoustic measurements during swallowing and looked for relationships to the viscous and viscoelastic rheology of the liquids. From acoustic analysis, the time required for the bolus thickened with locust bean gum to transfer through the pharyngeal phase was consistently longer than that for xanthan gum when compared at equivalent viscosity at 10 s^{-1}. They concluded that the results indicate that xanthan gum solutions flow as one coherent bolus through the pharyngeal phase with smaller variation of flow velocity than locust bean gum solutions, leading to a greater sensation of swallowing ease. They suggested that 'structured fluid', defined as fluid with yield stress such as xanthan gum solutions, is a rheological nature that allows the bolus to be swallowed in one go and that fluids possessing this nature are perceived as being easier to swallow.

Taken together, the studies so far seem to indicate that rheology does matter, or, at the very least, its relationship to safe swallowing warrants further investigation.

As discussed in Sect. 9.4.3, rheology certainly matters when it comes to developing test methods to characterise the consistency of thickened beverages.

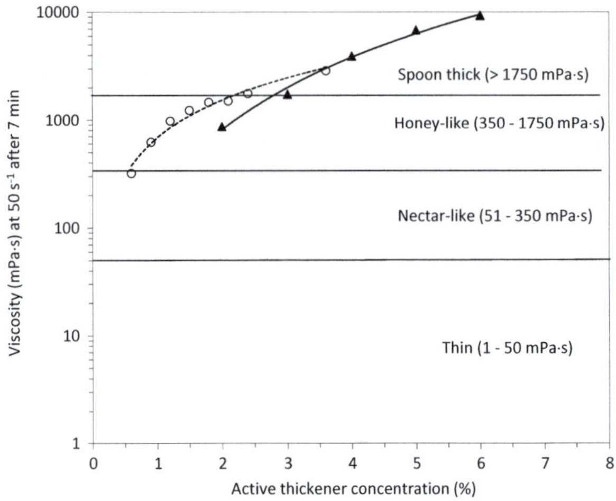

Fig. 9.8 Concentration dependence of viscosity at 50 s^{-1} xanthan (○) and modified starch (Δ)-based thickener for dysphagia at 20 °C measured 7 min after dispersion in tap water (Reproduced from Sworn et al. [53] with permission from the Royal Society of Chemistry)

The choice of test method can influence how the relative consistency of different products will be seen. The relationship between active thickener concentration and viscosity at 50 s^{-1} for xanthan gum and starch is shown in Fig. 9.8 in relation to the consistency criteria proposed by the National Dysphagia Diet project. The xanthan-based products are able to cover the nectar-like and honey-like consistencies, but due to the highly pseudoplastic nature of the viscosity, very high concentrations are needed to reach the spoon-thick consistency viscosity targets measured at 50 s^{-1}. Based on this measurement criterion, the starch products must be used at higher concentration than the xanthan to reach the honey-like consistency range but are more able to create the spoon-thick consistency.

In general, hospitals and nursing homes do not have sophisticated equipment for measuring the consistency of products, and so simple tests have been proposed. An example of this is the line spread test (LST) proposed by Paik et al. [39]. The test consists of placing the thickened solution in a metallic cylinder of 2 cm diameter and 2 cm height which is positioned on a graduated sheet (Fig. 9.9). The cylinder is then removed and the diameter (or the radius) of the spread is measured after a set time.

The LST has a number of practical advantages over the viscosity since it is very simple to perform, does not require viscometers or rheometers and can be made with a minimum of training.

In Japan proposals for classifying thickness of products for dysphagia have been made based on both viscosity at 50 s^{-1} and the line spread test (Table 9.5). The line spread test used cylinders of 2.8 cm height and 3.0 cm diameter. The Japanese proposal makes the arbitrary assumption that the viscosity at 50 s^{-1} is correlated with the LST, i.e. the higher the viscosity, the lower the radius in the LST. A study of

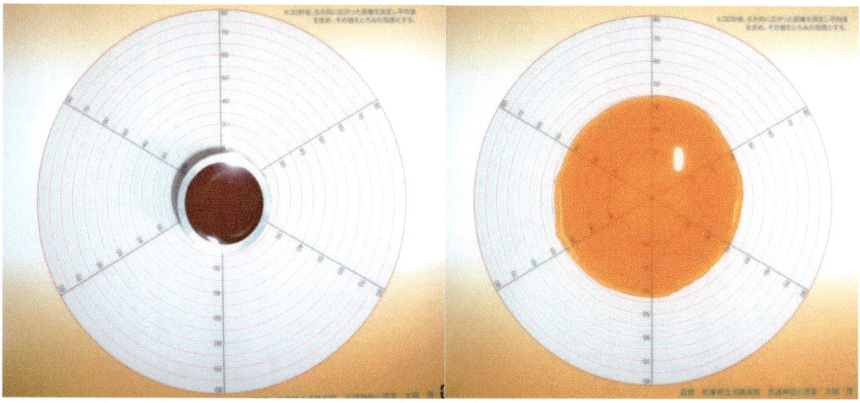

Fig. 9.9 Illustration of the LST test. Before (*left*) and after (*right*) removal of the cylinder (Reproduced from de Saint-Aubert et al. [44] with permission from the Royal Society of Chemistry)

Table 9.5 Proposed classifications of thickness in Japan and recommended dosages

Dosage of thickener system	Description	Viscosity at 50 s^{-1}(mPa·s)	LST value (radius in mm)
0.5–1 %	Mildly thick	50–100	43–40
1–2 %	Moderately thick	100–400	39–33
2–3 %	Extremely thick	400–600	32–30

five commercial thickeners for dysphagia containing different hydrocolloids using the two tests has shown that this assumption is not correct [44].

A plot of viscosity at 50 s^{-1} as a function of the LST diameter (Fig. 9.10) does not provide a universal correlation. In fact the correlation splits into two groups, one for xanthan-based products and one for the starch and starch/guar products. The LST value tends to be lower for the xanthan-based products compared with the starch- and guar-based products at equivalent viscosity at 50 s^{-1}. In other words, for the same viscosity, the starch type spreads more than the xanthan type. This is probably linked to the more gel-like viscoelastic properties of the xanthan gum solutions (Fig. 9.7). When solutions of commercial thickeners were prepared at the recommended viscosities given in Table 9.5 and the corresponding LST values were measured, only the pure xanthan-based products matched the recommendations for both viscosity and LST (Table 9.6). This suggests that different rheological phenomena are involved in the two tests and that the universal classification and concentration recommendations shown in Table 9.5 may not be relevant for all the thickener types. A classification per thickener type might be more appropriate.

This illustrates the complexity of finding a simple universal single point measurement that is applicable to different thickeners and relevant for the management of dysphagia. Even with a simple test such as the LST, factors such as the size of the cylinder and time allowed before measurement will influence the results. Whilst

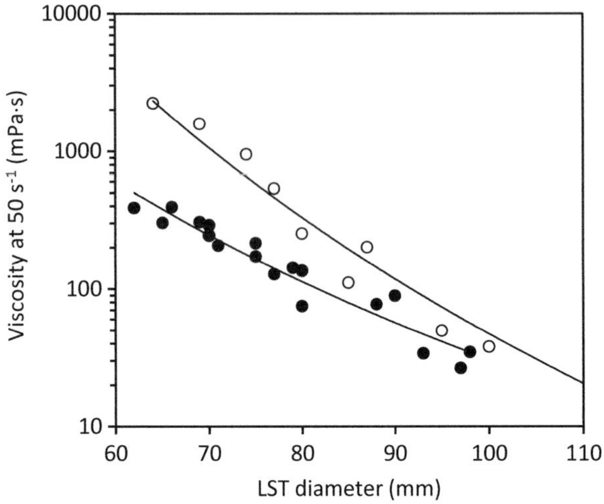

Fig. 9.10 Viscosity at 50 s^{-1} as a function of the LST value (biggest diameter of the spread) measured after full hydration in deionised water for xanthan (•) and modified starch (o)-based thickener (Reproduced from de Saint-Aubert et al. [44] with permission from the Royal Society of Chemistry)

Table 9.6 LST values of commercial thickeners prepared at the recommended viscosity at 50 s^{-1}

Recommendations from Table 9.5		Measured LST radius (mm) at corresponding viscosity at 50 s^{-1}			
Viscosity (mPa·s) @ 50 s^{-1}	LST radius (mm)	Xanthan product-1	Xanthan product-2	Starch and guar product	Starch product
50	43–40	43	43	52	51
100	43–40	40	42	44	43
200	39–33	37	37	43	42
400	39–33	33	33	39	41
600	32–30	32	31	38	40

such tests have potential for QC control and measurement of consistency to aid in the preparation of fluids for dysphagic patients, the fact that it is very difficult to define the exact rheological conditions means that, for R&D purposes, viscometers and rheometers that measure consistency under defined conditions of shear rate, shear stress, frequency or strain will continue to be required.

In 2013 the International Dysphagia Diet Standardisation Initiative (IDDSI) was created. Its stated aims are to 'develop global standardised terminology and definitions for texture modified foods and thickened liquids for individuals with dysphagia of all ages, in all care settings and all cultures'. A review of existing published national terminology and definitions conducted by the IDDSI has been published which also includes the rational aims and work plan for this initiative [5].

Table 9.7 IDDSI classification levels for drinks

Level	Name	IDDSI flow test
4	Extremely thick	If no liquid flows at all, the category is level 4 or above
3	Moderately thick	There is more than 8 ml remaining, but some liquid still flows through
2	Mildly thick	There is between 4 and 8 ml remaining
1	Slightly thick	There is between 1 and 4 ml remaining
0	Thin	All liquid has flowed through the syringe

The IDDSI [16] has recently introduced a framework for an eight-level classification of the consistency of foods and drinks together with recommended testing methods. It is intended that this framework should be adopted worldwide. The thickness of drinks is classified into four levels and is measured by how much liquid remains in a 10 mL slip tip syringe after being allowed to flow for 10 s. The classifications are outlined in Table 9.7.

The outcomes of this initiative will almost certainly have implications for the manufacture of commercial products for dysphagia.

9.6 Future Trends

'Does rheology matter?' and 'how thick is thick?' remain questions to be answered.

Most work involving rheological measurements to date has focused on simple flow and small amplitude viscoelastic properties. More complex rheological behaviour such as extensional viscosity, turbulent flow or response to large amplitude oscillatory shear (LAOS) may prove to be of value in understanding the complex process of swallowing. Extensional viscosity is a rheological technique in which the solution is subjected to extensional shear, that is to say it is stretched. Processes such as atomisation, turbulent flow, flow reduction and flow through porous media involve extensional flow. For Newtonian solutions the extensional viscosity is approximately three times the shear viscosity. Solutions of hydrocolloids have extensional viscosities much greater than three times the shear viscosity as a result of molecular deformation in the flow field. The ratio between shear viscosity and extensional viscosity is termed the Trouton ratio. Swallowing is a process that may also involve extensional flow. For example, Wendin et al. [55] found that in sensory tests carried out by persons with a normal swallow, the sensory attribute of ease of swallow was positively related to extensional viscosity. They suggested that this might be explained by the extensional flow during transport of the bolus, particularly when squeezed between the tongue and the palate. Standing et al. [50] have further explored the relevance of extensional viscosity by comparing the sensory properties of a Newtonian fluid, a Newtonian fluid with elasticity (Boger fluid) and a non-Newtonian (shear-thinning) fluid with elasticity. The elastic fluids had a Trouton ratio in the range of 15–20. The sensory panel, made up of 12 dysphagic patients, gave the non-Newtonian fluid with elasticity the highest score for perceived ease of swallowing (easiest to swallow) and the Newtonian fluid the lowest score. However the differences were not statistically significant. This branch of rheology and its relevance to dysphagia have received little attention to date, but these studies suggest that it merits further investigation.

The study of lubrication (tribology), common in the motor industry, is now gaining interest in the food industry in relation to sensory evaluation. This type of study is likely to yield new insights into the behaviour of foods during mastication and swallowing in the coming years that will benefit the development of products for dysphagia.

The complexity of making clinical trials with dysphagic patients and the difficulty in interpretation of the results have led some researchers to explore the possibility of using computer modelling to mimic the swallowing process [19, 20]. If such models can be shown to reflect reality, then there is no doubt that they will provide a very useful tool for the development of our understanding of the relationship between rheology and a safe swallow.

The patient should remain the focus point, and continued research into understanding the impact of bolus rheology on the swallowing experience could lead to the development of rheology-modified products with specific characteristics that maximise the safety of the swallow whilst maintaining a pleasant sensory experience for the patient.

Acknowledgements I thank my colleague Professor Niall Young for helpful discussions and critically reviewing the manuscript.

References

1. Brown, A., Mills, R.H., Daubert, C.R., and Casper, M.L. (1998). Establishing labels and standards for thickened liquids in the dysphagia diet. *The Consultant Dietician*, **23**, 1.
2. Coster, S.T., and Schwarz, W.H. (1987). Rheology and the swallow-safe bolus. *Dysphagia*, **1**, 93.
3. Cox, W.P. and Merz, E.H. (1958). Correlation of dynamic and steady flow viscosities. *J. Polym. Sci.*, **28**, 619.
4. Cichero, J.A.Y., Jackson, O., Halley, P.J., and Murdoch, B.E. (2000). How Thick Is Thick? Multicenter study of the rheological and material property characteristics of mealtime fluids and videofluoroscopy fluids. *Dysphagia*, **15**, 188–200.
5. Cichero, J.A.Y., Steele, C., Duivestein, J., Clavé, P., Chen, J., Kayashita, J., Dantas, R., Lecko, C., Speyer, R., Lam, P., & Murray, J. (2013). The need for international terminology and definitions for texture-modified foods and thickened liquids used in dysphagia management: Foundations of a global initiative. *Curr. Phys. Med. Rehabil. Rep.*,**1**, 280–291.
6. Cross, M.M. (1965). Rheology of non-Newtonian fluids: a new flow equation for pseudoplastic systems. *J. Colloid Sci.*, **20**, 417–437.
7. Garin, N., De Pourcq, J.N., Martın-Venegas, R., Cardona, D., Gich, I., & Mangues, M.A. (2014). Viscosity differences between thickened beverages suitable for elderly patients with dysphagia. *Dysphagia*, **29**, 483–488.

8. Germain, I., Dufresne, T., and Ramaswamy, H.S. (2006). Rheological characterisation of thickened beverages used in the treatment of dysphagia. *J. Food Eng.*, **73**, 64.
9. Glicksman, M. (1983). Food Hydrocolloids,Vols I, II and III. CRC Press Inc., FL.
10. Goulding, R., and Bakheit, A.M.O. (2000). Evaluation of the benefits of monitoring fluid thickness in the dietary management of dysphagic stroke patients. *Clinical rehabilitation*, **14**, 99.
11. Hanson, B., O'Leary, M.T., & Smith, C.H. (2012). The effect of saliva on the viscosity of thickened drinks. *Dysphagia*, **27** 10–19.
11. Hasegawa, A., Otoguro, A., Kumagai, H., & Nakazawa, F. (2005). *Nihon Shokuhin Kagaku Kogaku Kaishi* (in Japanese), **52**, 441–447.
13. Hasegawa, A., Nakazawa, F., & Kumagai, H. (2008). *Nihon Shokuhin Kagaku Kogaku Kaishi* (in Japanese), **55**, 330–337.
14. Hawdon, J.M., Beauregard, N., Slattery, J., & Kennedy, G. (2000). Identification of neonates at risk of developing feeding problems in infancy. *Dev. Med. Child Neurol.*, **42**, 235–9.
15. Hooke, R. (1678). *De PotentiaBestitutiva.*
16. IDDSI (2015) Detailed descriptors, testing methods and evidence. Drinks: Levels 0–4. www.iddsi.org
17. Imeson, A. (1999). Thickening and Gelling Agents for Food, 2nd edition. Aspen Publishers Inc.,MD.
18. Jansson,P-E., Kenne, L., & Lindberg, B. (1975). Structure of the extracellular polysaccharide from *Xanthomonas campestris. Carbohydrate Research*, **45**, 275–282.
19. Kamiya, T., Toyama, Y., Michiwaki, Y., & Kikuchi, T. (2013a) Development of a numerical simulator of human swallowing using a particle method (Part 1. Preliminary evaluation of the possibility of numerical simulation using MPS method). *35thAnnual International Conference of the IEEE EMBS* 4454–57.
20. Kamiya, T., Toyama, Y., Michiwaki, Y., & Kikuchi, T. (2013b) Development of a numerical simulator of human swallowing using a particle method (Part 2. Evaluation of the accuracy of a swallowing simulation using the 3D MPS method). *35thAnnual International Conference of the IEEE EMBS* 2992–95.
21. Kool, M.A., Harry Gruppen, H., Sworn, G., & Schols, H.A. (2013). Comparison of xanthans by the relative abundance of its six constituent repeating units. *Carbohydrate Polymers*, **98**, 914–921
22. Leonard, R.J., White, C., McKenzie, S., & Belafsky, P.C. (2014). Effects of bolus rheology on aspiration in patients with dysphagia. *J. Acad. Nutr. Diet*, **94**, 590–4.
23. Longmann, J.A. (1983). Evaluation and treatment of swallowing disorders. San Diego: College-Hill Pres Inc.
24. Mackley, M.R., Tock, C., Anthony, R., Butler, S.A., Chapman, G., & Vadillo, D. C. (2013). The rheology and processing behavior of starch and gum-based dysphagia thickeners. *J. Rheol.*, **57**, 1533.
25. Mann, L.L., and Wong, K. (1996). Development of an objective method for assessing viscosity of formulated foods and beverages for the dysphagic diet. *J. Am. Diet. Assoc.*, **96**, 585.
26. Matta, Z., Chambers IV, E., Garcia, J.M., & Helverson, J.M. (2006). Sensory characteristics of beverages prepared with commercial thickeners used for dysphagia diets. *J. Am. Diet Assoc.*, **106**, 1049–1054.
27. Melton, L.D., Mindt, L., Rees, D.A., & Sanderson, G.R. (1976). Covalent structure of the polysaccharide from *Xanthomonas campestris*: Evidence from partial hydrolysis studies. *Carbohydrate Research*, **46**, 245–257.
28. Mills, H. (1999). Rheology overview: control of liquid viscosities in dysphagia management. *Nutrition in Clinical Practice*, **14**, 52.
29. Mitchell, J.R. (1979). Rheology of polysaccharide solutions and gels. In,. *Polysaccharides in foods*. J.M.V. Blanshard, and J.R. Mitchell, (Eds), Butterworth & Co Ltd., London, pp 51–72.
30. Mitchell, J.R., Ferry, A.L., Desse, M., Hill, S.E., Hort, J., Marciani, L., & Wolf, B. (2008). Mixing hydrocolloids and water: Polymers versus particulates. In, *Gums and Stabilisers for the Food Industry 14*, P.A. Williams and G.O. Phillips (Eds.), RSC, Cambridge, pp 29–39.

31. Moret-Tatay, A., Rodríguez-García, J., Martí-Bonmatí, E., Hernando, I., & Hernandez, M.J. (2015). Commercial thickeners used by patients with dysphagia: Rheological and structural behaviour in different food matrices. *Food Hydrocolloids*, **51**, 318–326
32. Morris, E.R. (1984). Rheology of hydrocolloids. In, *Gums and Stabilisers for the Food Industry 2*. G.O. Phillips, D.J. Wedlock and P.A. Williams (Eds.), Pergamon Press, Oxford, pp 57–78.
33. Morris, E.R. (1989). Polysaccharide solution properties: Origin, rheological characterisation and implications for food systems. In, *Frontiers in Carbohydrate Research-1*. R.P. Millanc, J.N. BeMiller and R. Chandrasekaran, (Eds.); Elsevier Applied Science; London, pp 132–163.
34. Morris, E.R., Cutler, A.N., Ross-Murphy, S.B., Rees, D.A., and Price, J. (1981). The concentration and shear rate dependence of viscosity in random coil polysaccharide solutions. *Carbohydr. Polym.*, **1**, 5–21.
35. Nakauma, M., Ishihara, S., Funami, T., & Nishinari, K. (209). Swallowing profiles of food polysaccharide solutions with different flow behaviors. *Food Hydrocolloids*, **25**, 965–973.
36. Newman, L.A., Keckley, C., Petersen, M.C., & Hamner, A. (2001). Swallowing function and medical diagnoses in infants suspected of Dysphagia. *Pediatrics*, **108**, 6.
37. Newton, I. (1687). *Philosophiae Naturalis Principia Mathmatica.*
38. Nishinari,K., Takemasa, M., Sua, L., Michiwaki, Y., Mizunuma, H., & Ogoshi, H.(209). Effect of shear thinning on aspiration - Toward making solutions for judging the risk of aspiration. *Food Hydrocolloids,***25** 1737–1743.
39. Paik, N-J., Han, T.R., Park, J.W., Lee, E.K., Park M.S., & Hwang I-K. (2004). Categorisation of dysphagic diets with the line spread test. *Arch. Phys. Med. Rehabil.*, **85**, 857.
40. Payne, P., Methven, L., Fairfield, C., & Bell, A. (209). Consistently Inconsistent: Commercially Available Starch-Based Dysphagia Products. *Dysphagia*, **26**, 27–33.
41. Penney, B. (2014). Use of fluid thickener to reduce dysphagia risk. *Nursing Times*; **90**, 12, 16–18.
42. Phillips, G.O., & Williams, P.A. (2009). Handbook of Hydrocolloids, 2nd edition. Woodhead Publishing Ltd., Cambridge.
43. Ross-Murphy, S.B., Rheological Methods. In Chan, H.W.-S (ed) Biophysical Methods in Food Research, Critical Reports on Applied Chemistry Vol 5; SCI Blackwell; Oxford, 1984; pp 138–199.
44. de Saint-Aubert, C., Sworn, G., & and Jun Kayashita, J. (2012). Comparison of 2 tests used for the classification of food thickeners in the management of dysphagia. In, *Gums and Stabilisers for the Food Industry* 17, P.A. Williams and G.O. Phillips (Eds.), RSC, Cambridge, pp 359–368.
45. Sasaki, C.T., & Leder, S.B. (2015). Comments on selected recent dysphagia literature. *Dysphagia*, **30**, 482–487.
46. Seo, C-W., & Yoo, B. (2013). Steady and dynamic shear rheological properties of gum-based food thickeners used for diet modification of patients with dysphagia: Effect of concentration. *Dysphagia*, **28**, 205–29.
47. Shama, F., and Sherman, P. (1973). Identification of stimuli controlling the sensory evaluation of viscosity. II. Oral methods. *J Text. Stud.*, **4**, 91.
48. Sliwinski, E.L., La Faille, S., & Oudhuis, L.A.C.M. (2009). Effect of human saliva on the consistency of thickened foods for patients with dysphagia. *Clinical Nutrition Supplements*, **4**, 135.
49. Smith, T.L., Sun, M.M., and Pippin J. (2004). Characterising process control of fluid viscosities in nursing homes. *J. Am. Diet. Assoc.*, **104**, 969.
50. Standing, M., Qazi, A., Nystrom, M., Berta, M., Burlow, M., & Ekberg, O. (2015) Effect of shear vs extensional flow during swallowing. *Annual Transactions of the Nordic Rheology Society*, **23**, 63–65.
51. Steele, C.M., Alsanei, W.A., Ayanikalath, S., Barbon, C.E.A., Chen, J., Cichero, J.A.Y., Coutts, K., Dantas, R.O., Duivestein, J., Giosa, L., Hanson, B., Lam, P., Lecko, C., Leigh, C., Nagy, A., Namasivayam, A.M., Nascimento, W.V., Odendaal, I., Smith, C.H., & Wang, H. (2015). The influence of food texture and liquid consistency modification on swallowing physiology and function: A Systematic Review. *Dysphagia,***30**, 2–26.

52. Sworn, G. (2007). Natural thickeners. In, Handbook of Industrial Water Soluble Polymers, Edited by P. A. Williams, Blackwell Publishing Limited, Oxford, pp 10–31.
53. Sworn, G., Kerdavid, E. and Fayos, J. (2008). The role of hydrocolloids in the management of dysphagia. In, *Gums and Stabilisers for the Food Industry 14*, P.A. Williams and G.O. Phillips (Eds.), RSC, Cambridge, pp 392–401.
54. Tashiro, A., Hasegawa, A., Kohyama, K., Kumagai, H., & Kumagai, H. (2010). Relationship between the rheological properties of thickened solutions and their velocity through the pharynx as measured by ultrasonic pulse Doppler method. *Biosci. Biotechnlo. Biochem.*, **74**, 8, 1598–1605.
55. Wendin, K., Ekman, S., Bulow, M., Ekberg, O., Johansson, D., Rothenberg, E., & Stading, M. (2010). Objective and quantitative definitions of modified food textures based on sensory and rheological methodology. *Food & Nutrition Research*, **54**, 5134.
56. Wood,F.W. (1968). SCI Monograph No. 27: Rheology and texture of foodstuffs. 40.
57. Yamagata, Y., Izumi, A., Egashira, F., Miyamoto, K., & Kayashita, J. (2012). Determination of a suitable shear rate for thickened liquids easy for the elderly to swallow. *Food Sci. technol. Res.*, **18**, 3, 363–369.

Chapter 10
The Effect of Preparation Conditions on the Rheological Properties of *Gomatofu* (*Sesame Tofu*)

Emiko Sato

Abstract *Gomatofu*, a mixed gel consisting of *kudzu* starch and sesame seed, possesses unique textural characteristics. That is the Japanese traditional, healthy, and vegetarian food. Sato et al. (J Jpn Soc Food Sci Technol, Jpn 42:737–747, 1995) and Sato (HYDROCOLLOIDS, Edited by K. Nishinari, Elsevier Science B. V., Printed in the Netherlands, 269–274, 2000) investigated the effects of preparation conditions, mixing ratios, and cooking times on the physical properties of *gomatofu*. The previous papers indicated that *gomatofu* samples prepared by mixing sesame milk and *kudzu* starch at 250 rpm for 25 min had a uniform network structure (Sato et al., J Jpn Soc Food Sci Technol, Jpn 42:737–747, 1995). It was also found that *gomatofu* prepared with ingredients having a ratio of 40–50 g of *kudzu* starch to 40–60 g of sesame with 450 g of added water were palatable in softness, mouthfeel, and springiness (Sato et al., J Jpn Soc Food Sci Technol, Jpn 46:285–292, 1999). It can be seen that sesame components contribute to the strength and stability of *kudzu* gel starch (Sato et al., J Jpn Soc Food Sci Technol, Jpn 46:285–292, 1999), but a larger proportion of sesame contents decreased the cohesiveness due to the high level (55.9 %) of lipids in sesame seeds (Sato et al., J Jpn Soc Food Sci Technol, Jpn 42:871–877, 1995, J Jpn Soc Food Sci Technol, Jpn 46:367–375, 1999). Further, it was also found that favorable oil content was 3.4–6.4 % in the *gomatofu* (Sato et al., J Soc Rheol 33:101–108, 2003). The effects of roasting conditions of sesame seeds on the mechanical properties of *gomatofu* were investigated (Sato et al., J Home Econ, Jpn 58: 471–483, 2007). Overheating creates a burning smell, loss of flavor, and undesirable changes in the sesame components. The fracture stress increased at higher roasting temperatures because these fine particles of sesame have a filler effect, and they increased the elastic modulus of the gel. Sensory evaluation showed that *gomatofu* prepared with seeds roasted at 170 °C had best palatability. It was found that good mouthfeel is characterized by smoothness and is an important parameter in the texture of *gomatofu*.

E. Sato (✉)
Department of Health and Nutrition, Human Life Science, Prefecture of Niigata University, Niigata, Japan
e-mail: emi-sato@unii.ac.jp

I. Kaneda (ed.), *Rheology of Biological Soft Matter*, Soft and Biological Matter,
DOI 10.1007/978-4-431-56080-7_10

Keywords Preparation condition • Texture • Rheological properties • *Gomatofu*

10.1 Introduction

Gomatofu (sesame tofu), a mixed gel consisting of *kudzu* starch and sesame seed, possesses an extremely unique textural characteristics which is soft, smooth, and springy. The textural properties are greatly influenced by preparation, ingredients, and roasting conditions of sesame seed. The *gomatofu* was originally used in a kind of vegetarian dish in Chinese style (*Fucha*-cuisine) called "*mafu*" [10] mainly consisting of starch of and oil. In Japan, *gomatofu* (sesame tofu) is one of the traditional Japanese healthy foods consisting of *kudzu* starch and sesame seeds and is representative of all *shojin* (vegetarian) dishes. *Gomatofu* and sesame seeds are believed to have pharmacological actions [10, 11] in gastroenteritis disease and detoxification, reducing cholesterol and high blood pressure [11], and recovery from chronic peptic ulcer.

Bourne [12] stated that texture in the overall acceptability of food varies widely, depending upon the type of food, and may be affected by culture. Nishinari [13, 14] also stated that texture is classified into the physical factor of the palatability of food. Further, rheology is important in the understanding of textural properties of foods and food processing and mastication and deglutition in the eating process. Katsuta [15] mentioned that texture plays an important role in the food attribute, citing the study of Szczesniak and Kleyn [16] on the texture element analysis. On the other hand, Szczesniak [17] has classified "oily" in "other characteristics" in her texture profile. Matsumoto [18] has stated that oiliness seems to be affected by the texture and distribution of oils and fats. *Gomatofu* also possesses unique textural characteristics which are soft, smooth, and springy.

In this report is the effect of preparation conditions (mixing rate and cooking time [1, 5], ingredient ratio (*Kudzu* starch and sesame contents [3, 4]) and sensory evaluations [2, 4], different kinds of sesame materials [8, 9], the effect of sesame oil contents [6], and roasting conditions of sesame seed [7]) on the rheological and physical properties and microstructure of *gomatofu*.

10.2 Measurement Method

10.2.1 Materials

Highly pure (99.0 %, *Koki*, particle size: about 10–11 μm) *kudzu* (*Pueraria lobata*.) starch was purchased from *Inoue Tengyokudo* (*Nara*, *Yoshino*, 2003) as was shown in Fig. 10.1. The three kinds of sesame (*Sesamum indicum* DC) materials were produced in China (2004). The white sesame seeds from which hulls had been removed physically which shed their hulls are called "*Muki goma*" {length (L),

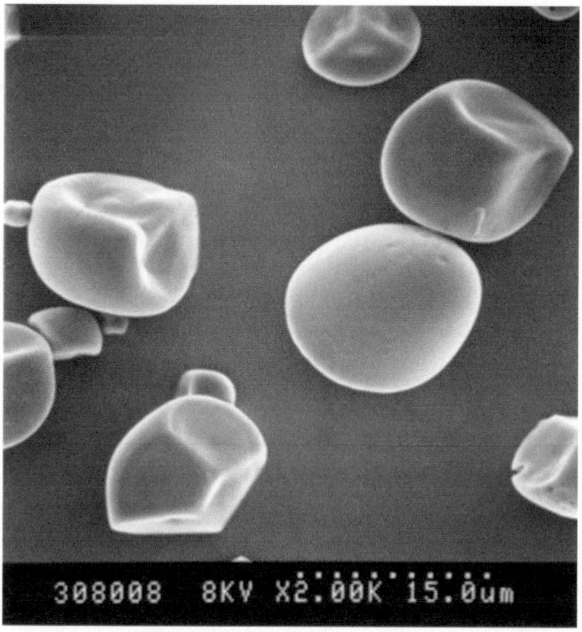

Fig. 10.1 Kudzu starch photograph

width (W), thickness (T): 3.1, 2.1, 0.8 mm}. The white and black sesame seeds with hulls which are dried after washing, called "*Arai-shiro-goma*" {L, W, T: 3.2, 1.9, 0.8 mm} and "*Arai-kuro-goma*" {L, W, T: 2.9, 1.8, 1.4 mm}, were purchased from Kadoya Oil Co. (Tonosho, Kagawa).

10.2.2 Chemical Analysis of Sesame Materials

The contents of lipids, proteins, and carbohydrates of sesame materials were analyzed quantitatively by the Soxhlet extraction method [19, 20], the Kjeldahl method of nitrogen determination [20, 21], and the phenol sulfuric acid method [22], respectively. In the determination of carbohydrate content, the sample was completely hydrolyzed by heating in 1.0 M sulfuric acid at 100 °C for 5 h. The hydrolyzed matter was neutralized by 1 M sodium hydroxide, and the sample was obtained by filtering. One milliliter of 5 % phenol solution was added to 2 ml of the sample solution. Five milliliter of 96 % sulfuric acid was added to each test tube so that the stream hit the liquid surface directly, and the sample was measured at 480–490 nm wavelengths quantitatively by colorimeter.

10.2.3 Preparation of Gomatofu

2.3.1.exp. (1) Preparing procedure of sample was shown in Fig. 10.2. Preparation procedures for *gomatofu* and the experimental apparatus are shown in Fig. 10.3.

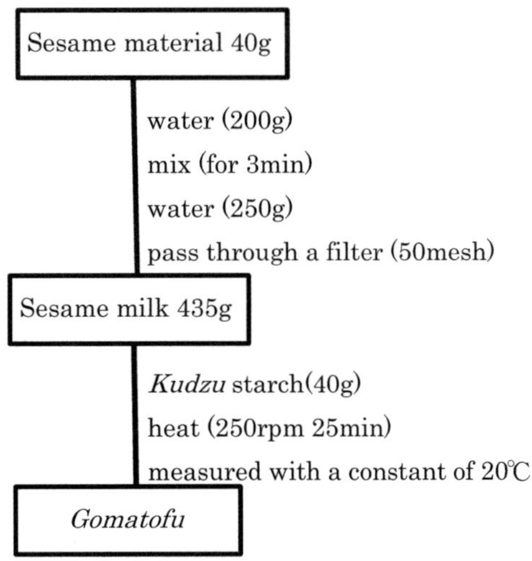

Fig. 10.2 Preparing procedure of gomatofu (sesame tofu)

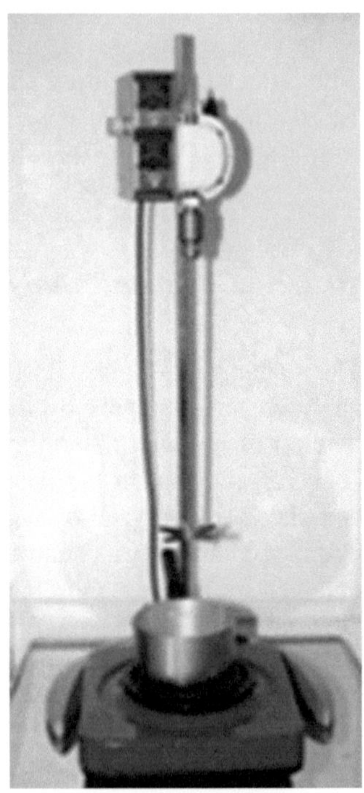

Fig. 10.3 System of mixing with cooking

Table 10.1 The composition of *Muki* white sesame seed and sesame milk. The 40 g of sesame and 450 g of water were mixed for 3 min and filtered using a sieve (50 mesh) to obtain about 435 g of sesame milk

	Moisture	Lipid	Protein	Carbohydrate
Sesame seeds (%)	4.9 ± 0.03	55.9 ± 0.30	19.8 ± 0.20	16.7 ± 0.30
Sesame seeds (g/40 g)	2.0 ± 0.10	22.4 ± 0.28	7.9 ± 0.35	6.7 ± 0.43
Sesame milk (%)	92.9 ± 0.02	4.3 ± 0.24	1.5 ± 0.08	1.0 ± 0.05
Sesame milk (g/435 g)	404.1 ± 0.12	16.5 ± 0.53	5.7 ± 0.32	4.4 ± 0.15

Briefly, 40 g of roasted sesame and 450 g of water were mixed for 3 min and filtered using a sieve (50 mesh = 297 μm) to obtain about 435 g of sesame milk and sesame residue. A suspension of sesame milk (435 g) and *kudzu* starch (40 g) was prepared by usual system A (simmering) and B (hot water dropping into cookpot to hold a constant quantity during cooking) for 25 min on an electric heater 450W (National NK-082) while mixing at a rate of 60, 150, 250, and 350 rpm using a mixing instrument (DC-3RT, Tokyo Rika Co., Japan) and the cooking times 15, 25, 35, and 45 min, respectively. Chemical composition of sesame seed and sesame milk was shown in Table 10.1. About 65–72 %of the composition of sesame materials transfers to the sesame milk. The hot suspension samples were immediately poured into a Teflon ring case (20 mm height × 20 mm inner diameter) and were kept in a 10 °C incubator (ASONE, PCI-301) for 24 h. Tests on the samples were performed at 20 °C. After samples were cooled at room temperature for 1 h, they were used for measurement by using a rheometer.

2.3.2.exp. (2) The suspension of sesame materials (sesame milk 435 g and *kudzu* starch 40 g) was prepared using a simmering method at the mixing rate of 250 rpm for 25 min in Teflon ring case (20 × 20 mm) [10.2 and 10.3]. The changes in the texture and viscoelasticities and microstructures of *gomatofus* (*kudzu* starch 40 g, water 450 g, sesame seed 0–80 g) prepared by simmering at a mixing rate 250 rpm for 25 min were investigated.

2.3.3exp. (3) The effect of different kinds of sesame materials (raw huskless white sesame seed, roasted white sesame seed, roasted black sesame seed, and defatted sesame flour) on the texture properties of *gomatofu* was investigated. The samples were named as follows: Ra-W (raw white tofu), Ro-B (roasted black *gomatofu*), and D.F (defatted sesame flour *gomatofu*), respectively. Three kinds of samples except D.F are called *Muki* (hull-less) *gomatofu*, *Arai* white (hull) *gomatofu*, and *Arai* black (hull) *gomatofu* hereafter. In the case of *kudzu*, the effect of *kudzu* starch amount (30–60 g) on the texture of *gomatofu* in the presence of 40 g sesame was investigated and in the case of sesame, the effect of sesame amount (0–80 g) on the texture of *gomatofu* in the presence of 40 g *kudzu*.

2.3.4. exp. (4) The effect of oil contents contained in *gomatofu* which was prepared with *Muki* white sesame seed on the texture was investigated. The five kinds of *gomatofu* were examined as followed; (a) defatted *gomatofu* (oil content 0 %), (b) defatted *gomatofu* (oil content 0.42 %), (c) control *gomatofu* (oil content

3.4 %), (d) 3 % sesame oil added (oil content 6.4 %), and (e) 6 % sesame oil added (oil content 9.4 %).

2.3.5. exp. (5) The effect of roasting conditions of sesame seeds and their effect on the mechanical properties of *gomatofu* were investigated [7]. *Gomatofu* was prepared with *Muki* (hull-less) and *Arai* (white and black, with hull) seeds (unroasted and roasted at 160, 170, 180, 190, and 200 °C) in an electric oven for 15 min.

10.2.4 Measurements of Gomatofu Samples

10.2.4.1 Rheological Measurement: Texture and Rapture and Viscoelasticity of *Gomatofu*

A creep meter (Rheoner RE-3305, Yamaden Co., Ltd. Tokyo, Japan), set by a thermostat chamber connected to a water bath to maintain the measuring temperature (20 °C), was used for the measurement of fracture properties of *gomatofu* under uniaxial compression. The samples were compressed to 80 % strain by a cylinder plunger (40 mm diameter, nonstick Teflon) at a compression speed of 1.0 mm/s using a 2 kg load cell. The cross-sectional area of the cylinder plunger used was larger than the sample area (20 mm diameter). Texture measurements were carried out at the same conditions as fracture measurements, but the samples were compressed to 60 % measurement strain. A method of the compression test by large deformation was the same as that described by Nishinari [14]. Care was taken so that the contact state between the plunger and cylindrical sample was fixed. Both the top and bottom surfaces of cylindrical gel samples were lubricated with silicon oil to make the frictional force as small as possible. Sherman [23] showed that lubrication, or lack of lubrication, of the contact surface between the food and compressing platen affects the force required to reach a given degree of compression. Thus, the samples were lubricated by silicon oil to ensure uniform deformation. The values of the fracture and texture measurements were obtained from the means and standard deviations of a series of ten tests, and significant difference between samples was examined by *t* test. The reason for carrying out a compression test rather than a penetration test in this study was to more closely approximate the texture of the *gomatofu*, i.e., the sensation of crushing the *gomatofu* on the tongue.

10.2.4.2 Scanning Electron Microscopy

Scanning electron microscopy (SME) (Chilled Natural SEM-3500, Hitachi, Co) observations were made at Niigata Water Works Bureau. As for *gomatofu* (gel), about 5 mm squares of the samples without any chemical treatment were observed at temperatures ranging from 0 to −10 °C. Air cell size (Heywood diameter) was analyzed by an image analyzer (LUZEX 500, Nireco, Co., Japan). The Heywood diameter equation used for the calculation was as follows:

$$\text{H.D. (Heywood diameter)} = \sqrt{(4/\pi)} \times \text{area}$$

The sample was observed on a stand in its natural state and was maintained at a low vacuum using a cool stage. It was possible to observe for long time by chilled SEM, even if the sample was susceptible to heat. The finer particles dispersed in the *gomatofu* were calculated according to the scale of SEM photographs.

10.2.4.3 Sensory Evaluation

Sensory evaluations of the texture of *gomatofu* made from two kinds of seeds (*Muki* and white *Arai*) and prepared at five kinds of roasting temperatures (unroasted, 160 °C, 170 °C, 180 °C, 190 °C) in the roasting experiment of sesame seeds were performed by ranking methods, sample numbers 5, panel [18] number 24 for each test. The panel was composed of students and teachers of Niigata Women's College. For each kind of *gomatofu* (*Muki*, *Arai*), hardness, springiness, and mouthfeel by discrimination test and total acceptance by palatable test were compared according to five samples; an order rating of 1, 2, 3, 4, or 5 was given to each sample, with 1 being the most or best, 2 being the second most/best, etc. The total sums of order were analyzed by means of Kramer's test [24] and further examined according to Newell and MacFarlane's test [25] to detect significant difference between samples. Kendall's agreement coefficient of concordance, sample numbers 5, panel numbers 24, $0 \leqq W \leqq 1$ was also used to judge the degree of agreement of the panel.

10.3 Result and Discussion

10.3.1 Effect on the Preparing Conditions: Mixing Rate and Cooking Time [1, 4]

The effects of mixing rates (60, 150, 250, 350 rpm) and the cooking times (15, 25, 35, 45 min) on the physical properties of *gomatofu* (sesame tofu) were studied by using a rheometer and a scanning electron microscope (SEM). *Gomatofu* was prepared using two different kinds of cooking methods: the usual A system (simmering) and B system (hot water dropping into cookpot to hold a constant quantity during cooking) were employed. Hardness of the (A) sample cooked for 25 min decreased with mixing time for all of the samples under various mixing rates and after 25 min of cooking increased with cooking time because of weight loss from the result of Fig. 10.4. Adhesiveness and viscosity of the Newtonian and the Voigt bodies ($\eta_N \eta_1$) of the sample mixed at 60 rpm increased with cooking time, while both of the samples mixed at 250 rpm and 350 rpm decrease with cooking

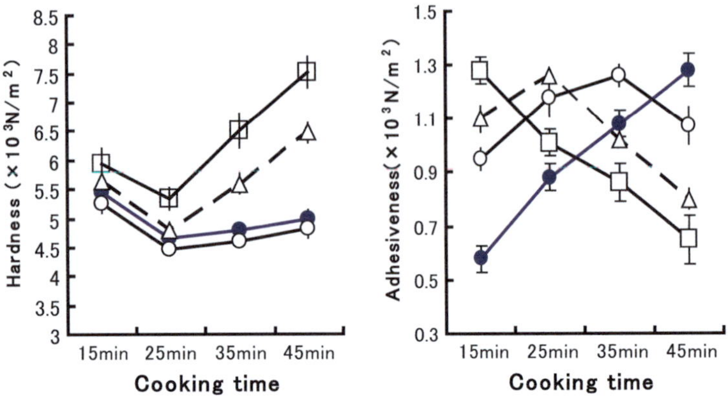

Fig. 10.4 Changes in the texture of gomatofu at various mixing rates of 60 (•), 150 (○), 250 (△), and 350 rpm (□) during cooking

time. Hardness, adhesiveness, and gumminess of the (B) sample for all mixing rates decreased exponentially. Adhesiveness showed different changes among mixing rates from Fig. 10.5. In other word, the hardness and gumminess of the sample cooked for 25 min decreased up to 25 min and turned to increase over 25 min of cooking increase with cooking time similar to the results obtained from penetration test in our previous paper [1]. Adhesiveness depended heavily on the mixing rate in the pertain test, but did lesser extent in the compression test. Cohesiveness did not change with the cooking time. The sample of 250 rpm for 25 min (2) was evaluated to have the best palatability, but the samples of 60 rpm for 15 min and 350 rpm for 45 min (3) had the worst texture organoleptically [6]. SEM observations as shown in Fig. 10.6 {the upper photographs are Cryo-SEM (×100) and the under photographs are Cryo-SEM (×10,000)} revealed that the sample cooked with a mixing rate of 60 rpm for 15 min (1) did not have a honeycomb structure, while the sample cooked with the mixing rate of 250 rpm for 25 min (2) had uniform-sized cells and a honeycomb structure and that the oil droplets got together and became lager in the sample of 350 rpm 45 min (Fig. 10.6(3)). It is assumed that *gomatofu* is a phase-separated model which has networks consisting of *kudzu* starch network and sesame protein by V. J. Morris [27].

10.3.2 Ingredient Ratio of Gomatofu [2,3]

The changes in the viscoelasticities and microstructures of *gomatofus* (*kudzu* starch 40 g, water 450 g, sesame seed 0–80 g) prepared by simmering at a mixing rate 250 rpm for 25 min as shown in Fig. 10.6(2) were investigated. This method of preparing *gomatofu* produced the most uniform size of air cells and honeycomb structure from Fig. 10.6(2). Creep meter (RE-3305), scanning electron microcopy

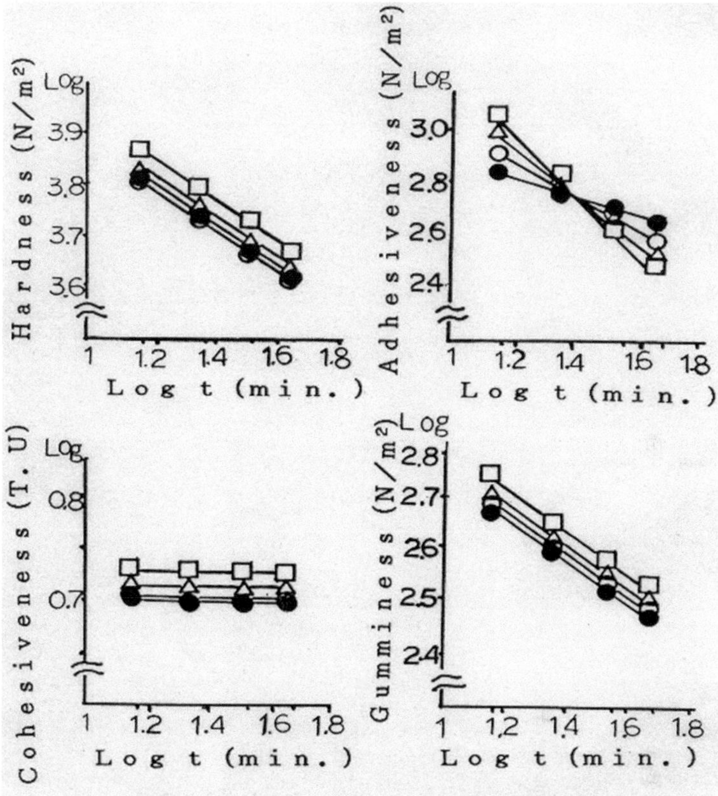

Fig. 10.5 Changes in the texture of gomatofu prepared by hot water dropping method at various mixing rates of 60 (●),150 (○), 250 (△), and 350 rpm (□) during cooking

(SEM,S-800), and syneresis observation were employed. The creep curve for the *gomatofu* was analyzed by using a four-element model, which consisted of Hookean and Voigt body elasticities (E_0,E_1) and Newtonian and Voigt body viscosities (η_N,η_1). The elasticities (E_0,E_1) of *gomatofu* increased, and its viscosities (η_N,η_1), as shown in Table 10.2, decreased with an increase of the sesame content. SEM observation (Fig. 10.8) revealed that (1) *kudzu* gel starch (sesame content 0 g) had a thick branched network microstructure, and (2) *gomatofu* (sesame contents 40 g) had a fibrous microstructure which surrounds globular sesame oil. It was found that 3 % of syneresis ratio from *kudzu* gel starch occurred during the first 2 h after preparation, while none occurred during the same period for *gomatofu* (sesame contents 60–80 g). From these results, it can be seen that sesame components contribute to the strength and stability of *kudzu* gel starch.

The effects of the ingredient ratio of *kudzu* starch and sesame on the texture of *gomatofu* (sesame tofu) were studied. The structure of *gomatofu* was observed with a scanning electron microscope (SEM). *Gomatofu* was prepared at a mixing

Upper photographs are Cryo-SEM x100),honeycomb stracture is 250rpm 25min.

(1) 60rpm 15min, (2)250rpm 25min, (3)350rpm 45min

Fig. 10.6 Effects of preparation conditions on the Cryo-SEM photographs obtained by fixing with glutaraldehyde and osmium acid for gomatofu (under photographs) (×10,000) kudzu starch (a), fibrous microstructure (f), globular sesame oils (o), and fibrous microstructure getting together (g)

Table 10.2 The viscoelasticity of *gomatofu* prepared with different *kudzu* starch and sesame

K (g)	W (g)	S (g)	SM (g)	E_0 (10^3 N/m^2)	E_1 (10^3 N/m^2)	η_1 (10^4 N/m^2)	η_N (10^7 N/m^2)	$\tau_{1(sec)}$
40	450	0	0	2.21 ± 0.15	2.12 ± 0.15	6.16 ± 0.25	3.36 ± 0.12	29
40	450	20	14	2.53 ± 0.19	2.35 ± 0.19	5.03 ± 0.24	3.03 ± 0.17	21
40	450	40	28	2.57 ± 0.16	2.60 ± 0.18	4.82 ± 0.30	2.64 ± 0.25	19
40	450	60	41	2.60 ± 0.18	2.68 ± 0.19	4.30 ± 0.28	2.58 ± 0.20	16
40	450	80	55	2.65 ± 0.21	2.86 ± 0.20	3.78 ± 0.30	2.34 ± 0.25	13

Contents *K kudzu* starch, *W* water, *S* sesame, *SM* solid material of sesame milk, E_0 elasticity of Hookean body, E_1 elasticity of Voigt body, η_1 viscosity of Voigt body, η_N viscosity of Newtonian body, τ_1 retardation time

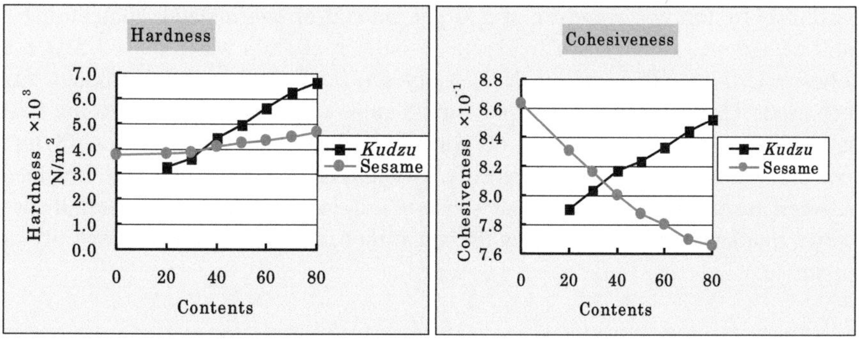

Fig. 10.7 Changes in the texture of gomatofu prepared with various amounts of sesame and kudzu starch at 250 rpm for 25 min. Kudzu: effect of kudzu starch amount (30–60 g) on the texture of gomatofu in the presence of 40 g sesame. Sesame: effect of sesame amount (0–80 g) on the texture of gomatofu in the presence of 40 g of kudzu starch

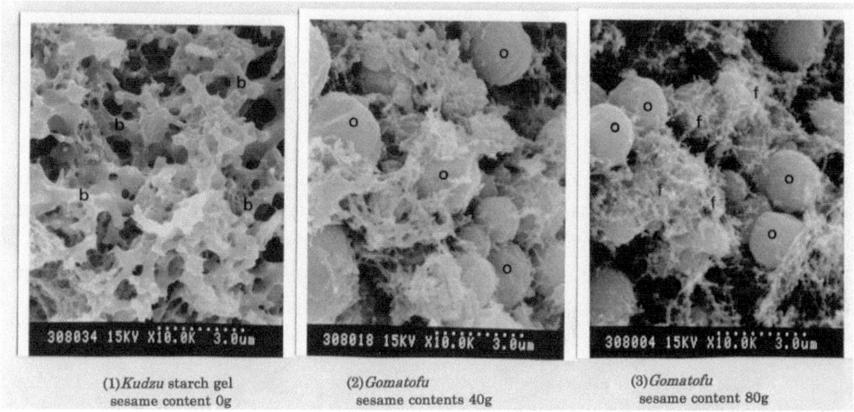

Fig. 10.8 SEM observation of gomatofu prepared with various amounts of sesame seeds at 250 rpm for 25 min by fixing with glutaraldehyde and osmium acid (×10,000)

rate of 250 rpm for 25 min. As a result of the texture measurement (Fig. 10.7), the more the *kudzu* starch and sesame contents increased, the more hardness, adhesiveness, and gumminess increased. Cohesiveness increased with an increase in the *kudzu* starch content, but decreased with increasing the sesame content. SEM observation (Fig. 10.8(3)) revealed that the larger amounts of sesame contained in the *kudzu* starch gel, the fibrous structure was formed. It was found that *gomatofu* prepared at a ratio 1:1–1.5:10–11 by weight of *kudzu* starch, sesame, and water was palatable in softness, mouthfeel, and springiness by the sensory test. A high correlation coefficient ($r = 0.94$, $p < 0.001$) was obtained between hardness by the texture measurement and hardness by the sensory score when the amounts of both *kudzu* starch and sesame were increased. A high correlation coefficient was also obtained between hardness and mouthfeel ($r = -0.96$,

$p < 0.001$) by the sensory score at a larger amount of *kudzu* starch content, but a high negative correlation coefficient ($r = -0.96$, $p < 0.001$) was obtained between cohesiveness and hardness by the sensory score when the sesame amount was increased. The sample of 250 rpm for 25 min was evaluated to have the best palatability, but the samples of 60 rpm for 15 min and 350 rpm for 45 min had the worst texture organoleptically. Good positive correlation was obtained between texture measurement and sensory test in most of the samples. It was shown that smooth mouthfeel was the most important factor in the palatability of *gomatofu*.

10.3.3 Effect on the Different Kind of Sesame Material

10.3.3.1 Chemical Composition

The chemical components of sesame seed and sesame milk are shown in Table 10.3. Sesame milk were obtained from 40 g sesame seed (in the case of 28 g defatted flour) and 450 g of water by mixing for 3 min and filtration. The amount of lipid in white sesame seed {*Kawamuki*: white raw (non-roast) huskless sesame seed and roasted white sesame seed} was greater than roasted black sesame seed, and their protein and carbohydrate contents were lower than those in roasted black sesame seed. Further, the amount of carbohydrate in *Muki* (raw huskless) white sesame seed was lower than roasted husk (white and black) sesame seed. About 60–75 % of the composition of sesame materials transfers to the sesame milk. Defatted sesame milk contains much protein and carbohydrate due to the removal of oil from sesame seed.

10.3.3.2 Texture

The results of texture parameters of *gomatofu* are shown in Fig. 10.9. Values for hardness of the *gomatofu* samples increased in the order of D.F > Ro-W > Ro-B > Ra-W. Cohesiveness increased in order of Ro-B > Ro-W > Ra-W > D.F, and adhesiveness of Ra-W and D.F were higher than those of Ro-W and Ro-B. From these results, it was found that cohesiveness of Ro-W and Ro-B was high and adhesiveness was low. The hardness of the sample Ra-W was the lowest, but the sample D.F had the highest. The cohesiveness of the sample D.F was the lowest. The elastic moduli (Eo) and the viscosities (η_N) of all the *gomatofu* samples increased in the order of Ro-W, Ro-B, and Ra-W according to their hardness, and the viscoelastic parameters increased with increasing number of days stored. Namely, values of the elastic moduli (Eo) and viscosities (η_N) for Ro-W and Ro-B were higher than those for Ra-W. The elastic moduli (Eo) of all of the *gomatofu* samples increased, and their viscosities (η_N) decreased with increasing sesame content.

Table 10.3 The chemical composition of four kinds of sesame seeds

		Moisture (g)	Fat (g)	Protein (g)	Carbohydrate (g)
Raw (unroasted)	Sesame seed (%)	4.0 ± 0.03	59.10 ± 0.42	19.72 ± 0.25	17.03 ± 0.29
White sesame seed (Ra-W)	Sesame seed (g/40 g)	1.40	23.60 (100 %)	7.89 (100 %)	68.7 (100 %)
	Sesame milk (%)	94.0 ± 0.05	4.11 ± 0.25	1.24 ± 0.08	0.94 ± 0.09
	Sesame milk 435 g	405.6	17.90 (75.6 %)	5.51 (73.9 %)	4.11 (60.0 %)
Roasted white	Sesame seed (%)	1.0 ± 0.03	59.80 ± 0.32	20.54 ± 0.23	20.40 ± 0.26
Sesame seed (Ro-W)	Sesame seed (g/40 g)	0.38	23.90 (100 %)	8.20 (100 %)	8.16 (100 %)
	Sesame milk (%)	93.60 ± 0.05	4.30 ± 0.45	1.27 ± 0.08	1.20 ± 1.09
	Sesame milk 435 g	402.0	17.40 (72.9 %)	5.43 (66.0 %)	5.12 (61.9 %)
Roasted black	Sesame seed (%)	1.80 ± 0.03	56.80 ± 0.52	20.54 ± 0.23	20.60 ± 0.29
Sesame seed (Ro-B)	Sesame seed (g/40 g)	0.70	22.80 (100 %)	8.20 (100 %)	8.24 (100 %)
	Sesame milk (%)	93.50 ± 0.15	3.77 ± 0.25	1.27 ± 0.08	1.41 ± 0.29
	Sesame milk 435 g	402.4	16.20 (70.5 %)	5.43 (66.0 %)	6.06 (59.0 %)
Defatted sesame	Sesame seed (%)	14.20 ± 0.09	11.70 ± 0.54	49.90 ± 0.55	34.30 ± 0.49
Flour (D.F)	Sesame seed (g/40 g)	1.17	3.26 (100 %)	13.90 (100 %)	9.72 (100 %)
	Sesame milk (%)	95.50 ± 0.25	0.23 ± 0.05	2.66 ± 0.02	1.77 ± 0.05
	Sesame milk 435 g	415.6	0.96 (29.4 %)	10.80 (77.0 %)	7.35 (75.5 %)

About 60–77 % of the composition of sesame materials (Ra-W, Ro-W, Ro-B) transfers to sesame milk

10.3.3.3 SEM Observations

SEM observation, as are shown in Fig. 10.10, revealed that air cell size in the samples was smaller and the cell number was larger in increasing order of Ro-W (13.3 μm) > Ro-B (14.3 μm) > Ra-W (20.5 μm) and both samples of Ro-W and Ro-B had a uniform cell size and fine structure. Conversely, the structures of Ra-W

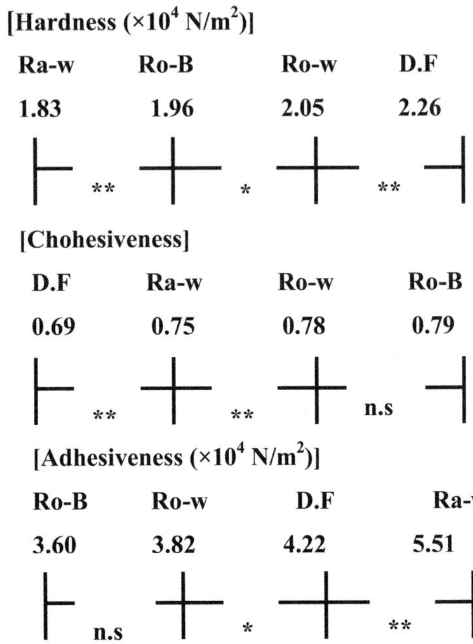

Fig. 10.9 Average of texture parameter of gomatofu

samples are coarser than Ro-W and Ro-B. It is assumed that *gomatofu* was a phase separated model [27] which had networks consisting of *kudzu* starch and sesame components (protein and lipid). Sesame (sesame protein, lipid, and water system) played an important role in a filler reinforcement effect (Matsumoto, 1979) of starch granules for *kudzu* starch gel. But in this study, since *gomatofu* was prepared under the same mixing rate and cooking time using different chemical components, it is assumed that the different air cell size is due to a change in the mechanical properties of *gomatofu*. It is considered that the soluble fiber (soluble carbohydrate) results in fine form structure. It also results in an increase of the value of hardness and of cohesiveness of *gomatofu* (Ro-W and Ro-B) which was prepared using roasted white sesame seed and black sesame seeds. These soluble fiber contents of the roasted sesame seeds were about 1.5 times larger than huskless sesame seeds. On the other hand, it is considered that air cell size of Ra-W (unroasted huskless (*Muki*) *gomatofu*) was larger and has a coarse structure. This is because the lipid of the huskless sesame seed acted as a gel inhibitor agent.

10.3.3.4 Conclusion

The sample *gomatofu* (Ra-W) prepared by using huskless (unroasted) sesame seed was investigated. Lipid content affecting "mouthfeel" of *gomatofu* was considered a possible factor. In this sensory test, mouthfeel of the sample Ra-W was the highest evaluated in all the samples. On the other hand, the sample of Ro-W and Ro-B (both

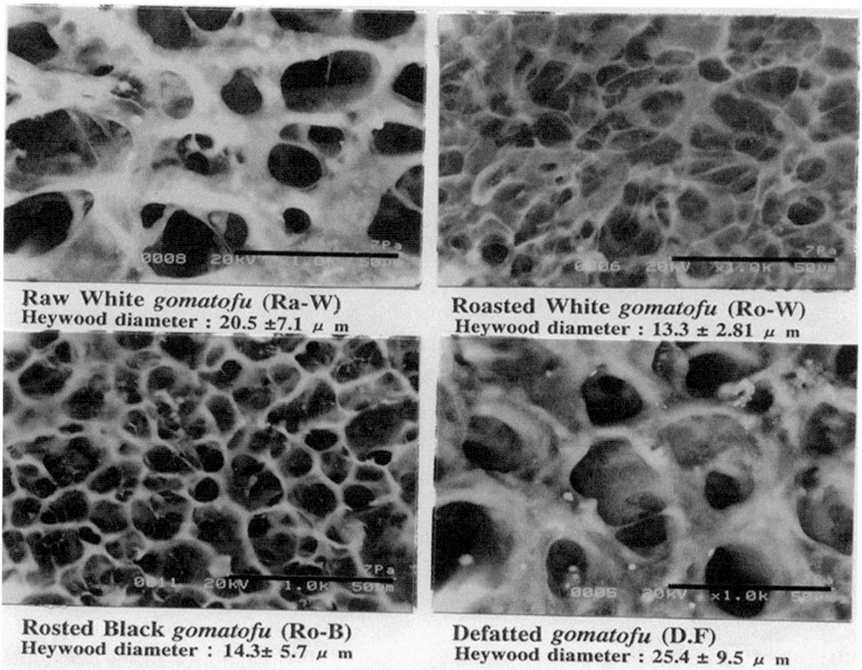

Fig. 10.10 Natural SEM photographs for gomatofu prepared with different kinds of sesame materials (×1,000)

using roasted sesame seed) had the best palatability because of springiness due to the high content of carbohydrate. From these results, it was clarified that different kinds of sesame materials affected the forming of structure and texture of *gomatofu*, because the chemical components were different from varieties of sesame materials. It is necessary to be clear about the effect of sesame components on the coagulation property of gel hereafter. It is considered that the texture of *gomatofu* prepared by using different sesame materials resulted in different textures and therefore these various kinds of *gomatofu* have affected regional food culture. All of the sensory parameters, except viscosity, and the judgment of the tasting panel agree highly with Kendall's coefficient. As a result of summarizing sensory test, these indicated that the Ra-W prepared with unroasted huskless sesame seed had the least hardness, but mouthfeel and viscosity of this sample were the highest. The Ro-B (*Arai* black) and Ro-W (*Arai* white) samples were evaluated to have the best palatability because of their superior springiness.

10.3.4 Effect on the Oil Contents Containing Gomatofu [6]

10.3.4.1 Ratio of Material

The ingredient ratio and oil content of the samples (*gomatofu*) are shown in Table 10.4. *Muki* white sesame seed of the sample (**a**) and (**b**) were defatted by Soxhlet extraction method. Preparation procedure of *gomatofu* and mixing apparatus are shown in Fig. 10.1. All of the samples of sesame milk were obtained by mixing sesame seed materials (**a**: 18 g, **b**;28 g, **c, d,** and **e**; 40 g) and 450 g of water for 3 min and filtration. About 60–75 % of the composition of sesame materials transfers to the sesame milk [8]. The samples are as follows: (a) defatted *gomatofu* (Soxhlet extraction, oil content 0 %), (b) defatted *gomatofu* (Soxhlet extraction, oil content 0.42 %), (c) control *gomatofu* (sesame seed 40 g, *kudzu* 40 g, water 450 g, oil content 3.4 %), (d) 3 % sesame oil added (oil content 6.4 %), and (e) 6 % sesame oil added (oil content 9.4 %). 435 g of sesame milk were obtained from (a) 18 g, (b) 28 g, and (c), (d), and (e) 40 g.

10.3.4.2 Texture

Matz [27] mentioned as follows: texture means those perceptions that constitute the evaluation of a food's physical characteristics by the skin or muscle senses of the buccal cavity, excepting the sensations of temperature or pain. Changes in the texture of *gomatofu* with various oil contents were shown in Fig. 10.11. From the results of the texture measurement, the hardness increased with increasing oil content, but cohesiveness in the sample (c), control *gomatofu*, showed the highest value on the first storage day. Conversely the adhesiveness of the sample (c) was the lowest. With increase in the number of storage days, hardness of the sample increased. Cohesiveness and adhesiveness decreased with increasing oil content. From the result of the creep measurement, the coefficient of elasticity and viscosity of the four-element model increased with the number of storage days.

10.3.4.3 SEM Observation

The size of air cells was found to become smaller with increasing content of oil by SEM (scanning electron microscopy) observation. Figure 10.12 presents the SEM photographs of the representative sample (**b**), (**c**), and (**e**) of *gomatofu* (×10,000) on the first storage day. It is revealed that the sample (**b**) had a thick branched structure of *kudzu* starch gel due to low oil contents. The sample (**c**) (control *gomatofu*, oil content 3.4 %) had a fibrous microstructure which surrounded globular droplets of sesame oils. The sample (**e**) had a finer fibrous microstructure which surrounded smaller globular sesame droplets than the sample (**c**), and the magnitude of hardness of the sample (**e**) was higher, but that of cohesiveness was lower than the sample (**c**). The larger amounts of sesame involved in the *kudzu* starch gels, the finer fibrous

Table 10.4 Ingredient ratio and oil content of the sample (*gomatofu*)

Sample (*gomatofu*)	Oil content (g)/sesame seed 100 g	Sesame seed (g)	Oil content (g)/sesame seed 40 g	Sesame oil added (g)	Water added (g)	Oil content (g)/sesame milk	Oil content (g)/*gomatofu* 100 g
(a) Defatted *gomatofu* (oil content 0 %)	0.0	18.0	0.00	0.0	450.0	0.00	0.00
(b) Defatted *gomatofu* (oil content 0.42 %)	10.3	28.0	2.88	0.0	450.0	2.19	0.42
(c) Control *gomatofu* (oil content 3.4 %)	58.0	40.0	23.20	0.0	450.0	17.90	3.38
(d) 3 % oil added (oil content 6.4 %)	58.0	40.0	23.20	15.9	434.1	33.80	6.38
(e) 6 % oil added (oil content 9.4 %)	58.0	40.0	23.20	31.8	418.2	49.70	9.38

[a–e] *Kudzu* starch of all sample was 40 g
[a] Defatted sesame seed was obtained from Soxhlet extraction
[b] Defatted sesame seed was obtained from compression grinder
[c] Basic ingredient (sesame seed 40 g, *kudzu* 40 g, water 450g)
[d] Three percent sesame oil added to basic ratio (**c** sample)
[e] Six percent sesame oil added to basic ratio (**c** sample)

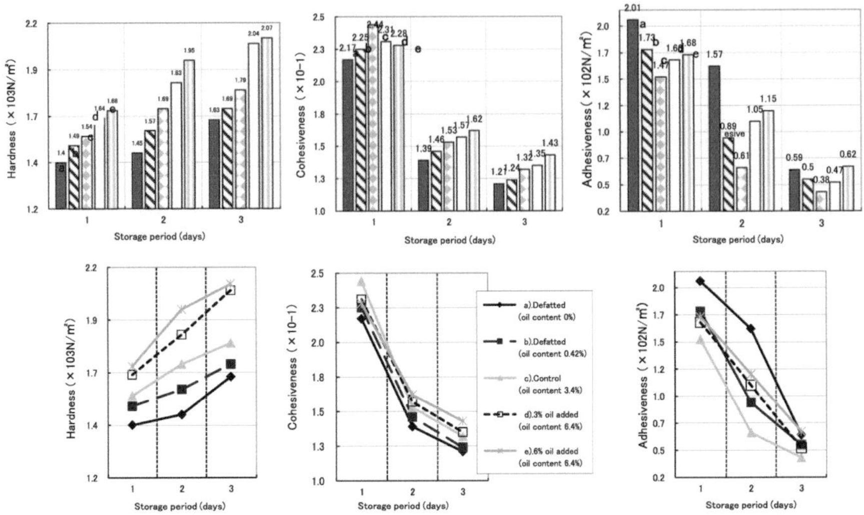

Fig. 10.11 Changes in the texture (hardness, adhesiveness, and cohesiveness) of gomatofu included different from oil contents

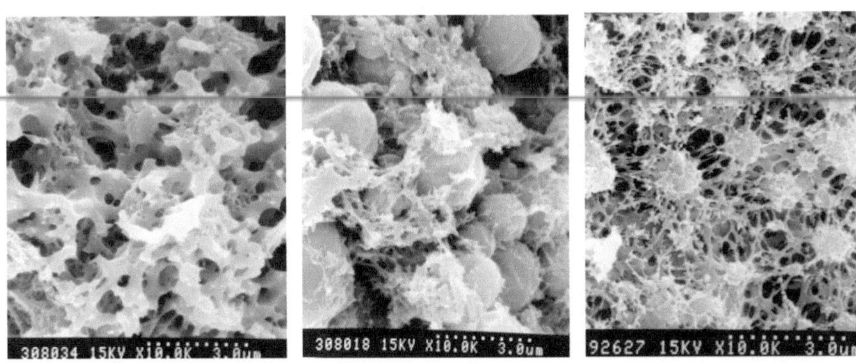

b. Defatted *gomatofu* (oil content 0.42%) **c. Control *gomatofu* (oil content 3.4%)** **e. 6% oil added (oil content 9.4%)**

Fig. 10.12 SEM photographs obtained by fixing with chemical treatment (glutaraldehyde) for gomatofu prepared with mixing rate at 250 rpm 25 min (×10,000)

structure was formed. In the present study, it was clarified that the larger amounts of sesame oil involved in the *kudzu* starch gels, the finer fibrous structure was formed. It is considered that the creep behavior reflects the structure of *gomatofu* observed by SEM.

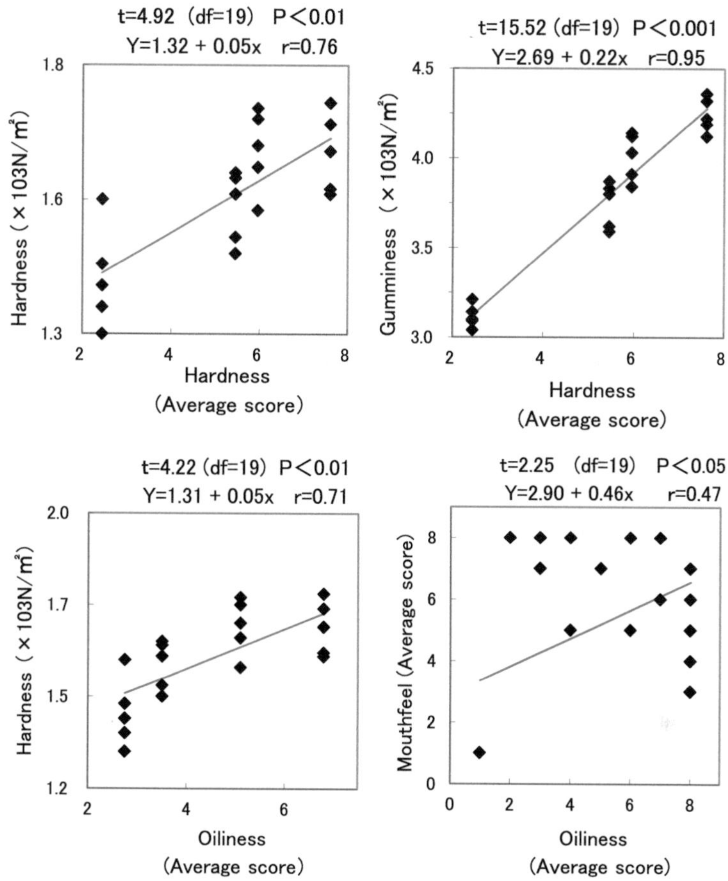

Fig. 10.13 Relationship between texture parameter and sensory score

10.3.4.4 Sensory Evaluation

Relationship between texture parameter and sensory scores is shown in Fig. 10.13. A positive correlation was obtained between instrumental texture measurement and sensory evaluation. The correlation between the hardness of the texture measurement and the hardness of the sensory score showed a significant value of 0.76 (P; probability <0.01). The correlation between the gumminess of the texture measurement and hardness of the sensory score showed a highly significant value of 0.95 ($P < 0.01$). The correlation between the hardness of the texture measurement and the oiliness of the sensory score showed a significant value of 0.71 ($P < 0.01$). In addition, the correlation between mouthfeel and oiliness of the sensory score showed a significant low value of 0.47 ($P < 0.05$), but mouthfeel and total acceptance showed a significant high value of 0.94 ($P < 0.001$, not shown in Fig. 10.9) of the sensory score. From the result of sensory evaluation, the samples (**c** and **d**) were highly evaluated for total acceptance. However, no significant difference was found in these samples by Newel and MacFarlane's test. From these results it was found that favorable oil content was 3.4–6.4 % in the *gomatofu*. Glicksman [28] has stated fatty or oily mouthfeel can be described as a combination of several basic parameters and these parameters were viscosity, lubricity, absorption, and others. In the sensory evaluation, the control sample (c) and the sample (d) were highly evaluated for total acceptance. From these results it was found that favorable oil content was 3.4–6.4 % in the sample (*gomatofu*).

10.3.5 Roasting Conditions of Sesame Seeds and Their Effect on the Mechanical Properties of Gomatofu [7]

10.3.5.1 Frequency Particle Size Measurement

Sesame materials were employed with *Muki* (hull-less) and *Arai* (white and black, with hull) seeds (unroasted and roasted at 160, 170, 180, 190, and 200 °C) in an electric oven for 15 min. The frequency particle size of sesame particles that passed through a 50 mesh sieve (=297 μm) in *Arai* white and *Muki* sesame milk was measured by a laser diffraction particle size analyzer (LDPA) (SALD-2100, Shimadzu Co., from *San-Ei Gen F.F.I., Inc. Japan*). Forty milliliter of deionized water was put into a cell of the LDPA instrument, and 0.5 ml of sesame milk diluted to one tenth (1/10) was dropped in the cell and measured three times. The average particle size was determined by the laser diffraction dispersion method.

10.3.5.2 Frequency Particle Size Measurement

Changes in the frequency of particle size with increasing of roasting temperature for *Arai* sesame milk are shown in Fig. 10.14. The average particle size increased remarkably with an increase in roasting temperature for both kinds of sesame seeds, ranging from 9.8 to 21.5 μm for *Arai* and from 27.8 to 40.4 μm for *Muki* when the temperature was equal to or lower than 170 °C and ranging from 85.9 to 86.4 μm for *Arai* when the temperature was 180 or 190 °C. The average particle size in the *Muki* sesame milk was larger than that of *Arai* for when compared at the same temperature. The frequency of particle size in the sesame milk of unroasted *Arai* had peaks at 2.4 μm, 8.3 μm, 29.0 μm, and 153.0 μm, but that of the samples roasted at 170 °C, 180 °C, and 190 °C had two peaks at 29.0 μm and 153.0 μm. The particle size of *Arai* sesame seeds roasted at 180 °C and 190 °C and passed through a sieve (297 μm) was most heavily concentrated in the range of 150.0–300 μm.

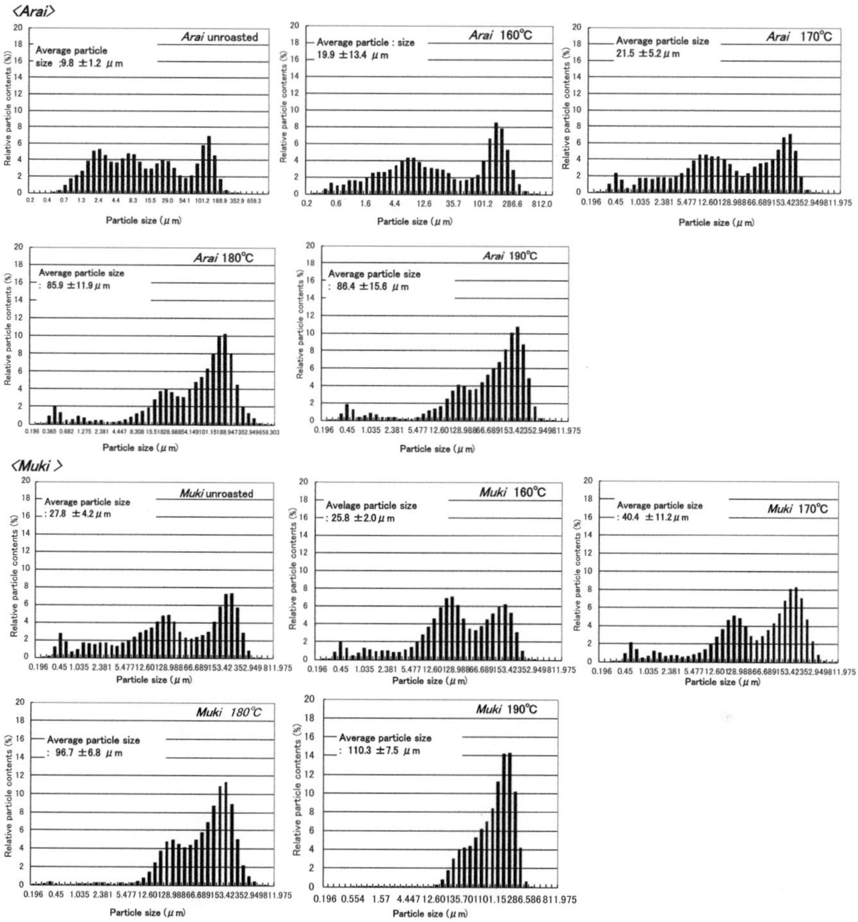

Fig. 10.14 Changes in the frequency particle size of Arai and Muki sesame milk with an increase of roasting temperature for sesame seeds. Forty milliliter of deionized water was put into a cell of the LDDA instrument, and 0.5 ml of sesame milk diluted to one tenth (1/10) was dropped in the cell and measured three times

10.3.5.3 Fracture Properties of *Muki Gomatofu* Prepared at Various Roasting Temperatures

Typical stress-strain curves of *Muki gomatofu* made from sesame seeds roasted at various temperatures are shown in Fig. 10.15. *Gomatofu* showed a ductile fracture, after passing through a plastic domain. The fracture stress of *gomatofu* was the lowest when seeds were roasted at 170 °C and the highest when seeds were roasted at 190 °C. There were significant differences ($p < 0.01$) in fracture stress between *gomatofu* prepared using seeds roasted at 170 °C and those roasted at 180 °C for two kinds of *gomatofu*. It is clear that *gomatofu* prepared using sesame seeds roasted at

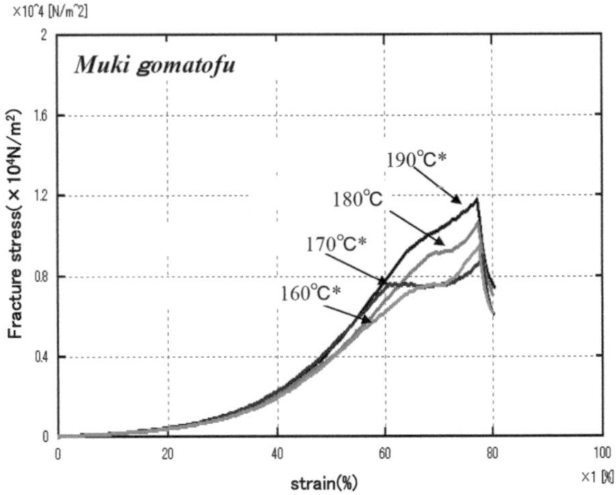

Fig. 10.15 Typical stress-strain curves at various roasting temperatures of Muki gomatofu *; sesame roasted at 160 °C, 170 °C, 180 °C, 190 °C

190 °C have higher fracture stress because their fracture stress and the fracture strain values (71–75 %) were higher than those of the other samples. It was considered that the values of fracture stress and strain of the *gomatofu* prepared using seeds roasted at 170 °C were the smallest.

10.3.5.4 The Comparison with the Fracture Properties of Agar Gels and That of *Gomatofu*

Both of fracture curves of agar gels and that of *gomatofu* were shown in Fig. 10.16. As the higher concentration of 0.3–0.9 % agar gel increased, the higher fracture stress ($0.12–4.5 \times 10^3$ J/m^3) and fracture energy of these agar gels also increased. The fracture strain of these agar gels was about 30–35 %, and no significant difference was recognized between agar gels. According to a previous report [26], fracture stain of 2.4 % (w/w) of agarose gel ranged from 818 % to 18 %, and the gel changes from a more "brittle" state to more extendable state with increasing molecular weight. In the present study, it is considered that the value of the fracture strain of agar gel increased due to 30 % content of agaropectin. On the other hand, fracture stress of unroasted sesame *gomatofu* was about 1.0×10^4 N/m^2 and 170 °C roasted sesame (*gomatofu*) was about 0.8×10^4 N/m^2. Therefore the value of fracture stress of *gomatofu* was very low compared with agar gel. As for a fracture strain of *gomatofu*, the value of unroasted sesame (*gomatofu*) was about 48 %, and that of 170 °C roast sesame (*gomatofu*) was about 63 %. Fracture strain of both *gomatofus* was shown to be much higher than that of agar gel. The fracture energy was almost the same, about 1.5×10^3 J/m^3 in both *gomatofus*. As a result of the findings mentioned above, it was clear that agar gel showed a brittle fracture in which the fracture stress was big, but a fracture strain was small, and it was firm and brittle. On the other hand, *gomatofu* showed a ductile fracture to break after passing

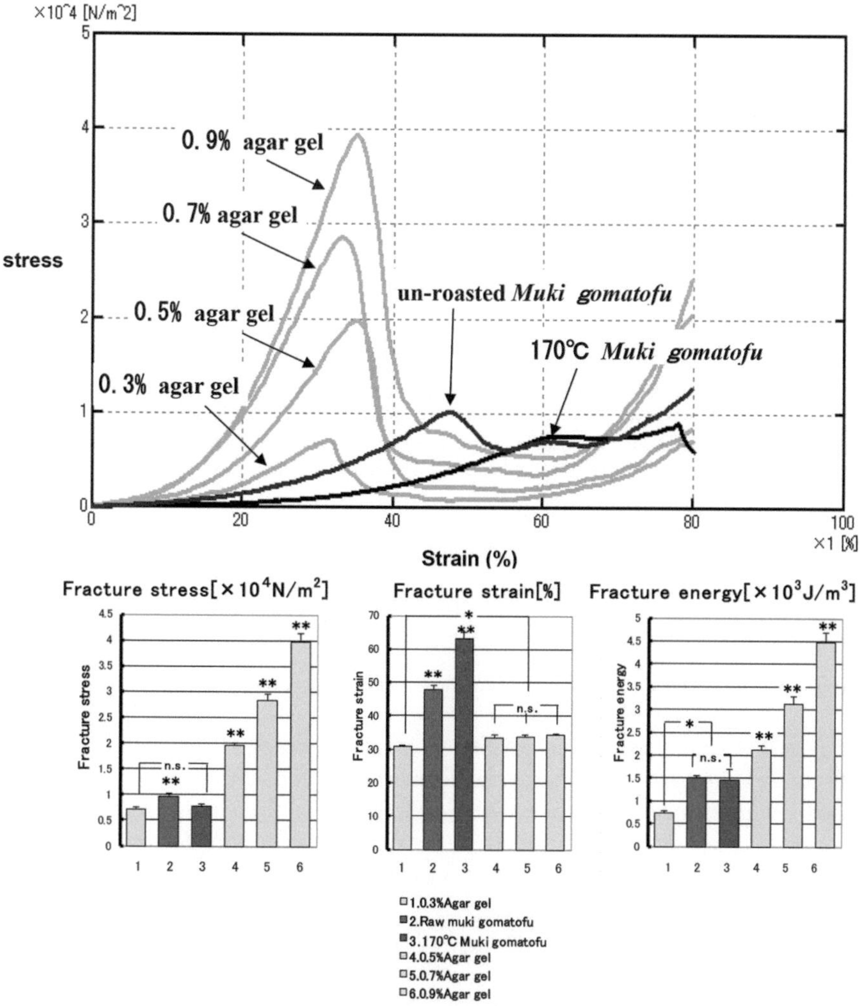

Fig. 10.16 The comparison with the fracture properties of agar gels and that of gomatofu. The means and SD of 10 ($n = 10$) for one sample were calculated, and significant difference between samples was examined by t test. **: $p < 0.01$, against the lowest mean value; *: $p < 0.05$, against the lowest mean value; *n.s.* no significant difference

through a plastic domain, and fracture stress was small. But a fracture strain had a greater fracture point compared with agar gel. Furthermore, the fracture stress of *gomatofu* had almost the same order of magnitude as that of 0.35 % agar gel. It was found that *gomatofu* was softer than agar gel, but that it was a tough gel. In addition, it was considered that *gomatofu* which used sesame roasted at 170 °C produced a tough gel which was hard to bite off in comparison with unroasted sesame *gomatofu*.

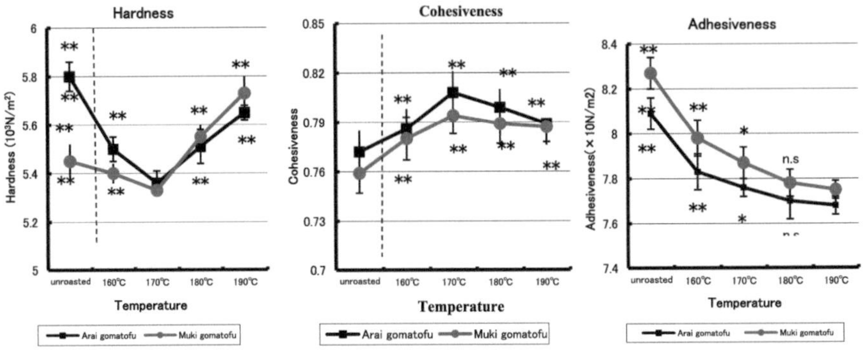

Fig. 10.17 Effect of roasting temperature on the texture properties of Arai white and Muki (*white*) gomatofu. The means and SD of 10 ($n = 10$) for one sample were calculated, and significant difference between samples was examined by *t* test. **: $p < 0.01$, against the lowest mean value; *: $p < 0.05$, against the lowest mean value; *n.s.* no significant difference

This is due to the fact that *gomatofu* roasted sesame at 170 °C was softer and had a slightly smaller fracture stress value.

10.3.5.5 Texture Properties of *Arai* and *Muki Gomatofu*

The average values (with standard deviation bar) of the ten measurements for each of three texture parameters (hardness, cohesiveness, and adhesiveness) of *Arai* and *Muki gomatofu* made from seeds at various roasting temperatures are shown in Fig. 10.17. The significant difference was examined against the lowest mean value by *t* test. The magnitude of the hardness H for *Muki gomatofu* was highest in order from $H_{190\,°C(p<0.01)} > H_{180\,°C(p<0.01)} > H_{unroasted\,(p<0.01)} > H_{160\,°C(p<0.01)} > H_{170\,°C}$; for *Arai Gomatofu*, the order starting with the highest was $H_{unroasted\,(p<0.01)} > H_{190\,°C(p<0.01)} > H_{180\,°C(p<0.01)} > H_{160\,°C(p<0.01)} > H_{170\,°C}$. For cohesiveness C, the order starting with the highest for *Arai* and *Muki gomatofu* was $C_{170\,°C(p<0.01)} > C_{180\,°C(p<0.01)} > C_{190\,°C} > C_{160\,°C} > C_{unroasted}$. For both kinds of *gomatofu*, the roasting temperature of 170 °C produced the lowest hardness but the highest cohesiveness. The cohesiveness of *Arai gomatofu* was higher than that of *Muki.* As for adhesiveness A, the magnitude for both *gomatofus* was highest in order from $A_{unroasted\,(p<0.01)} > A_{160\,°C\,(p<0.01)} > A_{170\,°C\,(p<0.05)} > A_{180\,°C\,(p=n.s)} = A_{190\,°C}$.

10.3.5.6 Sensory Evaluation of *Arai* white and *Muki Gomatofu*

Results of the ranking method are shown in Fig. 10.18. The number shows the total sum of order. Ranking of the hardness by sensory evaluation Hs for *Arai* was highest in order beginning with sample $Hs_{e,190\,°C(hard1)} > Hs_{a,unroasted} > Hs_{d,180\,°C} > Hs_{b,160\,°C} > Hs_{c,170\,°C(Muki,\,p<0.01)}$. These findings indicated that both *gomatofus* had the

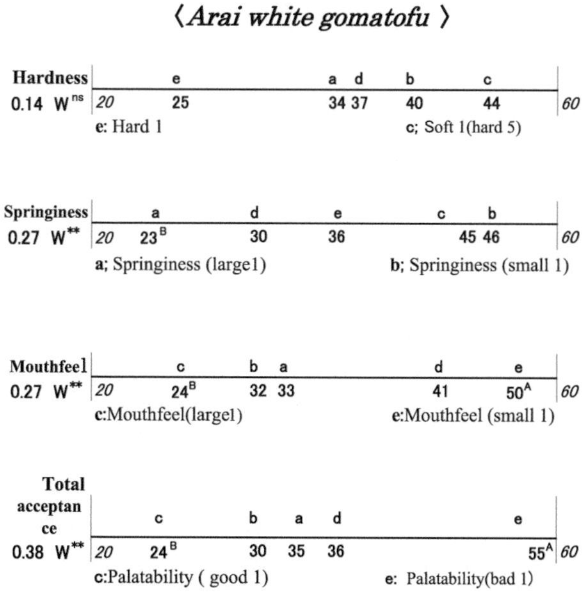

Fig. 10.18 Sensory evaluation of Arai white gomatofu prepared at various roasting temperatures by ranking method. The means and SD of 10 ($n = 10$) for one sample were calculated, and significant difference between samples was examined by *t* test. **: $p < 0.01$, against the lowest mean value; *: $p < 0.05$, against the lowest mean value; *n.s.* no significant difference

least hardness (i.e., maximum softness) when prepared with sesame seeds roasted at 170 °C. The Kendall coefficient was low ($W^{n.s} = 0.14$), indicating no significant difference. Ranking of the springiness S for *Arai* white was, in order from highest to lowest, sample $\mathbf{S_{a,\ (p<0.05)}}$>$\mathbf{S_d}$>$\mathbf{S_e}$>$\mathbf{S_c}$ and $\mathbf{S_b}$, and a significant difference ($p < 0.01$) was recognized between samples **a** and **b** by the N&M test. The ranking of the total acceptance TA for *Arai* was highest starting with $TA_{\ c,\ (p<0.05)}$>$TA_{\ b}$, >$TA_{\ a}$,>$TA_{\ d}$,>$TA_{\ e,\ (p<0.01)}$; for both *Arai* ($p < 0.05$) and *Muki* ($p < 0.01$), **e** or the sample roasted at 170 °C, were the highest although the difference between *Arai* **c** and **b** was not significant according to N&M test. Namely, the sample of *Muki* **c**, 170 °C ($P < 0.01$), was also evaluated to have the best palatability, respectively. For both *Arai and Muki*, the ranking of total acceptance was lowest for the sample **e** or the sample roasted at 190 °C ($P < 0.01$). *Arai* **e** 190 °C ($P < 0.01$) had the lowest overall evaluation among the *Arai* because of having slight bitterness or "*zaratuki*" and powdery. In the parameters of springiness ($p < 0.01$), mouthfeel ($p < 0.01$), and total acceptance ($p < 0.01$), the judgment of the tasting panel showed high agreement based on Kendall's coefficient (W); the exception was hardness.

10.3.5.7 Conclusion

It was reported that *gomatofu* follows a phase-separation model [27] having networks of *kudzu* starch and sesame components as shown in Fig. 10.19. According to empirical knowledge of cooks, heating of sesame seeds should be stopped when three seeds pop open. Overheating creates a burning smell, loss of flavor [29], and undesirable changes in the sesame components. Sesame seeds are usually roasted until the hull is burst by internal vapor pressure; in this study, sesame seeds started popping at 170 °C. The difference in the physical properties of sesame due to roasting greatly affected the color, sesame components and texture of the seeds, and the ease of grinding. *Arai* white sesame roasted at 170 °C were structurally well expanded, became thicker [28], and were easily broken by roasting. *Gomatofu* prepared using seeds roasted at 170 °C had appropriate color and a lower level of protein, a smaller average particle size (*Arai*,21.5 μm; *Muki*, 40.4 μm), and peak of relative particle content in sesame milk (*Arai*, 7.8 %; *Muki*, 8.2 %) than those using seeds roasted at 190 °C (*Arai*, 86.4 μm and 11.6 %; *Muki*,110.3 μm and 14.3 %). The values of the fracture stress and strain (Fig. 10.15) and hardness (Fig. 10.17) for *Arai* white and *Muki gomatofu* were lowest for 170 °C (softness), and the magnitude of cohesiveness was large. The magnitude of adhesiveness for *gomatofu* decreased with roasting temperatures; it was guessed that this trend was due to the influence of the particle size. The particle size of *gomatofu* became smaller (2.0–6.0 μm) by mixing during preparing of *gomatofu*. It was guessed that most of the finer sesame particles (comparatively soft) were dispersed in the network of the *gomatofu* by mixing the suspension of sesame milk and *kudzu* starch. The fracture stress increased at higher roasting temperatures because these fine particles of sesame have a filler effect [30], and they increased the elastic modulus of the gel. Sensory evaluation showed that *gomatofu* prepared with seeds roasted at 170 °C had best palatability (*Arai gomatofu*, $p < 0.05$; *Muki gomatofu*, $p < 0.01$) due to their softness and mouthfeel. It was found that good mouthfeel is characterized by smoothness and is an important parameter in the texture [5]. But *gomatofu* prepared with seeds roasted at 190 °C became hard and burnt taste, and those of mouthfeel was rated the worst ($p < 0.01$) because it was not smooth. It was considered that oral perception of grittiness [31] was influenced not only by average particle size and the density of the dispersion particles but also by the physical properties of the starch gel and the variety of starch. In addition, it was difficult for panel to detect grittiness when the particle size was small and soft, and the viscosity of the dispersion medium was high [32]. Oral perception of grittiness for *gomatofu* was slightly strong and was evaluated as "*zaratuki*" and powdery in this study, though the sesame particle size in *gomatofu* roasted at a high temperature (190 °C) was not so large, and those particles were comparatively hard.

Acknowledgments Author would like to thank H. Watanabe working in Niigata Water Works Bureau where was examined by using SEM (Chilled Natural SEM-3500, Hitachi, Co) for observations, and T. Funami working in San-Eigen F.F.I., Inc. Osaka where was made by using

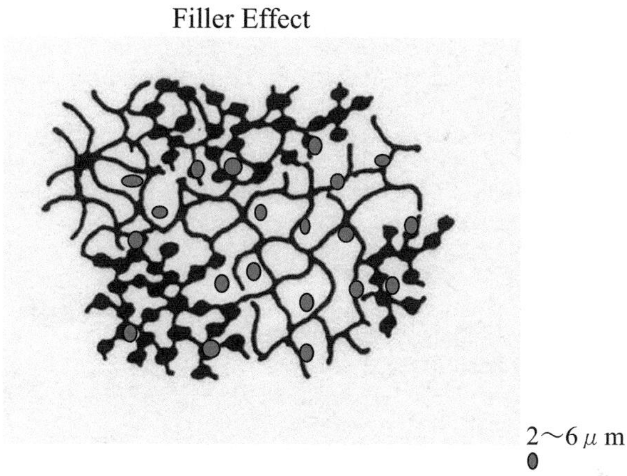

Fig. 10.19 Phase-separated networks (by V.J. Morris) [27]. Fine particles of sesame seeds roasted at higher temperature (180, 190 °C) have a filler effect, and they increased the elastic modulus of the gomatofu

Laser Diffraction Particle Size Analyzer (SALD-2100, *Shmadzu*Co), for *San-Ei Gen F.F.I., Inc.* Japan.

References

1. Sato, E., Miki, E., Gohtani, S., & Yamano, Y.: *Nippon Shokuhin Kagaku Kogaku Kaishi. (J. Japanese Society for Food Science and Technology, Jpn)***42**, 737–747. (1995)
2. Sato, E.: *HYDROCOLLOIDS*, Edited by K. Nishinari, Elsevier Science B. V. by Printed in the Netherlands, 269–274 (2000).
3. Sato, E., Ito, R., Yamano, Y.: *J. Japanese Society for Food Science and Technology, Jpn*) **46**, 285–292. (1999)
4. Sato, E., Miki, E., Gohtani, S., & Yamano,Y.. *Nippon Shokuhin Kagaku Kogaku Kaishi. (J. Japanese Society for Food Science and Technology, Jpn*) **42**, 871–877. (1995)
5. Sato, E., Ito, R., Yamano, Y.. *Nippon Shokuhin Kagaku Kogaku Kaishi. (J Japanese Society for Food Science and Technology, Jpn*) **46**, 367–375. (1999).
6. Sato, E., Shinbo, Y,, Watanabe,H., and Nishinari, K.: *J. Society of Rheology*, **33**, 101–108 (2003)
7. Sato,E., Masahide Watanabe, H., Noda S and Nishinari, K.: *Journal of Home Econ-mics,Japan,***58**, 471–483 (2007)
8. Sato, E.,.*Nihon Chori kagakukaishi*, (*J. Cookery Science Japan)*, **34,** 295–300. (2001)
9. Sato, E.:, *Food Hydrocolloids,***17**, 901–906. (2003).
10. *Shanghai kagakugijyutu* syuppansha; *Cyuyaku Daijiten,* Syougakukan hen, Tokyo, 2452 (1998)
11. *Shanghai kagakugijyutu* syuppansha; *Cyuyaku Daijiten,* Syougakukan hen, Tokyo 801 (1998)
12. Bourne M, "Food texture and Viscosity Concept and Measurement" 2 nded, Academic Press, NY. (2002)
13. Nishinari K, *J Texture Studies,***35,** 113 (2004)
14. Nishinari K, *Nihon Reoroji Gakkaishi* (*J Soc Rheol, Jpn*), **31**, 41 (2003)
15. Katsuta K, " Shinn Shokkann Jiten", 36 (1999), Nishinari K, Nakazawa F, Katsuta K. & Toda J, Science Forum, Tokyo.
16. Szczesniak AS and Kleyn DH, *Food Technol*, **17,** 74 (1963)
17. Szczesniak AS, *J Food Sci*, **28,** 385, (1963).

18. Matsumoto F, *Nihon Yu Kagaku Kaishi (J Soc Oil Chem, Jpn)*, **17,** 657 (1968)
19. Kandate, M.,: *Food analytical method,* Dobunsyoin Tokyo, 90 (1986)
20. Sato, E., Honma, N., Shibuya, U., Ishihara, K.: *Nihon Kasei gakkaishi*, (*J. Home economics Jpn.)*, **30,** 775–779. (1979)
21. Kjeldahl: Method of Analysis of AOAC, 10th. Ed.774 (1965).
22. Hodge,J.E.&Hofreiter, B.T.: Method in Carbohydrate Chemistry, Vol.1 (Whistler,R.L.,& Wolfrrom, M.L., eds.), 388, Academic Press, New York and London (1962).
23. Culioli,J.,and Sherman. P.: *J. Texture Stud.* **7,** 353–372 (1976)
24. Kramer, A.: *Food Tecnol.,* **14** (11), 576–582 (1960).
25. Newell, G. J & Mac Farlane, J. D.: Expanded tables for multiple comparison procedures in the analysis of ranked data. *J. Food. Sci.,* **53**,1721–1725 (1978).
26. Watase.M, Nishinari. K (1983) Rheological properties of agarose gels with different molecular weights, Rheol. Acta 22, p580–87.
27. V.J. Morris: *"Gums and Stabilisers for the Food Industry 3*" 87–99 (1985).
28. Glicksman M, *Food Technol*, **45,** 94 (1991)
29. Takeda, T., Aono, H., Fukuda,Y., Hatae,K., and Shimada, A*Nihon Kasei Gakkaisi* (*J. Home Econ, Jpn).* **51**.1115–1125, (2000)
30. Yamane.H.: *Reoroji Deta Handobukku, Nihon Reoroji gakkaihen,* Maruzen, Tokyo,77 (2006)
31. Imai,E.,Hatae,K.,Shimada,A.: *J. Texture Studies*, **26,** 561,(1995)
32. Engelen, L., Van der bilt, A., Schipper, M., Bosman F.; *J. Texture Studies*, **36,** 373 (2005)

Part III
Applications: Cosmetics and Personal Care Products

Chapter 11
Rheology Control Agents for Cosmetics

Isamu Kaneda

Abstract "Usage feeling" is one of the most important characteristics of cosmetics, particularly skin-care products. Because the usage feeling of cosmetics is strongly related to their rheological properties, the ingredients, which affect the rheological properties of the product, are key factors in the development of skin-care products. In this chapter, two types of novel rheology control agents for cosmetics that have been developed are described. The first is hydrophobically ethoxylated urethane, which contains a relatively large C24 hydrophobe. This telechelic polymer forms a transient network structure in aqueous systems. Because the transient network structure is due to physical interactions, it is easily destroyed and reconstructed. This interesting physical property creates a unique usage feeling for use in skin-care products. The second agent is a water-swellable microgel that was polymerized in a W/O microemulsion system. Although water-soluble polymers are widely used in cosmetics as viscosity thickeners, the thickeners often suffer a serious problem, so-called spinability, which is due to the entanglement of polymer chains. In contrast, microgels avoid such problems. The details of these rheology control agents, their syntheses, physicochemical properties, and rheological properties are reviewed.

Keywords Cosmetics • Rheology control agent • Telechelic polymers • Microgels

11.1 Introduction

For cosmetic development, a very important challenge is the control of rheological properties. Because cosmetic products, especially skin-care products, are used every day and continuously, usage feeling, which includes "spreadability," "freshness," and "moistness," is an important evaluation criterion used by consumers to select a product. Therefore, the usage feeling of cosmetic products must be considered from the first stages of formulation. Undoubtedly, the usage feeling is strongly related to the rheological properties of the ingredients. Therefore, rheological

I. Kaneda (✉)
Department of Food Science and Wellness, Rakuno Gakuen University, Ebetsu, Japan
e-mail: kaneda-i@rakuno.ac.jp

I. Kaneda (ed.), *Rheology of Biological Soft Matter*, Soft and Biological Matter,
DOI 10.1007/978-4-431-56080-7_11

measurements and analyses are indispensable tools for cosmetic formulators and scientists. Cosmetic products contain many kinds of ingredients, and the cosmetic formulators must make the best choice to formulate a product that is preferred by consumers.

Most skin-care cosmetics are O/W emulsions. The oil phase of the emulsion is comprised of emollient ingredients that act on the stratum corneum of the skin. In the aqueous phase, polyols such as glycerin have been used as a moisturizing ingredient. A high inner-oil phase emulsion has "body" without any viscosity thickeners [30]. For example, a semisolid O/W cream, using swollen lamellar gels, is widely utilized in skin-care products [33]. There is little demand for rheology control agents for these semisolid emulsions; however, they are needed for the formulation of emulsions that contain less inner-oil phase.

Because most skin-care products are O/W emulsions, rheology control agents are added to the continuous phase, that is, the aqueous phase. Consequently, most are water-soluble or water-swellable polymers. Undoubtedly, nowadays, the most popular rheology control agents for cosmetics are cross-linked polyacrylic acids and their analogues [10, 38]. Cross-linked polyacrylic acid is considered a microgel. Carbopol, a well-known cross-linked polyacrylic acid, forms a porous structure in the aqueous phase. This structure results in a good "body" feeling, and the product exhibits shear-thinning, non-Newtonian flow [24, 31]. This behavior is preferred in rheology control agents for cosmetic products.

Polysaccharides are also used in cosmetics, and those used are known to be safe and nontoxic. For example, xanthan gum, which is an extracellular polysaccharide secreted by the bacterium *Xanthomonas capestris*, is widely utilized not only in cosmetics but also in foods, in personal care products, and in other fields. In particular, because this polysaccharide has a rigid structure, the formation of a porous 3D structure, which occurs at a certain critical concentration, gives "body" to the product.

"Shear-thinning, non-Newtonian flow" is a key term in the consideration of the rheological properties of cosmetic products [7, 27]. In the next section, we review the current demands in rheological properties for cosmetic products.

11.2 Importance of Rheology Control for Cosmetic Products

11.2.1 Usage Feeling and Rheology Control

There are two characteristic factors that determine the usage feeling of cosmetics. These are the "primary feeling" and the "secondary feeling" [6]. Yielding characteristic and shear thinning are strongly related to "primary feeling," which is the usage feeling during the removal and application of the cosmetic formulation. These formulations require "body" when they are taken out of the bottle; otherwise they drip or run down the skin. Subsequently, the applied formulation is spread and rubbed on the skin. The most important usage feeling in this situation is spreadability. Therefore, products exhibiting shear-thinning, non-Newtonian flow are preferable because of their good spreadability.

Rheological properties, such as shear-thinning flow accompanied by yield stress, are not only important for reasons of consumer palatability but, also, from a psychological view point. The psychological effect of such non-Newtonian flow behavior creates the feeling of cosmetic penetration. Generally, because the skin surface is hydrophobic, the aqueous phase of the cosmetic does not penetrate the skin immediately. The emollient effect of skin-care cosmetic products is due to moisture retention in the outer part of the skin surface. However, consumers often desire that cosmetic products applied to the skin penetrate the skin. A shear-thinning, non-Newtonian fluid with finite yield stress may satisfy these consumers because changes in viscosity create the sensation of product penetration.

In addition to "primary feeling," both the "secondary" and "after feelings" are important. "Secondary feeling" is the usage sensation felt when rubbing the cosmetic product on the skin. Because cosmetic products are spread in a thin film on the skin, the "secondary feeling" relates to tribological properties between the skin and the hands or fingers. The feeling after application of the cosmetic product is called the "after feeling." A sticky "after feeling" is unacceptable for consumers. This unpleasant feeling is strongly affected by the polymers added as a rheology control agents; therefore, careful consideration of rheology control agent is necessary.

11.2.2 The Properties of the Polymer Solutions

Here, we review the fundamental properties of polymer solutions to aid understanding of the following sections. Although polymer chains are isolated in dilute solutions, over their critical concentration, the chains are in contact. This is the so-called overlap concentration, c^*. The value of c^* depends on the type of polymer and its molecular weight. Generally, high molecular weight polymers and rigid polymers have small c^* values.

Polymer solutions that have concentrations in excess of c^* are called "semi-dilute solutions" or "concentrated solutions." In such solutions, entanglement of the polymer chains occurs. Because the entanglement points act like cross-link points in gels, the solutions are viscoelastic. However, entanglement is temporary; consequently, the elastic properties disappear after a finite time length. The lifetime of the temporary cross-links is estimated by determining the relaxation time of the transient network structure using a rheological measurement method.

The temporary viscoelastic properties of cosmetic emulsions containing polymers in a semi-dilute regime affect the tribological characteristics, that is, those concerning the "secondary feeling." Too much entanglement causes "spinability," an undesirable usage feeling. The "after feeling" is also affected by the polymers. The nonvolatile ingredients of cosmetic products remain on the skin after evaporation of water. Consequently, polymers in the cosmetic products condense on the skin. It is easy to imagine that the polymer entanglement increases during this process and the firmly entangled polymer chains can cause an unpleasant sticky feeling on the skin. Therefore, if too high a concentration of polymer is used in the cosmetic, various problems concerning the usage feeling can arise.

As mentioned above, the polymer concentration should be minimized to avoid unfavorable usage feelings. Moving forward, we will show practical issues for the use of polymers in cosmetic products, in particular, skin-care cosmetic products.

11.2.3 Practical Issues for the Application of Polymers as a Rheology Control Agent

11.2.3.1 Acid and Salt Resistance

Because the addition of a small amount of polymer is beneficial, polyelectrolytes that have large radii of gyration or intrinsic viscosities are preferred, for example, polyelectrolytes that have groups that can dissociate such as carboxylic acid groups. Because these dissociated groups are more polarized, in aqueous solutions, the polymers are hydrated and dissolve in water. Moreover, the dissociated groups act as fixed points of charge on the polymer chains, and the repulsive interactions between these charged groups cause expansion of the polymer chains. This effect causes high viscosity, thickening the emulsion. Cross-linked polyacrylic acid, one of the most popular rheology control agents, contains carboxyl groups. The solubility of the polyacrylic acids is low in acidic media and the turbidity is high; that is, their aqueous solution never become completely transparent. In contrast, in alkaline solutions, the polymers dissolve well, and the resultant aqueous solutions are transparent. The appearance of products, such as lotions, that contain almost no oil is important. Therefore, if a polyacrylic acid is used in such a product, the alkalinity of the formulation must be carefully controlled.

When the active ingredient is a salt, increasing the concentration of the salt increases the ionic strength. The increasing counterion concentration shields electrostatic repulsion around the dissociating groups, even at a suitable pH, and the viscosity reduces. Consequently, the polymer concentration must be increased to obtain the desired viscosity. However, as mentioned previously, increasing the polymer concentration can lead to unfavorable effects, such as an unpleasant "after feeling."

11.2.3.2 Realization of Drastic Shear Thinning

Shear thinning is a common non-Newtonian flow behavior that occurs in many materials, including polymer solutions, emulsions, liquid crystals, and suspensions [5]. As mentioned in Sect. 11.2.1, it is desirable that shear thinning is drastic;

that is, the difference between the viscosity at the low shear rate and high shear rate is large. The ultimate goal of increasing the viscosity at low shear rates is to produce a finite yield stress. For example, percolated microgels may exhibit yield stress and shear-thinning, non-Newtonian flow properties if there are sufficient soft particles [9, 41]. Carbopols, a class of polyacrylic acid, are rheology control agents that enable this drastic shear thinning, but they have poor acid resistance. However, microgels polymerized with a monomer containing a strong acid, such as sulfonic acid, may overcome this disadvantage.

Polymers applied in sufficient concentrations exhibit shear-thinning, non-Newtonian flow due to entanglement. However, polymers in cosmetic emulsions may remain on the skin after application and subsequent evaporation of moisture. Remaining polymer condenses and becomes more firmly entangled. Such firmly entangled polymer chains can cause a "sticky feeling" on the skin. In the development of cosmetic formulations, the "after feeling" must be considered. The degree of the polymer chain entanglements depends on the length of the polymer chains; therefore, the use of relatively short polymer chains may avoid such a problem. However, short polymers do not cause sufficient viscosity thickening. To avoid this contradictory situation, temporary 3D network structures can be used. Hydrogels consisting of temporary 3D network structures, made by physical interactions such as hydrophobic interactions, are promising materials. Because the network structures are fragile, the hydrogels flow easily under shear deformation, yet they have "body" when not stressed.

Two practical examples of these novel rheology control agents developed using concepts mentioned are described in the following sections. One example is a hydrogel that consists of telechelic polymers. In this hydrogel, temporary 3D network structures are formed in the hydrogel. The second example is a water-swelling microgel, which is polymerized in a confined space leading a spherical particle shape.

11.3 C24HEUR as a Rheology Control Agent for Cosmetics

11.3.1 Telechelic Polymer

Telechelic polymers are a class of associating polymers. The polymers are linear chains containing reactive end groups. When the polymers dissolve in water, the hydrophobes form self-assembled structures. Because the assemblies act as a temporary cross-linker, the solution forms hydrogels. A conceptual illustration of the 3D-network structure for the temporary hydrogels is shown in Fig. 11.1.

As shown in Fig. 11.1, hydrophobic alkanes form self-assemblies called "flower micelles." The flower micelles are connected together by "elastic active chains" and some polymers exist as "dangling chains." The flower micelles work as cross-link points in hydrogels comprised of telechelic polymers. Because the molecular

Fig. 11.1 Conceptual illustration of the self-assemblies of associating telechelic polymers

assemblies form via hydrophobic interactions, the binding energy is lower than that of covalent bonds. Therefore, cross-links are easily broken under deformation or stress. Therefore, aqueous solutions of telechelic polymers are shear thickening at low shear rates and shear thinning at higher shear rates [28, 29].

Hydrophobically modified ethoxylated urethanes (HEURs) are typical examples of these raw materials that are currently used. The first industrial application of HEURs was as rheology control agents in water-based paints [11], which contain polymer latex. Paint formulations containing HEURs cause fewer spatters when rolled onto surfaces; in addition, they show good leveling characteristics. Consequently, HEURs are promising rheology control agents for formulations that require shear-thinning, non-Newtonian flow on yield stress, for example, cosmetics. Because the required flow performance is dependent on the product, the flow properties of the material must be adjusted by changing its chemical structure. The flow properties of HEURs strongly depend on the size of the water-soluble chains and hydrophobic end chains [3, 8, 42]. Because it is easy to chemically modify HEURs, they can be easily adapted for different applications.

11.3.2 The Fundamental Characteristics of C24HEUR

Yoshida et al. [42] developed a HEUR for use in cosmetic formulations. The end-cap groups contain C24 alkanes; therefore, the polymer is named C24HEUR. The chemical structure of C24HEUR is shown in Fig. 11.2.

The fundamental characteristics of C24HEUR have been investigated. The most important characteristic is the critical concentration for formation of "flower micelles." The dependence of polymer concentration on hydrodynamic size was investigated by dynamic light scattering (DLS). The CONTIN routine was used to analyze the time correlation function. The hydrodynamic radius was calculated

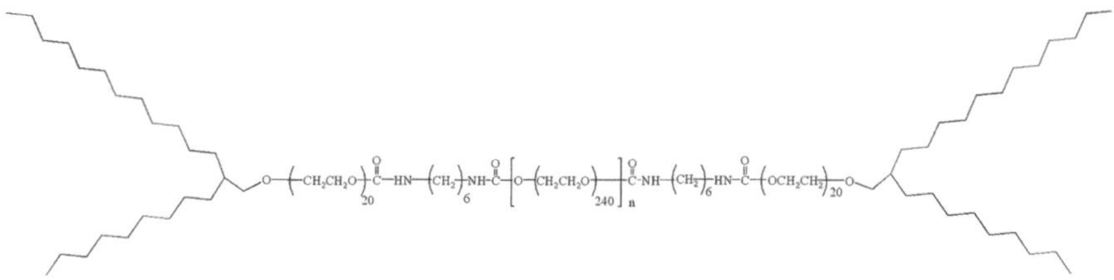

Fig. 11.2 The chemical structure of C24HEUR

using the Stokes-Einstein equation, Eq. 11.1:

$$R_h = \frac{k_B T}{6\pi \eta D} \tag{11.1}$$

k_B is the Boltzmann constant, T the absolute temperature, η the viscosity of the solvent (water), and D the mutual diffusion coefficients determined by DLS. Figure 11.3 shows the size distribution of the associated structure at various concentrations.

A monomodal peak appeared at 0.1 %, indicating that micelles comprised of C24HEUR are present at that concentration. The R_{h} at this concentration is approximately 10^2 nm. Although this value is larger than the expected size of micelles comprising C24 alkenes, the hydrodynamic size of the micelles is larger due to the presence of "loop chains." When the polymer concentration increases, R_{h} also increases. In particular, large associated structures appear at concentrations greater than 0.7 %, as shown schematically in Fig. 11.1. This indicates that a connected porous structure forms above this critical concentration.

The most distinguishing rheological properties of HEURs in aqueous solution are their linear viscoelastic properties [2]. It is well known that the dynamic modulus is well described by a single-mode Maxwell model. Figure 11.4 shows the frequency dependence of the dynamic modulus for a 1 wt% C24HEUR aqueous solution.

The curves are the calculated results using the Maxwell model (Eqs. 11.2 and 11.3).

$$G' = G_0 \frac{\omega^2 \tau^2}{1 + \omega^2 \tau^2} \tag{11.2}$$

$$G'' = G_0 \frac{\omega \tau}{1 + \omega^2 \tau^2} \tag{11.3}$$

G' and G'' are storage and loss moduli, respectively. G_0, ω, and τ are equilibrium modulus, angular velocity, and relaxation time, respectively. A HEUR solution has a single relaxation time, i.e., it exhibits elastic (rubber-like) properties under rapid deformation, yet it flows under slow deformation. The relaxation time of a HEUR solution depends on the length of the middle chain and the size of its hydrophobic

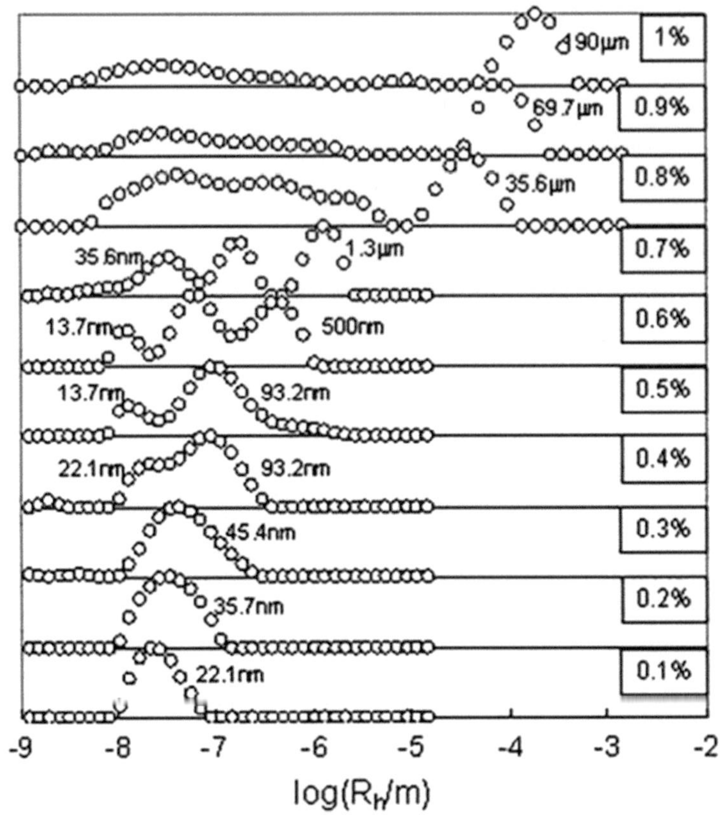

Fig. 11.3 Size distributions, obtained from DLS, for various concentrations of C24HEUR. The weight percentage of polymer and the mean hydrodynamic radius are indicated

residues. Tanaka and Edwards [35] devised the transient network theory to explain this behavior. The relaxation time of a HEUR hydrogel is strongly related to the lifetime of flower micelles formed from HEUR molecules [35, 36].

The steady-state flow behavior of an aqueous solution of HEURs is also unique. The steady-state flow curve for a 1.5 wt% C24HEUR aqueous solution is shown in Fig. 11.5

As mentioned above, Fig. 11.5 shows the initial shear thickening at low shear rates, followed by shear thinning at higher shear rates (the dotted line in the figure is an eye guide). In particular, the viscosity drops abruptly after peak shear thickening. According to modified transient network theory, this phenomenon is due to the nonlinear stretching effect of the effective chains in the transient network. Moreover, in this situation, it is expected that shear banding occurs [36].

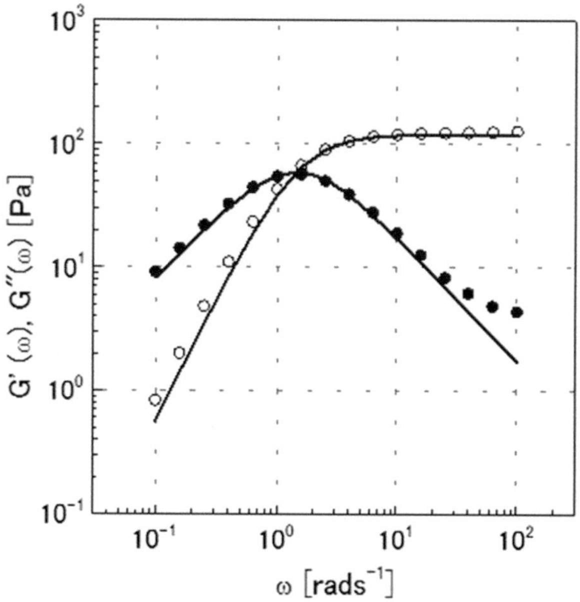

Fig. 11.4 The dynamic modulus against frequency for a C24HEUR 1 wt% aqueous solution. *Open circles* and *closed circles* denote storage and loss modulus, respectively. The *curves* denote the prediction of a Maxwell model (Eqs. 11.2 and 11.3)

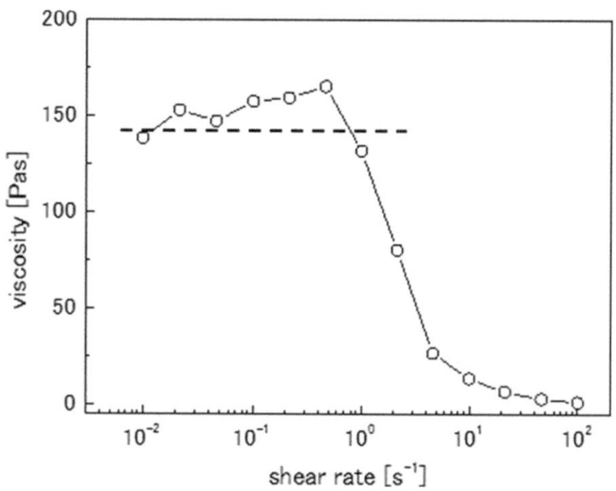

Fig. 11.5 Flow curve for C24HEUR 1.5 wt% aqueous solution (The *dotted line* is an eye guide as the zero-shear viscosity)

11.3.3 Nonlinear Flow Properties for C24HEUR Under Large Deformation

Recently, studies of the behavior of HEUR under large deformations have been reported [4, 12]. Because the "first feeling" for cosmetics is related to start-up flow, we focused in these nonlinear rheological properties for aqueous HEUR solutions. When these viscoelastic fluids are used in cosmetics, their rheological behavior under large deformations must be considered because, under typical usage, cosmetics are subject to large deformation. As an example of nonlinear deformation, stress growth behavior under the step shear flow was examined.

In general, the stress under a certain strain is defined by the balance of the tensile strength between the end of the active chain and the lifetime and dissociation rate of the transient cross-link point. If the tension and the dissociation rate for the system can be estimated, the mechanical properties can also be estimated. In particular, the behavior in the nonlinear region is quite interesting [25, 26].

The shear stress growth behavior can be modeled by Eq. 11.4:

$$\eta^{+}(t) = \eta^{(1)}t + \eta^{(2)}\frac{t^2}{2!} + \eta^{(3)}\frac{t^3}{3!} + \cdots \tag{11.4}$$

Here $\eta^{(1)}$, $\eta(2)$, and η (3) can be explained as follows:

$$\eta^{(1)} = G'(\omega_{\to\infty}) \tag{11.5}$$

$$\eta^{(2)} = -\lim \omega G''(\omega) \tag{11.6}$$

$$\eta^{(3)} = -\left(A + B\dot{\gamma}^2\right) \tag{11.7}$$

The third coefficient is dependent on the shear rate. In particular, coefficient B includes a factor showing the effect of nonlinear stretching of the elastic active chains. If the factor is equal to zero, the chain reduces to a Gaussian series and B becomes zero. In other words, any unusual behavior during the time scale is contained in the third term of Eq. 11.4. The dynamics of the network contains the effect of nonlinear stretching, and this nonlinear stretching effect can be observed as an unusual increase in the viscosity growth curve under the initial shear flow.

Here, we will introduce a typical example of this unusual increase in stress. The viscosity growth curves for a 1.5 wt% C24HEUR aqueous solution at several shear rates are shown in Fig. 11.6. Because, in a transient network, stress under a particular level of strain is determined by the balance between the tension between the ends of the active chains and the lifetime dissociation rate of the transient cross-link points [25, 26], the viscosity growth curve was checked at three constant shear rates. For easier comparison of samples, we have introduced the concept of normalized shear

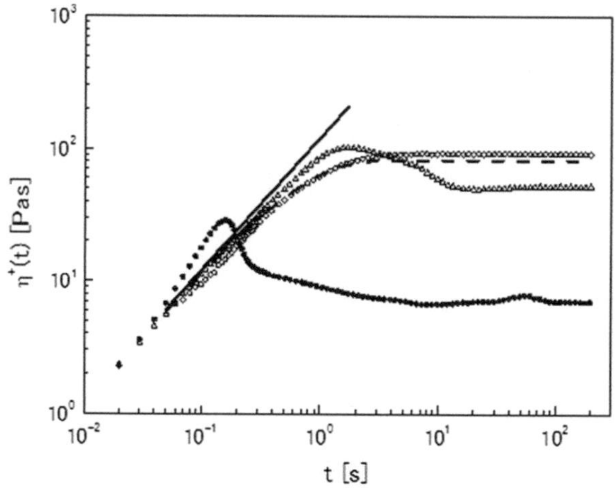

Fig. 11.6 Viscosity growth *curves* for a C24HEUR 1.5 % aqueous solution. *Open circles*, *squares*, and *closed circles* denote the results at $W_i = 0.1$, 1.0, and 10, respectively. The *solid line* and *dotted line* denote the elastic response at very short time periods (Eq. 11.8) and expected viscosity growth curve estimated by Eq. 11.9, respectively

rare, $\dot{\gamma}\tau$. The open circles, open squares, and closed circles in Fig. 11.6 denote the viscosity growth curves for $\dot{\gamma}\tau = 0.1$, 1.0, and 10, respectively. The solid lines denote the linear elastic response calculated using the following equation:

$$\eta = G_0 t \tag{11.8}$$

The dotted lines denote the viscoelastic response predicted by the Maxwell model, as described in Eq. 11.9:

$$\eta^+(t) = G_0\tau\left[1 - \exp\left(-\frac{t}{\tau}\right)\right] \tag{11.9}$$

An upturn in stress is clearly evident for the curve under highest shear rate flow.

As shown, a stress upturn, which is the overshoot from the elastic response according to Eq. 11.8 (the straight line in Fig. 11.6), appears at high shear rate flow ($W_i = 10$). It is interesting to consider the influence of this unusual increase in stress on the "primary feeling" because consumers may sense this unusual response when they start to spread the cosmetic product on their skin. This may be felt as an initial resistance followed by decreasing resistance. Therefore, C24HEUR is a promising rheology control agent not only because of its acid resistance but also because of its preferable usage feeling.

Concerning cosmetic applications, the effect of the various additives on the rheological properties of products must be examined because, generally, cosmetics contain a variety of ingredients. In particular, alcohols and polyols are important ingredients, used as preservatives and emollients. As mentioned above, the mechanical properties of the transient networks in HEUR hydrogels are strongly related to the physicochemical properties of the formed micelles, that is, the cross-links formed in the hydrogels. Therefore, the effects of these additives on the rheological properties of HEUR aqueous solutions must be investigated, in particular, their nonlinear flow behavior. Therefore, we will discuss studies on the effects of additive glycerol [22] and methanol [23].

11.3.4 Effect of Glycerol on the Rheological Properties of C24HEUR

Figure 11.7 shows the dependence of the mechanical characteristics on C24HEUR hydrogels on the glycerol concentration. Upon addition of glycerol, the relaxation time shortened. In contrast, G_0 slightly increased with increasing glycerol concentration. This result suggests that glycerol affects the structure of the polymer network at high concentrations. The effects of glycerol on the micellar structure of a nonionic surfactant [32] and PEO-PPO-PEO block copolymers [1] have been reported. In general, upon addition of glycerol, PEO chains shrink as a result of dehydration. From linear viscoelastic measurements, glycerol seems to change the cross-link points of the transient network. As shown in Fig. 11.7, shorter relaxation times reduce the duration that the hydrophobic end chains of C24HEUR remain at the micellar junctions.

The effect of glycerol on the viscosity growth behavior of aqueous solutions of C24HEUR at 3 wt% containing 30 wt% glycerol is shown in Fig. 11.8. Because the

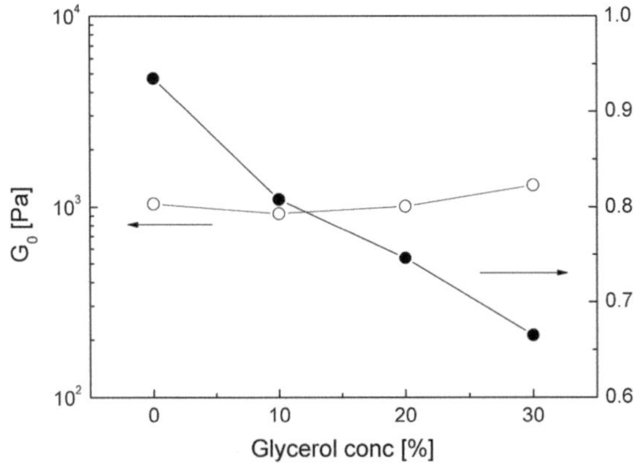

Fig. 11.7 Variation of estimated equilibrium shear modulus (G_0) and relaxation time (τ) vs. glycerol concentration. *Open* and *closed symbols* denote G_0 and τ, respectively (From Ref. [22], with permission from Springer)

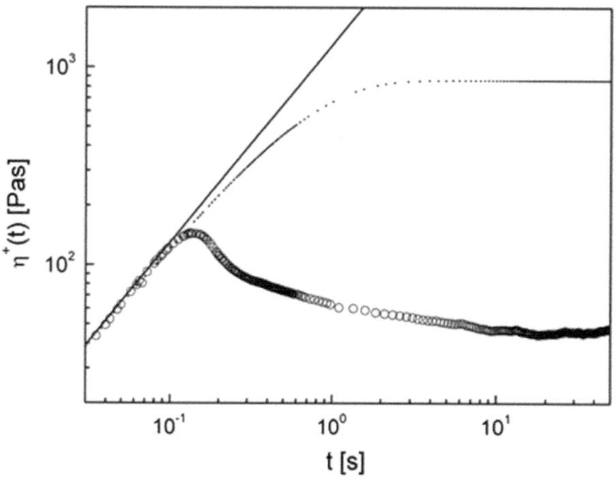

Fig. 11.8 The viscosity growth *curves* at various shear rates for a C24HEUR 3 wt% aqueous solution containing 30 wt% glycerol at 10 s^{-1}. The *lines* and *dotted lines* denote the elastic response at very short time periods (Eq. 11.8) and expected viscosity growth curve estimated by Eq. 11.9, respectively

relaxation times for these samples are approximately 0.7 s, values of W_i are greater than one.

The unusual increase in stress shown in Fig. 11.6 is not present in the sample containing 30 wt% glycerol. This indicates that glycerol affects the dynamics of the network, particularly the mechanical response of the active chains in the networks. Furthermore, the disappearance of the unusual upturn in stress means that the sample containing 30 % glycerol does not involve nonlinear stretching. As mentioned above, glycerol reduced the relaxation time. Consequently, we believe that glycerol affected the structure of the cross-links of the transient network in this system. Moreover, glycerol affected the manner of nonlinear deformation. A possible hypothesis is described as follows: because glycerol is a water-soluble solvent containing three hydroxyl groups, it may affect the hydration of the hydrophilic chains of C24HEUR, that is, the PEO chains. In the presence of a small amount of water-miscible solvent, a cononsolvency effect occurs [17, 19, 37, 39, 40]. Consequently, polymers shrink as a result of dehydration. If the active chains, connected at both ends to the cross-link point, shrink, the active chains between the cross-link points collapse. When shear flow is applied to the system, the collapsed chains reel out. Because the polymer chains are reeled out, they cannot be further stretched. Therefore, samples containing high concentrations of glycerol do not show the unusual increase in stress.

11.3.5 Effect of Methanol on the Rheological Properties of C24HEUR

Methanol also affects the rheological properties of aqueous solutions of C24HEUR [23], and Fig. 11.9 shows the dependence of the linear viscoelastic characteristics of these solutions on the methanol concentration.

In this case, the polymer concentration was fixed at 1.5 wt%. These results show that addition of methanol both reduces the relaxation time and increases G_0, with G_0 increasing slightly with methanol concentration. The relaxation time behavior is similar to that induced by the addition of glycerol, although the effect of G_0 is different. According to the classical theory for gels, the modulus of a gel is given simply by the following equation, Eq. 11.10:

$$G = \nu k_B T \tag{11.10}$$

ν is the number of effective chains in the gel network, k_B is the Boltzmann constant, and T is temperature. Because our polymer concentration was constant (1.5 wt.%), the G_0 behavior suggests that the number of effective chains in the gel network increases when methanol is added. This may be because methanol reduces the aggregation number of the flower micelles.

Because the steady-state flow behavior changes with micelle properties, the addition of methanol is expected to affect the steady-state flow curve. We determined the steady-state viscosity at various shear rates in the range 0.05–500 s^{-1} for samples containing methanol (0–10 mol%). We analyzed our data in terms of normalized viscosities, η/η_{init}, and normalized shear rates, $\dot{\gamma}\tau$, where η_{init} is the viscosity measured at 0.05 s^{-1} and τ is obtained from measurements of dynamic

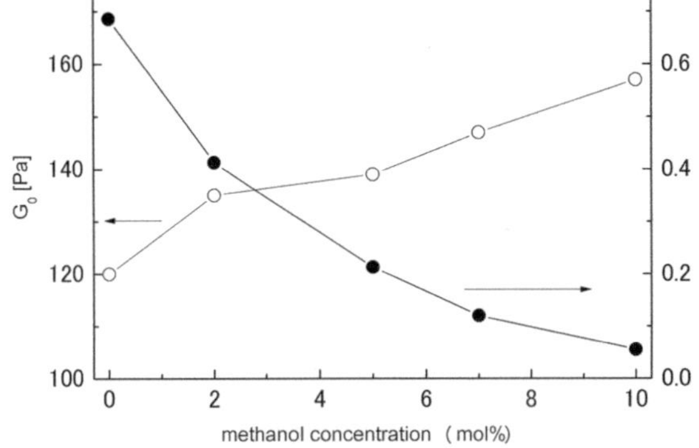

Fig. 11.9 Variation of estimated equilibrium shear modulus (G_0) and relaxation time (τ) vs. methanol concentration. *Open* and *closed symbols* denote G_0 and τ, respectively (From Ref. [23], with permission from Springer)

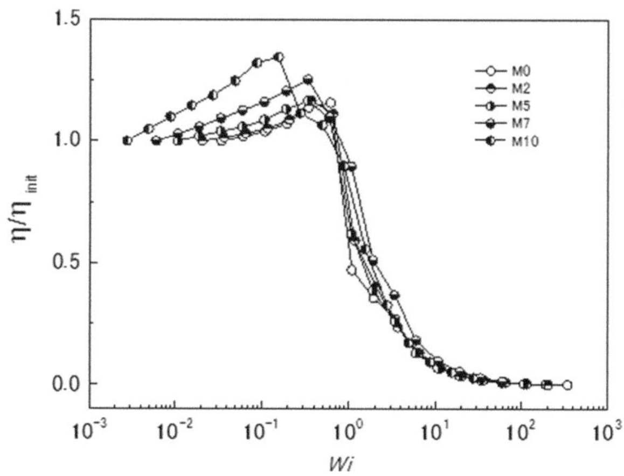

Fig. 11.10 Flow curves for C24HEUR 1.5 wt% solutions containing methanol in concentrations ranging from 0 to 10 mol% (M0, M2, M5, M7, and M10). Normalized viscosity and shear rate are employed in the coordinate system. Viscosity is normalized as η_{init}, which is the apparent viscosity at 0.05 s^{-1}. Shear rate is shown as Weissenberg number ($W_i = \tau\dot{\gamma}$). The values of τ for these samples were estimated by a Maxwell model (Eqs. 11.2 and 11.3) (From Ref. [23], with permission from Springer)

moduli. Such a normalized coordinate system enables the comparison of the shear-thickening behavior at different methanol concentrations.

Figure 11.10 shows flow curves (normalized viscosity, η/η_{init}, vs. normalized shear rate, $\dot{\gamma}\tau$) for the measured samples. The labels, M0, M2, M5, M7, and M10, denote the methanol concentration (mol%) in each sample. The normalized viscosity increases gradually at first; then, at a normalized shear rate < 1, it decreases abruptly. The intensity of shear thickening (η/η_{init}) depends on methanol concentration and becomes quite significant. Moreover, the viscosity peak shifts to s shorter time (lower shear rate) with increasing methanol concentration. Although shear-thickening behavior is affected by methanol concentration, the shear-thinning behavior of the samples at flow rates above the critical shear rate is similar, and the flow curves of the samples essentially overlap beyond $\dot{\gamma}\tau = 1$. This phenomenon is expected, and we speculate that it is caused by shear-banding transitions.

The unusual stress increase that occurs at the step shear flow shown in Fig. 11.6 disappears on addition of both methanol and glycerol. Figure 11.11 shows the viscosity growth curve for C14HEUR 1.5 wt.% solution containing 10 mol% of methanol.

The origin of this phenomenon is as the same that for glycerol. As shown in Fig. 11.8, the effect of glycerol is similar to that of methanol concerning relaxation time, in particular, concerning the lifetime of the temporary cross-link points. Therefore, methanol also affects the properties of the cross-link point and the elastic active chains.

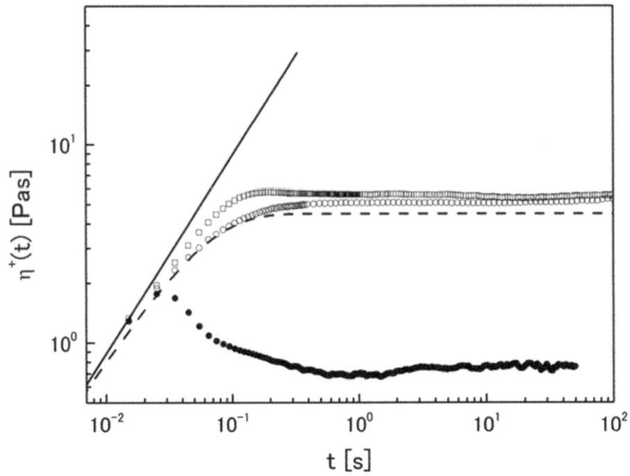

Fig. 11.11 Viscosity growth curves for a C24HEUR 1.5 % aqueous solution containing 10 mol% methanol. *Open circles*, *squares*, and *closed circles* denote the results at $W_i = 0.1$, 1.0, and 10, respectively. The *solid line* and *dotted line* denote the elastic response at very short time periods (Eq. 11.8) and expected viscosity growth curve estimated by Eq. 11.9, respectively (From Ref. [23], with permission from Springer)

11.4 Water-Swellable Microgel

Microgels are promising candidates for cosmetic rheology control agents that do not cause undesirable usage feelings such as the "sticky after feeling" or "spinability" (see Sect. 11.2). Carbopol is one of the most famous microgel-type rheology control agents for cosmetics [10, 38]. However, this class of cross-linked polyacrylates has several issues that must be overcome. The most serious issue is their pH dependence because Carbopol contains carboxyl groups in its main chains, and it forms transparent gels only under alkali conditions, thus reducing the formulation options for cosmetics. One solution for this issue is the application of neutral polymer microgels. For example, agar microgels have been used in cosmetics [16]. However, agar microgels have unsatisfactory usage feelings due to their poor swelling and brittle mechanical properties.

Polyacrylamides are considered to be suitable polymers for cosmetic rheology control agents due to their usage feeling. Acrylamido-2-methyl-1-propanesulfonic acid (AMPS) is an interesting monomer. Because the monomer is sulfonic acid, a polymer copolymerized with AMPS can swell over a wide pH range. In fact, several kinds of copolymers that incorporate AMPS have been developed as cosmetic rheology control agents [13–15].

However, a problem exists, that is, how to polymerize microgels incorporating AMPS? Polymer microgels are commonly prepared by either emulsion polymerization or precipitation polymerization. However, because both AMPS and the resultant polymer are soluble in water, it is difficult to employ precipitation polymerization.

Moreover, AMPS is practically insoluble in organic solvents; consequently, emulsion polymerization cannot be used.

To make water-swellable polyAMPS microgels, we have developed a novel inverse microemulsion polymerization (IMEP) system [18]. It is well known that a ternary system using a nonionic surfactant exhibits a "phase inversion phenomenon." The temperature at which the phase inversion occurs is called the hydrophilic-lipophilic balance (HLB) temperature. A transparent and low viscosity phase appears in the vicinity of this temperature. The phase is an O/W microemulsion (Wm) or W/O microemulsion (Om). Because Om consists of reverse micelles, swollen by water, this phase can be used as a polymerization "container" for polymers containing AMPS. Moreover, since the Om phase is determined by both the composition of the system and the temperature, it is easy to control the size of the "container." Therefore, the size of the swellable microgels can be controlled to be very small, almost the same size as that of the microemulsion particles.

11.4.1 Inverse Microemulsion Polymerization System

To characterize the pseudo-ternary system consisting of a monomer aqueous solution, nonionic surfactant, and n-hexane, a phase diagram study was performed. The aqueous phase, monomer solution consisted of dimethylacrylamide (DMAA) and AMPS. The monomer concentration of the aqueous phase was 20 wt%, and the molar mixing ratio of DMAA and AMPS was 80:20. The weight ratio of the aqueous phase and n-hexane was 50:50. The surfactant was added at 5 % of the total weight of the system. C18E_n-type nonionic surfactants were used in the study. Here, n indicates the mean stoichiometry of ethylene oxide, which ranged from 4 to 6 M. The n value was adjusted through variation of the mixing weight ratio of C18E3 and C18E6, which are industrial grade surfactants. The phase inversion temperatures of the systems were determined by conductivity measurements. The phase diagram is shown in Fig. 11.12.

The curve in Fig. 11.12 indicates the phase inversion temperature. The system was observed to be transparent and had low viscosity at around this temperature. This implies that the reversed micelle solution phase exists at this temperature. From this result, we can construct an inverse microemulsion polymerization system. If the temperature for the polymerization is known, the components of the pseudo-ternary system can be adjusted to yield the desired HLB temperature. For example, if we want to perform a thermal radical polymerization at 65 °C, the necessary surfactant is C18E5.

We polymerized water-swellable microgels consisting of DMAA and AMPS with additional methylenebisacrylamide (MBA) as a cross-linker. In this section, these microgels are labeled as D-A microgels.

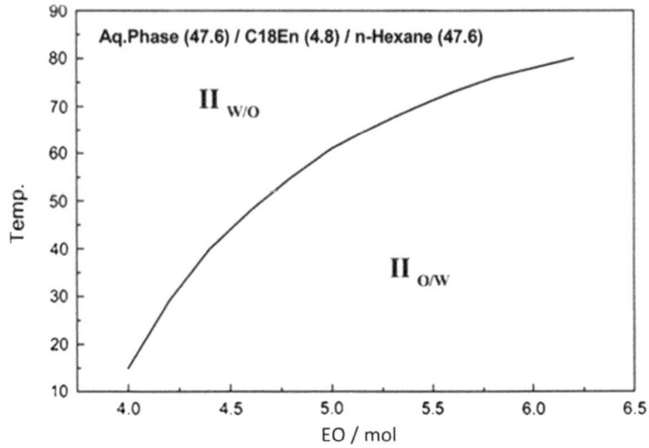

Fig. 11.12 The phase diagram of the pseudo-ternary inverse microemulsion polymerization system. The composition ratio of the system is shown in the panel. The *curve* shows the phase conversion temperature (From Ref. [18], with permission from Elsevier)

11.4.2 Spherical Shape of the Water-Swellable Microgels Polymerized in a Confined Space

To produce a favorable "primary feeling," skin-care cosmetic products must have "body" under low deformation speeds, yet they should flow easily under higher deformation speeds, as mentioned in Sect. 11.2. If sufficient microgel is added to the aqueous phase of an emulsion, percolation through the microgels gives "body" to the emulsion; moreover, the resultant product gains shear-thinning, non-Newtonian flow properties if the microgel particles are sufficiently soft. As an initial step in the study of the rheological properties of water-swellable microgels, we investigated the properties of dilute microgel solutions.

It is important to know the critical volume fraction at which the microgels form a percolating network in aqueous solution. Because the microgels swell greatly in aqueous systems, it is difficult to estimate their volume fraction from the weight ratio (concentration). However, the intrinsic viscosity can be used. The intrinsic viscosity is a characteristic concerning the spatial size of colloidal particles and is defined by Eq. 11.11:

$$[\eta] = l\,im_{c \to 0} \frac{(\eta - \eta_s)}{c\eta_s} \tag{11.11}$$

η is the viscosity of the microgel dispersion, ηs is the viscosity of the solvent, and c is the microgel concentration. We measured the intrinsic viscosities in 3 mM sodium chloride aqueous solution for a series of cross-linked DMAA-AMPS copolymers, which were polymerized in aqueous solution and IMEP. The intrinsic viscosity of the microgels polymerized in solution was ca. 25 dL/g. In contrast, for microgels polymerized in IMEP, it was ca. 7 dL/g, indicating that the cross-linked polymers polymerized in IMEP, that is, a confined space, were more compact than those polymerized in solution. A conceptual scheme is shown in Fig. 11.13.

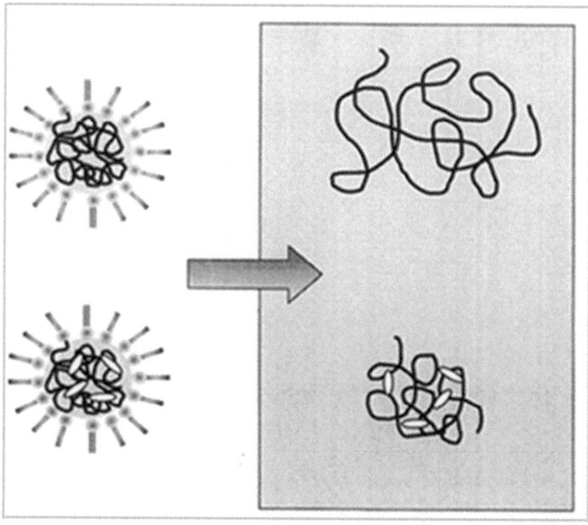

Fig. 11.13 Conceptual illustration for polymerization in a confined space (From Ref. [18], with permission from Elsevier)

11.4.3 Flow Behavior of the Microgel Dispersions

To understand the overlap concentration (c*) for the microgels, the relationship between the relative viscosity and the microgel concentration was studied. The microgel concentration was normalized by its $[\eta]$ value. Generally, $[\eta]$ is related to c* and the volume fraction occupied by colloidal particles, (ϕ), by assuming that the colloidal particles behaving as impenetrable spheres according to Eq. 11.12:

$$\frac{\eta}{\eta_s} = 1 + 2.5\phi \tag{11.12}$$

From Eqs. 11.11 and 11.12, the relationship between $[\eta]$ and ϕ is given by the following equation:

$$[\eta] = \frac{2.5\phi}{c} \tag{11.13}$$

Because it is expected that spherical colloid particles begin to overlap when ϕ is around 0.4, c* can be described as follows:

$$c^* \approx \frac{1}{[\eta]} \tag{11.14}$$

The inflection point in the viscosity versus the concentration plot is considered to be c*.

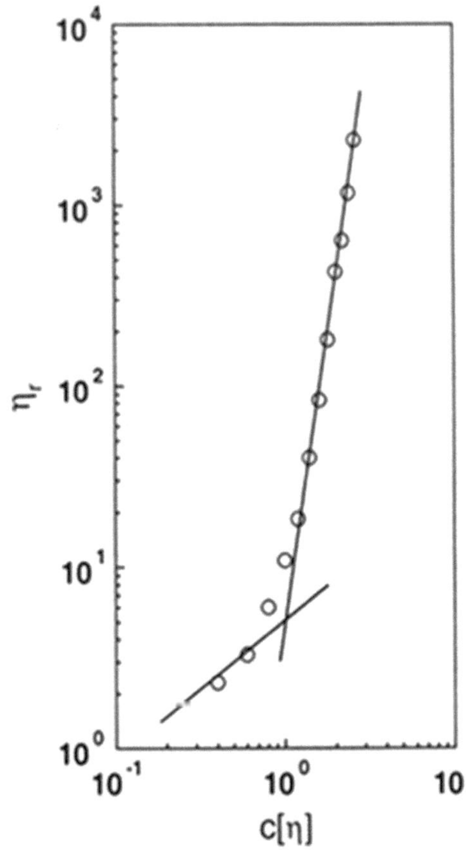

Fig. 11.14 Concentration dependency of the relative viscosities for D-A microgel dispersions at various concentrations ranging from $c[\eta]=0.4$ to 2.6 (From Ref. [20], with permission from Elsevier)

Figure 11.14 shows the relationship between the relative viscosities and the normalized microgel concentration, ($c[\eta]$). It is clear that the inflection point is at around $c[\eta]=1$. This indicates that the microgels can be considered as hydrodynamically impenetrable spheres and that, as such, the impenetrable spheres do not have dangling chains.

11.4.4 Shear-Thinning, Non-Newtonian Flow of the Microgel Dispersions

The yield stress is one of the important rheological properties for "primary feeling" of skin-care cosmetics (see Sect. 11.2.1). We investigated the yielding properties from the steady-state flow curve. A typical flow curve for an aqueous dispersion of a microgel is shown in Fig. 11.15. The flow curves show that the aqueous dispersions of the microgel exhibit pseudoplastic flow, that is, they have finite

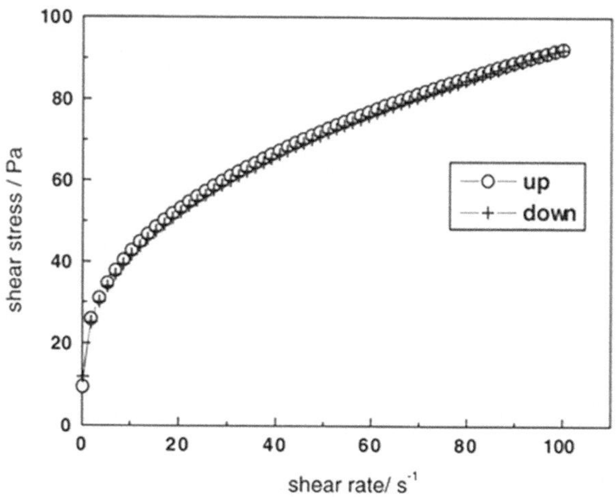

Fig. 11.15 Hysteresis of the flow *curve* for the D-A microgel dispersion. The concentration of the sample was c[η] = 2.0. *Circles* and *crosses* denote upward and downward procedures, respectively (From Ref. [20], with permission from Elsevier)

yield stresses and exhibit shear-thinning flow. The circles and crosses denote the results for the upward (increasing shear stress) and downward (decreasing shear stress) procedures, respectively. Although most condensed colloidal dispersions show hysteresis, in this case, there is none. The lack of hysteresis may be due to the softness of the microgel particles.

To estimate the yield stress from the flow curve, we used the Herschel-Bulkley equation, an empirical power-law equation.

$$\sigma = \sigma_y + k(\dot{\gamma})^n \tag{11.15}$$

σ and γ are the shear stress and shear rate, respectively. The three parameters are characteristic of non-Newtonian flow, and k is a consistency parameter, σy is yield stress, and n is the H-B index. This empirical power-law model has been used in various studies [16, 18, 34].

This empirical equation is useful for analysis of this type of flow curve. The relationship between the yield stress and $c[\eta]$ is shown in Fig. 11.16.

The yield stress increases above $c[\eta] = 1.8$. The volume fraction occupied by the microgels for this concentration is 0.72, derived from Eq. 11.13. If the particles are hard spheres, the dispersion has infinite viscosity at around $\phi = 0.5$, the random close packing limit. A value of 0.72 means that the microgels packed each other with large deformation. Because the microgel was only slightly cross-linked (0.5 mol% of MBA was copolymerized), the microgels may become progressively deformed as the microgel concentration increases. Also, this is likely to be the apparent yield stress due to the friction between microgel particles.

The H-B index is also interesting concerning the flow properties, in particular, the shear-shinning behavior. Shear-thinning behavior is associated with $0 < n < 1$ and the

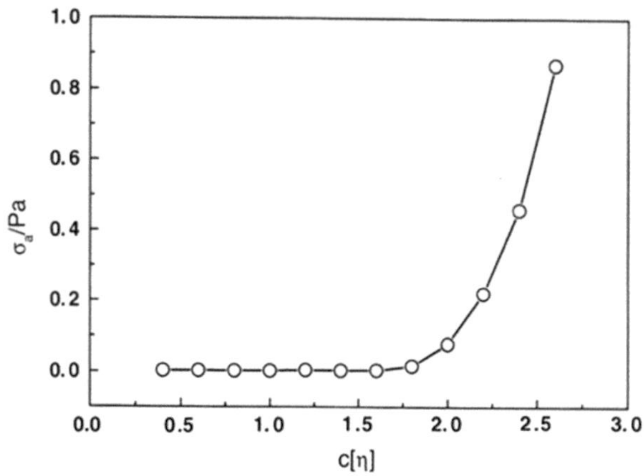

Fig. 11.16 Concentration dependence of the apparent yield stress evaluated from Eq. 11.15 for the A-D microgel dispersions (From Ref. [20], with permission from Elsevier)

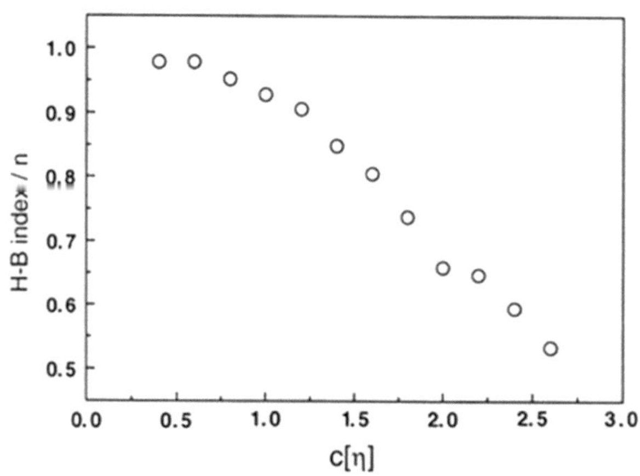

Fig. 11.17 Concentration dependence of the H-B index evaluated by Eq. 11.15 for the A-D microgel dispersions (From Ref. [20], with permission from Elsevier)

unusual shear-thickening behavior at $n > 1$. Figure 11.17 shows the concentration dependence of n for the aqueous microgel dispersions. It shows that n for the lower concentration samples is nearly 1, indicating the samples almost behave as Newtonian fluids. However, at higher concentrations, the value n becomes smaller, eventually becoming close to 0.5, according to the Casson's theory for spherical particle dispersions. Interestingly, the relationship between the food texture and H-B index shows that foods having a "fresh" texture, such as tomato juice, have $n = 0.5$. Therefore, a cosmetic product that has $n = 0.5$ can be expected to have a "fresh" primary feeling [16].

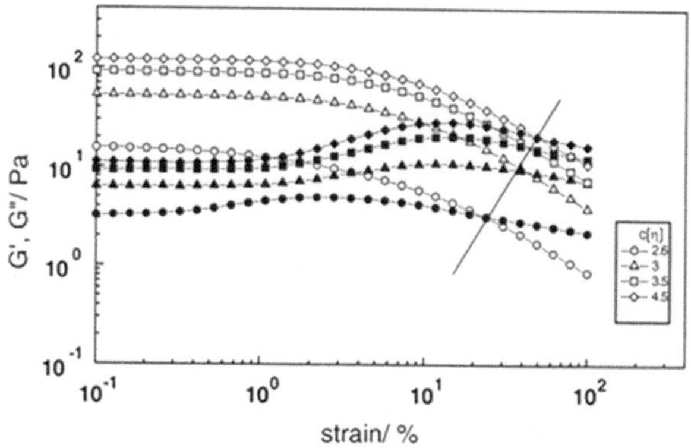

Fig. 11.18 Strain dependence of the shear modulus for the A-D microgel dispersions. *Open* and *closed symbols* denote storage and loss modulus, respectively. All measurements were performed at 1 Hz (From Ref. [20], with permission from Elsevier)

Shear-thinning non-Newtonian flow properties for water-swellable microgels consisting of DMAA-AMPS, described above, strongly contribute to the "primary feeling" of skin-care cosmetic products. In particular, because the microgels are polymerized in a confined space, the swollen shape is expected to be spherical, as shown by several measurements. Furthermore, these microgels have a distinct "primary feeling" that stands out from the other rheological control agents.

11.4.5 Condensed Microgel Aqueous Dispersions

The yielding behavior of condensed microgel dispersions can also be studied by dynamic viscoelasticity measurements [20]. The dynamic modulus for D-A microgel aqueous dispersions during strain-sweep measurements is shown in Fig. 11.18. The storage modulus (G') is always larger than loss modulus (G'') in the linear viscoelastic region. Although the microgels in the dispersion are essentially discrete, the condensed dispersion acts as a continuous network structure. The condensed microgel dispersion, which appears as a "bulk gel," collapses under large shear strains. The crossover point of G' and G'' is one of the criteria for this collapse. The solid line denotes the crossover point for the dispersion at various concentrations ranging from $c[\eta] = 2.5$ to 4.5. The shear strain at collapse points is around 10 %. The values are somewhat larger than a condensed emulsion or dispersion that consists of hard spheres, and this may be due the "softness" of the microgels.

It is shown that D-A microgels are closed packing and they behave like "bulk gels" in highly concentrated dispersions. We attempted to observe this close packing using electron microscopy [21]. The freeze-fracture transmittance electron microscopy (FF-TEM) method is good for such wet samples. Figure 11.19 shows

Fig. 11.19 Freeze-fracture TEM image of the A-D microgel dispersion (c[η] = 3.0)

an FF-TEM image of the microgel aqueous dispersion ($c[\eta] = 14$). It is clear that the microgels are packed with deformation.

The condensed state of the D-A microgels after drying was also observed by atomic force microscopy (AFM). A small amount of the aqueous A-D microgel dispersion was spread on a mica plate and dried for 18 h at room temperature. The dried sample was observed by AFM, as shown in Fig. 11.20. Dome-shaped projections are visible in the image, indicating that the shapes of the microgel particles in aqueous dispersion, shown in Fig. 11.19, are retained after drying. Therefore, when a cosmetic product containing the microgels is applied to the skin, the surface of the skin may show this uneven shape after evaporation of moisture. It is well known the frictional resistance of uneven surfaces is smaller than that of flat and smooth surfaces. Therefore, the surfaces observed in Fig. 11.20 may influence the "after feeling" of cosmetic products comprising A-D microgels. Moreover, the observed uneven surface means that the microgels are swollen even after drying for 18 h. Consequently, we expect that the swollen microgels would function well as a moisturizing agent because the swollen microgels may supply necessary moisture to the stratum corneum.

11.5 Summary

Because the 3D structures formed in aqueous solutions of C24HEUR are temporary, they dissociate easily under deformation; however, hydrogels consisting of C24HEUR have "body" in their resting state. The aqueous solution of C24HEUR

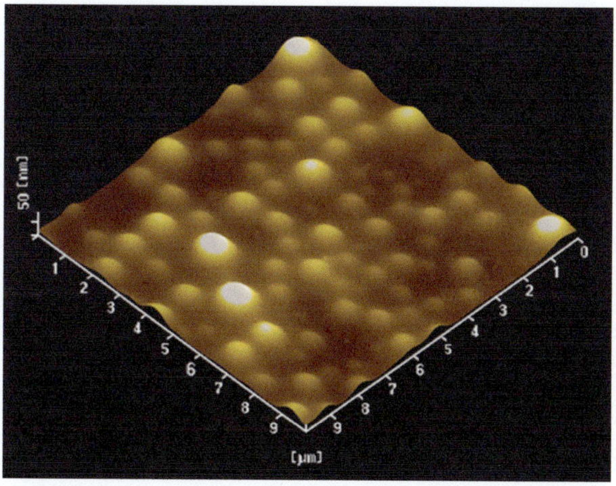

Fig. 11.20 AFM image of the A-D microgel in the dry state

exhibits shear-thinning, non-Newtonian flow: a desirable feature for a cosmetic rheology control agent. Moreover, because the main part of the polymer is comprised of polyethyleneoxide, a nonionic polymer, the polymer has both acid and salt resistance.

Unusual nonlinear flow properties, the topic of this section, namely, the unusual stress upturn under step shear flow, are interesting because friction at the start of rubbing is expected to be related to the "primary feeling." However, it should be noted that, in some cases, coexisting ingredients may affect these features.

A novel microgel that can be swollen, even in acidic aqueous phases, has been developed. The microgel exhibits excellent performance as a cosmetic rheology control agent not only compensating for the disadvantages of existing ingredients. In particular, because the microgels are polymerized in a confined space, they have a completely spherical shape, and this shape is the origin of its unique features as a cosmetic rheology control agent.

References

1. Alexandridis, P., and Yang, L. (2000) Micellization of polyoxyalkylene block copolymer in formamide. Macromolecules 33: 3382–3391
2. Annable T, Buscall R, Ettelaie R, Whittlestone D (1993) The rheology of solutions of associating polymers: comparison of experimental behavior with transient network theory. J. Rheol. 37 (4):695–726
3. Barmar, M., Ribitsch, V., Kaffashi, B., Barikani, M., Srreshtedari, Z., and Pfragner, J. (2004) Influence of prepolymers molecular weight on the viscoelastic properties of aqueous HEUR solution. Colloid Polym. Sci. 282:454–460

4. Barmar, M., Barikani, M., and Kaffashi, B. (2005) Steady shear viscosity study of various HEUR models with different hydrophilic and hydrophobic size. Colloids and Surfaces A 253:77–82
5. Barnes, H.A., Hutton, J.F., and Walters, K. (1991) An introduction to rheology Elsevier
6. Bekker, M., Webber, G.V., and Louw, N.R. (2013) Relating rheological measurements to primary and secondary skin feeling when mineral-based and Fischer-Tropsch wax-based cosmetic emulsions and jellies are applied to the skin. International J. Cosmetic Sci. 35:354–361
7. Brummer, R., and Godersky, S. (1999) Rheological studies to objectify sensations occurring when cosmetic emulsions are applied to the skin. Colloids and Surfaces A 152:89–94
8. Calvet, D., Collet, A., Viguier, M., Berret, J.-F., and Serero, Y. (2003) Perfluoalkyl end-capped poly (ethylene oxide). Synthesis, Characterization, and Rheological behavior in Aqueous Solution. Macromolecules 36:449–457
9. Fridrikh, S., Raquois, C., Tassin, J.F., and Rezaiguia, S. (1996) Rheological behavior of concentrated suspensions of soft spheres. J. Chim. Phys. 93:941–959
10. Gutowski, I.A., Lee, D., de Bruyn, J.R., and Frisken, B.J. (2012) Scaling and mesostructure of Carbopol dispersions. Rheol. Acta 51:441–450
11. Glass, J.E. (Ed.) Polymers in aqueous media: Performance through association. Advance in Chemistry Series ACS 1993, p248
12. Kaffashi, B., Barmar, M., and Eyvani, J. (2005) The steady state and dynamic rheological properties of telechelic associative polymer solutions. Colloids and Surfaces A 254: 125–130
13. Kaneda, I., Miyazawa, K., and Yanaki, T. (2001a) Japanese Patent 2001–114641
14. Kaneda, I., Miyazawa, K., and Yanaki, T. (2001b) Japanese Patent 2001–114642
15. Kaneda, I., Miyazawa, K., and Yanaki, T. (2001c) Japanese Patent 2001–115135
16. Kaneda, I., and Yanaki, T. (2002) Rheology of agar microgel dispersion. Nihon Reoroji Gakkaishi 30 (2):89–94
17. Kaneda, I., and Vincent, B. (2004) Swelling behaviour of PMMA-g-PEO microgel particles by organic solvents. J. Colloid Interface Sci. 274:49–54
18. Kaneda, I., Sogabe, A., Nakajima H. (2004) Water-swellable polyelectrolyte microgels polymerized in an inverse microemulsion using a nonionic surfactant. J. Colloid Interface Sci. 275: 450–457
19. Kaneda, I., and Vincent, B. (2004) Swelling behavior of PMMA-g-PEO microgel particles by organic solvent. J. Colloid Interface Sci 274: 49–54
20. Kaneda, I., and Sogabe, A. (2005) Rheological properties of water swellable microgel polymerized in a confined space. Colloids and Surfaces A 270–271:163–170
21. Kaneda, I. (2006) The yield stress of a soft and water swellable microgel aqueous suspension in semi-dilute regime. Nihon Reoroji Gakkaishi 34 (2):77–81
22. Kaneda, I., Koga, T., and Tanaka, F. (2009) Time-dependent flow properties of transient hydrogels with temporal network junctions. Progr. Colloid Polym. Sci. 136:31–38
23. Kaneda, I., Koga, T., Tand anaka, F. (2012) Rheological properties of physical gel formed by hydrophobically modified urethane ethoxylate (HEUR) associative polymers in methanol-water mixtures. Rheol. Acta 51:89–96
24. Ketz, R.J., Prud'homme, R.K., Graessley, W.W. (1988) Rheology of concentrated microgel solutions. Rheol. Acta 27: 531–539
25. Koga, T., Tanaka. F., and Kaneda, I. (2009a) Stress growth in transient polymer networks under startup shear flow. Progr. Colloid Polym. Sci. 136:39–46
26. Koga, T., Tanaka, F., Kaneda, I., and Winnik, F.M. (2009b) Stress buildup under strat-up shear flows in self-assembled transient networks of thelechelic associating polymers. Langmuir 25 (15):8626–8638
27. Lukic, N., Jaksic, I., Krstonosic, V., Cekic, N., and Savic, S. (2012) A combined approach in characterization of an effective w/o hand cream: the influence of emollient on texture, sensorial and in vivo skin performance. International J. Cosmetic Sci. 34:140–149
28. Ma, S., and Cooper S.L. (2001) Shear thickening in aqueous solution of hydrocarbon end-capped poly (ethylene oxide). Macromolecules 34:3294–3301

29. Marrucci, G., Bhargava, S., and Cooper, S.L. (1993) Models of shear-thickening behavior in physically cross-linked networks. Macromolecules 26:6483–6488
30. Mason, T.G., Bibette, J., and Weitz, D.A. (1995) Elasticity of compressed emulsions. Phys. Rev. Letters 75 (10): 2051–2054
31. Ozkan, S., Gillece, T.W., Senak, L., and Moore, D.J. (2012) Characterization of yield stress and slip behaviors of skin/hair care gels using steady flow and LAOS measurements and their correlation with sensorial attributes. International J. Cosmetic Sci. 34:193–201
32. Penfold, J., Staples, E., Tucker, I., and Cummins, P. (1997) The structure of nonionic micelles in less polar solvents. J. Colloid Interf. Sci. 185:424–431
33. Riberio, H.M., Morais, J.A., and Eccleston, G.M. (2004) Structure and rheology of semisolid o/w creams containing cetyl alcohol/non-ionic surfactant mixed emulsifier and different polymers. International Cosmetic Sci. 26:47–59
34. Saramito, P. (2009) A new elastoviscoplastic model based on the Herschel-Bulkley viscoplastic model. J. Non-Newtonian Fluid Mech. 158:154–161
35. Tanaka, F. and Edwards, S.F. (1992) Viscoelastic properties of reversibly crosslinked polymer networks – Transient Network Theory. Macromolecules 25 (5): 1516–1523
36. Tanaka F and Koga T (2006) Nonaffine transient network theory of associating polymer solutions. Macromolecules 39:5913–5920
37. Tanaka, F., Koga, T., and Winnik, F.M. (2008) Competitive hydrogen bonds and cononsolvency of poly (N-isopropylacrylamide)s in mixed solvents of water/methanol. Physical Rev. Letters 101:028302
38. Wang, S., Kislaliglu, M.S., and Breuer, M. (1999) The effect of rheological properties of experimental moisturizing creams/lotions on their efficacy and perceptual attributes. International Cosmetic Sci. 21:167–188
39. Winnik, F.M., Ringsdorf, H., and Venzmer, J. (1990) Methanol-water as a co-nonsolvent system for poly(N-isopropylacrylamide). Macromolecules 23:2415–2416
40. Winnik, F.M., Ottaviani, M.F., Bossmann, S.H., Garcia-Garibay, M., and Turro, N.J. (1992) Cononsolvency of poly(N-isopropylacrylamide) in mixed water-methanol solutions: A look at spin-labeled polymers. Macromolecules 25:6007–6017
41. Wolfe, M.S., and Scopazzi, C. (1989) Rheology of swellable microgel dispersions: Influence of crosslink density. J. Colloid Interface Sci. 133:265–277
42. Yoshida, K., Nakamura, A., Nakajima, Y., Fukuhara, T., Inoue, H., and Kaneda, I. (2007) Use of associating polymers as multifunctional thickeners: Studies of their structure in aqueous solutions via NMR, QELS,fluorescence, and rheology measurements. IFSCC Magazine 10:35–39

Chapter 12
Rheological Properties of Personal Lubricants

Aaron S.M. Goh, Beng Sim Chuah, and K.C. Nguyen

Abstract In this study, a number of lubricants were assessed for their rheological properties, and qualitative correlation with sensory experience was made. The lubricants contained various structurants including xanthan, hydroxyethyl cellulose, carboxymethyl cellulose and carbomer. A range of shear thinning behaviour was found, with the samples containing xanthan gum showing the highest degree of shear thinning behaviour and the highest degree of gel-like behaviour. The non-water-based lubricants which contained silicon oil or glycerin showed the lowest viscosities and relatively little shear thinning. On cooling, samples containing hydroxyethyl cellulose showed a crossover temperature where G' became higher than G'', suggesting a transition from a more liquid-like behaviour to a more solid-like behaviour. The crossover temperature was higher for samples which exhibited higher steady shear viscosities. The rheological measurements correlated with sensory data that the xanthan-based lubricant could be easily spread and had a more 'solid' feeling with little tendency to drip or flow.

Keywords Lubricant • Rheology • Friction

12.1 Introduction

Personal lubricants are applied topically during intimate activities. The lubricants serve to enhance pleasure and reduce trauma at sensitive body areas by reducing skin-to-skin friction [10]. When applied onto condoms, the lubricants may facilitate unrolling of the condoms as well as preventing the surfaces of the rolled condom from sticking together during storage [6]. With added microbicides and spermicides, the lubricants may also be used to prevent transmission of pathogens and to act as a birth control method [7]. These lubricants are closely related to vagina gels which are used primarily for medical and clinical purposes (e.g. [25]).

A.S.M. Goh (✉) • B.S. Chuah • K.C. Nguyen
Singapore Institute of Technology, Singapore, Singapore
e-mail: aaron.smgoh@gmail.com; bengsim.chuah@gmail.com; kc.nguyen@ansell.com

I. Kaneda (ed.), *Rheology of Biological Soft Matter*, Soft and Biological Matter,
DOI 10.1007/978-4-431-56080-7_12

Personal lubricants can be categorised as either water-based, oil-based or silicone-based. The lubricants need to meet certain rheological requirements – for example, the lubricants should have the ability to spread yet does not travel over and away from desired mucosal tissues during usage. As a result, thickeners or rheology modifiers are added to a number of lubricant products. Whereas comprehensive rheological characterisation has been applied to vagina gels (e.g. [9, 15, 21, 25]), the characterisation of personal lubricants tends to be simpler and often relies on a single-point measurement of viscosity (e.g. [10]).

The simple point measurement is quite sufficient for lubricants with a predominantly Newtonian behaviour such as those having typically low viscosity. However, for structured lubricants that exhibit shear thinning, thixotropy or gelation behaviour, a better understanding of the lubricants requires a more diverse range of tools from the rheology toolbox. For example, a representative shear rate during coitus has been estimated to be around 100 s^{-1} [21], whereas the shear rate, on standing after initial application, can be of a much smaller value. Using a single-point viscosity measurement thus does not provide an adequate picture of what the lubricant experiences from the time of application to during coitus.

The aim of this study was to compare the rheological properties of some commercially available personal lubricants. In addition, preliminary measurement of friction properties of the lubricants was also explored.

12.2 Materials and Methods

12.2.1 Materials

Eleven commercially available lubricants containing different structurants were selected for characterisation. Three lubricants were water-based and were structured by carboxymethyl cellulose (CMC), one by carbomer, four by hydroxyethyl cellulose (HEC) and one by xanthan. Some of the lubricants also contained co-solvents such as propylene glycol and glycerin. These co-solvents may have been used to increase the lubricity of the lubricants [1]. In addition, one of the lubricants consisted of silicon oil and another contained glycerin as the main component. For the rest of this chapter, the lubricants will be referred to as according to Table 12.1.

12.2.2 Rheological Tests

The rheological properties were measured using a TA AR1500EX rheometer with a cone-plate configuration (diameter 60 mm and angle 2°). Steady shear behaviour was measured by varying the shear rate from 0.01 to 1,000 s^{-1}. The resulting

Table 12.1 List of commercially available lubricants tested

Sample reference	Structurant
CMC1	Carboxymethyl cellulose
CMC2	Carboxymethyl cellulose
CMC3	Carboxymethyl cellulose
CAR	Carbomer
HEC1	Hydroxyethyl cellulose
HEC2	Hydroxyethyl cellulose
HEC3	Hydroxyethyl cellulose
HEC4	Hydroxyethyl cellulose
XAN	Xanthan gum
SIL	Silicon oil
GLY	Glycerin

multipoint flow curve shows how the viscosity of the lubricants would change as a function of the shear rate. The shear rate represents the rate at which the lubricants were sheared.

Oscillation tests were also conducted to obtain the storage (G') and loss (G'') moduli. The oscillation tests initially comprised strain sweep at a frequency of 1Hz to determine the region of linear viscoelasticity. Subsequently, frequency sweep tests from 0.1 to 100 Hz were performed at a strain of 0.1 % which was within the linear viscoelastic region. In addition, temperature ramp tests between 10 and 80 °C were conducted at a strain of 0.1 % and frequency of 1 Hz to determine the dependence of G' and G'' on temperature.

All tests were conducted at 37 $\pm$ 0.1 °C. Two replicates were performed for each test condition.

12.2.3 Friction Tests

The measurement of lubricity of personal lubricants had previously been attempted by Ahmad et al. [1, 2] and Fevola and Vochecowicz [11]. The lubricity was defined as inversely proportional to the coefficient of friction [1], i.e. the lower the friction, the higher the lubricity of the lubricant. The test rigs used by the authors are shown schematically in Fig. 12.1. In the first attempt by the authors, the lubricant was applied between a metal disc of known weight and a glass plate. The metal disc was attached to a force transducer. By sliding the metal disc across the glass plate, the friction force was measured and the coefficient of friction was calculated by dividing the friction force by the weight of the metal disc. In the second attempt, a universal testing machine was utilised in the horizontal direction together with a friction rig. A non-lubricated condom was mounted over a glass sledge and was moved in a linear reciprocating fashion across a metal plate which was covered by a polyethylene/foil liner. The friction measurements were recorded over a 5-min period.

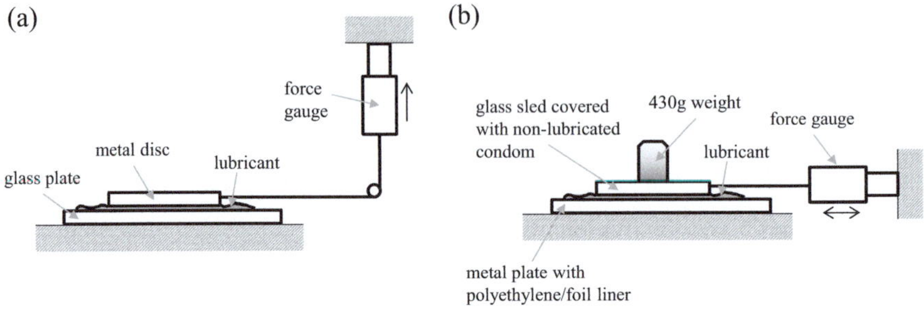

Fig. 12.1 Friction measurement methods used by (**a**) Ahmad et al. [1] and (**b**) Ahmad et al. [2]

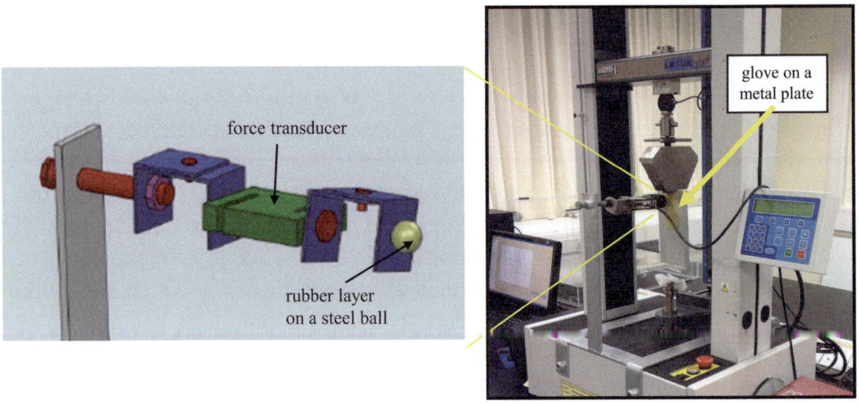

Fig. 12.2 Friction test rig for this study. The insert shows a schematic of the arm used to measure the normal force

Based on the methods described above, a vertical test rig was set up for the current study (see Fig. 12.2). The first part of the rig consisted of a metal plate attached vertically to the crosshead of a universal test machine (Lloyd Instruments LR10kPlus). The metal plate could be moved in a linear reciprocating motion up and down following the movement of the crosshead. A rubber glove, cut to a rectangular shape, was attached to the metal plate and used to represent a soft substrate. A new glove was used for each test. The second part of the rig consisted of a roller screw mechanism to apply a normal force to the metal plate. The roller screw mechanism was based on a fixed arm, to which a load cell was attached. The contacting surface was made of rubber which was fixed on to the surface of a steel ball. The normal force from the force transducer was recorded on a data logger. Concurrently, the friction force was recorded on the universal test machine. Tests were performed at 20 °C on selected lubricants.

12.3 Results

12.3.1 Rheological Measurements

Figures 12.3 and 12.4 compare the relationship between viscosity and shear rate for the lubricants tested. A wide range of viscosities were observed. The lubricant based on glycerin exhibited the lowest range of viscosities. The CMC-based lubricants showed a similar range of viscosities which suggests that the concentrations of the structurant used were rather similar. The HEC-based lubricants showed a larger

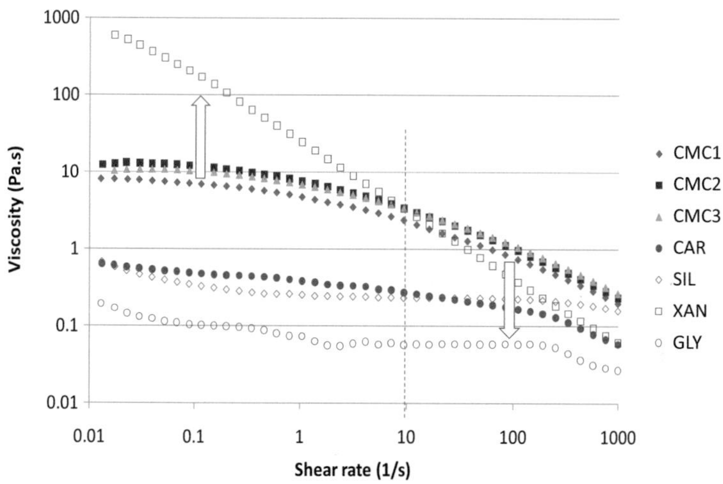

Fig. 12.3 Relationship between viscosity and shear rate for different lubricants at 37 °C

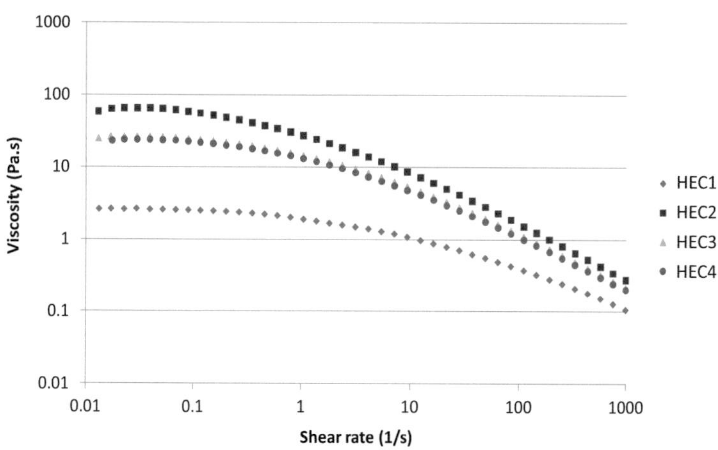

Fig. 12.4 Relationship between viscosity and shear rate for HEC-based lubricants at 37 °C

degree of differences which may be due to differences in the concentration or the molecular weight of the structurant.

Generally, the lubricants did not have a constant viscosity when the shear rate was varied. Shear thinning, which refers to the decrease in viscosity with increasing shear rate, was exhibited by all the lubricants with varying degrees. The degree of shear thinning was the highest for the xanthan-based lubricant whose viscosity dropped by four orders of magnitude between the shear rate range of 0.01–1,000 s^{-1}. The lubricants based on HEC showed the second highest degree of shear thinning, with viscosities dropping by more than two orders of magnitudes within the same shear rate range. The lubricants based on silicon oil and glycerin, which showed a relatively lower range of viscosities, also showed the lowest degrees of shear thinning.

The marked shear thinning of the xanthan-based lubricant is typical of xanthan solutions. Xanthan's rigid, rodlike structures form reversible entangles which lead to a weak network in solution when at rest (see Fig. 12.5). On shearing, the degree of aggregation of the xanthan molecules is reduced, and the molecules are aligned to the direction of shear, leading to a large reduction in resistance to flow [16]. The interactions between xanthan molecules are not permanent and are shear-reversible [23].

The shear thinning behaviour observed in the CMC- and HEC-based lubricants is also due to the progressive mechanical disruption of intermolecular entanglements and deformation of polymer coil [3, 4, 18]. At low strain rates, the molecules are entangled with each other resulting in a high viscosity. As the strain rate increases, the molecules become oriented to the direction of shear, and there is a lower resistance to flow leading to lower viscosities. Benchabane and Bekkour [4] suggested that two ranges of rheological behaviour can be observed for CMC solutions – the first is a Newtonian plateau at low shear rates and the second is the shear thinning region at higher shear rates. The authors also suggested that when the concentrations of the structurants used are higher than a critical concentration called the overlap concentration, a shear thickening behaviour is apparent at the lowest shear rates. The shear thickening is thought to be a result of an increase in intermolecular interactions as the shear rate rises. These three types of behaviour can be observed in the lubricant tested for both the CMC- and HEC-based lubricants, with the shear thickening behaviour apparent in the lubricants with relatively higher viscosities.

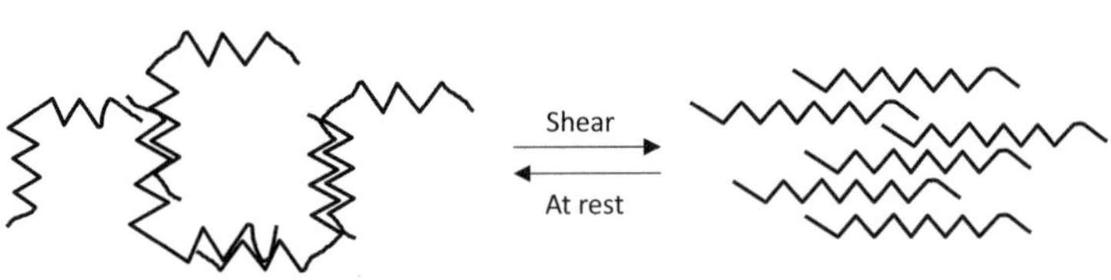

Fig. 12.5 Schematic diagram of xanthan molecules at rest and under shearing (From Ref. [23])

Carbomer dispersions consist of discrete soft polymeric particulates (microgels) that, on swelling, fill up space. The degree of interactions between individual microgels (such as due to concentration) and the properties of the polymer within each microgel govern the rheological behaviour of the dispersion [12]. When sheared, the carbomer microgels as a whole and the polymer chain segments within each microgel deform and align themselves in the direction of flow, leading to a shear thinning behaviour [13]. Although carbomer solutions can show a high degree of shear thinning (e.g. [13]), the carbomer-based lubricant in this study showed a relatively low range of viscosities and degree of shear thinning.

The measurement of the steady shear viscosities shows the advantage of performing a multipoint viscosity measurement as compared to a single-point viscosity measurement. Referring to Fig. 12.3, if a single-point steady shear viscosity is taken at $10 s^{-1}$, the xanthan-based lubricant would appear to have a similar viscosity as the CMC-based lubricants. If a lower shear rate is taken as a reference, e.g. at $0.1 s^{-1}$, then the steady shear viscosity of the xanthan-based lubricant would be significantly higher than the other lubricants. On the contrary, if a higher shear rate is used as the reference, e.g. at $100 s^{-1}$, then the xanthan-based lubricant would have a lower steady shear viscosity compared to the CMC- and HEC-based lubricants. Using a single-point viscosity measurement may therefore lead to an incorrect interpretation of the viscosity of the lubricants.

The dependence of the storage and loss moduli as a function of frequency is shown in Figs. 12.6 and 12.7. For the carbomer-based lubricant, the loss modulus G'' was higher than the storage modulus G' for the whole frequency range. In addition, both G' and G'' increased with increasing frequency. These observations

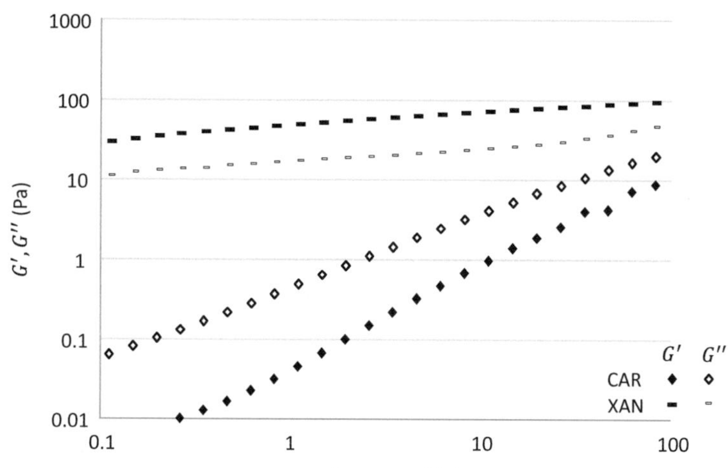

Fig. 12.6 Variation of storage and loss moduli for xanthan- and carbomer-based lubricants as a function of frequency at 37 °C

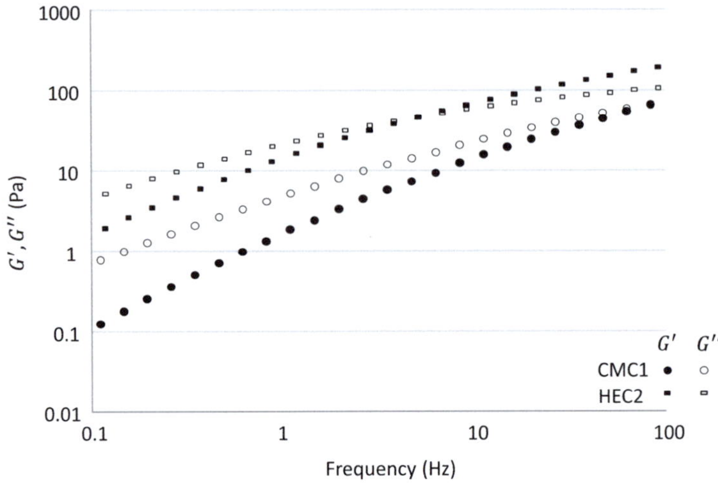

Fig. 12.7 Variation of storage and loss moduli for CMC- and HEC-based lubricants as a function of frequency at 37 °C

were also found for the lubricants based on silicone oil and glycerin and are not shown here. On the contrary, for the xanthan-based lubricant, G' was higher than G'' for all frequencies, and a relatively low dependence on frequency was observed. The mechanical spectra for the xanthan-based lubricant suggest that the concentration of the xanthan used exceeds the overlap concentration [8]. For the HEC- and CMC-based lubricants, G' was lower than G'' at low frequencies, but a crossover point may be noticeable where G' was then higher than G'' at higher frequencies.

The mechanical spectra are commonly used to aid in classifying the rheological behaviour of solutions. Four categories may be distinguished [5, 20]:

- Strong gel – G' is significantly higher than G'', and both moduli are independent of frequency.
- Weak gel – G' is slightly higher than G'', and both moduli are slightly dependent on frequency.
- Entangled polymer solution – G' is lower than G'' at low frequencies, but both moduli increase with increasing frequency. There is a crossover point such that, at higher frequencies, G' is higher than G''.
- Non-entangled polymer solution – G' is much lower than G'' at all frequencies, and both moduli are strongly dependent on frequency.

From the above classification conditions, it can be seen that the xanthan-based lubricant was exhibiting a weak gel behaviour. Both the CMC- and HEC-based lubricants exhibit the behaviour of an entangled polymer solution, while the carbomer-based lubricant was exhibiting a behaviour similar to a non-entangled

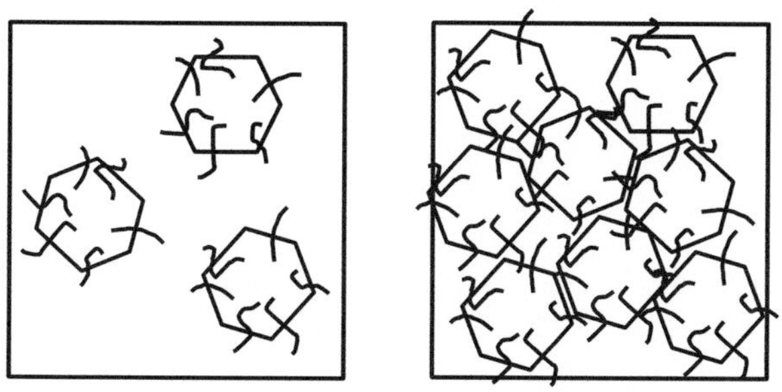

Fig. 12.8 Schematic of microstructures of a carbomer-based dispersion at low and high concentrations of the structurant

polymer solution. The relatively weaker properties of the carbomer-based lubricant may be due to a low concentration of the carbomer such that the discrete microgels have low interactions with each other (see Fig. 12.8). If the concentration of the microgel was increased, the particles would increasingly be confined sterically by their neighbours which would lead to stronger rheological properties.

For all lubricants with the exception of xanthan- and HEC-based lubricants, temperature sweep tests between 10 and 80 °C showed decreasing G' and G'', with G'' larger than G', as the temperature was increased (results not shown). For the xanthan-based lubricant, G' was larger than G'' for the whole range of temperature tested (results not shown).

The temperature sweep data for three of the HEC-based lubricants are shown in Fig. 12.9. HEC3 and HEC4 lubricants showed a similar behaviour, so only data for HEC3 are shown. For HEC1, G'' was always higher than G' within the range of temperatures tested. For HEC2 and HEC3, a crossover point was observed where G' was higher than G'' at lower temperatures than the crossover temperature. The crossover temperature was higher for HEC2 (~40 °C) than for HEC3 (~25 °C). From the steady shear viscosity data in Fig. 12.4, HEC1 showed the lowest viscosities, while HEC2 showed the highest viscosities. HEC 3 and HEC4 showed intermediate values of viscosities amongst the lubricants structured by HEC. These viscosity and oscillatory test data suggest that HEC2 has the highest degree of internal structuring, possibility due to a higher concentration of HEC compared to the other HEC-based lubricants.

It should be noted that the presence of co-solvents in the lubricants such as propylene glycol and glycerol may affect the influence of the structurants on the rheological properties. The presence of the co-solvents may result in a lower concentration of available water to hydrate the structurant or a reduction in the swelling ability of the structurant [14].

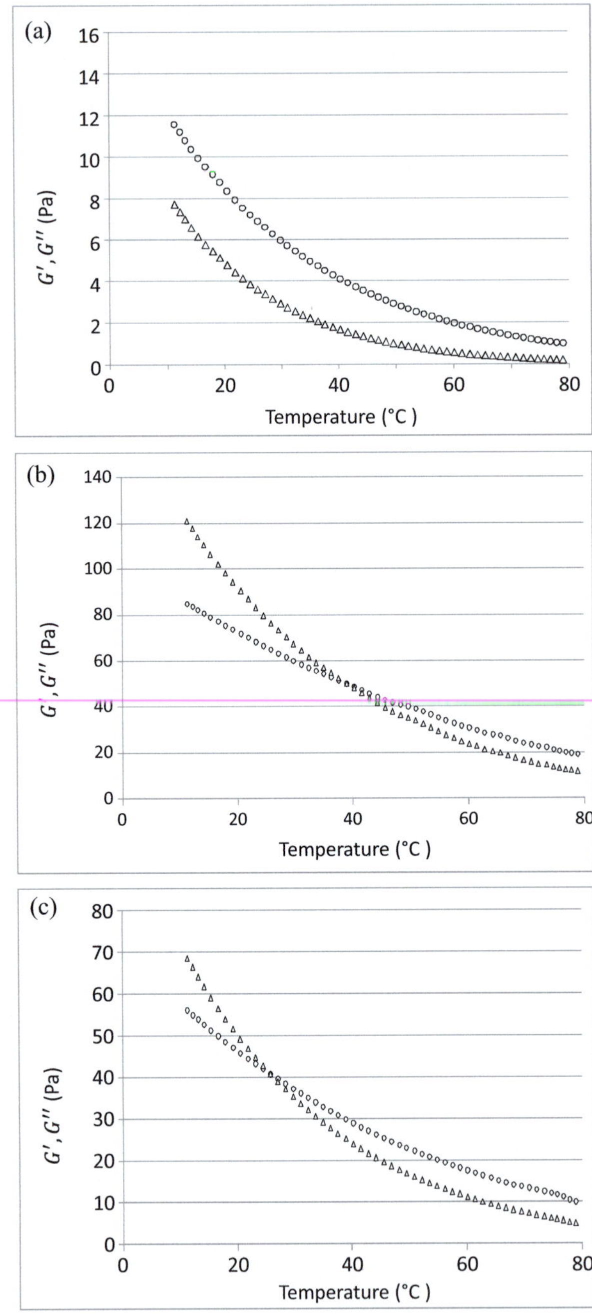

Fig. 12.9 Variation of G' (Δ) and G' (⬠) with temperature for (**a**) HEC1, (**b**) HEC2 and (**c**) HEC3 lubricants

12.3.2 Friction Measurements

An example of a friction trace recorded on the universal test machine is shown in Fig. 12.10. Initially no lubricant was applied to the rubber surface, and a relatively higher friction force was measured. After the lubricant was applied, the friction force decreased. Due to the reciprocating motion of the platform, the friction force was recorded either as positive or negative values corresponding to the direction of motion.

A difference in the friction force between the two directions of motion can be observed from the data. The difference was caused by the compliance of the arm used to measure the normal force which resulted in a lower normal force for motion in one direction than in the other direction. Since the normal force was measured concurrent to the frictional force measurement, the calculation of the coefficient of friction (= friction force/normal force) was not affected by the compliance of the arm. The normal force recorded was in the range of 5–8 N.

Figure 12.11 shows the variation of the average coefficient of friction with sliding speed for five representative lubricants. For HEC2 and CAR, three sliding speeds corresponding to 30, 200 and 400 mm/min were tested. These lubricants generally showed little variation in the coefficient of friction within the sliding speed range tested. For XAN, CMC1 and SIL, two more addition sliding speeds corresponding

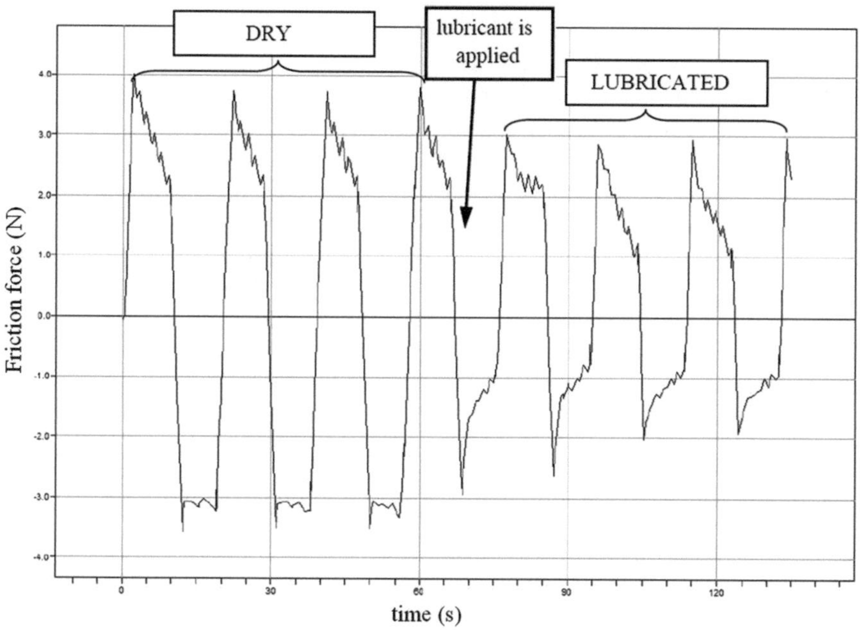

Fig. 12.10 Variation of friction force with time before and after lubricant is applied. The positive and negative values correspond to the reciprocating (*up* and *down*) movement of the metal plate

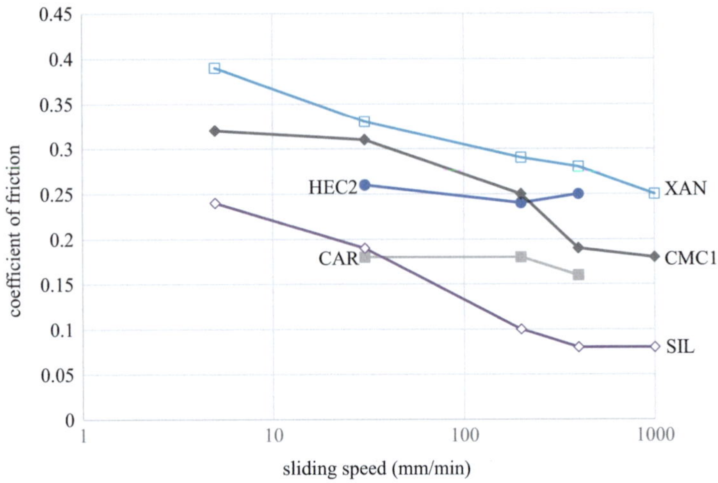

Fig. 12.11 Variation of coefficient of friction with sliding speed for selected lubricants

to 5 and 1,000 mm/min were also tested. For these three samples, the coefficient of friction decreased with increasing sliding speed, i.e. the lubricity increased when the shearing rate was faster. The trend in the data suggests that the lubrication provided by the lubricants could be within the boundary and/or mixed lubrication regimes where the surfaces are in close contact with each other or are partly in contact and partly separated by the lubricants, respectively [19].

The different lubricants exhibited different values of coefficient of friction. The xanthan-based lubricant gave the highest coefficient of friction compared to the other lubricants, whereas the silicon oil-based lubricant gave the lowest coefficient. The differences may be due to the different degrees of adsorption of the structurants to the surface. Within the boundary and mixed lubrication regimes, the adsorption of a polymer film on the surface has been found to strongly determine the coefficient of friction [22].

The values of coefficient of friction obtained from this study are in agreement with the range obtained by Ahmad et al. [1] but not those obtained by Ahmad et al. [2]. The latter obtained values lower than 0.1 for most of the lubricants tested at a sliding speed of 120 mm/min. It should be noted that the measurement of coefficient of friction is sensitive to many parameters related to the experimental setup such as surface conditions, mechanical properties of substrates as well as compliance of the system. The values obtained from different samples may be compared relative to each other within an experimental setup but may not necessarily agree with tests done under a different setup. To have the best prediction of the lubricants in practice would probably require a more realistic test rig such as those used to simulate condom breakage (e.g. [24]).

12.4 Relation Between Instrumental and Sensory Properties

A diverse range of shear rates is encountered during physical use of personal lubricants. During application of the lubricants, the lubricants might be squeezed out from the packaging and should be sufficiently 'thin' to allow easy spreading. These activities involve relatively medium range of shear rates. Once spread, the lubricants should remain in place and not drip due to gravity or bulk movement of the body. The shear rate involved here is very low. At fast shearing, such as during coitus, high shear rates may be involved.

The large degree of shear thinning, involving a high viscosity at low shear rates and a low viscosity at high shear rates, suggests that the xanthan-based lubricant would be easily spread but with little tendency to drip or flow. The high shear thinning behaviour may also lead to increased spreadability of the lubricant during coitus [21]. In addition, the weak gel behaviour (solid-like) may give rise to a more solid feeling. Internal user studies (results not shown) suggest that between the xanthan-, CMC- and HEC-based lubricants, the xanthan-based lubricant showed a more 'pleasant' texture prior to application, was easier to spread once applied and was able to last long during coitus. These user perceptions are consistent with the rheological findings.

Compared to microbicide gels, research into the relationship between rheological/tribological properties and sensorial attributes of lubricants has been very limited (e.g. [17]). This study shows that the rheological properties are useful in explaining how users perceive the lubricants, but more work is necessary to develop quantitative models of the relationships between rheological/tribological properties and sensory attributes.

Acknowledgement The authors would like to thank Andrew Lai Eng Khiong and Tee Kok Wee from Curtin University Sarawak for performing the experiments.

References

1. Ahmad N, Junction M, Koll GE, et al. (1999) Personal lubricant compositions. US Patent 5885591
2. Ahmad N, Junction M, Patel K, et al. (2000) Personal lubricant compositions. US Patent 6139848
3. Arfin N, Bohidar HB (2012) Concentration selective hydration and phase states of hydroxyethyl cellulose (HEC) in aqueous solutions. International Journal of Biological Macromolecules, 50, 759–767
4. Benchabane A, Bekkour K (2008) Rheological properties of carboxymethyl cellulose (CMC) solutions. Colloid Polymer Science, 286, 1173–1180
5. Clark AH, Ross-Murphy SB (1987) Structural and mechanical properties of biopolymer gels. Advances in Polymer Science, 83, 57–192
6. Cojocariu C, Harrison MJ, Woodward DC, et al. (2014) Aqueous lubricant composition. WO 2014152154

7. Cunha AR, Machado RM, Palmeira-de-Oliveira, et al. (2014) Characterisation of commercially available vaginal lubricants : a safety perspective. Pharmaceutics, 6, 530–542
8. Cuvelier G, Launay B (1986) Concentration regimes in xanthan gum solutions deduced from flow and viscoelastic properties. Carbohydrate Polymers, 6, 321–333
9. Das Neves J, da Silva MV, Goncalves MP, et al. (2009) Rheological properties of vaginal hydrophilic polymer gels. Current Drug Delivery. 2009, 6, 83–92
10. Dezzutti CS, Brown ER, Moncla B, et al. (2012) Is wetter better? An evaluation of over-the-counter personal lubricants for safety and anti-HIV-1 activity. PLoS One, 17, e48328
11. Fevola MJ, Vochecowicz S (2009) Lubricious, non-tacky personal lubricant. EP 2080509 A1
12. Gutowski IA, Lee D, de Bruyn JR, et al. (2012) Scaling and mesostructure of Carbopol dispersions. Rheologica Acta, 51, 441–450
13. Islam MT, Rodriguez-Hornedo N, Ciotti S, et al. (2004) Rheological characterization of topical carbomer gels neutralized to different pH. Pharmaceutical Research, 21, 7
14. Jones DS (2006) Thermorheological (dynamic oscillatory) characterization of pharmaceutical and biomedical polymers. In Thermal Analysis of Pharmaceuticals. Craig DQM, Reading M (eds.), CRC Press, 311–358
15. Kieweg S, Katz D (2006) Squeezing flows of vaginal gel formulations relevant to microbicide drug delivery. Journal of Biomechanical Engineering, 128, 540–553
16. Launay B, Doublier JL, Cuvelier G (1986) In Functional Properties of Food Macromolecules. Hill SE, Mitchell JR, Ledward DA (eds.), Elsevier Applied Science Publishers, London, 1–78
17. Mahan ED, Zaveri T, Ziegler GR, et al. (2014) Relationships between perceptual attributes and rheology in over-the-counter vaginal products: a potential tool for microbicide development. PLoS One, 9, e105614
18. Meadows J, Williams PA, Kennedy JC (1995) Comparison of the extensional and shear viscosity characteristics of aqueous hydroxyethylcellulose solutions. Macromolecules, 28, 2683–2692
19. Moore D F (1975). The elastohydrodynamic transition speed for spheres sliding on lubricated rubber. Wear, 35, 159–170
20. Nishinari K (2009) Some thoughts on the definition of a gel. In Tokita M, Nishinari K (eds.), Progress in Colloid and Polymer Science, vol. 136, Springer-Verlag Berlin, Heidelberg, 87–94
21. Owen DH, Peters JJ, Katz DF. (2000) Rheological properties of contraceptive gels. Contraception. 2000, 62, 321–326
22. Stokes JR, Macakova L, Chojnicka-Paszun A, et al. (2011) Lubrication, adsorption and rheology of aqueous polysaccharide solutions. Langmuir, 27, 3474–3484
23. Urlacher B, Noble O (1999) Xanthan gum. In Thickening and Gelling Agents for Food. 2nd ed. Imeson A (ed.), Blackie Academic and Professional., London, 284–311
24. White ND, Hill DM, Bodemeier (2008) Male condoms that break in use do so mostly by a "blunt puncture" mechanism. Contraception, 77, 360–365
25. Yu T, Malcolm K, Woolfson D, et al. (2011) Vaginal gel drug delivery systems : understanding rheological characteristics and performance. Expert Opinion on Drug Delivery, 8, 1309–1322

Part IV
Applications: Biological Tissues

Chapter 13
Development of PVA Hydrogels with Superior Lubricity for Artificial Cartilage

Atsushi Suzuki, Saori Sasaki, and Teruo Murakami

Abstract Physically cross-linked poly(vinyl alcohol) (PVA) gels are versatile biomaterials with their excellent biocompatibility and mechanical strength. Since the late 1970s, the semicrystalline PVA gels prepared by a freeze-thawing method (FT gel) have been extensively studied in characterizing the structural and functional properties for practical applications, such as artificial hydrogel cartilage. Recently, a simple preparation method for a physically cross-linked PVA gel by a cast-drying method (CD gel) was reported. Although the network nanostructures are similar, CD gels are transparent and elastic, while FT gels are opaque and less elastic. The crystallization conditions of these systems have been investigated; the gels become highly swollen and rigid by the selection of optimum preparation conditions. In this chapter, the mechanical properties of FT and CD gels, such as tearing energy, sliding friction, and abrasion loss, are reviewed in connection with the nano- and microstructures of physical PVA gels. The tribological properties of FT and CD gels are compared with that of natural cartilage. Based on experimental results, simple preparation methods to improve the mechanical and lubrication properties of physically cross-linked PVA gels are presented, and the mechanism of superior lubricity in newly developed PVA gels for artificial hydrogel cartilage is discussed.

Keywords Poly(vinyl alcohol) gel • Hybrid gel • Superior lubricity • Artificial hydrogel cartilage

---

A. Suzuki (✉)
Department of Materials Science & Research Institute of Environment and Information Sciences, Yokohama National University, 79-7 Tokiwadai, Hodogaya-ku, Yokohama 240-8501, Japan
e-mail: asuzuki@ynu.ac.jp

S. Sasaki • T. Murakami
Research Center for Advanced Biomechanics, Kyushu University, 744 Motooka, Nishi-ku, Fukuoka 819-0395, Japan

I. Kaneda (ed.), *Rheology of Biological Soft Matter*, Soft and Biological Matter,
DOI 10.1007/978-4-431-56080-7_13

13.1 Introduction

Many attempts have been made to develop artificial cartilage from soft materials with superlubricity based on bionic designs by elucidating the lubrication mechanisms of natural synovial joints [1–5]. Among candidate soft materials, poly(vinyl alcohol) (PVA) gel is frequently studied for such applications [4, 6–11]. Some PVA gels show low coefficients of friction under specific conditions and high water contents, comparable to natural articular cartilage. For example, PVA gels prepared by a freeze-thawing method (FT gel) [12, 13] were reported to show very low friction coefficient under walking conditions when lubricated with a hyaluronate solution containing protein [14]. When focusing only on PVA gels, the use of PVA powders with high molecular weight [6] and lubricant alterations [8, 9] has been reported to enhance the mechanical and tribological properties of PVA. However, these hydrogels lack mechanical durability and wear resistance. A simple method to prepare a physically cross-linked PVA gels by cast-drying (CD gel) was recently introduced [15, 16]; this method is promising for PVA applications.

For the application of hydrogels as artificial articular cartilage, several essential properties are required, such as biocompatibility, low friction, high wear resistance, suitable water content, and high mechanical strength (i.e., stiffness, fracture toughness, and fatigue resistance). CD and FT gels both have high water absorbance and excellent mechanical properties compared to other hydrogels. However, the mechanical strength of each gel is insufficient for practical use, and FT or CD gels alone cannot satisfy all required properties. The mechanical strength and wear resistance of these gels are particularly low compared to those of natural articular cartilage; further improvements are necessary for practical uses. In addition, the elution of uncross-linked PVA [17, 18], an inevitable phenomenon in physical gels, should be prevented or minimized.

New preparation methods for PVA gels have been studied, such as the combination of different types of PVA gels with different structures, elastic moduli, and characteristics of permeability. A lamination method and hybrid techniques using FT and CD gels were examined in order to improve the mechanical and tribological properties [19, 20] of the PVA gel as a candidate for practical use as artificial cartilage.

In this chapter, recent developments in the swelling and mechanical properties of PVA gels are presented in connection with the nano- and microstructures of physical PVA gels. The text focuses on materials prepared from an aqueous PVA solution without additional chemical reagents. The results are discussed based on gel science and technology, as well as biphasic finite element (FE) analysis [20]. A method for preparing biomimetic artificial PVA cartilage is presented for use as an artificial cartilage with superior lubricity, based on bionic design principles.

13.2 Preparation of Physical PVA Gels

It has been reported that a high degree of polymerization is required to obtain firm gels with strong amorphous networks with homogeneously distributed microcrystallites, while a high degree of hydrolysis is required to form large microcrystallites cross-linking the amorphous regions [15, 21]. Therefore, it is important to use a starting PVA powder with high degrees of both polymerization and hydrolysis to obtain a swollen gel using repeated freeze-thawing and cast-drying methods. The starting PVA solution is generally an aqueous solution with 10–15 wt.% PVA obtained by dissolving PVA powder at high temperatures, sometimes using an autoclave.

In this chapter except Sect. 13.6, one sort of PVA powder was kindly supplied by Kuraray Co., Ltd. (Cat. No. PVA117) and used without further purification for most experiments. The average degree of polymerization was 1,700, and the average degree of hydrolysis was between 98 and 99 mol%. The concentration of the starting PVA solution was 15 wt.% PVA, obtained by dissolving PVA powder in distilled water over 90 °C under stirring for more than 2 h.

13.2.1 *FT Gels*

The outline of the typical freeze-thawing method is as follows [22–25]: The aqueous solution is frozen at around the freezing temperature, $T_{\text{freeze}} = -20$ °C, for a specified period between 8 and 24 h. The frozen solutions are subsequently thawed at a temperature, T_{thaw}, above 0 °C, usually set to room temperature ~23 °C for a specified period more than 8 h. This freezing-thawing cycle is repeated several times.

In the present experiments, T_{freeze} was fixed to −20 °C, and the duration of frozen storage was 24 h. Thawing was performed at $T_{\text{thaw}} = 4$ °C either with or without a constant heating rate, typically 1 °C·min^{-1}, for a specified period up to 24 h.

13.2.2 *CD Gels*

The standard cast-drying method used here was as follows [26]: The pre-gel PVA aqueous solution was cast into a dish and left to dry at room temperature in air until the weight stabilized. The ambient relative humidity, *RH*, ranged between 15 % and 70 % (depending on the time of day, or season; typically $RH = 50$–60 %) under uncontrolled humidity conditions. The conditions adopted for the preparation of the standard sample were the gelation (drying) temperature, $T_{\text{gel}} = \sim 25$ °C (room temperature), and the weight, $W_{\text{p}} = 3$ g of the solution in the polystyrene dish of 33 mm in diameter. The thickness of the films thus obtained was ~0.3 mm.

Recently, CD gels were prepared under different drying conditions, namely, T_{gel}, *RH*, and W_p [26–28]. The elution of polymer from the gels was quantitatively measured, as well as the swelling ratio, crystallinity, and characteristic size of microcrystallites. Some results are presented here in addition to the experimental results.

13.2.3 Network Structures

The network structures of both physical PVA gels are composed of microcrystallites and amorphous zones, as schematically shown in Fig. 13.1. Although the network nanostructures of FT and CD gels are similar, CD gels are transparent and elastic, whereas FT gels are opaque and less elastic. The FT and CD methods of gelation are based on phase separation induced by the natural solidification and vaporization of water, respectively, both of which cause the formation of microcrystallites in the PVA phase. The networks consist of dense and dilute regions of polymers, which can be introduced permanently by these gelation processes. The size of the dense polymer regions from the FT method should be much larger than that from the CD method, causing the opacity of FT gel. The degree of inhomogeneity can be controlled between transparency and opacity by changing the water content of the initial pre-gel solution, w_i in the FT method. By decreasing w_i from 85 wt.% to a selected value, microcrystallites are formed by the CD mechanism prior to the FT process, to an extent dependent on the value of w_i. As shown in Fig. 13.2, the gel becomes transparent as w_i is reduced from 85 % (standard FT gel) to several percent (standard CD gel) [29]. A reduction in w_i decreases W_t/W_d and increases crystallinity, causing changes in the properties of gels related with the network structure, such as the mechanical strength and water flow.

The water flows in the two gels are completely different. Considering a report on the water flow in chemically cross-linked poly(*N*-isopropylacrylamide) gel (NIPA gel) [30], fluid flows rapidly in the opaque hydrogel, but extremely slow in the transparent hydrogel. Similar models can be proposed for both NIPA and PVA networks, since the opacity of both samples is attributed to the aggregation of cross-links, despite the different mechanisms of network formation. Since the flow of water and the permeability affect the surface friction, the average pore size of the polymer network and the distribution of pores in the hydrogels are extremely important.

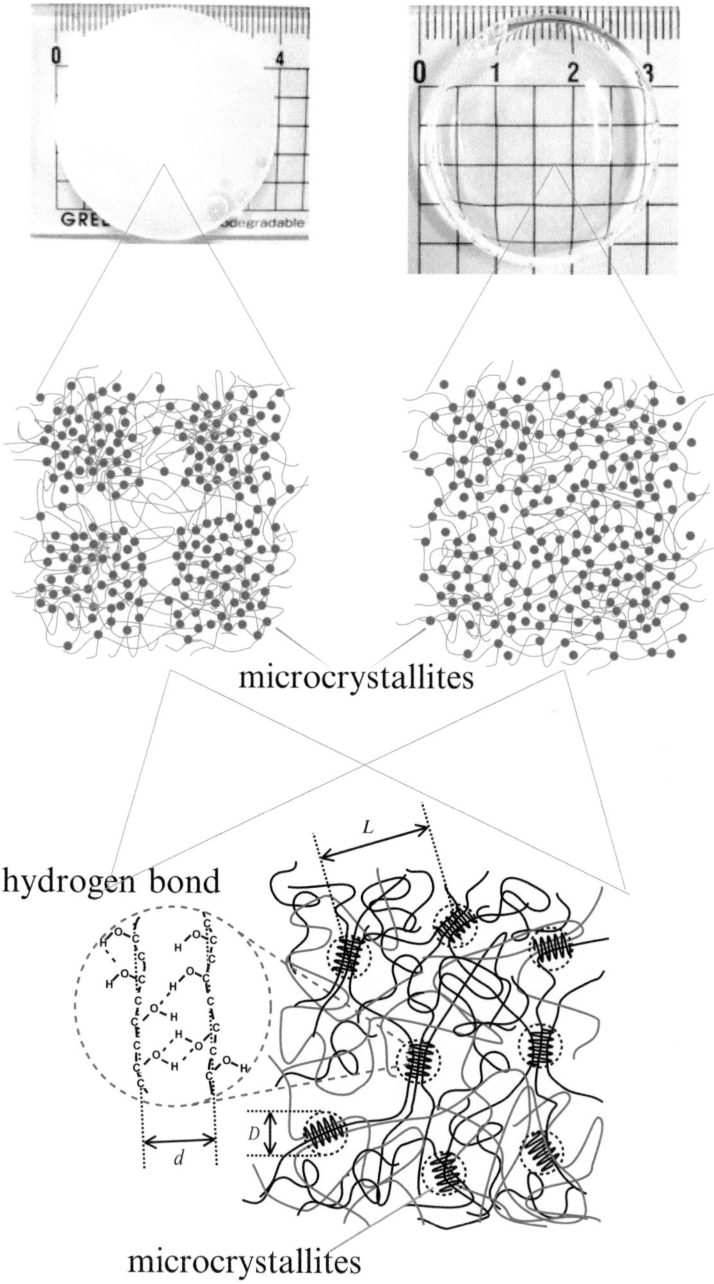

Fig. 13.1 Schematic illustration of network structure of physical PVA gel in multi-scales (milli-, micro-, and nanoscales), which consists of a swollen amorphous network of PVA physically cross-linked by microcrystallites

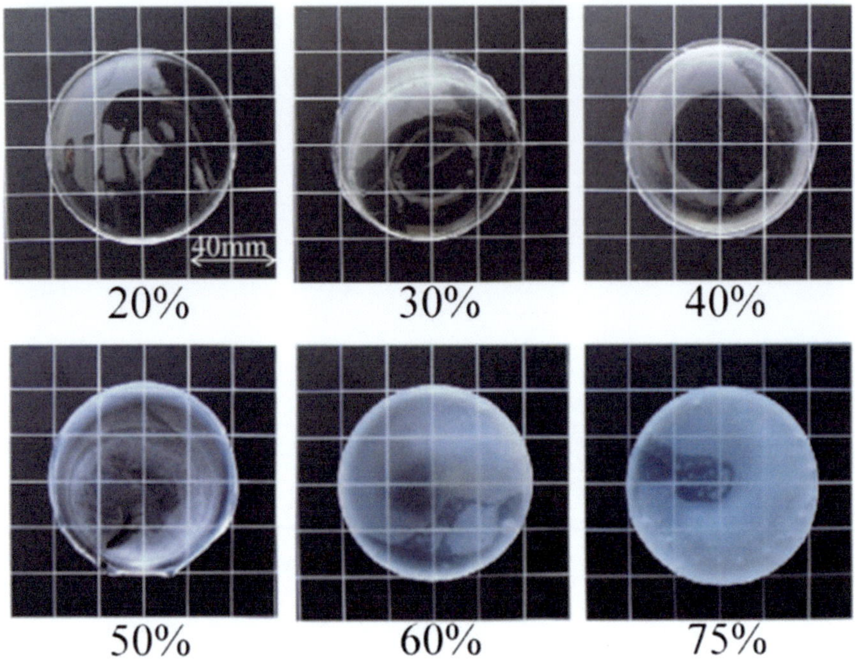

Fig. 13.2 Pictures of CD FT composite gels prepared using the initial PVA solution with different water content, w_i wt % (From Ref. [29], with permission from Wiley)

13.3 Experimentals and Evaluation Methods

13.3.1 Swelling and Elution Ratios

For experimental studies of hydrogels, the measurement of the swelling ratio by measuring the changes in volume is a fundamental technique [31]. This is because the observed volume should be reflected by the average network structure, and the macroscopic volume change should be determined by the interactions between the monomers; the swelling ratio can exaggerate the microscopic conformation at the molecular level. In other words, a small change in the external conditions can induce a dramatic macroscopic volume change [32].

After gelation, the total weight of the initial dried sample was measured and defined as W_0. The dried sample film was immersed in pure water at room temperature. The weight ratio of pure water to dried gel was kept constant (water, $W_0 = 695{:}3$) [17, 21, 27, 28, 32]. Here, the swelling ratio is defined as W_t/W_d, where W_t is the weight of the swollen sample in its equilibrium state and W_d is the dehydrated weight after measuring W_t. Therefore, the water content, w, can be expressed as $1 - (W_t/W_d)^{-1}$. Notably, W_d was smaller than the total weight of the initial dried sample, W_0, because of elution.

For the quantitative evaluation of the eluted polymer into the external medium (water), the concentration of total carbon (TC) was measured using a Total Organic Carbon (TOC) analyzer (TOC-V_{CSN}, Shimadzu) with a high accuracy of 50 $\mu g \cdot L^{-1}$, and the weight of the eluted polymer, W_e, was calculated from the TOC by subtracting the Total Inorganic Carbon (TIC) contained in carbon dioxide from the TC measurement [32]. For the calculation of W_e, the TIC in pure water was measured in advance, and a calibration curve was obtained by considering the aging effects in air and the glass bottle. The elution ratio, e, was defined as follows:

$$e = \frac{W_e}{W_0} \times 100\,(\%) \tag{13.1}$$

The following three experiments were conducted to determine the swelling ratio of physical gels [32]:

1. Simple swelling process (designated as "simple swelling"):
   A sample was placed in a container with a limited amount of pure water at room temperature and left to reach the equilibrium state. After the respective waiting period, the sample was removed from the water, and the weight, W_t, was measured. Then, the sample was dried for 24 h at 60 °C, and the weight of the dried sample, W_d, was measured. Using W_t and W_d, the swelling ratio, W_t/W_d, was obtained.
2. Water exchange process without drying (designated as "simple exchange"):
   A sample was placed in pure water and left for a designated time period at room temperature. After W_t was measured, the sample was placed into fresh pure water and kept at room temperature for 24 h. According to the literature [17], this waiting period was sufficient for the gel to reach the equilibrium state, as will be shown in Sect. 13.4.1. Although complete elution requires more time, almost all polymers were eluted after 24 h. This process was repeated N times, where N denotes the number of water exchanges and W_e was calculated from the measured TOC of the respective samples of residual solvent water.
3. Water exchange process with drying (designated as "exchange with drying"):
   A sample was placed in pure water and left for a designated time period at room temperature. After W_t was measured, the sample was dried at 60 °C for over 24 h. The sample was then placed in fresh pure water and kept at room temperature. This process was repeated N times, and W_e was calculated from the measured TC of the respective samples of residual water. Concurrently, the weight of the sample dried at 60 °C after water exchange, W_d, was determined and was found to be smaller than W_0 because of elution.

The molecular weight distribution of the eluted polymers was obtained using size-exclusion chromatography (SEC). The molecular weight distribution of the sample was measured via SEC using a high performance liquid chromatography (HPLC) instrument (Agilent 1100 Series, Agilent) and column (TSKgel alpha-M, TOSOH). In this study, standard poly(ethylene oxide) (PEO) solutions with varying molecular weights were used as calibration standards for the analysis with an eluent of methanol + 0.1 M NaCl solution. The oven temperature was 40 °C, and the flow velocity was 0.5 mL·min^{-1}. The weight-average molecular weight, M_w, and polydispersity index, *d*, defined as the ratio of M_w to the number-average molecular weight, M_n, were calculated.

13.3.2 Properties of Microcrystallites

To study the network microstructures, the formation of hydrogen bonds and the characteristic properties of PVA physical gels, Fourier-transform infrared (FT-IR) spectroscopy, X-ray diffraction (XRD), and differential scanning calorimetry (DSC) were applied.

The mid-IR spectra of the dried plate gels were measured using an FT-IR spectrophotometer (Jasco, FT-IR-610) equipped with an attenuated total reflectance (ATR) attachment with a zinc selenide crystal in order to investigate the conformation, dissociation, and bonding state of the functional groups of the swollen gels in the solvent. An appropriate amount of gel was placed on the zinc selenide crystal, and the FT-IR spectra were then measured at room temperature.

Small-angle X-ray scattering (SAXS) measurements were performed at room temperature using a SAXS system with a wavelength of 0.1488 nm at BL10C in the Photon Factory, High Energy Accelerator Research Organization (KEK) [26, 33]. A small- and wide-angle X-ray scattering (SWAXS) system with a wavelength of 0.1542 nm, operated at 40 kV and 50 mA (SAXSess, Anton Paar, PANalytical), was additionally used. The average crystallite size *D* was estimated by performing wide-angle X-ray scattering (WAXS) measurements using RINT-2000 (Rigaku Corp., Tokyo, Japan) diffractometer, operated at 40 kV and 200 mA at room temperature. From the WAXS profiles, *D* was calculated using the Scherrer equation [34], $D = k\lambda/(\beta\cos\theta)$, where $k = 1$ (the shape of the crystal is assumed to be a sphere), β is the width at the half-maximum intensity of the reflection, θ is the Bragg scattering angle, and $\lambda = 1.54$ Å is the wavelength of the X-ray radiation.

The crystallinity was measured using a DSC (DSC8000, PerkinElmer) and a thermogravimetric analyzer (TGA, TGA4000, PerkinElmer). Two identical samples were cut from the dried sample. One sample of weight W_0 was heated from 25 to 250 °C at a heating rate of 10 °C min^{-1} using the DSC. Simultaneously, the true weight of the total polymer was calculated by performing TGA on the other sample. The sample was heated and maintained at 120 °C for 30 min. The rate of weight loss, w^*, was measured and regarded as the residual water content of the initial dried sample. From these measurements, the crystallinity, X_c, was calculated using the following equation [27]:

$$X_c = \frac{\Delta H}{\Delta H_{PVA}(1 - w^*)} \tag{13.2}$$

where ΔH and ΔH_{PVA} indicate the amount of heat required to melt the microcrystallites in the dried sample and in 100 % crystalline PVA (=138.6 J·g^{-1} [35]), respectively. The extrapolated onset temperature, T_{eo}, was also measured from the DSC curve and regarded as the approximate melting point of the microcrystallites.

A microscopic image of the swollen PVA gel surface was obtained by PeakForce Tapping mode atomic force microscopy (AFM, Dimension Icon AFM, Bruker) in pure water at room temperature, with a silicon cantilever with spring constant of 0.7 N·m^{-1} and a probe tip radius of curvature of 2 nm (ScanAsyst Fluid+, Bruker). The images were processed only by flattening to remove the background slope.

13.3.3 Mechanical Strength

For tensile strength testing, the gels in their equilibrium states were cut using a dumbbell cutter (JIS K-6251-8), and the dumbbell-shaped test samples (0.5–0.6 mm in thickness, 50 mm in total length, the central parallel portion width and length being 8.5 mm and 16 mm, respectively) were used to obtain stress-strain (σ-ε) curves in air for their respective gels by a uniaxial loading test. The rate of elongational stretch between the grips was 1.0 mm·s^{-1}. The forces generated during deformation were measured using two testers. The gel was stretched in pure water at room temperature using a uniaxial loading tester of a double-leaf spring with four strain gauges. A tensile tester (INSTRON 5965, Instron Japan Co. Ltd.) was also used for measurements in air at room temperature. Using a food dye, two markers were applied to the gel by hand in the gauge section, comprising the central parallel portion. Images of the gel and markers were captured during the tests by a video camera and saved to a computer, from which the distance between the markers was analyzed to determine the strain. From the measured values of force and elongation, the nominal stress, σ, and strain, ε, were calculated; from these, the breaking stress was obtained. The initial tensile strength, corresponding to the Young's modulus, E, was evaluated using the data in the low-strain region below $\varepsilon = \sim 0.2$. The data was plotted in log-log scale, and a solid line with a slope of 1 was fitted to the linear region by the least-square method.

The tear force was obtained by a trouser tear test in water. For this, the samples were first cut into rectangular shapes (0.5–0.6 mm (CD gel) or 2.0 mm (FT gel) in thickness, 75 mm in length and 30 mm in width), and initial slits (25 mm in length) were cut in the center of one of the short sides of the samples, from which the tear propagated. The force generated during deformation was measured using a double-leaf spring with four active strain gauges to maintain the accuracy of the measurements. In this test, the rate of separation of the grips holding the gel was held constant at 0.1 mm·s^{-1}. The tear energy, G, was defined as the fracture energy released per unit area of crack surface growth (J·m^{-2}), which was obtained by dividing the tear force by the gel thickness.

The compression force was measured using a TA.XTPlus Texture Analyzer (Texture Technologies Co.) with a cylindrical probe of 25 mm in diameter [26].

The 30 % compression stress, $\sigma_{30\,\%}$, was measured at room temperature for each gel.

A ball-on-disk reciprocating friction test was conducted in water at room temperature except Sect. 13.6, at a sliding speed of 20 mm·s^{-1} and distance of 100 m (stroke, 25 mm; 2,000 total cycles) at a constant load of 5.88 N, using a friction tester (TriboGear TYPE:38, HEIDON). In the reciprocating friction test lubricated with saline and simulated synovial fluid in Sect. 13.6, the frictional behavior of the ellipsoidal PVA gel specimens was examined against a glass plate, corresponding to a simplified knee prosthesis model.

13.4 Typical Characteristic Properties of Physical PVA Gels

13.4.1 Swelling and Elution Properties

The swelling and elution ratios were measured during a simple swelling. Figure 13.3a shows the time revolutions of e and W_t/W_d of standard CD gel. The measurements were repeated three times for each sample, and the average of the data is shown in the figure; the error bar is less than that of the size of the symbols. The samples differed with waiting periods, resulting in scattered data points. With increasing time, both e and W_t/W_d increased and became constant after 48 and 24 h, respectively. This observation confirmed the elution from the gel into the outer water, showing that it stopped after the gel reached equilibrium. Furthermore, as shown in Fig. 13.3b, the short chains are suggested to be the predominant eluted species in the solvent; the peak position of the molecular weight distribution curve shifts to a higher molecular weight as a result of the elution of polymers with lower molecular weights. The M_w of the gel increased from 6.86×10^4 (PVA117 powders) to 7.73×10^4 after elution, while the polydispersity index, d, decreased from 2.92 to 2.22. This is caused by the elution of the short chains $\left(M_w = 3.00 \times 10^4\right)$.

The elution ratio was measured during both the simple exchange and exchange with drying processes. Figure 13.3c shows the cumulative e of all CD gels during the repeated water exchange processes. The cumulative e value gradually increased after a steplike increase at $N = 1$ until e approached a constant value. The elution ratio approached zero when sufficient water exchange had occurred. Notably, elution during the exchange with drying process stopped earlier than that during the simple exchange process, indicating that elution is restricted by the formation of additional hydrogen bonds during the drying process. In the simple exchange experiments, more than 50 % eluted at $N = 1$. In contrast, more than 80 % of the total polymers were eluted during the first swelling under exchange with drying. From the figure, it is evident that elution stops after ~12 water exchanges for the exchange with drying process, but continues over several dozen exchanges for the simple exchange process under the present experimental conditions, including temperature and weight ratio of dried gel to water.

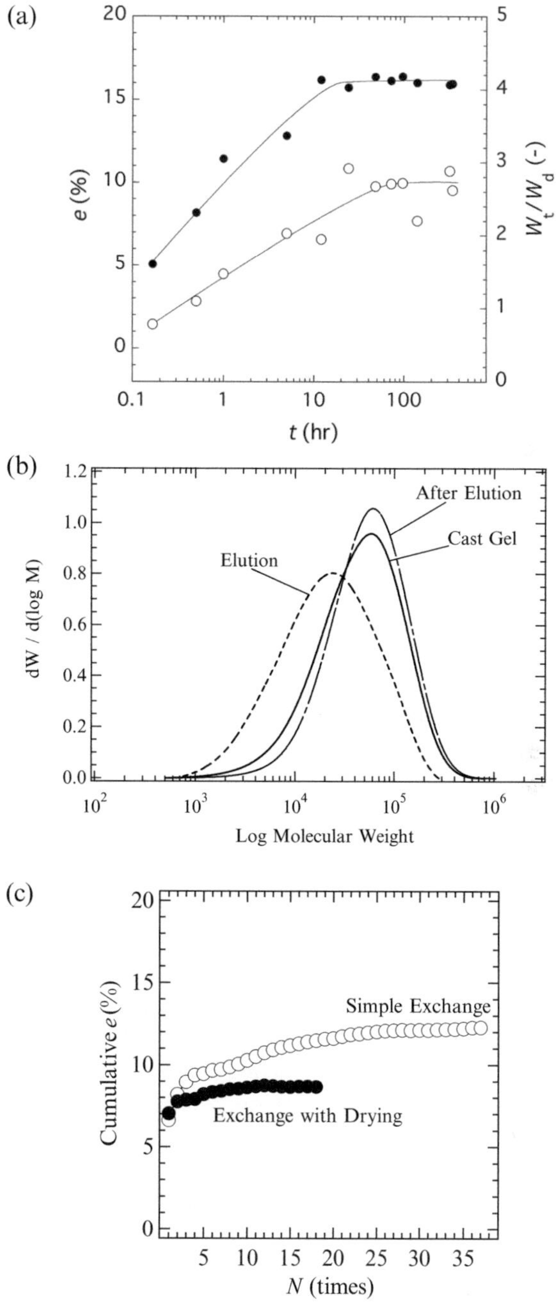

Fig. 13.3 (**a**) Time evolutions of the elution ratio, *e* (*open circles*), and the swelling ratio, W_t/W_d (*closed circles*), of PVA117 CD gel in the swelling process after gelation (simple swelling). (From Ref. [17], with permission from RSC) (**b**) Molecular weight distributions of original gel, eluted PVA, and gel after elution. (**c**) The cumulative elution ratio, *e*, of CD gels during the simple exchange (*open circles*) and exchange with drying (*closed circles*) processes (From Ref. [18], with permission from Elsevier)

To examine the material parameters of PVA powders, which determine the elution amount, the following experiments were conducted: After sufficient water exchange, a CD gel (called first cast gel) was redissolved, and the solution was decanted to obtain a redissolved CD gel (second cast gel). The solvent water of second cast gel was repeatedly exchanged and was redissolved again to obtain third cast gel. The second and third cast gels were prepared after the cumulative e became almost constant. The changes in the swelling and elution behavior of the redissolved CD gels were evaluated quantitatively. Figure 13.4a shows the changes in the cumulative elution ratio (e) during the water exchange processes. From this figure, it is apparent that e is largest at $N = 1$ in each sample, and the cumulative e gradually increases as N is increased until finally becoming constant. Although e at $N = 1$ decreases in the second and third redissolved CD gels, the figure confirms that elution still occurs. Moreover, the cumulative e of the redissolved CD gels stabilizes much earlier than that of the first cast gel. The fact that the cumulative e of the redissolved CD gels apparently decreases indicates that the elution is suppressed by changes in the network structure from the redissolution process. From the chromatograms at $N = 1$ of the first, second, and third solvents, the molecular weight distributions of the PVA in the gels and outer solvent were calculated by the method described in the experimental section. As shown in Fig. 13.4b, the molecular weight distribution curve of the PVA of the gel network changes; the peak position of the molecular weight distribution curve shifts to a higher molecular weight as a result of the elution of polymers with lower molecular weights. In Fig. 13.4c, the molecular weight distribution of the eluted PVA varies correspondingly; the peak position shifts to a higher molecular weight compared with that of the PVA eluted from the first cast gel. This observation is consistent with our previous work, in which the M_w of the eluted polymer became larger as the M_w of the polymer in the gel increased [26].

In a previous report [17], three mechanisms that occur during swelling after drying were presented. One involves the release of some of the polymers that were not cross-linked during gelation; this elution stops when the eluted polymer reaches a critical concentration in a limited amount of solvent. Therefore, the residual non-cross-linked polymers are eluted after the next water exchange. Another mechanism is the destruction of microcrystallites during the swelling process, which results in the partial release of polymers that were cross-linked during gelation. The third mechanism is the result of the positive or negative osmotic pressure that may be induced by the release of the cross-linked polymers, which leads to a change in the swelling ratio by the decrease of polymers in gels; this occurs because the elution of polymers results in fewer polymer chains, which decreases the attractive forces between the polymer chains. Concurrently, polymer elution might lead to increased attractive forces by the increased number of hydrogen bonds, introduced to the gels as a result of the increase in the number of free hydroxyl groups by the decrement in dangling bonds by elution. Notably, the X_c value of the film dried at 60 °C after swelling (i.e., after elution) was 0.538, almost 150 % larger than that of the as-prepared standard PVA cast film ($X_c = 0.345$). It was much larger than that of cast film dried at 60 °C ($X_c = 0.385$), which was only about 10 % larger than that

Fig. 13.4 (**a**) Cumulative elution ratio, *e*, of first eluted (*open circles*), second eluted (*closed triangles*), and third eluted (*open squares*) solvents. The molecular weight distribution of PVA (**b**) from redissolved gels and (**c**) from outer solvent water (From Ref. [27], with permission from Springer)

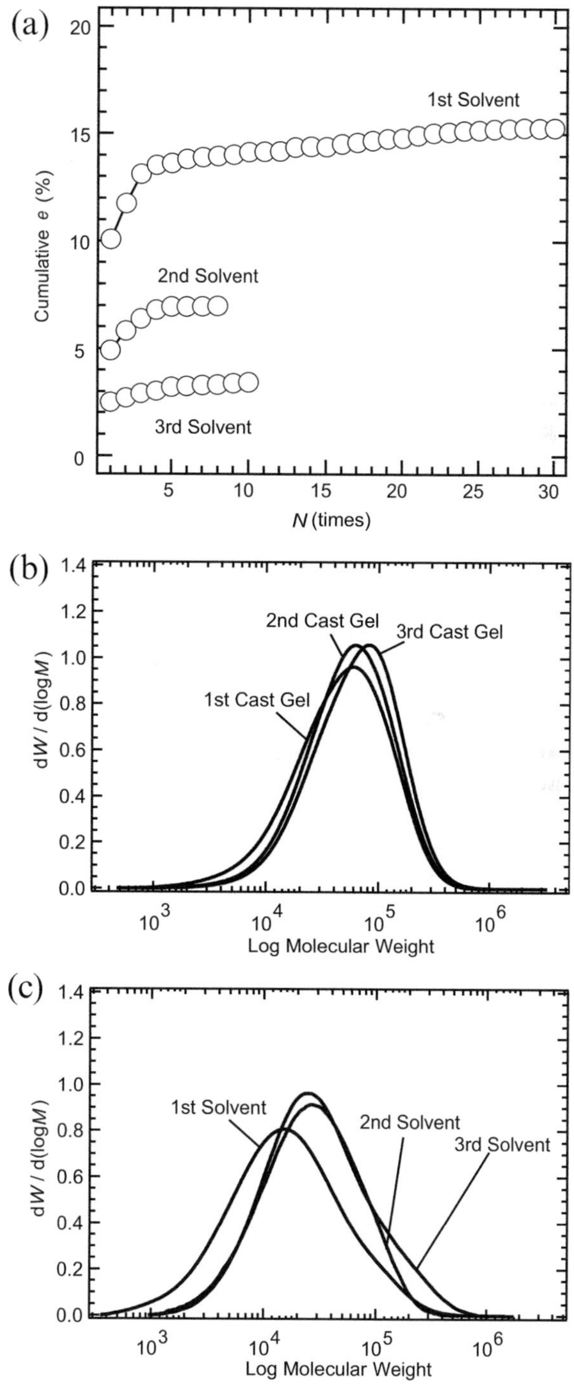

of the as-prepared standard PVA cast film. This result suggests that the swelling of the standard gel after gelation assisted the formation of many more additional microcrystallites than that of CD gel directly prepared at 60 °C.

With these competing factors, repeated water exchange results in further elution into freshwater, regardless of whether the sample is dried between exchanges. After a sufficient number of water exchanges, the amount of polymers eluted decreases and appears to approach zero; however, we cannot exclude the possibility that elution could continue for the duration of the experimental time (i.e., a year) if the resolution of the apparatus was higher than that used for the present experiment.

13.4.2 Formation and Destruction of Microcrystallites

The swelling ratio of standard CD gel depends on the amount of residual water, which was obtained by interrupting the drying instead of completing drying [16]. In this study, sufficiently dried PVA cast films were exposed to ambient air or maintained under different humidity conditions at room temperature to incorporate certain amounts of water. Figure 13.5a shows the X_c and T_{eo} values of PVA cast films with different rates of weight loss, w^*, as measured by TG as defined in Eq. 13.2. With increased w^*, X_c decreases and T_{eo} increases. The decrease in X_c indicates a decrease in microcrystallite structure as a result of the swelling pressure by the adsorption of moisture. Furthermore, the increase in T_{eo} corresponds to the increase in the average melting temperature of microcrystallites in PVA cast films. These observations suggest that small microcrystallites were destroyed by the swelling pressure created by the addition of water, whereas larger microcrystallites grew sufficiently to remain intact during the swelling process. These results suggest that the microstructure of a standard CD gel after gelation can be changed by the amount of residual water, which depends on the experimental conditions of gelation, such as T_{gel} and the surrounding *RH*.

As described in the earlier experimental section, the weight ratio of solvent to W_0 was fixed to 231.7 (695/3) in the present experiments. Figure 13.5b shows W_t/W_d and e of the standard CD gels immersed in different weight ratios of solvent water. With increases in the weight of solvent water, decreases occurred in W_t/W_d as well as increases in e. This indicates the origin of an attractive force during the elution. The observed decrease in swelling ratio could be attributed to the additional formation of hydrogen bonds between free –OH groups generated by elution. According to these results, the same standard CD gels, after sufficient drying, show different swelling and elution ratios depending on the changes in the environment conditions. This change could be attributed to the changes in the number and size of microstructures, related to the elution phenomenon.

Figure 13.6a shows SWAXS spectra of the PVA cast gel film in a simple swelling process. The intersection of the two extrapolated lines before and after the broad peak in the dried gel with $W_t/W_d = 1.05$ is $q = 0.65$ nm^{-1} ($L = 9.7$ nm). With an increase in W_t/W_d (from 1.05 to 2.6) by slowly adding water, the peak position shifts

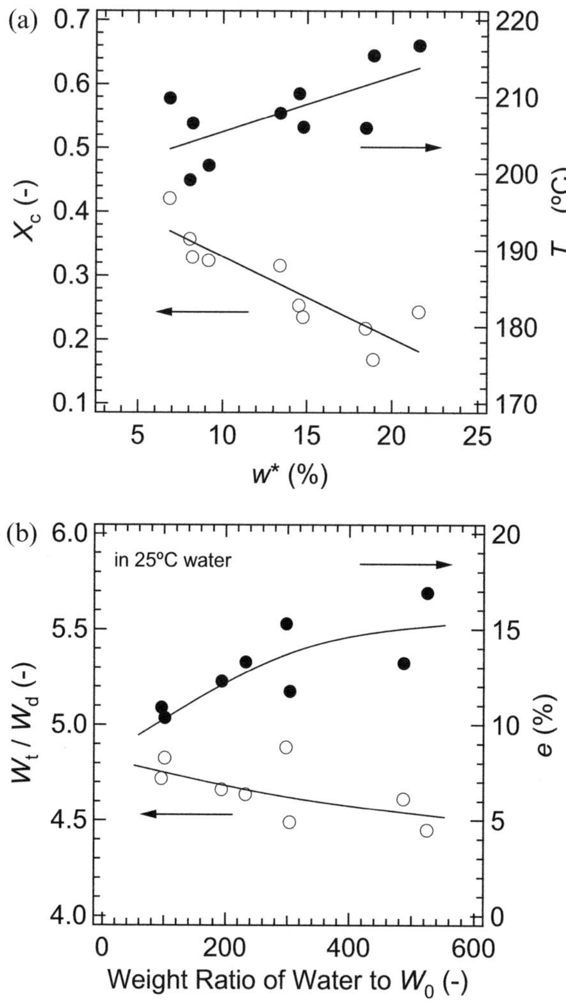

Fig. 13.5 (**a**) X_c and (**b**) T_{eo} of cast dried PVA films with different w^* and (**b**) changes in e and W_t/W_d with ratio of solvent weight to W_0

to a lower wavelength and finally disappears above $W_t/W_d = 3.3$ as the scattering from water becomes dominant in the spectrum. As is shown in Fig. 13.6b, L increases at first and gradually increases when W_t/W_d exceeds 1.8. Therefore, we can consider L of the gel with $W_t/W_d = 2$ to be approximately L of the swollen gel.

Notably, these observations of the microstructures were performed in dried or half-dried gels, because it is difficult to analyze the network structures of fully swollen hydrogels with high water content, w, as in the case in the CD and FT gels (where w >80 % of the total weight). Recent experiments have indicated, however,

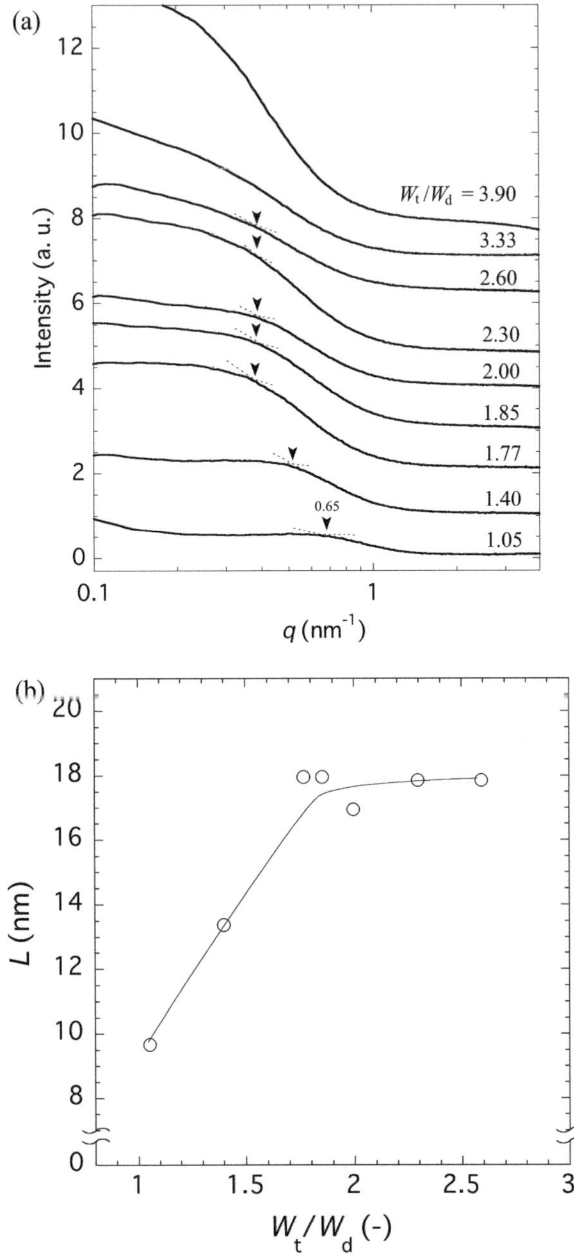

Fig. 13.6 (**a**) SWAXS spectra of as-prepared CD gels with increasing W_t/W_d. The vertical axis of each spectrum is shifted for clarity. (**b**) L of CD gels during increasing W_t/W_d (From Ref. [33], with permission from IOP)

that the observations of D, X_C, and L of gels in the dried state would correlate qualitatively with those of gels in the swollen state [18, 27, 28].

13.4.3 Factors Affecting the Swelling and Elution Ratios

The swelling and elution properties of physically cross-linked CD gels depend on the network structure of the PVA and cross-linking, which is characterized by the size, number, and distribution of microcrystallites. Therefore, the swelling and elution ratios can be manipulated by the adept control of the conditions adopted for the preparation of gels. Among the various factors that influence the formation of microcrystallites, the temperature and *RH* at gelation play an important role. In addition, the size of the gel is also crucial in determining the network structure of the gels.

The weight of PVA powders, W_p, was changed in addition to T_{gel} and *RH*. Figure 13.7 shows W_t/W_d, e, X_c, and D of CD gels prepared using $C_{PVA} = 15$ wt.% PVA solution with different W_p values (0.15–0.6 g), which correspond to the solution weight, 1–4 g, under different T_{gel} and *RH*. As seen from Figs. 13.7a, b, both the ratios of W_t/W_d and e decrease with increase in W_p under different combinations of T_{gel} and *RH*. For values of $W_p < 0.45$ g, as shown in Figs. 13.7c, d, we can observe a decrease in X_c and an increase in D with an increase in W_p. Comparing the changes in the microscopic parameters, X_c and D, with the macroscopic observations, W_t/W_d and e, the effect of the change in D is more prevalent than that of X_c for W_t/W_d. This is consistent with the results of the mechanical strength testing, described earlier. These results suggest that the effects of increases in W_p on the W_t/W_d ratio are similar to that of *RH* and that the increase in D is necessary to understand the swelling and elution behaviors.

Finally, the relation between W_t/W_d (or w) and e was examined to determine the optimum condition required to prepare CD gels. As shown by a solid line in Fig. 13.8a, the elution ratio is well correlated with the water content; all data follow this single master curve. High T_{gel} or *RH* values result in low swelling and elution ratios, whereas low T_{gel} or W_p values result in high swelling and elution ratios. The elution ratio can be correlated to the swelling ratio, because the microstructure determines both the ratios. According to the experimental results, the elution ratio can be effectively minimized with high T_{gel}, high *RH*, and large W_p. Elution is accelerated when the water content exceeds 70 %. Three different data points from CD gels prepared under different combinations of T_{gel}, *RH*, and W_p are presented in Fig. 13.8a. As shown, all three data sets exhibit similar swelling and elution ratios. This observation indicates that the state of microcrystallites could be similar in swollen gels, although the X_c and D values of the cast dried films differed, given that weak microcrystallites are destroyed during the swelling process (Fig. 13.8b). The size and number of microcrystallites clearly influence the elution ratio.

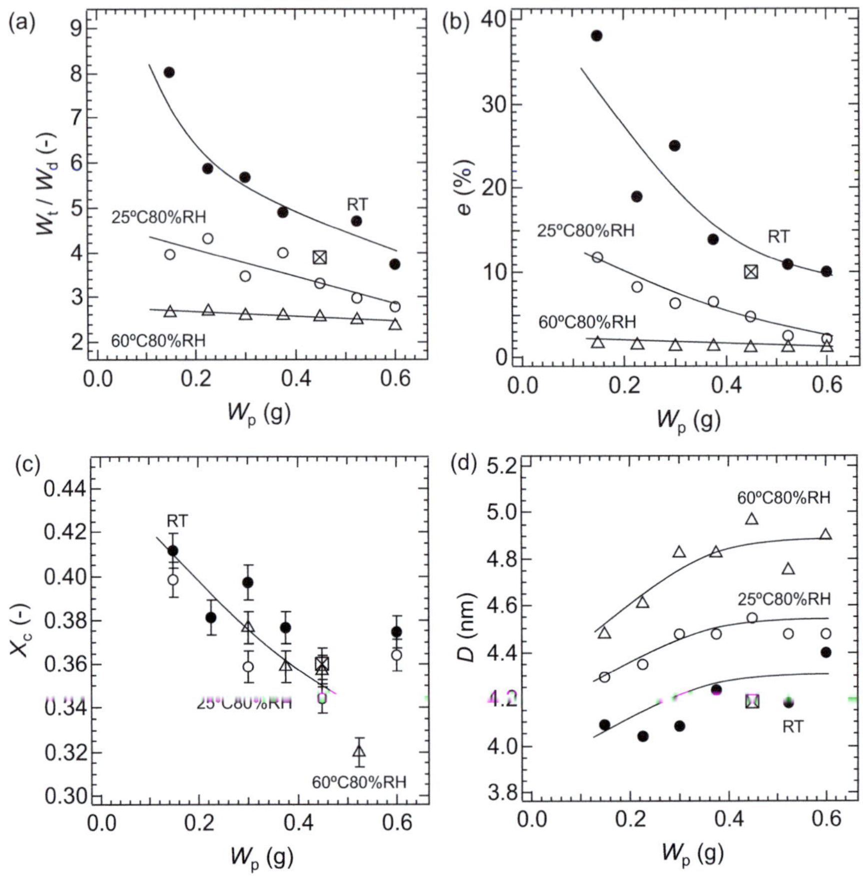

Fig. 13.7 (**a**) Swelling ratio, (**b**) elution ratio, (**c**) X_c, and (**d**) D of PVA cast film with different W_p prepared using 15 wt.% PVA solution. The cross symbol corresponds to standard CD gel (From Ref. [28], with permission from Wiley)

13.5 Mechanical Properties and Nano- and Microscaled Network Structures

13.5.1 Improvements of Tensile Strength by Mild Treatments

Figure 13.9a shows the nominal stress-strain (σ-ε) curve of each swollen gel, measured in water during a water exchange process without drying. As discussed in Sect. 13.4.1, the tensile strength, equal to the breaking stress in the case of the present PVA gels, increased with increasing N. Polymer elution may caused increased attractive forces by additional hydrogen bonds introduced in the gels as a result of the increase in the number of free hydroxyl groups resulting from the decrease in dangling bonds by elution. Correspondingly, the slope in the low ε range,

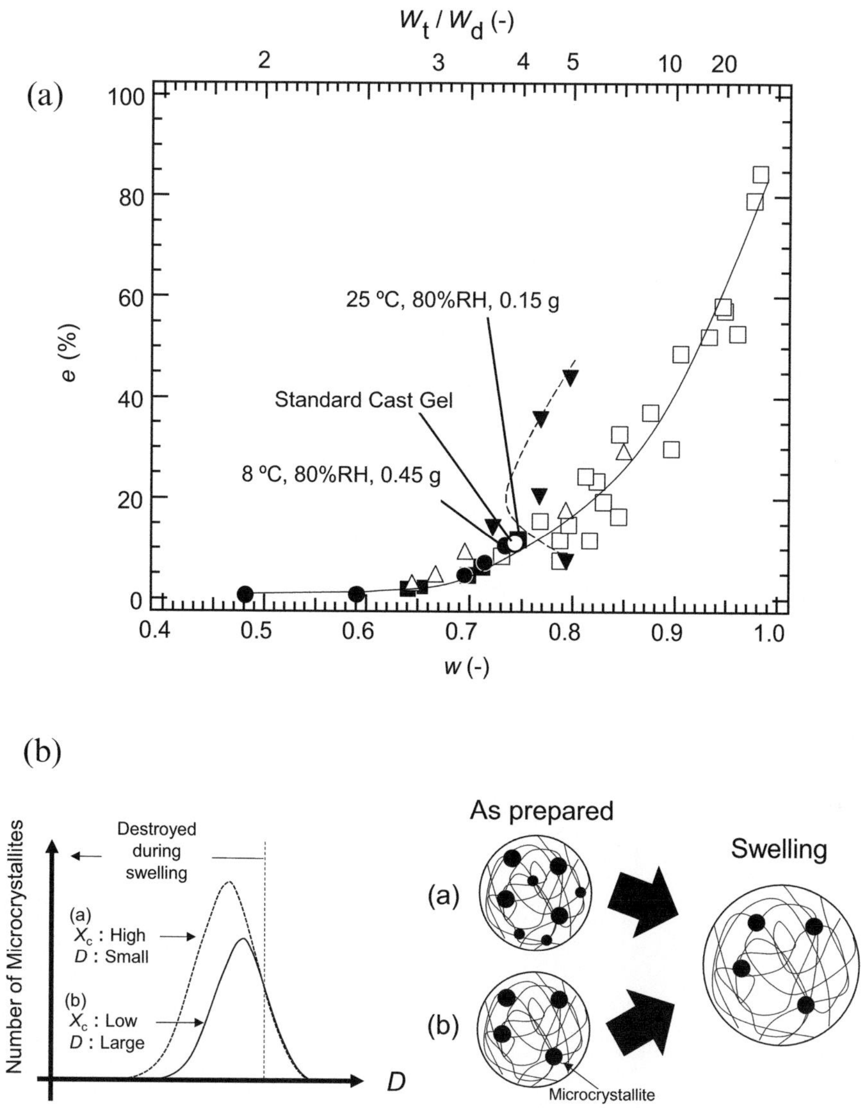

Fig. 13.8 (**a**) Relation between swelling ratio and elution ratio of CD gels: *open circle*, standard gel; *open square*, different W_p; *open triangle*, different T_{gel}; *closed circle*, different T_{gel} with 80 %RH; *closed square*, different W_p and T_{gel} with 80 %RH; and *inverted triangle*, different degrees of polymerization. (**b**) Schematic of the distribution of microcrystallites and the network structure of PVA cast gel (From Ref. [28], with permission from Wiley)

corresponding to the Young's modulus, E, increased apparently with increasing N (Fig. 13.9b). The tensile strain, equal to the breaking strain in the case of the present PVA gels, on the other hand, was maintained at $\varepsilon = \sim 4$.

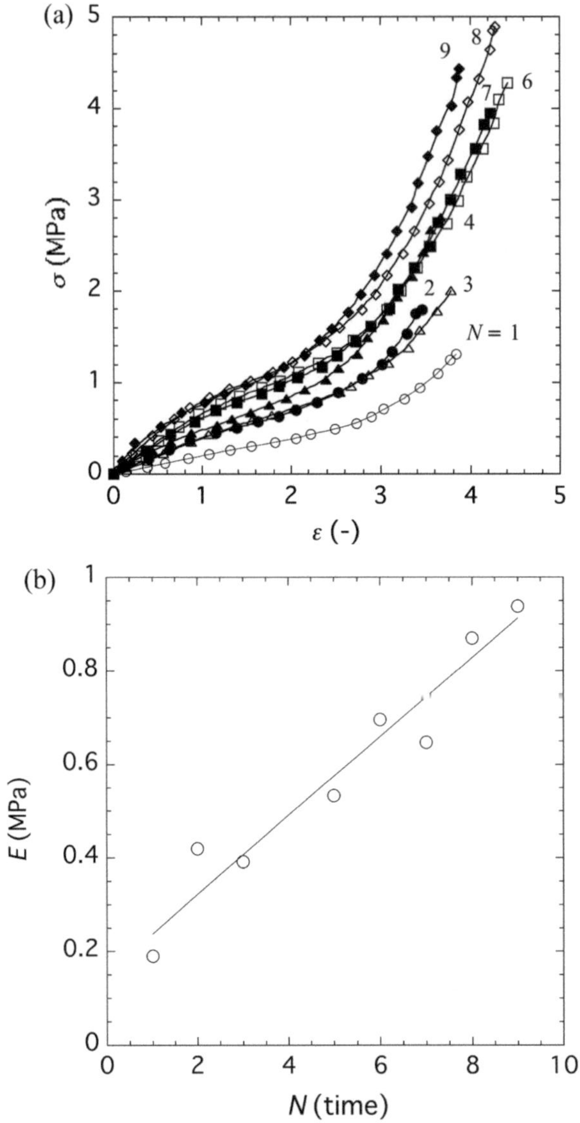

Fig. 13.9 (**a**) Stress-strain (σ-ε) curves during the simple elongation test of CD gels during water exchange with drying, where the gels were ruptured at the maximum ε, and (**b**) Young's modulus, E, of the samples, determined from the σ-ε curve in (**a**) (From Ref. [33], with permission from IOP)

Figure 13.10a shows the nominal σ-ε curve of swollen gels prepared at different T_{gel} values, measured in water. The breaking stress increased with an increase in T_{gel}, while the slope in the low ε range (linear portion of a σ-ε curve), corresponding to the Young's modulus, E, increased rapidly with an increase in T_{gel} above 50 °C

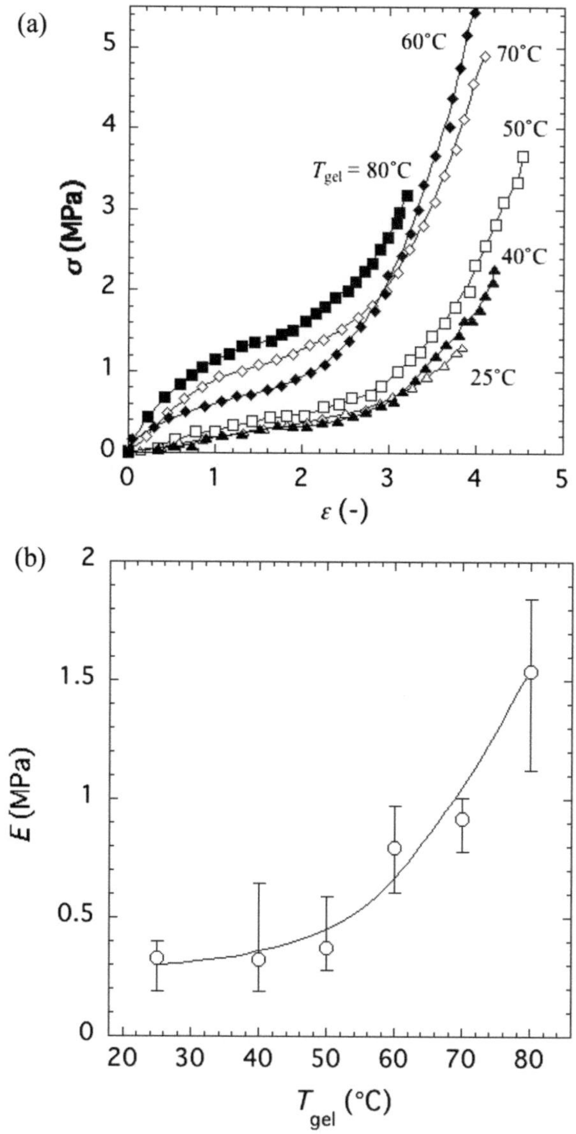

Fig. 13.10 (**a**) Stress-strain curve and (**b**) Young's modulus of CD gels with different T_{gel}. The error bar denotes the range of error among three samples (From Ref. [26], with permission from RSC)

(Fig. 13.10b). According to the literature [33], the increments of the breaking stress and E of CD gel with increasing T_{gel} resulted from increments of the crystallinity. The tensile strain was also maintained at $\varepsilon = \sim 4$.

Originally, the PVA cast films were prepared at room temperature and under an uncontrolled atmosphere. Therefore, even in gels prepared by the same method, the absolute values of the swelling ratio deviated by ±10 %. This is because environmental factors such as temperature and humidity were not controlled during

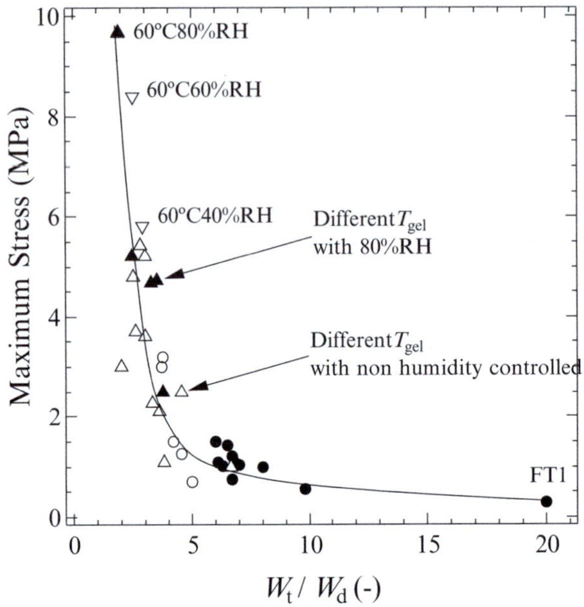

Fig. 13.11 Relationship between the breaking stress and the swelling ratio. *Open circles* denote CD gels, prepared by the standard recipe with non-humidity controlled, and *closed circles* denote FT gels, prepared by different number of freeze-thawing cycles. *Open*, *closed*, and *inverse open triangles* indicate CD gels, prepared by different T_{gel} and/or humidity (From Ref. [36]) DOI: 10.1177/0954411915615469

preparation. Recently, the effects of drying temperature as well as humidity on the swelling properties and mechanical strength of the gels were investigated. The swelling ratio was found to change with adjustments to both factors.

By these methods of mild water exchange treatments or temperature and humidity changes, it is possible to control the swelling behavior of the gel to some extent. However, according to swelling theory and rubber elasticity, presented in Eqs. 13.1 and 13.2, ϕ decreases (the volume increases) and F increases as ν increases, and vice versa. In Fig. 13.11, the breaking stresses of FT and CD gels prepared using a 15 wt.% PVA117 pre-gel solution by different conditions are well correlated with the swelling ratios and following a master curve [36]. Therefore, the realization of both high water absorbability and high mechanical strength is difficult in gels prepared by either FT or CD methods.

13.5.2 Tearing Resistance

Figure 13.12a shows the variation in G with time for both FT4 and CD gels. From this, it is evident that in the case of the CD gel, G propagates steadily after first

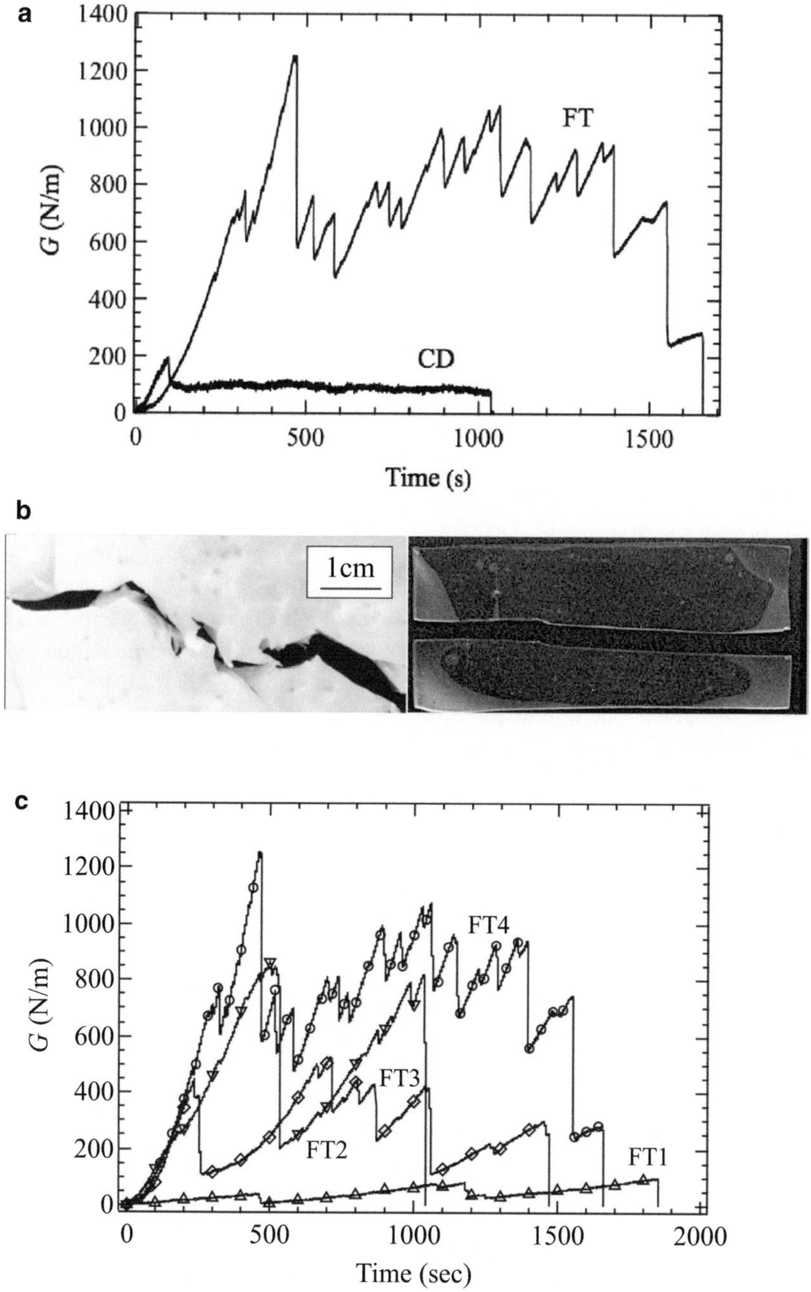

Fig. 13.12 (**a**) Fracture energy, *G*, of FT and CD gels during a tear test with a pulling velocity of $v = 0.1$ mm/s. (**b**) Fracture surfaces of FT (*left*) and CD (*right*) gels. (**c**) *G* of FT gels prepared by different numbers of the repeated freeze-thawing cycles (From Ref. [29], with permission from Wiley)

peaking during the initial stage. This larger peak is attributed to the initial fracture energy required to initiate propagation in the vicinity of the slit end, a region in which the polymer network is reinforced by the turbulence induced by cutting the slit. In the case of the FT gel, however, G exhibits a stick-slip behavior, and the average G is $\sim$10 times larger than that of a CD gel in the flat region. Pictures of typically fractured gels by tear tests are presented in Fig. 13.12b, in which a characteristic fracture pattern is observed. This appearance of stick-slip behavior in the fracture pattern of FT gel corresponds visually to the oscillatory tear curve. These results indicate that it is the microscopic network structure as characterized by fibrils that determines the tear energy of the FT gels. The appearance of stick-slip behavior and the large tear energy results from the formation of a microscale network of fibrils (bundled polymer) and from network inhomogeneity on the microscale.

Several factors can affect the fracture force of hydrogels, including the degree of cross-linking, total concentration of the polymer, and fracture velocity; all must be examined to fully understand the tear mechanism. In the case of FT gels, the swelling ratio decreased, and the crystallinity increased with an increasing number of repeated cycles and a subsequent increase in G (Fig. 13.12c). This was also attributed to the increased formation of microcrystallites in the network by repeating the FT process.

As shown in Fig. 13.13a, the G of CD gels increases with gelation temperature, T_{gel}, from 25 to 60 °C. This suggests that a change in T_{gel} affects the network structure, especially regarding the size, number, and distribution of microcrystallites, determined at gelation. According to the literature [26], the growth of microcrystallites is enhanced by reducing the water content (i.e., evaporating bound water) of polymers with increasing T_{gel}. In actuality, the present results shown in Fig. 13.13b are consistent with the preliminary result [26], in which W_t/W_d was reduced and the crystallinity increased with increasing T_{gel}. The measurements were conducted on three different samples prepared using the same method and at the same time, except for $T_{gel} = 30$ °C, which were prepared using the same method and at the same time.

As shown in Fig. 13.2a, the gel becomes transparent as the initial water content is reduced; accordingly, the tear energy is also reduced (Fig. 13.14a). This is understandable given the decrease in W_t/W_d and increase in crystallinity associated with a decrease in w (Fig. 13.14b). These results therefore indicate that the network structure, characterized by microscale fibrils, can increase the tear energy. In other words, these fibrils can determine the tear energy of the FT gels.

13.5.3 Friction in Water

First the frictional behavior was tested for physical PVA gels. In addition to FT and CD gels, hybrid gels were tested, which were prepared as follows. For the first FT gel layer (bottom layer), the PVA solution (15 wt.% PVA117 aqueous solution) was

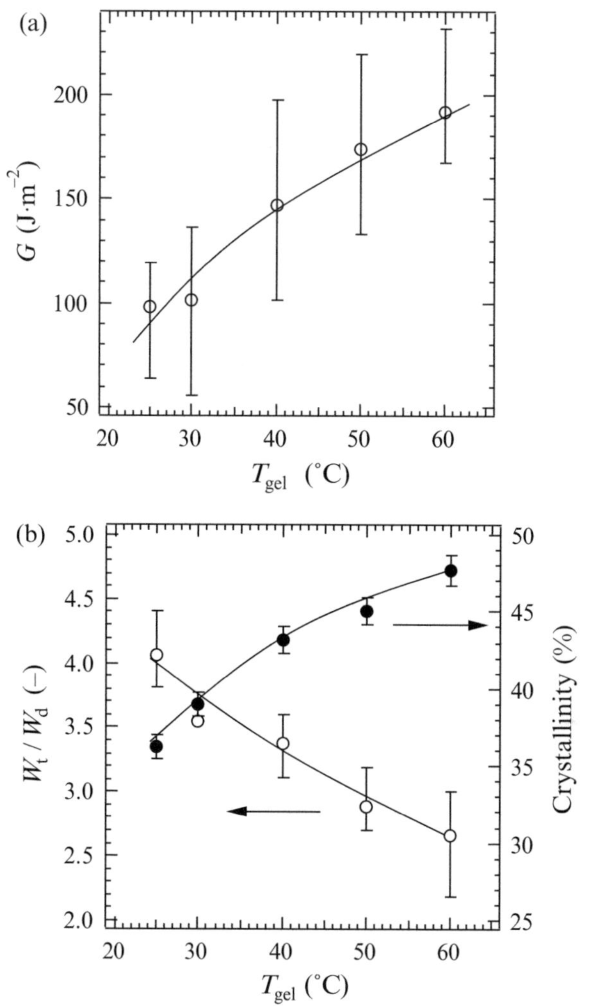

Fig. 13.13 Effects of the drying temperature, T_{gel}, on the fracture energy, G, (**a**) and the swelling ratio, W_t/W_d, of CD gels (From Ref. [29], with permission from Wiley)

frozen at −20 °C for 24 h, and then thawed at 4 °C for 16 h. This cycle was repeated four times. After the freezing-thawing process, for the second CD gel layer (upper layer), the same PVA solution was poured on the FT gel layer and dried at room temperature. Cross-sectional images of the hybrid gel are shown in Fig. 13.15. In the reciprocating friction test, the upper CD layer was used as a rubbing surface, and the bottom FT layer was fixed to the specimen holder with a cyanoacrylate adhesive.

To measure the lubrication properties of the physical PVA gels, a ball-on-plate reciprocating friction test was conducted under constant loading, corresponding to the migrating contact area on the biphasic PVA hydrogel. The reciprocating tests were conducted under lubrication with water at room temperature. As shown in

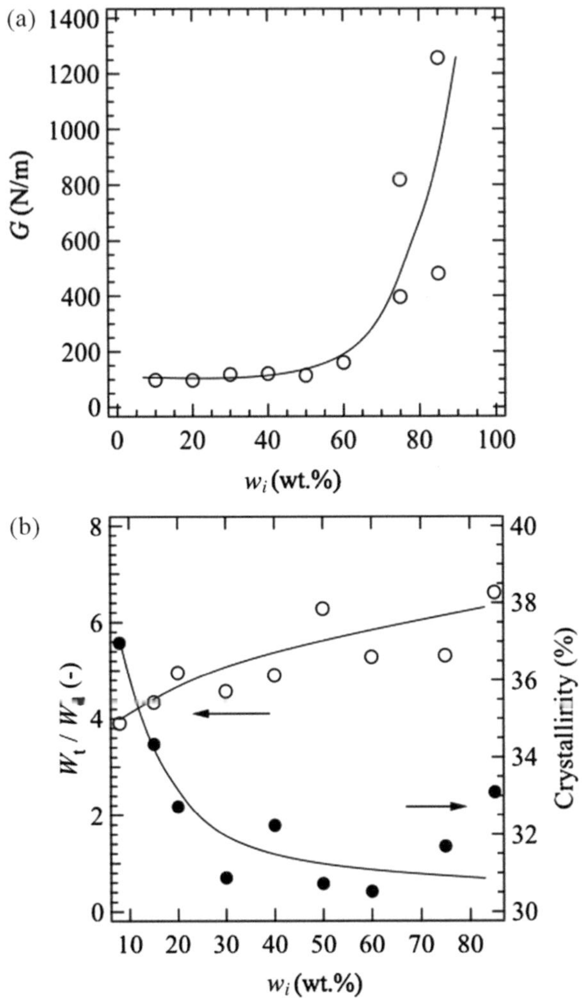

Fig. 13.14 (**a**) w dependence of the fracture energy, G, and (**b**) the swelling ratio, W_t/W_d, and the crystallinity of CD-FT composite gels prepared using the initial PVA solution with different water content, w_i wt% (From Ref. [29], with permission from Wiley)

Fig. 13.16a, the FT gel shows a large initial coefficient of friction, μ_k, that gradually increases, a change that is similar to that of natural articular cartilage. In contrast, a lower friction coefficient is displayed by the CD gel, experiencing a slight increase over time. CD or FT gels alone each exhibit a clear time-dependent change from a medium initial friction coefficient to a high coefficient, exceeding 0.1 or 0.2, respectively. However, the hybrid gel shows an extremely low μ_k of ~0.01 with no increase throughout the test. In addition, the compressive force was measured during the reciprocating tests using a texture analyzer with a cylindrical probe of 25 mm in

(a)

(b)

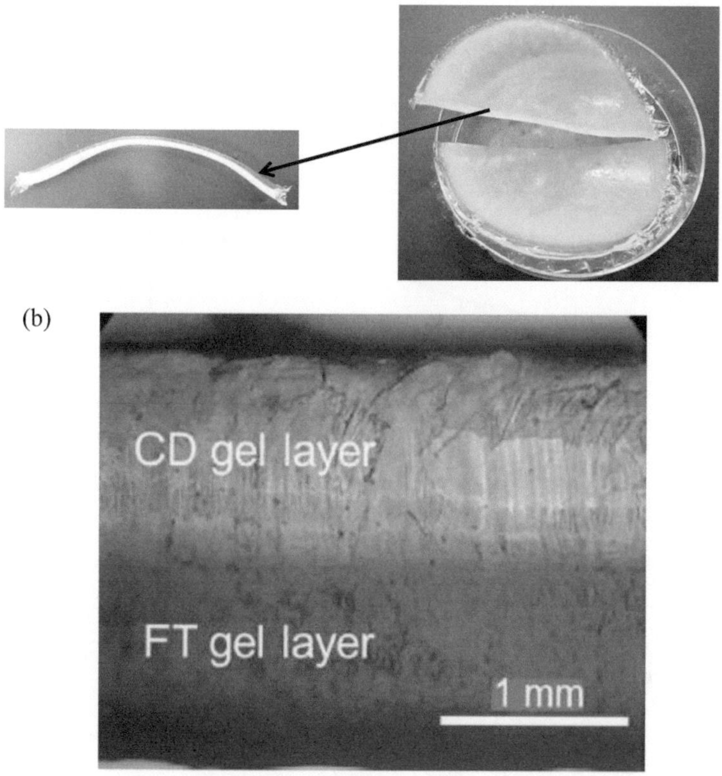

Fig. 13.15 (**a**) Pictures of a hybrid gel of laminated layers of two different hydrogels (*upper* CD and *lower* FT gels), which was prepared in a petri dish of polystyrene with an inner diameter of 85 mm. (From Ref. [36]) DOI: 10.1177/0954411915615469 (**b**) Cross section of a laminated hybrid gel of CD on FT gel in its swollen state

diameter [26]. The swelling ratio and the 30 % compression stress, $\sigma_{30\,\%}$, measured at room temperature for the respective gels, are presented in Fig. 13.16b. To fabricate FT dry gels, the FT gels were sufficiently dried in air at room temperature and then immersed in pure water again, identical to the treatment for the bottom layer of the hybrid gel. The FT dry gel shows the smallest W_t/W_d and the largest $\sigma_{30\,\%}$ of the tested gels. When $\sigma_{30\,\%}$ increases, W_t/W_d decreases correspondingly because of the decrease in the number of microcrystallites. The hybrid gel shows intermediate values of both characteristics.

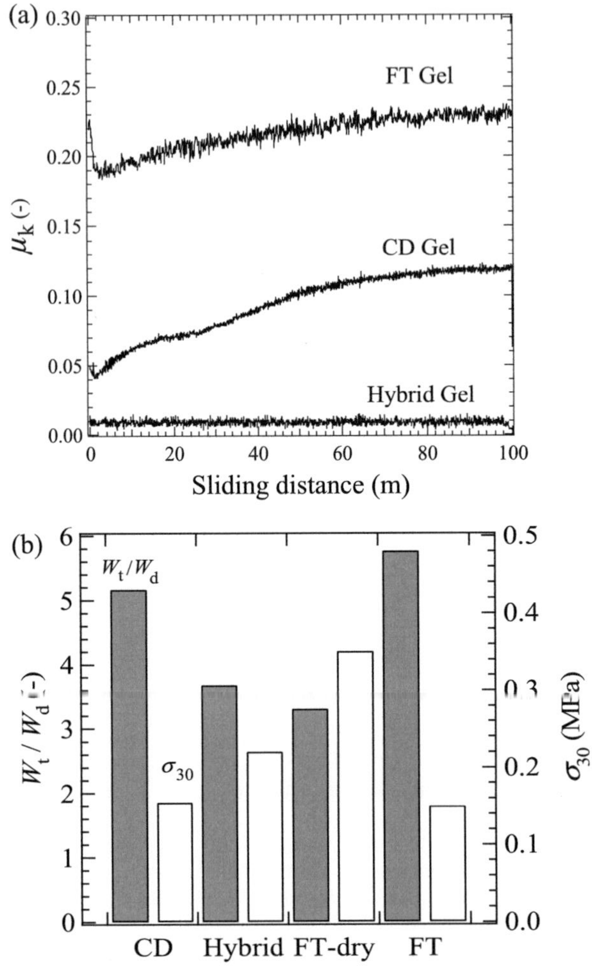

Fig. 13.16 (**a**) The coefficient of friction of FT, CD, and hybrid gels in water. (**b**) Swelling ratio and compression stress for 30 % reduction, σ_{30}, of the respective gel (From Ref. [36]) DOI: 10.1177/0954411915615469

13.6 Superior Lubricity in Saline and Simulated Synovial Fluid

In the previous section, three PVA gels fabricated by FT, CD, or hybrid methods were prepared as biocompatible artificial hydrogel cartilage materials and were tested by a ball-on-disk (or ball-on-plate) reciprocating test with a migrating contact area in water. The gels were physically cross-linked by microcrystallites by hydrogen bonding, with different nano- and microscale network structures depending on the fabrication method.

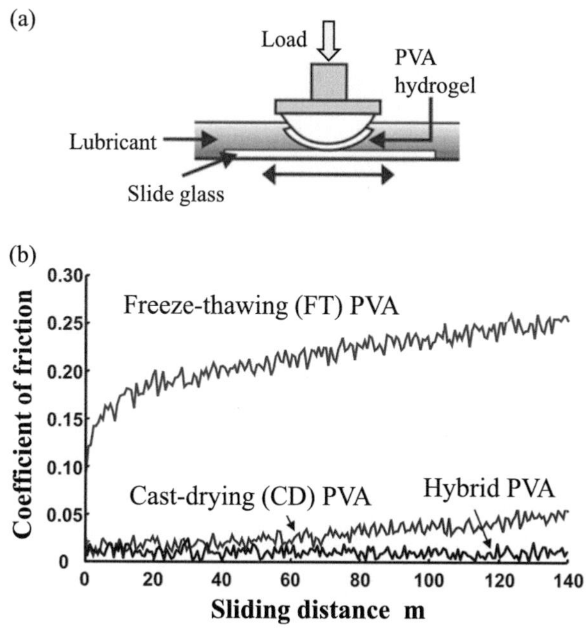

Fig. 13.17 (**a**) A reciprocating test with an ellipsoidal PVA specimen against a glass plate, conducted under conditions of continuous and constant loading. (**b**) Frictional behaviors of PVA hydrogels in saline in reciprocating tests (From Ref. [20], with permission from Elsevier)

To evaluate the lubrication properties of these gels under severe operating condition, reciprocating tests with ellipsoidal PVA specimens against a glass plate were conducted under conditions of continuous and constant loading, corresponding to a stationary contact area on the biphasic PVA hydrogel (Fig. 13.17a) [20, 37–39]. The frictional behaviors of PVA gel specimens against a flat glass plate were analyzed at a continuous loading of either 2.94 N or 9.8 N. The sliding speed and stroke length were 20 mm·s^{-1} and 35 mm, respectively, and the lubricants were saline or simulated synovial fluid.

13.6.1 Friction in Saline Solution

As shown in Fig. 13.17b, the three hydrogels exhibit distinct frictional behaviors in saline solution. The FT gel was prepared with five repeated freezing-thawing treatments from a 20 wt.% solution of a different PVA powder with a polymerization degree of 2,000 and an average degree of saponification of 98.4–99.8 mol% [20]. The FT gel shows a medium initial friction coefficient with a gradual increase, while the CD gel shows a much smaller friction coefficient but a similar gradual increase. In the case of the hybrid gel, a very low friction coefficient is maintained throughout

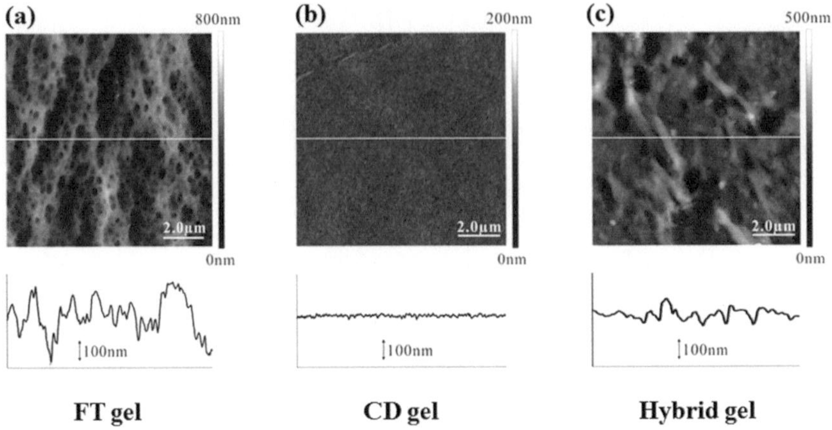

Fig. 13.18 AFM images and profiles of PVA hydrogel surfaces in pure water (From Ref. [20], with permission from Elsevier)

the test. The differences in the frictional behaviors of the three hydrogels can be examined from a perspective of material properties and structures.

Figure 13.18 shows AFM images and profiles of the PVA hydrogel surfaces in pure water [20]. The FT gel has a rough surface with a dendritic and porous structure, but the CD gel has a very smooth surface of ~10–20 nm in height. These differences appear to be caused by the bulk structures of the gels, which contain both crystalline and amorphous regions as discussed in previous studies [20]. The material properties and structures in both PVA hydrogels appear to depend on the structures, which are controlled by hydrogen bonding. The FT gel has a heterogeneous network structure with a milky white appearance, composed of microcrystallites and considerable amorphous zones, and has high permeability and porosity. The CD gel has a uniform network structure with transparency and low permeability. In contrast, the hybrid gel has medium surface roughness of ~100 nm in height. The surface layer of the hybrid gel is laminated by the CD method, but the surface structure appears to have been affected by the roughened interfacial structure of the bottom FT layer. The influence of the biphasic properties of PVA gels on interstitial fluid pressurization which controls the friction was investigated as described below.

13.6.2 Biphasic Analysis of PVA Hydrogels

The material properties of PVA hydrogels were estimated by compressive relaxation tests [20, 38–40]. As suggested above, the FT gel has a high permeability of 2.0×10^{-13} m^4·(Ns)$^{-1}$, while the CD gel has a low permeability of 2.4×10^{-15} m^4·(Ns)$^{-1}$ similar to that of articular cartilage. The Young's modulus and Poisson's ratio for the FT and CD gels were estimated as 110 kPa and 0 and 190 kPa and 0.41,

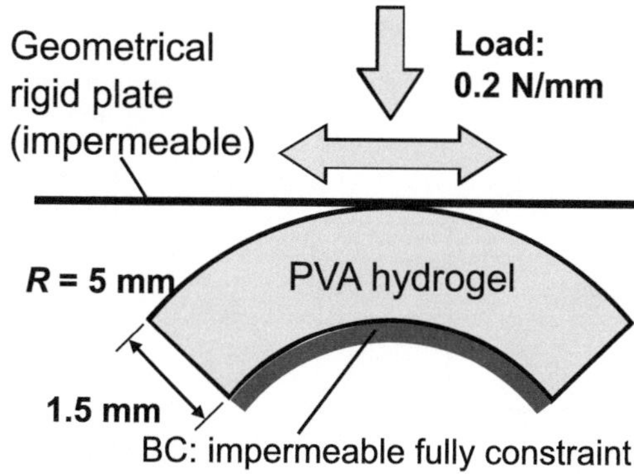

Fig. 13.19 Finite element model for a continuous loading to cylindrical PVA hydrogel (From Ref. [20], with permission from Elsevier)

respectively. In order to evaluate the biphasic lubrication mechanisms for the three kinds of PVA hydrogels in reciprocating tests, biphasic finite element (FE) analysis using the commercial package ABAQUS (6.8–4) was applied, based on the previous FE analysis for articular cartilage in reciprocating motion by Sakai et al. [41]. A two-dimensional FE analysis was conducted for a cylinder with PVA hydrogel layer against an impermeable rigid plate in reciprocating motion under continuous loading for the model, shown in Fig. 13.19. The thickness of the PVA hydrogel layer is 1.5 mm; the outer radius is 5 mm. The hybrid gel is composed of 0.7 mm (outer) CD layer and 0.8 mm of FT layer. The biphasic tissue was modeled by CPE4RP (four-node bilinear displacement and pore pressure, reduced integration with hourglass control) elements; the mesh was designed with 0.1 mm thickness with 0.1 mm on the surface. The bottom of the model was set to be fixed and impermeable. The other surfaces were not fixed and basically permeable, with the exception of the contact region. A load of 0.2 N·mm^{-1} was applied at the center of the cylindrical PVA hydrogel surface with a ramp time of 1 s; the load was then held constant for 292 s (73 cycles) of reciprocation motion. For a stroke of 8 mm and a period of 4 s, the sliding speed is 4 mm·s^{-1}, selected to suppress hydrodynamic action. The changes in interstitial fluid pressure, stress in the solid phase, and fluid flow distribution during the reciprocating motion were examined. The friction coefficient for solid-to-solid contact, μ_{solid}, between the geometrical rigid plate and the solid phase was assumed according to experimental data, as described later. The total friction was estimated from the summation of the horizontal forces, including adhesive and ploughing forces, where the ploughing force appears to be infinitesimal under continuous constant loading condition for the flattened hydrogel surface.

The distributions of interstitial fluid pressure and fluid load partitioning (FLP) for the three PVA hydrogels are shown immediately after loading and after 292 s in Fig. 13.20. No distinct difference appears in the interstitial fluid pressure and FLP immediately after loading among the FT, CD, and hybrid gels. However, after 292 s, clear differences are observed. The low permeability CD gel maintains a

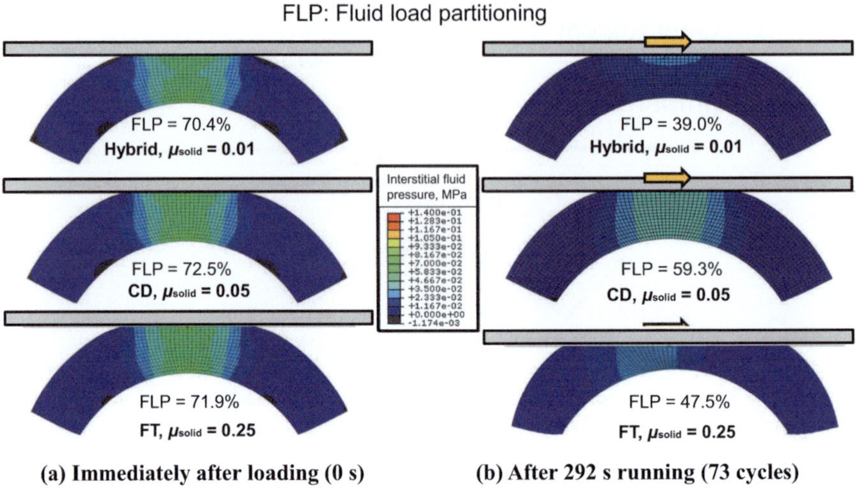

Fig. 13.20 Changes in interstitial fluid pressure in PVA hydrogels (From Ref. [20], with permission from Elsevier)

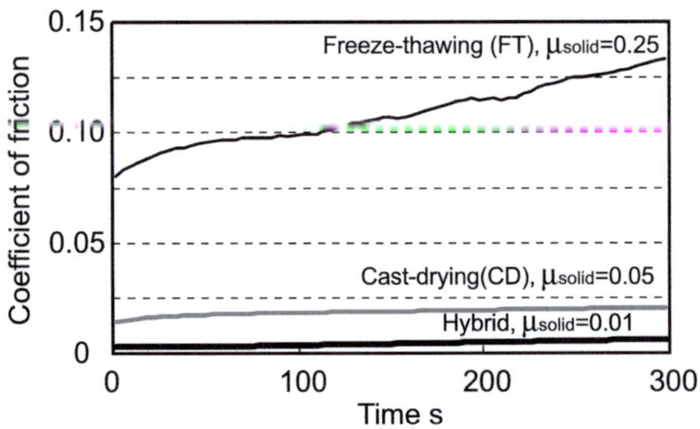

Fig. 13.21 Estimation of changes in friction based on biphasic FE analysis (From Ref. [20], with permission from Elsevier)

similar pressure distribution and an FLP of 59.3 % with a somewhat lower level. In the higher permeability FT gel, fluid pressurization almost diminishes except in the central zone, where FLP = 47.5 % and the elastic deformation becomes larger. In contrast, in the hybrid gel, the fluid pressure distribution is modified to a lower level as FLP = 39.0 % after 292 s, but the elastic deformation is restricted to an area similar to that of the CD gel. Notably, the interstitial fluid pressure for the hybrid gel at 50 s has a similar distribution to, and FLP = 69.4 % of, those immediately after loading.

The estimated coefficient of friction based on the biphasic FE analysis is shown in Fig. 13.21 [20]. The comparison of this biphasic estimation with the experimental data in Fig. 13.17 indicates a clear correspondence for the FT and CD gels. For the low friction of the hybrid gel, however, another lubrication mechanism is required for surface lubricity in the lower FLP condition, although the biphasic lubrication mechanism is effective for the high FLP maintained until 50 s after loading in this case.

13.6.3 Effect of Synovia Constituents on Friction and Wear of PVA Hydrogels

For PVA-FT gel lubricated with a saline solution, high friction and significant wear were observed. Therefore, the effects of adding lubricating synovia constituents on the friction and wear were examined. In this study, four main constituents were selected: 0.5 wt.% hyaluronic acid (HA; molecular weight, 9.2×10^5), 1.4 wt.% albumin, 0.7 wt.% γ-globulin, and 0.01 wt.% Lα-dipalmitoylphosphatidylcholine (DPPC) to evaluate the tribological performances of these gels in physiological conditions. As shown in Fig. 13.22, the addition of single constituents could not provide behaviors of both low friction and low wear. It is considered that a simulated synovial fluid containing 0.5 wt.% HA, 1.4 wt.% albumin, 0.7 wt.% γ-globulin, and 0.01 wt.% DPPC may improve both the friction and wear properties of the PVA-FT gel. This simulated synovial fluid was effective in the improvement of the characteristics of the hybrid gel, with a friction coefficient of 0.003 and minimal

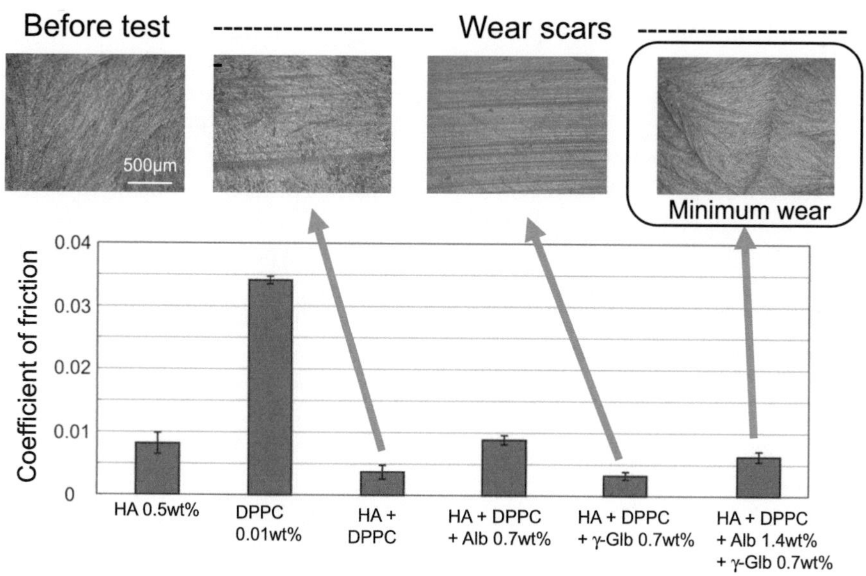

Fig. 13.22 Influence of synovia constituents on friction and wear of PVA-FT hydrogel (From Ref. [39]) DOI: 10.1177/1350650114530273

wear [42]. These results indicate that the synergistic combination of the biphasic and boundary lubrication mechanisms is important, but the elucidation of the detailed lubrication mechanism for PVA hydrogels lubricated by simulated synovial fluid remains for future study.

13.7 Summary and Future Visions

For the application of hydrogels as artificial articular cartilage, several essential properties are required, as mentioned. PVA gels used for artificial cartilage must combine superior mechanical strength and excellent lubricating properties, comparable to those of cartilage at natural synovial joints. The CD and FT gels both have high water absorbance and excellent mechanical properties compared to other hydrogels. As shown in the present typical examples of physical gels, the CD gel showed better performance, characterized by a high swelling ratio and low wear, of the physical gels. However, CD gel alone cannot satisfy all required properties for artificial cartilage. Of the required properties, the mechanical strength and wear resistances of the FT and CD gels were poor compared to those of natural synovial joints; further improvements are necessary for practical uses.

It may be difficult to significantly improve the mechanical properties using conventional fabrication methods. The conventional FT and CD fabrication methods are based on phase separations induced by the natural phenomena of the solidification and vaporization of water. In contrast to these simple and convenient methods, it is possible to improve the properties of gels by introducing artificial fabrication processes. For example, some attempts were made to control the polymer networks of FT and CD gels on multiple scales. In the current study, a lamination method and hybridization techniques using FT and CD gels were presented in order to improve the mechanical properties of the composite gels. The lamination allowed the control of the surface and bulk mechanical properties independently, which improved the mechanical and tribological properties. Better mechanical performances were achieved by the synergistic interaction between the two types of physical gels, partly attributed to the additional microcrystallites formed at the interface between the networks. It was shown that the swelling ratio, mechanical properties, and wear behavior could be controlled by laminating FT and CD gels.

A unidirectional freezing method is another example of an artificial fabrication technique, by which a microscale fibril network structure is formed in the network of FT gels. This anisotropic network structure can promote better mechanical performance along the fibril direction. These techniques provide hierarchical network structures that cannot be obtained by conventional single-step processing.

For the next research step, the authors are developing hybrid hydrogels containing nanoscale structurally controlled CD gels on microstructurally controlled FT gels as another candidate material for artificial cartilage with superior lubricity. Natural synovial joints have multiple layers, each of which has a specific function. For this reason, high performance is expected from the hybridization of single layers of hydrogels with different properties. Based on fundamental biological and materials research, we are applying such a hybrid technique to real materials systems. The tribological properties of the real system were significantly improved in our preliminary experiments, which will be reported elsewhere in the near future.

Acknowledgement The author would like to thank Kuraray Co., Ltd. for supplying PVA powders. This work was supported by the Grant-in-Aid for Specially Promoted Research of Japan Society for the Promotion of Science (Kaken: 23000011).

References

1. Dowson D and Jin ZM. Micro-elastohydrodynamic Lubrication of Synovial Joints. *Eng. Med.*, 1986; 15:65–67.
2. Murakami, T. JSME International Journal, Series III 1990, 33, 465–474.
3. Murakami T, Higaki H, Sawae Y, Ohtsuki N, Moriyama S, Nakanishi Y. *Proc. Inst. Mech. Eng. Pt. H J. Eng. Med.* 1998, *212*, 23–35.
4. Oka M, Ushio K, Kumar P, Ikeuchi K, Hyon SH, Nakamura T, Fujita H. *Proc. IMechE Part H: J. Eng. Med.* 2000, *214*, 59–68.
5. Arakaki K, Kitamura N, Fujiki H, Kurokawa T, Iwamoto M, Ueno M, Kanaya F, Osada Y, Gong JP, Yasuda K. *J. Biomed. Mat. Res. Part A* 2010, *93*, 1160–1168.
6. Cha W-I, Hyon S-H, Oka M, Ikada Y. *Macromol. Symp.* 1996, *109*, 115–126.
7. Kobayashi M, Hyon S.H. *Materials* 2010, *3*, 2753–2771.
8. Nakashima K, Sawae Y, Murakami T. *Tribo. Intern.* 2007, *40*, 1423–1427.
9. Yarimitsu S, Nakashima K, Sawae Y, Murakami T. *Tribology Online* 2007, 2, 114–119.
10. Bodugoz-Senturk H, Choi J, Oral E, Kung JH, Macias CE, Braithwaite G, Muratoglu OK. The effect of polyethylene glycol on the stability of pores in polyvinyl alcohol hydrogels during annealing. *Biomaterials*, 2008; 29: 141–149.
11. Bodugoz-Senturk H, Macias CE, Kung JH, Muratoglu OK. Poly(vinyl alcohol)–acrylamide hydrogels as load-bearing cartilage substitute. *Biomaterials* 2009; 30: 589–596.
12. Peppas NA. Turbidimetric studies of aqueous poly(vinyl alcohol) solutions. *Makromolekulare Chemie* 1975; 176: 3443–3440.
13. Nambu M. *Japanese Patent Kokai* 1982; No. 57/130543. Nambu M. Rubber-like poly(vinyl alcohol) gel. *Kobunshi Ronbunshu* 1990; 47: 695–703 (in Japanese).
14. Murakami T, Sawae Y, Higaki H, Ohtsuki N, Moriyamaa S. The Adaptive Multimode Lubrication in Knee Prostheses with Artificial Cartilage during Walking. In: Dowson D et al. (eds) *Elastohydrodynamics '96: Fundamentals and applications in lubrication and traction.* Elsevier Science 1997, 383–394.
15. Otsuka E and Suzuki A. A Simple Method to Obtain a Swollen PVA Gel Crosslinked by Hydrogen Bonds. *J. Appl. Polym. Sci.* 2009, 114, 10–16.
16. Otsuka E and Suzuki A. Swelling properties of physically cross-linked PVA gels prepared by a castdrying method. *Prog. Colloid Polym. Sci.* 2009, *136*, 121–126.
17. Otsuka E, Sasaki S, Koizumi K, Hirashima Y, Suzuki A. Elution of polymers from physically cross-linked poly(vinyl alcohol) gels. *Soft Matter* 2010; 6: 6155–6159.
18. Sasaki S and Suzuki A. Effects of repeated water exchange and the molecular-weight distribution of PVA cast gels on the elution of polymers. *React Func Polym* 2013; 73: 878–884.
19. Suzuki A, Sasaki S, Sasaki S, Noh T, Nakashima K, Yarimitsu S, Murakami, T. Elution and Wear of PVA Hydrogels by Reciprocating Friction. Extended Abstract for World Tribology Congress 2013, 2013; Torino, in USB.

20. Murakami T, Sakai N, Yamaguchi T, Yarimitsua S, Nakashima K, Sawae Y, Suzuki A. Evaluation of a superior lubrication mechanism with biphasic hydrogels for artificial cartilage. Tribology International 2015; 89: 19–26.
21. Sasaki S, Otsuka E, Hirashima Y, and Suzuki A. Elution of Polymers from PVA Cast Gels with Different Degrees of Polymerization and Hydrolysis. *J. Appl. Polym. Sci.* 2012, 126, E233–E241.
22. Hassan CM and Peppas NA. Structure and applications of poly(vinyl alcohol) hydrogels produced by conventional crosslinking or by freezing/thawing methods. *Adv Polym Sci* 2000; 153: 37–65.
23. Stauffer SR and Peppas NA. Poly(vinyl alcohol) hydrogels prepared by freezing-thawing cyclic processing, Polymer, 33 (1992) 3932–3936.
24. Peppas NA and Stauffer SR. Reinforced uncrosslinked poly (vinyl alcohol) gels produced by cyclic freezing-thawing processes: a short review, J. Controlled Release, 16 (1991) 305–310.
25. Lozinsky VI. A Brief History of Polymeric Cryogels. *Adv Polym Sci* 2014; 263: 1–48.
26. Otsuka E, Komiya S, Sasaki S, Xing J, Bando Y, Hirashima Y, Sugiyama M, Suzuki A. *Soft Matter* 2012, 8, 8129–8136.
27. Sasaki S and Suzuki A. Suzuki, Change in molecular weight distribution by elution of polymers from PVA cast gel. *Polym. Bull.* 2014, 71, 2383–2394.
28. Sasaki S and Suzuki A. Polym. Adv. Tech. 2016, 27, 318–324.
29. Noh T, Bando Y, Ota K, Sasaki S, Suzuki A. Tear force of physically crosslinked poly(vinyl alcohol) gels with different submicrometer-scale network structures. *J Appl Polym Sci* 2015; 136: 121–126.
30. Suzuki A and Yoshikawa M. Water flow in poly(*N*-isopropylacrylamide) gels. *J. Chem. Phys.* 2006; 125: 174901.
31. Suzuki A. Phase transition in gels of sub-millimeter size induced by interaction with stimuli. *Adv Polym Sci*, 1993; 110: 199–240.
32. Sasaki S and Suzuki A. Effects of repeated water exchange and the molecular-weight distribution of PVA cast gels on the elution of polymers. *React Func Polym* 2013; 73: 878–884.
33. Otsuka, E, Sugiyama, M, Suzuki, A. Network Microstructure of PVA Cast Gels Observed by SAXS Measurements. *J. Phys.: Conf. Ser.* 2010, *247*, 012043.
34. Costa, H. S., Rocha, M. F., Andrade, G. I., Barbosa-Stancioli, E. F., Pereira, M. M., Orefice, R. L., Vasconcelos, W. L. and Mansur, H. S. Sol–gel derived composite from bioactive glass–polyvinyl alcohol. *J. Mater. Sci. 2008;*43, 494–502 (2008).
35. Peppas NA and Merrill EW. Crosslinked poly(vinyl alcohol) hydrogels as swollen elastic networks. *J Appl Polym Sci* 1977; 21: 1763–1770.
36. Suzuki A and Sasaki S. Swelling and mechanical properties of physically crosslinked poly(vinyl alcohol) hydrogels. Proc IMechE Part H: Journal of Engineering in Medicine, 2015, 229, 828–844
37. Murakami T, Yarimitsu S, Nakashima K, Sawae Y, Sakai N, Araki T, Suzuki A. Time-dependent frictional behaviors in hydrogel artificial cartilage materials. Proc. 6th International Biotribology Forum, 2011, 65–68.
38. Murakami T, Importance of adaptive multimode lubrication mechanism in natural and artificial joints, *Proc IMechE, J. Eng. Tribol.*, 2012; 226, 2012, 827–837.
39. Murakami T, Yarimitsu S, Nakashima K, Yamaguchi T, Sawae Y, Sakai N, Suzuki A. Superior Lubricity in Articular Cartilage and Artificial Hydrogel Cartilage. *Proc IMechE Part J: J Engineering Tribology*, 2014; 228, 1099–1111.
40. Yamaguchi T, Murakami T. Surface friction and bulk transport properties in hydrogels. Preprint JAST Conference (in Japanese) 2013-9; 199–200.
41. Sakai N, Hagihara Y, Furusawa T, Hosoda N, Sawae Y, Murakami T. Analysis of biphasic lubrication of articular cartilage loaded by cylindrical indenter. *Tribol Int* 2012; 45: 225–36.
42. Murakami T, Yarimitsu S, Nakashima K, Sakai N,Yamaguchi T, Sawae Y, Suzuki A. Synergistic Lubricating Function with Different Modes for Artificial Hydrogel Cartilage, Proc. The 8th International Biotribology Forum and The 36th Biotribology Symposium, 2015, 1–2.

Chapter 14
Physical Properties of Pig Vitreous Body

Masahiko Annaka and Toyoaki Matsuura

Abstract Microscope laser light scattering spectroscopy is applied to investigate the structural and dynamical properties of the gel network in the pig vitreous body. It is natural to consider the collagen motion is coupled with the dynamics of Na-hyaluronate; therefore, we develop the equation for the mode coupling of flexible Na-hyaluronate and semirigid network of collagen. The unique physical properties of a gel arise from its structure. The gel is characterized by two kinds of bulk coefficients: (1) the elastic constants of the gel network and (2) the friction coefficients between gel network and the fluid. We determine the friction coefficients between vitreous gel network and gel fluid using a specially designed apparatus. Together with experimental and theoretical results, the dynamics of collagen and Na-hyaluronate explains two relaxation modes of the fluctuation. The diffusion coefficient of collagen obtained from D_{fast} and D_{slow} is very close to that in aqueous solution, which suggests the vitreous body is in the swollen state. The diffusion coefficients were found to be dependent on the position (surface or central part) of the vitreous body from which the scattered light sampled. The inhomogeneous distribution of Na-hyaluronate and collagen and the shell structure of the vitreous body were suggested.

Keywords Vitreous body • Na-hyaluronate • Collagen • Dynamic light scattering • Friction coefficient

M. Annaka (✉)
Faculty of Science, Department of Chemistry, Kyushu University, 744 Motooka, 819-0395 Fukuoka, Japan

International Research Center for Molecular Systems (IRCMS), Kyushu University, 744 Motooka, 819-0395 Fukuoka, Japan
e-mail: annaka@chem.kyushu-univ.jp

T. Matsuura
Department of Ophthalmology, Nara Prefecture General Medical Center, 1-30-1, Hiramatsu, 631-0846 Nara, Japan

I. Kaneda (ed.), *Rheology of Biological Soft Matter*, Soft and Biological Matter,
DOI 10.1007/978-4-431-56080-7_14

14.1 Introduction

The vitreous body is a tenuous gel that contains collagen and Na-hyaluronate [1]. The fraction of the polymer network is only about 1–2 %, and the remaining is water. Therefore, large amount of water is sustained within the dilute polymer network. The vitreous body is located between the lens and the retina that comprises 80 % of the overall volume of the eye. The functions of the vitreous body are supposed to keep the shape of the eyeball, to absorb the external mechanical shock, to maintain the homeostasis of the eye, and to regulate the position of the lens. The appearance of fresh vitreous body is transparent, and, hence, the vitreous body is considered a uniform tissue. These functions of the vitreous body are still under controversy.

A critical juncture in the advancement of gel research was the realization that scattered light intensity fluctuations, arising from concentration or density fluctuations within gels, represented thermally excited acoustic and elastic vibrations of the gel matrix (phonons) that are rapidly dampened by frictional forces [2]. Subsequent light scattering experiments were able to verify that (1) the magnitude of the scattered light intensity fluctuations was dependent upon the compressibility of the gel network, (2) the ratio of the elastic modulus and frictional coefficient of the network in its fluid medium, or the effective pore size of the network, can be determined by the collective diffusion coefficients of gels obtained from the decay times of the intensity correlation functions, and (3) as the gel approaches the critical points, critical phenomena can be evidenced in the form of the divergent behavior in the observed scattered light intensities and the collective diffusion coefficients of gels [3].

The vitreous body has a complex structure, in which the highly flexible Na-hyaluronate is considered to be interwoven by semirigid network of collagen threads. Many studies performed to date have suggested that Na-hyaluronate, which has a coil shape, is uniformly distributed throughout the three-dimensional network of collagen fibers that form the triple helix in the vitreous body [4]. However, the structural properties of the vitreous body have not been determined. And essentially no investigations on the dynamics and phase equilibrium properties of the vitreous body, however, have been performed to verify indisputably that the vitreous body is indeed a gel network.

Some diseases affect changes in the complex structure of the vitreous body. The collapse of the vitreous body may cause many diseases such as posterior vitreous detachment, vitreous bleeding, and retinal detachment [5, 6]. In the search for the underlying principle of the functions of the vitreous body, it is crucial to understand its physical properties, which leads to promote better understanding of the mechanism of diseases of the vitreous body.

In this study, the microscope laser light scattering spectroscopy (MLLSS) is applied to investigate the structural and dynamical properties of the gel network in the calf vitreous body. It is natural to consider the collagen motion is coupled with the dynamics of Na-hyaluronate; therefore, we develop the equation for the mode coupling of flexible Na-hyaluronate and semirigid network of collagen. The unique physical properties of a gel arise from its structure. The gel is characterized by two kinds of bulk coefficients: (1) the elastic constants of the gel network and (2) the friction coefficients between gel network and the fluid. We determine the friction coefficients between vitreous gel network and gel fluid using a specially designed apparatus [7]. Together with experimental and theoretical results of dynamic light scattering (DLS), we discuss the elastic properties of the calf vitreous body.

14.2 Theory for the Density Fluctuation of the System

The situation that Na-hyaluronate polymers fill in the meshes of collagen fiber network to prevent them from collapsing can be modeled as a complex system of polymers interacting with the network meshes. The DLS measures the time correlation of density fluctuation of the scattering entities, which are the segments of the Na-hyaluronate polymer and the collagen mesh. The concentration fluctuation of the Na-hyaluronate polymer segment, $\delta C_{\mathrm{HA}}(\boldsymbol{r},t)\left(=C_{\mathrm{HA}}(\boldsymbol{r},t)-C_{\mathrm{HA}}^{0}\right)$, where $C_{HA}(\boldsymbol{r},t)$ and C_{HA}^{0}, respectively, are the local and the averaged concentrations of the segment, can be described by the following diffusion equation [8].

$$\frac{\partial\delta C_{\mathrm{HA}}(\boldsymbol{r},t)}{\partial t}=D_{\mathrm{HA}}\nabla^{2}\delta C_{\mathrm{HA}}(\boldsymbol{r},t)+L_{\mathrm{HA-Col}}D_{\mathrm{HA}}\frac{C_{\mathrm{HA}}^{0}}{C_{\mathrm{Col}}^{0}}\nabla^{2}\delta C_{\mathrm{Col}}(\boldsymbol{r},t) \quad (14.1)$$

where D_{HA} and $L_{\mathrm{HA-Col}}$, respectively, are a diffusion coefficient defined in the interparticle interaction-free condition and a phenomenological coefficient, which is significant in the case that the acting force of collagen matrix on the Na-hyaluronate polymer is large. The subscripts, HA and Col, denote hyaluronic acid and collagen, respectively. Here, C_{Col} and $\delta C_{\mathrm{Col}}(\boldsymbol{r},t)$, respectively, are the averaged concentration and the concentration fluctuation of the collagen segment. The collagen network can be regarded as an elastic body. According to the linear theory for the elastic body, a force balance among elastic and the external forces can be described by the following Newton equation:

$$\rho\frac{\partial^{2}\boldsymbol{u}}{\partial t^{2}}=\mu\nabla^{2}\boldsymbol{u}+\left(K+\frac{1}{3}\mu\right)\nabla(\nabla\cdot\boldsymbol{u})-f\frac{\partial\boldsymbol{u}}{\partial t}+\boldsymbol{F}_{\mathrm{Col}} \quad (14.2)$$

where ρ, $\boldsymbol{u}$, μ, K, f, and $\boldsymbol{F}_{\mathrm{Col}}$, respectively, are the density of the network, a displacement vector of the collagen segment, a shear modulus, a bulk modulus, a friction coefficient of a unit volume, and the force acted by the hyaluronate polymer. The inertia term is negligibly small in the fluctuation, that is, $\rho\partial^{2}\boldsymbol{u}/\partial t^{2}=0$. Then Eq. (14.2) leads to

$$\frac{\partial\delta C_{\mathrm{Col}}(\boldsymbol{r},t)}{\partial t}=D_{\mathrm{S}}\nabla^{2}\delta C_{\mathrm{Col}}(\boldsymbol{r},t)+L_{\mathrm{Col-HA}}D_{\mathrm{Col}}\frac{C_{\mathrm{Col}}^{0}}{C_{\mathrm{HA}}^{0}}\nabla^{2}\delta C_{\mathrm{HA}}(\boldsymbol{r},t) \quad (14.3)$$

where

$$D_S = \frac{K + 4\mu/3}{f} \tag{14.4}$$

$$D_{Col} = \frac{T}{f} C_{Col}^0 \tag{14.5}$$

L_{Col-HA} is a phenomenological coefficient related to the acting force of Na-hyaluronate polymer on collagen matrix, and T is the Boltzmann temperature. Here, we use units where Boltzmann's constant k_B is unity to use energy units for temperature. In deriving Eq. (14.3), the following relations are used:

$$C_{Col}^0 = \nabla \cdot \boldsymbol{u}(\boldsymbol{r}) + \delta C_{Col}(\boldsymbol{r}, t) \tag{14.6}$$

$$\boldsymbol{F}_{Col} = -L_{Col-HA} T C_{Col}(\boldsymbol{r}, t) \nabla \ln C_{HA}(\boldsymbol{r}, t) \tag{14.7}$$

and

$$f = C_{Col}^0 \zeta_{Col} \tag{14.8}$$

where ζ_{Col} is a friction coefficient of the collagen segment. Equations (14.1) and (14.3) can be rewritten in the form of Fourier transform of the concentrations as

$$\frac{\partial \delta C_{HA}(q, t)}{\partial t} = -q^2 \left\{ D_{HA} C_{HA}(q, t) + L_{HA-Col} D_{HA} \frac{C_{HA}^0}{C_{Col}^0} C_{Col}(q, t) \right\} \tag{14.9}$$

$$\frac{\partial \delta C_{Col}(q, t)}{\partial t} = -q^2 \left\{ D_S C_{Col}(q, t) + L_{Col-HA} D_{Col} \frac{C_{Col}^0}{C_{HA}^0} C_{HA}(q, t) \right\} \tag{14.10}$$

where

$$\delta C_{HA}(\boldsymbol{r}, t) = \left(\frac{1}{2\pi}\right)^3 \int\int\int dr^3 C_{HA}(\boldsymbol{q}, t) \exp(i\boldsymbol{q} \cdot \boldsymbol{r})$$

and

$$\delta C_{Col}(\boldsymbol{r}, t) = \left(\frac{1}{2\pi}\right)^3 \int\int\int dr^3 C_{Col}(\boldsymbol{q}, t) \exp(i\boldsymbol{q} \cdot \boldsymbol{r})$$

Equations (2.9) and (2.10) can be converted into matrix form as

$$\frac{\partial \boldsymbol{C}(\boldsymbol{q})}{\partial t} = -q^2 \widetilde{\boldsymbol{A}} \boldsymbol{C}(\boldsymbol{q}) \tag{14.11}$$

where

$$\boldsymbol{C}(\boldsymbol{q}) = \begin{bmatrix} C_{\mathrm{HA}}(\boldsymbol{q}, t) \\ C_{\mathrm{Col}}(\boldsymbol{q}, t) \end{bmatrix}$$

and

$$\widetilde{\boldsymbol{A}} = \begin{bmatrix} D_{\mathrm{HA}} & L_{\mathrm{HA-Col}} D_{\mathrm{HA}} \dfrac{C_{\mathrm{HA}}^0}{C_{\mathrm{Col}}^0} \\ L_{\mathrm{Col-HA}} D_{\mathrm{Col}} \dfrac{C_{\mathrm{Col}}^0}{C_{\mathrm{HA}}^0} C_{\mathrm{HA}} & D_{\mathrm{S}} \end{bmatrix}$$

The solution of Eq. (14.11) is given as

$$\begin{bmatrix} C_{\mathrm{HA}}(\boldsymbol{q}, t) \\ C_{\mathrm{Col}}(\boldsymbol{q}, t) \end{bmatrix} = \boldsymbol{B}(\boldsymbol{q}) \begin{bmatrix} D_{\mathrm{HA}}(\boldsymbol{q}) \exp\left(-q^2 \lambda_1 t\right) \\ D_{\mathrm{Col}}(\boldsymbol{q}) \exp\left(-q^2 \lambda_2 t\right) \end{bmatrix} \tag{14.12}$$

where

$$\widetilde{\boldsymbol{B}}^{-1}(\boldsymbol{q}) \widetilde{\boldsymbol{A}}(\boldsymbol{q}) \widetilde{\boldsymbol{B}}(\boldsymbol{q}) = \begin{bmatrix} \lambda_1 & 0 \\ 0 & \lambda_2 \end{bmatrix} \tag{14.13}$$

with

$$2\lambda_{1/2} = D_{\mathrm{HA}} + D_{\mathrm{S}} \pm \sqrt{(D_{\mathrm{HA}} - D_{\mathrm{S}})^2 + 4 D_{\mathrm{HA}} L_{\mathrm{Col-HA}} D_{\mathrm{Col}} L_{\mathrm{HA-Col}}} \quad (\lambda_1 \geq \lambda_2) \tag{14.14}$$

and

$$\widetilde{\boldsymbol{B}} = \frac{1}{\sqrt{D_{\mathrm{HA}} L_{\mathrm{Col-HA}} D_{\mathrm{Col}} L_{\mathrm{HA-Col}} + \Delta D^2}} \begin{bmatrix} L_{\mathrm{Col-HA}} D_{\mathrm{Col}} \dfrac{C_{\mathrm{Col}}^0}{C_{\mathrm{HA}}^0} C_{\mathrm{HA}} & \Delta D \\ -\Delta D & L_{\mathrm{HA-Col}} D_{\mathrm{HA}} \dfrac{C_{\mathrm{HA}}^0}{C_{\mathrm{Col}}^0} \end{bmatrix}$$

$$\widetilde{\boldsymbol{B}}^{-1} = \frac{1}{\sqrt{D_{\mathrm{HA}} L_{\mathrm{Col-HA}} D_{\mathrm{Col}} L_{\mathrm{HA-Col}} + \Delta D^2}} \begin{bmatrix} L_{\mathrm{HA-Col}} D_{\mathrm{HA}} \dfrac{C_{\mathrm{HA}}^0}{C_{\mathrm{Col}}^0} & -\Delta D \\ \Delta D & L_{\mathrm{Col-HA}} D_{\mathrm{Col}} \dfrac{C_{\mathrm{Col}}^0}{C_{\mathrm{HA}}^0} C_{\mathrm{HA}} \end{bmatrix}$$

$$\Delta D = \frac{|D_{\mathrm{HA}} - D_{\mathrm{S}}|}{2} \left[1 - \sqrt{1 + \frac{4 D_{\mathrm{HA}} L_{\mathrm{Col-HA}} D_{\mathrm{Col}} L_{\mathrm{HA-Col}}}{(D_{\mathrm{HA}} - D_{\mathrm{S}})^2}} \right]$$

Therefore, the concentration fluctuation can be given by

$$C_{\mathrm{HA}}(\boldsymbol{q},t) \propto L_{\mathrm{Col-HA}} D_{\mathrm{Col}} \frac{C^0_{\mathrm{Col}}}{C^0_{\mathrm{HA}}} C_{\mathrm{HA}} \exp\left(-q^2\lambda_1 t\right) + \Delta D \exp\left(-q^2\lambda_2 t\right) \quad (14.15)$$

$$C_{\mathrm{Col}}(\boldsymbol{q},t) \propto -\Delta D \exp\left(-q^2\lambda_1 t\right) + L_{\mathrm{HA-Col}} D_{\mathrm{HA}} \frac{C^0_{\mathrm{HA}}}{C^0_{\mathrm{Col}}} \exp\left(-q^2\lambda_2 t\right) \quad (14.16)$$

Finally, Eq. (14.14) yields a double exponentially decaying function for the time correlation of the electric field of scattering light as follows:

$$g^{(1)}(q,\tau) = A_{\mathrm{fast}} \exp\left(-q^2\lambda_1 t\right) + A_{\mathrm{slow}} \exp\left(-q^2\lambda_2 t\right) \quad (14.17)$$

where A_{fast} and A_{slow} are constant.

14.3 Experimental

14.3.1 Materials

The vitreous body of the calf eye was chosen for our study because it is easily obtained and handled. Calf vitreous body was isolated from the sclera of the eyeball. The choroid membrane was also carefully removed by the standard method [9]. The sample preparation was performed within 8 h after extraction of the eye at a local slaughterhouse.

14.3.2 Microscope Laser Light Scattering Spectroscopy

Since the vitreous body is difficult to hold in a sample cell, we used the technique of the microscope laser light scattering spectroscopy (MLLSS). The technique differs from the conventional dynamic laser light scattering technique in that the scattering volume in the MLLSS in some 105 times is smaller, making it as low as 2 μm^3. It, therefore, becomes possible to analyze the motion of particles inside the vitreous body. The schematic diagram of the MLLSS setup is illustrated in Fig. 14.1 [10, 11]. The beam of a He-Ne laser (632.8 nm, Spectra Physics 124 A) was focused onto the equational plane of the vitreous body through an optical fiber connected to the fiber coupler that is equipped with a condensing lens. The beam is sharply focused onto the various positions of the vitreous body placed under an upright microscope (Nikon Optiphot). The cross section at the focal point within the vitreous body was approximately 2 × 2 μm. The focal region was imaged onto a photomultiplier tube (EMI 9863B-350) by a 50 μm diameter optical fiber embedded in one of the eyepieces of the microscope (Gamma Scientific 700-10-36A). The fluctuations

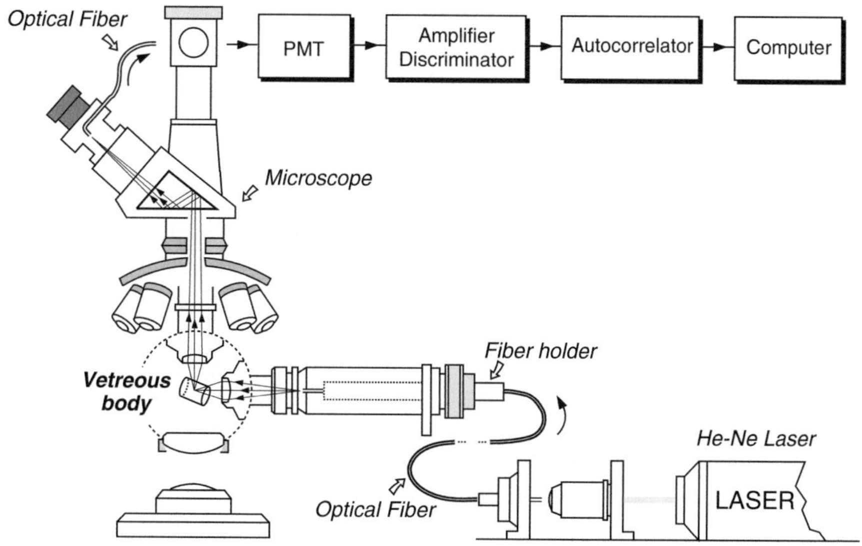

Fig. 14.1 Schematic illustration of the microscope laser light scattering spectroscopy (MLLSS)

of the number of photons were analyzed using the autocorrelator (Brookhaven Instrument, Model BI-2030A). The temporal fluctuations of the scattered light intensity $I(t)$ were analyzed in terms of intensity autocorrelation functions [12, 13]:

$$C(\tau) = \langle I(t) \cdot I(t+\tau) \rangle_t \tag{14.18}$$

where $\langle \cdots \rangle$ stands for the time average over t. The rate of the fluctuations of the scattered light intensity that represents the density fluctuations of the vitreous gel is proportional to the rate of local swelling and shrinking of the gel via molecular Brownian motions. There are also permanent and static inhomogeneities within the vitreous that also contribute to light scattering. The light intensity scattered by these inhomogeneities does not fluctuate with time. The scattered light intensity is, thus, the superposition of contributions from scattering elements that are static from those that dynamically fluctuate:

$$I(t) = I_S + I_D(t) \tag{14.19}$$

In dynamic light scattering, the time correlation of the intensity of scattered light is recorded. Assuming the Gaussian nature of the scattered light photons, the correlation function of the intensity of scattered light is rewritten in terms of the autocorrelation function $g(\tau)$ of the scattered electric field $E(t)$ that is related to the scattered light intensity: $I(t) = E(t)E^{\star}(t)$

$$g(\tau) = \frac{\langle E(t) \cdot E(t+\tau) \rangle_t}{\langle E(t) \cdot E(t) \rangle_t} \tag{14.20}$$

Then $C(\tau)$ is written as

$$C(\tau) = (I_{\mathrm{S}} + I_{\mathrm{D}})^2 + A\left\{I_{\mathrm{D}}^2 \cdot g^2(\tau) + 2I_{\mathrm{S}} \cdot I_{\mathrm{D}}\, g(\tau)\right\} \tag{14.21}$$

where A is the efficiency parameter of the apparatus, which is uniquely determined by the optical configuration of the setup, the value, I, and the average intensities scattered by the gel fluctuations and the static inhomogeneities. For the present experiments, A was determined to be 0.8.

The electric field autocorrelation function $g(\tau)$ can be easily extracted from the intensity autocorrelation function $C(\tau)$ from knowledge of the initial value $C(0)$, the baseline $C(\infty)$, and the coefficient A, using the above relations. As we shall see later and as shown in Fig. 3, the autocorrelation functions can have a distinct double-exponential feature. This indicates the presence of two different modes within the gel. Therefore, the correlation functions $g(\tau)$ are analyzed using the following relationship:

$$g(\tau) = \frac{A_{\mathrm{fast}}}{A_{\mathrm{fast}} + A_{\mathrm{slow}}} \exp\left(-D_{\mathrm{fast}} q^2 \tau\right) + \frac{A_{\mathrm{slow}}}{A_{\mathrm{fast}} + A_{\mathrm{slow}}} \exp\left(-D_{\mathrm{slow}} q^2 \tau\right) \tag{14.22}$$

where A_{fast} and A_{slow} are the amplitudes and D_{fast} and D_{slow} the diffusion coefficients of the fast and slow components in the bimodal distribution, respectively. The wave number q is defined by the scattering angle θ , the wavelength λ of the laser beam in the vitreous body, and the refractive index n:

$$q = \frac{4\pi n}{\lambda} \sin\frac{\theta}{2} \tag{14.23}$$

In the present experiments, q was approximately $6.3 \times 10^5\ \mathrm{cm}^{-1}$.

14.3.3 Measurement of Friction Coefficient

The principle of the measurement of the friction coefficient of a vitreous body is schematically shown in Fig. 14.2. The apparatus was originally designed by Tokita and Tanaka, and they determined the friction coefficients of poly(acrylamide) gel and poly(N-isopropylacrylamide) gel accurately under various conditions [7]. A vitreous body of thickness d is fixed to the wall by silicon glue, and water is pushed through the vitreous body with a small pressure P. The average velocity of the water flow through the openings inside the gel, v, is determined by measuring the rate at

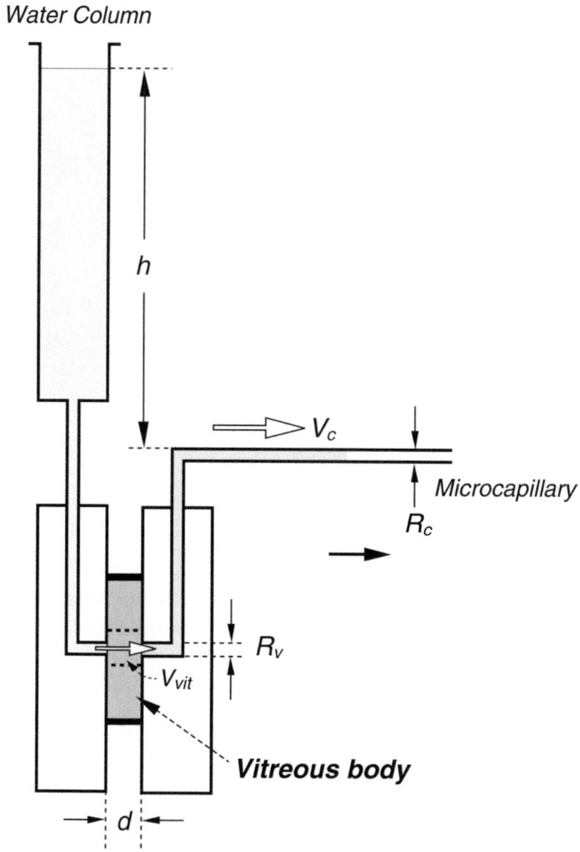

Fig. 14.2 Schematic diagram of the apparatus for the measurement between the gel network and water. The sample mold is tightly held between two plexiglass plates. The velocity of the fluid flow in the micropipette is measured by a movable microscope with an accuracy of 0.001 mm

which water flows out of the vitreous body in a steady state. The friction coefficient, f, is defined as

$$f = \frac{P}{d \cdot v} \tag{14.24}$$

The chromatography column with 50 cm length is used as a reservoir of water to generate the hydrostatic pressure. The range of the height of the water column can be changed from 20 to 60 cm, which corresponds to the pressure from 2×10^4 to 6×10^4 dyn/cm^2. The temperature of the sample is controlled within an accuracy 0.1 °C. The apparatus is set on the vibration-free table to avoid any external mechanical disturbances. The rate of water flow through the gel was determined by measuring the movement of the water meniscus in a micropipette under a microscope.

14.4 Results

14.4.1 Dynamic Light Scattering

Figure 14.3 shows a correlation function obtained by MLLSS from the core region (position B in Fig. 14.4), observed at a scattering angle 135° and at a temperature of 23 °C. From nonlinear least squares fit to data using Eqs. (14.21), (14.22), and (14.23) (the solid line in Fig. 14.3), the essential parameters for the diffusion coefficients $D_{\text{fast}} = 5.9 \times 10^{-8}\,\text{cm}^2/\text{s}$ and $D_{\text{slow}} = 3.0 \times 10^{-9}\,\text{cm}^2/\text{s}$ that represent the two apparent, collective diffusion motions of the vitreous body, designated as fast and slow motions, respectively, were derived. The relative contributions of the fast and slow components to the overall dynamic scattered light intensity I_{D} were designated as $\%A_{\text{fast}}$ and $\%A_{\text{slow}}$, respectively, where $\%A_{\text{fast}} = 100 \times A_{\text{fast}}/\,(A_{\text{fast}} + A_{\text{slow}})$. The static $\%I_{\text{S}}$ and dynamic $\%I_{\text{D}}$ components of scattered light intensities, where $\%I_{\text{D}} = 100 \times I_{\text{D}}/\,(I_{\text{D}} + I_{\text{S}})$, to the total scattered light intensity $I\,(t)$ observed at particular wave vector, were also determined. For the data shown in Fig. 14.3, $\%A_{\text{fast}} = 40$ and $\%A_{\text{slow}} = 50$.

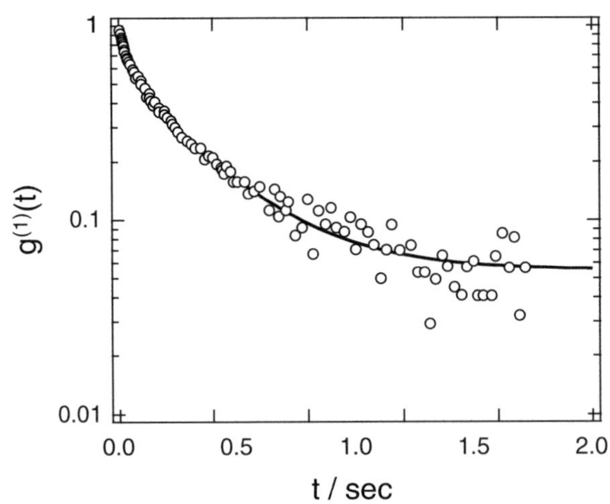

Fig. 14.3 A correlation function from light scattered by the calf vitreous body (position B indicated in Fig. 14.4) collected at scattering angle 135 °C and at 23 °C. The solid line represents a computer-generated fit to the data using Eqs. (14.21), (14.22), and (14.23). The results of the fit yielded for $D_{\text{fast}} = 5.9 \times 10^{-8}\,\text{cm}^2/\text{s}$, $D_{\text{slow}} = 3.0 \times 10^{-9}\,\text{cm}^2/\text{s}$, and $\%A_{\text{fast}} = 40$

Table 14.1 Position dependence of dynamics of vitreous body

Sampled position	D_{fast} cm²/s	D_{slow} cm²/s	$\%A_{fast}$	I_D	I_S
A	$(10 \pm 0.98) \times 10^{-8}$	$(10 \pm 0.98) \times 10^{-8}$	40	45	55
B	$(5.9 \pm 0.83) \times 10^{-8}$	$(3.0 \pm 0.78) \times 10^{-8}$	40	45	55
C	$(7.8 \pm 1.52) \times 10^{-8}$	$(3.8 \pm 0.60) \times 10^{-8}$	40	45	55

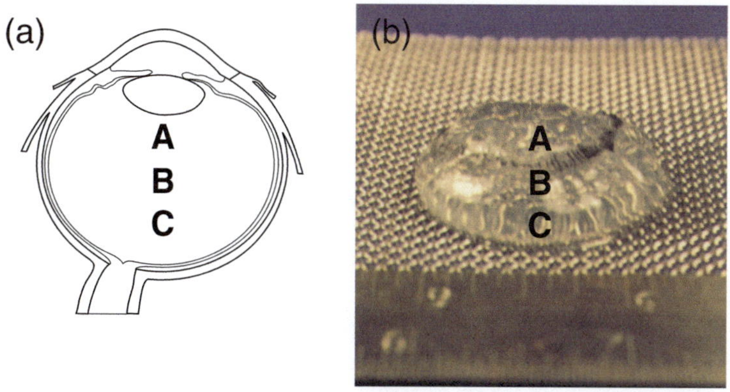

Fig. 14.4 Cross section of the eye and the photograph of the calf vitreous body. A, B, and C indicate the sampled positions for MLLSS experiment. Positions A and C: the outer region of the vitreous body (position A corresponds to the region near the ciliary body) and position B: the central region of the vitreous body

In the normal vitreous body, the values of the diffusion coefficients exhibited the position dependency. Table 14.1 summarizes the result of DLS regarding the dependence of the observed dynamics upon the position within the vitreous body as indicated in Fig. 14.4a. Each data is the average value of five measurements. In the surface part of the vitreous body (positions A and C), the diffusion coefficient is larger than that in the core region (B). Swann and Constable [14] reported the inhomogeneous distribution of Na-hyaluronate and collagen, and the vitreous body has the shell structure: the surface part has higher concentration comparing with its central part, which serves the mechanical stabilization of the vitreous body. These facts well correspond to the experimental results obtained here.

14.4.2 *Friction Coefficient*

The velocity of the water in the micropipette at the stationary state, V_C, is obtained by measuring the shift of water meniscus for a given period of time using a movable microscope with an accuracy of 0.001 mm. The velocity thus obtained is plotted as a function of the pressure in Fig. 14.5. At relatively high pressures, the water

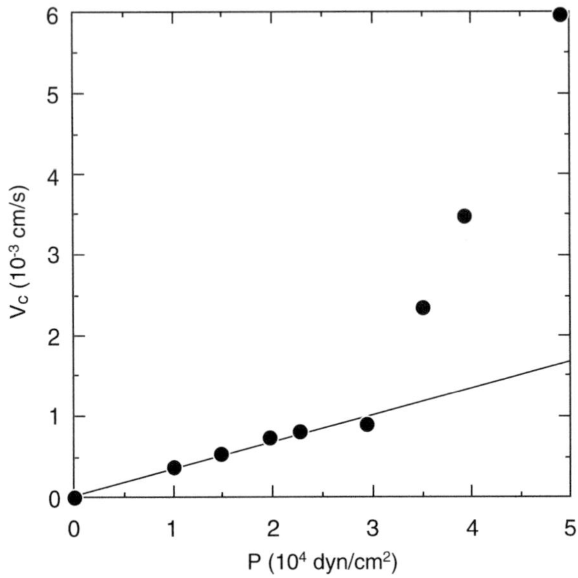

Fig. 14.5 The velocity of water in the micropipette is plotted as a function of the water pressure. The *solid line* in this figure is obtained by the least squares fit

flow in the vitreous body is high. Presumably, the network structure of the vitreous body is broken under high pressure; therefore, we chose five points measured at lower pressure to calculate the friction coefficients. The relationship between the applied pressure and the velocity is linear at lower pressures as predicted. The friction coefficient between vitreous body and water, $f = 6.4 \times 10^9\,\mathrm{dym \cdot s/cm^4}$, was obtained from the slope of the straight line in Fig. 14.5 using the following equation:

$$f = -\left(\frac{dV_C}{dp}\right)^{-1} \cdot \left(\frac{1}{d}\right) \cdot \left(\frac{R_V}{R_C}\right)^2 \qquad (14.25)$$

where $dV_C/dp = 3.1 \times 10^{-8}\,\mathrm{cm^3/dyn \cdot s}$ is determined from the slope of the straight line as given in Fig. 14.5. The thickness of the sample d is $d = 0.5\,\mathrm{cm}$. The factor $(R_V/R_C)^2$ is the ratio of the area of the micropipette of radius $R_C = 3.4 \times 10^{-2}\,\mathrm{cm}$ and the area of the hole in the vitreous body with radius $R_V = 3.3 \times 10^{-1}\,\mathrm{cm}$, which is necessary to convert V_C to V_{vit}. The value of f measured here is from the total water flow along the orbital axis in the vitreous body. More precise measurements at different positions are needed, but it is very difficult to measure water flow in a high-water-content vitreous body without destruction of its network structure. Therefore, we use this friction coefficient in the following discussions.

14.5 Discussions

The vitreous body can be regarded as a gel composed of the highly flexible Na-hyaluronate polymers. The fluid gel is interwoven by semirigid network of collagen threads which serve the mechanical stabilization of the body. As mentioned in the theoretical section, the DLS measures the time correlation of density fluctuation of the scattering entities, which are the segments of the Na-hyaluronate polymer and the collagen mesh. Synthetic gels, such as poly(acrylamide) gel, have been known to show single diffusion coefficient [3]. The vitreous body, however, is found to show two different diffusion coefficient modes, one relatively fast and the other relatively slow mode. This may be due to the complex structure of the vitreous body, in which the highly flexible Na-hyaluronate is interwoven by semirigid collagen threads [9]. In this case, it is natural to consider the coupling of the collagen motion with the dynamics of Na-hyaluronate can explain the fast mode. According to Eqs. (14.14) and (14.17), D_{fast} $(= \lambda_1)$ and D_{slow} $(= \lambda_2)$ are given by

$$D_{\text{fast}} = D_{\text{HA}} - \Delta D \quad \text{and} \quad D_{\text{slow}} = D_{\text{S}} + \Delta D \quad (D_{\text{HA}} > D_{\text{S}}) \tag{14.26}$$

and

$$D_{\text{fast}} = D_{\text{S}} - \Delta D \quad \text{and} \quad D_{\text{slow}} = D_{\text{HA}} + \Delta D \quad (D_{\text{HA}} < D_{\text{S}}) \tag{14.27}$$

$D_{\text{HA}} \simeq 2.4 \times 10^{-8}\,\text{cm}^2/\text{s}$ has been reported for the Na-hyaluronate diluted in the 0.2 M NaCl aqueous solution, molecular weight of which is similar to that $\left(M_{\text{W}} \sim 1.4 \times 10^6\right)$ in the calf vitreous body [15]. Using this result, the values of D_{S}, ΔD, and $D_{\text{Col}} L_{\text{Col-HA}} L_{\text{HA-Col}}$ are evaluated from the obtained D_{fast}- and D_{fast}-values and are tabulated in Table 14.2. It should be mentioned that the value of $D_{\text{Col}} L_{\text{Col-HA}} L_{\text{HA-Col}}$ is very close to the reported diffusion constant of collagen with a contour length 280 nm in 0.1N HCL aqueous solution $\left(D_{\text{Col}} \sim 8.9 \times 10^{-8}\text{cm}^2/\text{s}\right)$ [16]. It is plausible that the $L_{\text{Col-HA}} L_{\text{HA-Col}}$-value for the collagen strongly coupled to Na-hyaluronate in the vitreous body is an order of magnitude 1. It can be inferred, therefore, that the collective diffusion coefficient of collagen in the vitreous body is very close to that in the aqueous solution. This can be realized in the highly swollen vitreous body which has enough space for the polymer segments to move freely.

Table 14.2 Position dependence of calculated values of ΔD, the collective diffusion coefficient D_{S}, the longitudinal modulus M, and $D_{\text{Col}} L_{\text{Col-HA}} L_{\text{HA-Col}}$

Sampled position	ΔD cm^2/s	D_{S} cm^2/s	$M (= K + 4\mu/3)$ Pa	$D_{\text{Col}} L_{\text{Col-HA}} L_{\text{HA-Col}}$ cm^2/s
A	-2.0×10^{-8}	8.0×10^{-8}	510	6.3
B	-2.1×10^{-8}	3.9×10^{-8}	250	3.2
C	-2.0×10^{-8}	6.0×10^{-8}	380	4.7

It should be mentioned that the q-dependency experiment must be needed to confirm the observed two modes are diffusive. However, unfortunately, we could not conduct further experiments using the calf vitreous body in order to avoid the danger of infection of bovine spongiform encephalopathy (BSE). We, therefore, conducted q-dependency experiments by using the pig vitreous body [17]. Since a pig and a cow are mammals, the chemical composition and the structure of both vitreous bodies are basically the same. For a polydisperse system, the decay time of the correlation function can be described by a weighted sum of exponentials [18]:

$$g^{(1)}(\tau) = \int_0^{+\infty} G(\Gamma) \exp(-\Gamma\tau)\, d\Gamma \tag{14.28}$$

In order to clarify the nature of the decay rate Γ^{-1}, the q-dependence of $G(\Gamma)$ was examined. Figure 14.6 shows (a) the scattering angle dependence of normalized electric field autocorrelation function, $g^{(1)}(\tau)/g^{(1)}(0)$ vs. t, and (b) the q^2-dependence of the Γq^{-2}. The distribution of relaxation rates was obtained by an inverse Laplace transformation of Eq. (14.28) with CONTIN program [19]. For a diffusional mode, one expects that Γ is proportional to q^2. For all scattering angles, the Γq^{-2} for both fast and slow modes is almost q independent as shown in Fig. 14.6b. This indicates that the observed two modes are diffusive. It should be mentioned that the apparent q^2 dependence of the Γq^{-2} is within experimental error. Very slow relaxation with decay time ranging from 1 to 10 s is observed as shown in Fig. 14.6a. The vitreous gel has a very dilute network-like structure of collagen whose interstices are filled with Na-hyaluronate. There seems to exist a very slow accordion-like movement of semirigid network of collagen threads in the vitreous gel. This could originate in the observed very slow relaxation. It is

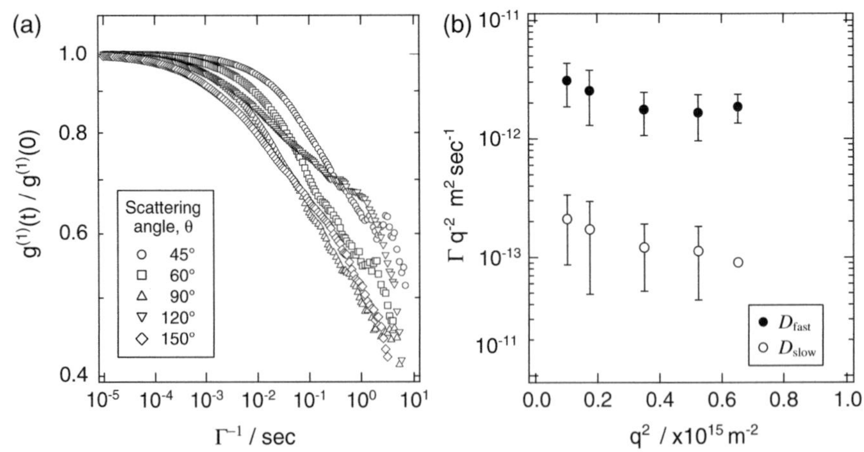

Fig. 14.6 The scattering angle dependence of normalized electric field autocorrelation function, $g^{(1)}(\tau)/g^{(1)}(0)$ vs. t (a), and the q^2-dependence of the Γq^{-2} for the pig vitreous body

important to note that this slow relaxation, not yet well understood, is possibly generated by a different physical mechanism and is independent of the other faster mode. Therefore, for the CONTIN analysis, we introduced a cutoff at 0.1 s and subtracted the nonzero baseline [20]. It was verified for the correlation curves that the cutoff did not influence the decay rates and amplitudes of the faster modes. But no satisfactory explanation for this slow relaxation is given here. The detailed study on the pig vitreous body is in progress now.

The longitudinal elastic modulus of the vitreous network $M\,(=K+4\mu/3)$, estimated from $D_S\left(=3.7\times10^{-12}\,\mathrm{m^2/s}\right)$ and $f\left(=6.4\times10^{9}\,\mathrm{N\cdot s/m^4}\right)$ using Eq. (14.4), is $2.4\,\mathrm{N/m^2}$. Assuming Poisson's ratio of the network is zero, the longitudinal elastic modulus is readily $G=2\mu$ [2]. Therefore, the shear modulus of the vitreous network is calculated to be $\mu=0.012\,\mathrm{N/m^2}$. Tokita et al. determined the share modulus μ of calf vitreous body to be in the order of $0.05\,\mathrm{N/m^2}$ by using torsion pendulum apparatus [21]. Although this value is in the same order of our result, there still remain several factors we have to consider. The shear modulus of the vitreous gel may be variable between species. It is worthy to mention that Nickerson et al. recently reported the time-dependent modulus change of the bovine vitreous body [22]. The drop in the shear modulus indicates a molecular-level change from a relatively rigid *in oculo* state, which is not stable outside of the constraints of the eye, to a measurably softer state *ex oculo*. The viscoelastic properties of the vitreous body are still under controversy; however, we could determine the elastic constant μ from the independently obtained values: the time correlation of the scattered light intensity and the friction coefficient between polymer network and solvent, f from macroscopic method.

The elastic moduli, M, at different positions are also listed in Table 14.2. The intraocular pressure is created by the continual renewal of fluids within the interior of the eye and is known to be within the range of 1300–2600 Pa ($\simeq$10–20 mmHg) under normal physiological condition. It is clinically known that the intraocular pressure is reduced to less than 670 Pa ($\simeq$5 mmHg) by artificially blocking the production of aqueous humor. In this case, only the elastic modulus of the gel network is considered to contribute the intraocular pressure; therefore, our experimental results are consistent with the clinical observation.

This study represents new structural and dynamic properties of the vitreous body. The simple consideration presented here may be helpful toward understanding of physicochemical aspects of diseases. To develop a more thorough understanding of the vitreous gel, however, much more conclusive investigations are warranted under various conditions, which is the subject of future studies.

14.6 Conclusion

Dynamic light scattering spectroscopy was applied to investigate the structural and dynamical properties of calf vitreous body. From the observations of the dynamics of scattered light scattered by the calf vitreous body, two different diffusion coefficient modes, one relatively fast and the other relatively slow mode, were observed. The diffusion coefficients were found to show position dependency (surface or central part), which suggested the inhomogeneous distribution of Na-hyaluronate and collagen and the vitreous body has the shell structure. We developed the coupling theory in vitreous gel system describing the coupling of the collagen motion with the dynamics of Na-hyaluronate under the assumption that Na-hyaluronate polymers was filled in the meshes of collagen fiber network. The calculated value of diffusion coefficient of collagen is very close to that in aqueous solution, which suggests the vitreous body is in the swollen state and the validity of our model. In the quest of the principle behind diseases of vitreous body, more precise study, however, must be needed to reveal the structural and dynamical properties of the vitreous body at molecular level. This will open the door for the quantitative understanding of the structure and physical properties of vitreous body.

References

1. E. R. Berman, M. Voaden, Biochemistry of the Eye; G. K. Smelser Ed. Academic Press, London, p. 373 (1970).
2. T. Tanaka, L. O. Hocker, and G. B. Benedek, J. Chem. Phys., **59**, 5151 (1973).
3. T. Tanaka, D. J. Fillmore, S. -T. Sun, I. Nishio, G. Swislow, and A. Shah, Phys. Rev. Lett., **45**, 1636 (1980).
4. E. A. Balazs, New and Controversial Aspects of Retinal Detachment, A. McPherson Ed. Academic Press, Philadelphia, Vol 1, p. 3 (1968).
5. S. Norman, J. Arch. Ophthal. **79**, 568 (1968).
6. S. T. William, Trans. Amer. Acad. Opthalmol. & Otolaryngol. **72**, 217 (1968).
7. M. Tokita and T. Tanaka, J. Chem. Phys., **95**, 4613 (1991).
8. S. Sasaki, F. M. J. Schipper, J. Chem. Phys. **115**, 4349 (2001).
9. J. G. F. Worst, L. I. Los, Cisternal Anatomy of the Vitreous, Kugler Publications, Amsterdam, p. 1 (1995).
10. I. Nishio, T. Tanaka, S. -S. Sun, Y. Imanishi, S. T. Ohnishi, Science **220**, 1173 (1983).
11. J. Peetermans, I. Nisio, S. T. Ohnishi, T. Tanaka, Biophysics **83**, 352 (1986).
12. J. P. Munch, S. Candau, J. Herz, and G. Hild, J. Phys. (Paris) **38**, 971 (1977).
13. J. P. Munch, P. Lemarechal, and S. Candau, J. Phys. (Paris) **38**, 1499 (1977).
14. D. A.Swann, I. Constable, Invest. Ophthal. **11**, 164 (1972).
15. R. Takahashi, K. Kubota, M. Kawada, A. Okamoto, Biopolymers **50**, 87 (1999).
16. K. Claire, R. Pecora, J. Phys. Chem. B **101**, 746 (1997).
17. The DLS experiments on pig vitreous body was conducted using 22mW Uniphase He-Ne laser ($\lambda = 632.8$ nm) (San Jose, USA) as a light source and an ALV-5000/EPP correlator (Langen, Germany) at 25°C. Pig vitreous body was treated by the same method as the vitreous body of calf as described in experimental section. The sample preparation was performed within 8 h after extraction of the eye at a local slaughterhouse.
18. B. J. Berne, R. Pecora, Dynamic Light Scattering with Applications to Chemistry, Biology, and Physics, Wiley, New York (1976).
19. S. W. Provencher, Biophys. J. **16**, 27 (1976).
20. C. M. Papadakis, K. Almdal,K. Mortensen, F. Rittig, P. Stepanek. Macromol. Symp. **162**, 275 (2000).
21. T. Tokita, Y. Fujita, K. Hikichi, Biorheology **21**, 751, (1984)
22. C. S. Nickerson, J. Park, J. A. Kornfield, H. Karageozian, J. Biomechanics **41**, 1823 (2008).

MIX
Papier aus verantwortungsvollen Quellen
Paper from responsible sources
FSC® C105338

If you have any concerns about our products,
you can contact us on
ProductSafety@springernature.com

In case Publisher is established outside the EU,
the EU authorized representative is:
Springer Nature Customer Service Center GmbH
Europaplatz 3, 69115 Heidelberg, Germany

Printed by Libri Plureos GmbH
in Hamburg, Germany